D0906468

INTRODUCTION
TO PHYSICAL
POLYMER SCIENCE

INTRODUCTION TO PHYSICAL POLYMER SCIENCE

SECOND EDITION

L. H. Sperling

Lehigh University
Bethlehem, Pennsylvania

A WILEY-INTERSCIENCE PUBLICATION

JOHN WILEY & SONS, INC.

New York / Chichester / Brisbane / Toronto / Singapore

Copyright © 1992 by John Wiley & Sons, Inc.

Library of Congress Cataloging in Publication Data:
Sperling, L. H. (Leslie Howard), 1932–
    Introduction to physical polymer science /L. H. Sperling—2nd
ed.
        p.   cm.
    "A Wiley-Interscience publication."
    Includes bibliographical references and index.
    ISBN 0-471-53035-2 (cloth: alk. paper)
    1. Polymers.   2. Polymerization.   I. Title.
QD381.S635   1992   91-44021
668.9—dc20   CIP

Printed in the United States of America

10 9 8 7 6 5

This book is dedicated to my grandsons, Ryland Everett Sweigard, born October 13, 1988, and Tresten Andrisen Sweigard, born February 18, 1991. They think that plastics are to play with.

PREFACE TO THE SECOND EDITION

When I was small, the books and songs I knew just *were*, with no particular authorship. I got my lesson on the rights and duties of authorship from my mother one day, discussing the Bible. "If we don't think that a given passage is worded in the best way for today, why don't we just rewrite it?" I asked.

My mother was horrified. "That Book was written thousands of years ago! It was handed down from generation to generation as it was set down on paper by its authors!" Books were sacred to her.

So it was that I learned that all books, songs, plays, and operas are written by particular men and women who have a strong and clear message to be expressed. Only they have the privilege of revising their works. If someone wasn't satisfied, they could write their own book!

Now, after a period of still less than a decade, I find that many new ideas have come to the fore in physical polymer science. A few important concepts somehow got left out of the first edition of this book. A handful of pages and paragraphs wandered in that should not have been there. I, the author of this work, joyfully take the privilege that is only mine to prepare this second edition. Thank you for giving me the opportunity.

Major specific changes include the following: three new chapters were added. The first one resulted from a split of the chapter titled, "The Bulk State," into two chapters, "The Amorphous State" and "The Crystalline State." The second one resulted from the vast increase in importance of the liquid crystalline state, resulting in a chapter entitled "Polymers in the Liquid Crystalline State." This chapter absorbed some of the material originally in the chapter entitled "Mechanical Behavior of Polymers." The third new chapter, at the end of the book, is entitled, "Modern Topics."

Other major changes include the addition of the macromolecular hypothesis and historical development to Chapter 1, photophysics and fluorescence to Chapter 2, movement of the sections on thermodynamics of mixing and phase separation and fractionation from Chapter 4 to Chapter 3, ahead of molecular weight determination, and additional material on thermodynamics of blending polymers and polymer–polymer phase diagrams to Chapter 4.

The following changes refer to the new chapter numbers. Additional material on self-diffusion of polymers was added to Chapter 5, and fiber spinning and structure were added to Chapter 6. As mentioned previously, Chapter 7 is an entirely new chapter on polymer liquid crystals. Greater discussion of dynamic mechanical behavior is given in Chapter 8. A discussion of gelatinous materials was added to Chapter 9. To Chapter 10, a discussion of rheology and a new demonstration were added (in the appendix). Chapter 11 was given a complete reorganization, because of the new material on stress–strain behavior and crack healing. Chapter 12 is an entirely new chapter, concerned with a variety of topics, emphasizing surface and interfacial behavior, electrical behavior of polymers, nonlinear optics, and high-temperature materials.

Again, I want to take this opportunity to thank the many students who helped in proofreading the manuscript. In proofing the first edition, a student said to me: "Dr. Sperling! You always taught us to write and rewrite manuscripts until they read like a thousand violins playing in the night! This paragraph here, however, reads like 50 violins playing in the day, and two of them are squeaking!" In preparing the second edition, a student again said to me: "Look at this equation! All three terms must always be positive, yet you have equated the sum of them to zero!" The result in each case, of course, is one fewer error that future students will have to endure.

Many thanks must also be given to the Department of Chemical Engineering and the Department of Materials Science and Engineering, as well as the Materials Research Center and the Center for Polymer Science and Engineering at Lehigh University. Dr. D. A. Thomas, long my research partner, contributed some juicy study problems. Special thanks must be given to Ms. Gail Kriebel and her staff at the E. W. Fairchild–Martindale Library, Ms. Andrea Pressler, photographer, and Ms. Virginia Newhard, my secretary.

L. H. Sperling

Bethlehem, Pennsylvania
June 1992

PREFACE TO THE FIRST EDITION

Research in polymer science continues to mushroom, producing a plethora of new elastomers, plastics, adhesives, coatings, and fibers. All of this new information is gradually being codified and unified with important new theories about the interrelationships among polymer structure, physical properties, and useful behavior. Thus the ideas of thermodynamics, kinetics, and polymer chain structure work together to strengthen the field of polymer science.

Following suit, the teaching of polymer science in colleges and universities around the world has continued to evolve. Where once a single introductory course was taught, now several different courses may be offered. The polymer science and engineering courses at Lehigh University include physical polymer science, organic polymer science, and polymer laboratory for interested seniors and first-year graduate students, and graduate courses in emulsion polymerization, polymer blends and composites, and engineering behavior of polymers. There is also a broad-based introductory course at the senior level for students of chemical engineering and chemistry. The students may earn degrees in chemistry, chemical engineering, metallurgy and materials engineering, or polymer science and engineering, the courses being both interdisciplinary and cross-listed.

The physical polymer science course is usually the first course a polymer-interested student would take at Lehigh, and as such there are no special prerequisites except upper-class or graduate standing in the areas mentioned above. This book was written for such a course.

The present book emphasizes the role of molecular conformation and configuration in determining the physical behavior of polymers. Two relatively new ideas are integrated into the text. Small-angle neutron scattering is doing for polymers in the 1980s what NMR did in the 1970s, by providing an entirely new perspective of molecular structure. Polymer blend science now offers thermodynamics as well as unique morphologies.

Chapter 1 covers most of the important aspects of the rest of the text in a qualitative way. Thus the student can see where the text will lead him or her, having a glimpse of the whole. Chapter 2 describes the configuration of polymer chains, and Chapter 3 describes their molecular weight. Chapter 4 shows the interactions between solvent molecules and polymer molecules. Chapters 5–7 cover important aspects of the bulk state, both amorphous and crystalline, the glass transition phenomenon, and rubber elasticity. These three chapters offer the greatest depth. Chapter 8 describes creep and stress relaxation, and Chapter 9 covers the mechanical behavior of polymers, emphasizing failure, fracture, and fatigue.

Several of the chapters offer classroom demonstrations, particularly Chapters 6 and 7. Each of these demonstrations can be carried out inside a 50-minute class and are easily managed by the students themselves. In fact, all of these demonstrations have been tested by generations of Lehigh students, and they are often presented to the class with a bit of showmanship. Each chapter is also accompanied by a problem set.

The author thanks the armies of students who studied from this book in manuscript form during its preparation and repeatedly offered suggestions relative to clarity, organization, and grammar. Many researchers from around the world contributed important figures. Dr. J. A. Manson gave much helpful advice and served as a Who's Who in highlighting people, ideas, and history.

The Department of Chemical Engineering, the Materials Research Center, and the Vice-President for Research's Office at Lehigh each contributed significant assistance in the development of this book. The Lehigh University Library provided one of their carrels during much of the actual writing. In particular, the author thanks Sharon Siegler and Victoria Dow and the staff at Mart Library for patient literature searching and photocopying. The author also thanks Andrea Weiss, who carefully photographed many of the figures in this book.

Secretaries Jone Susski, Catherine Hildenberger, and Jeanne Loosbrock each contributed their skills. Lastly, the person who learned the most from the writing of this book was...

L. H. SPERLING

Bethlehem, Pennsylvania
November 1985

CONTENTS

9   Cross-linked Polymers and Rubber Elasticity   382

Values of Often Used Constants[†]

Avogadro's number	N_A	6.022×10^{23} molecules/mol
Boltzmann's constant	K	1.380×10^{-16} erg/K
		1.380×10^{-23} J/K
Gas constant, molar	R	8.314 J/mol · deg K
		82.05 cm^3 · atm/mol · deg K
		1.987 cal/mol · deg K
		8.31×10^7 dynes · cm/mol · K
Planck's constant	h	6.626×10^{-34} J · sec
Speed of light in vacuum	c	2.997×10^8 m/sec

Useful Conversion Factors

1 dyne/cm^2 = 1.450×10^{-5} lb/in.^{2} = 1.02×10^{-6} kg/cm^2

1 Pa = 10 dynes/cm^2 = 7.5×10^{-3} mm Hg = 10^{-5} bar
    = 1.02×10^{-5} kg/cm^2

1 Pa = 1 N/m^2

1 bar = 1 atm pressure

1 J = 2.387×10^{-1} cal = 1×10^7 ergs

1 Pa · sec = 10 poise

1 MPa = 1×10^7 dynes/cm^2 = 145 lb/in.^{2}

1 nm = 10 Å

1 N = 10^5 dynes = 1.02×10^{-1} kgf = 2.248×10^{-1} lbf

[†]J. A. Dean, Ed., *Lange's Handbook of Chemistry*, 12th ed., McGraw-Hill, New York, 1979, pp. 2–3.

INTRODUCTION
TO PHYSICAL
POLYMER SCIENCE

1

INTRODUCTION TO POLYMER SCIENCE

Polymer science was born in the great industrial laboratories of the world of the need to make and understand new kinds of plastics, rubber, adhesives, fibers, and paints. Only much later did polymer science come to academic life. Perhaps because of its origins, polymer science tends to be more interdisciplinary than most, combining chemistry, chemical engineering, materials, and other fields as well.

Chemically, polymers are long-chain molecules of very high molecular weight, often measured in the hundreds of thousands. For this reason, the term "macromolecules" is frequently used when referring to polymeric materials. The trade literature sometimes refers to polymers as resins, an old term that goes back before the chemical structure of the long chains was understood.

The first polymers used were natural products, especially cotton, starch, proteins, and wool. Beginning early in the 20th century, synthetic polymers were made. The first polymers of importance, Bakelite and nylon, showed the tremendous possibilities of the new materials. However, the scientists of that day realized that they did not understand many of the relationships between the chemical structures and the physical properties that resulted. The research that ensued forms the basis for physical polymer science.

This book develops the subject of physical polymer science, describing the interrelationships among polymer structure, morphology, and physical and mechanical behavior. Key aspects include molecular weight and molecular weight distribution, and the organization of the atoms down the polymer chain. Many polymers crystallize, and the size, shape, and organization of the crystallites depend on how the polymer was crystallized. Such effects as annealing are very important, as they have a profound influence on the final state of molecular organization.

1

Other polymers are amorphous, often because their chains are too irregular to permit regular packing. The onset of chain molecular motion heralds the glass transition and softening of the polymer from the glassy (plastic) state to the rubbery state. Mechanical behavior includes such basic aspects as modulus, stress relaxation, and elongation to break. Each of these is relatable to the polymer's basic molecular structure and history.

This chapter provides the student with a brief introduction to the broader field of polymer science. Although physical polymer science does not include polymer synthesis, some knowledge of how polymers are made is helpful in understanding configurational aspects, such as tacticity, which are concerned with how the atoms are organized along the chain. Similarly, polymer molecular weights and distributions are controlled by the synthetic detail. This chapter starts at the beginning of polymer science, and it assumes no prior knowledge of the field.

1.1 FROM LITTLE MOLECULES TO BIG MOLECULES

The behavior of polymers represents a continuation of the behavior of smaller molecules at the limit of very high molecular weight. As a simple example, consider the normal alkane hydrocarbon series

$$
\begin{array}{ccc}
\text{H} & \text{H}\ \ \text{H} & \text{H}\ \ \text{H}\ \ \text{H} \\
| & |\ \ \ | & |\ \ \ |\ \ \ | \\
\text{H}-\text{C}-\text{H} & \text{H}-\text{C}-\text{C}-\text{H} & \text{H}-\text{C}-\text{C}-\text{C}-\text{H} \\
| & |\ \ \ | & |\ \ \ |\ \ \ | \\
\text{H} & \text{H}\ \ \text{H} & \text{H}\ \ \text{H}\ \ \text{H} \\
\text{Methane} & \text{Ethane} & \text{Propane}
\end{array}
\qquad (1.1)
$$

These compounds have the general structure

$$ \text{H}(\!-\!\text{CH}_2\!-\!)_n\text{H} \qquad (1.2) $$

where the number of $-\text{CH}_2-$ groups, n, is allowed to increase up to several thousand. The progression of their state and properties is shown in Table 1.1.

At room temperature, the first four members of the series are gases. Pentane boils at 36.1°C and is a low-viscosity liquid. As the molecular weight of the series increases, the viscosity of the members increases. Although commercial gasolines contain many branched-chain materials and aromatics as well as straight-chain alkanes, the viscosity of gasoline is markedly lower than that of kerosene, motor oil, and grease because of its lower average chain length.

These latter materials are usually mixtures of several molecular species, although they are easily separable and identifiable. This point is important because most polymers are also "mixtures"; that is, they have a molecular

Table 1.1 Properties of the alkane series

Number of Carbons in Chain	State and Properties of Material	Use, Dependent on Chain Length
1–4	Simple gas	Bottled gas for cooking
5–11	Simple liquid	Gasoline
9–16	Medium-viscosity liquid	Kerosene
16–25	High-viscosity liquid	Oil and grease
25–50	Simple solid	Paraffin wax candles
1000–3000	Tough plastic solid	Polyethylene bottles and containers

weight distribution. In high polymers, however, it becomes difficult to separate each of the molecular species, and people talk about molecular weight averages.

Compositions of normal alkanes averaging more than about 20–25 carbon atoms are crystalline at room temperature. These are simple solids known as wax. It must be emphasized that at up to 50 carbon atoms the material is far from being polymeric in the ordinary sense of the term.

The polymeric alkanes contain 1000–3000 carbon atoms and are known as polyethylene. Polyethylene has the chemical structure

$$\peglet CH_2 - CH_2 \pegret_n \tag{1.3}$$

which originates from the structure of the monomer ethylene, $CH_2 = CH_2$. In some places the structure is written

$$\peglet CH_2 \pegret_{n'} \tag{1.4}$$

or polymethylene. (Then, $n' = 2n$.) The relationship of the latter structure to

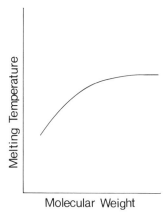

Figure 1.1 The molecular weight – melting temperature relationship for the alkane series. An asymptotic value of about 140°C is reached for very high molecular weights (polyethylene).

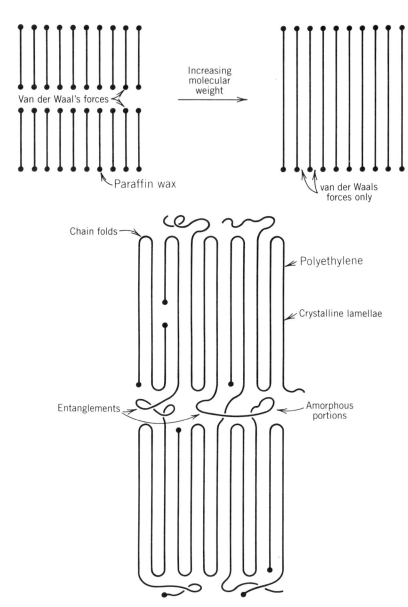

Figure 1.2 Comparison of wax and polyethylene structure and morphology.

the alkane series is clearer. While true alkanes have CH_3 — as end groups, most polyethylenes have initiator residues.

Even at a chain length of thousands of carbons, the melting point of polyethylene is still slightly molecular-weight-dependent, but most linear polyethylenes have melting temperatures, T_f, near 140°C. The approach to the theoretical asymptote of about 145°C at infinite molecular weight (1) is illustrated schematically in Figure 1.1.

The greatest differences between polyethylene and wax lie in their mechanical behavior, however. While wax is a brittle solid, polyethylene is a tough plastic. Comparing resistance to break of a child's birthday candle with a wash bottle tip, both of about the same diameter, shows that the wash bottle tip can be repeatedly bent whereas the candle breaks on the first deformation.

Polyethylene is a tough plastic solid because its chains are long enough to connect individual stems together within a lamellar crystallite by chain folding (see Figure 1.2). The chains also wander between lamellae, connecting several of them together. These effects add strong covalent bond connections both within the lamellae and between them. On the other hand, only weak van der Waals forces hold the chains together in wax.

In addition, a certain portion of polyethylene is amorphous. The chains in this portion are rubbery, imparting flexibility to the entire material. Wax is 100% crystalline, by difference.

The long chain length allows for entanglement (see Figure 1.3). The entanglements help hold the whole material together under stress. In the melt state, chain entanglements cause the viscosity to be raised very significantly also.

The long chains shown in Figure 1.3 also illustrate the coiling of polymer chains in the amorphous state. One of the most powerful theories in polymer science (2) states that the conformations of amorphous chains in space are

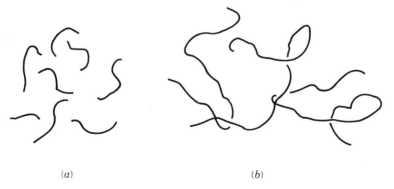

(a)   (b)

Figure 1.3 Entanglement of polymer chains. (a) Low molecular weight, no entanglement. (b) High molecular weight, chains are entangled. The transition between the two is often at about 600 backbone chain atoms.

random coils; that is, the directions of the chain portions are statistically determined.

1.2 MOLECULAR WEIGHT AND MOLECULAR WEIGHT DISTRIBUTIONS

While the exact molecular weight required for a substance to be called a polymer is a subject of continued debate, often polymer scientists put the number at about 25,000 g/mol. This is the minimum molecular weight required for good physical and mechanical properties for many important polymers. This molecular weight is also near the onset of entanglement.

1.2.1 Effect on Ultimate Tensile Strength

The effect of molecular weight on the tensile strength of polymers is illustrated in Figure 1.4. At very low molecular weights, the ultimate tensile stress, σ_b, is near zero. As the molecular weight increases, the tensile strength rapidly increases, then gradually levels off. Since a major point of weakness at the molecular scale is the chain ends, which do not transmit the covalent bond strength, it is predicted that the tensile strength also reaches an asymptotic value at infinite molecular weight. Then, a large part of the curve in Figure 1.4 can be expressed (3, 4)

$$\sigma_b = A - \frac{B}{M_n} \tag{1.5}$$

where M_n is the number-average molecular weight (see below) and A and B are constants.

1.2.2 Molecular Weight Averages

The same polymer from different sources may have different molecular weights. Thus polyethylene from source A may have a molecular weight of 150,000 g/mol, whereas polyethylene from source B may have a molecular weight of 400,000 g/mol (see Figure 1.5). To compound the difficulty, all

Figure 1.4 Effect of polymer molecular weight on tensile strength.

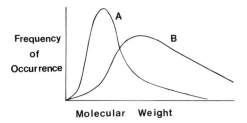

Figure 1.5   Molecular weight distributions of the same polymer from two different sources, A and B.

synthetic polymers and most natural polymers (except proteins) have a distribution in molecular weights. That is, some molecules in a given sample of polyethylene are larger than others. The differences result directly from the kinetics of polymerization.

However, these facts led to much confusion for chemists early in the 20th century. At that time, chemists were able to understand and characterize small molecules. Compounds such as hexane all have six carbon atoms. If polyethylene with 2430 carbon atoms were declared to be "polyethylene," how could that component having 5280 carbon atoms also be polyethylene? How could two sources of the material having different average molecular weights both be polyethylene, noting A and B in Figure 1.5?

The answer to these questions lies in defining average molecular weights and molecular weight distributions (5, 6). The two most important molecular weight averages are the number-average molecular weight, M_n,

$$M_n = \frac{\sum_i N_i M_i}{\sum_i N_i} \tag{1.6}$$

where N_i is the number of molecules of molecular weight M_i, and the weight-average molecular weight, M_w,

$$M_w = \frac{\sum_i N_i M_i^2}{\sum_i N_i M_i} \tag{1.7}$$

For single-peaked distributions, M_n is usually near the peak. The weight-average molecular weight is always larger. For simple distributions, M_w may be 1.5–2.0 times M_n. The ratio M_w/M_n, sometimes called the polydispersity index, provides a simple definition of the molecular weight distribution. Thus all compositions of $+CH_2—CH_2+_n$ are called polyethylene, the molecular weights being specified for each specimen.

For many polymers, a narrower molecular distribution yields better properties. The low end of the distribution may act as a plasticizer, softening the material. Certainly it does not contribute as much to the tensile strength. The high-molecular-weight tail increases processing difficulties, because of its enormous contribution to the melt viscosity. For these reasons, great emphasis is placed on characterizing polymer molecular weights.

1.3 MAJOR POLYMER TRANSITIONS

Polymer crystallinity and melting were discussed previously. Crystallization is an example of a first-order transition, in this case liquid to solid. Most small molecules crystallize, an example being water to ice. Thus this transition is very familiar.

A less classical transition is the glass–rubber transition in polymers. At the glass transition temperature, T_g, the amorphous portions of a polymer soften. The most familiar example is ordinary window glass, which softens and flows at elevated temperatures. Yet glass is not crystalline, but rather it is an amorphous solid. It should be pointed out that many polymers are totally amorphous. Carried out under ideal conditions, the glass transition is a type of second-order transition.

The basis for the glass transition is the onset of coordinated molecular motion in the polymer chain. At low temperatures, only vibrational motions are possible, and the polymer is hard and glassy (Figure 1.6, region 1) (7). In the glass transition region, region 2, the polymer softens, the modulus drops three orders of magnitude, and the material becomes rubbery. Regions 3, 4, and 5 are called the rubbery plateau, the rubbery flow, and the viscous flow

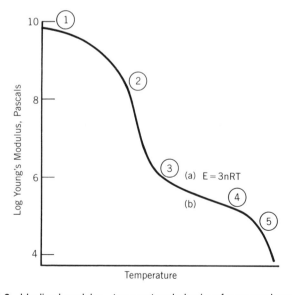

Figure 1.6 Idealized modulus – temperature behavior of an amorphous polymer.

Table 1.2 Typical polymer viscoelastic behavior at room temperature

Region	Polymer	Application
Glassy	Poly(methyl methacrylate)	Plastic
Glass transition	Poly(vinyl acetate)	Latex paint
Rubbery plateau	*Cross*-poly(butadiene–*stat*–styrene)	Rubber bands
Rubbery flow	Chicle[a]	Chewing gum
Viscous flow	Poly(dimethyl siloxane)	Lubricant

[a]From the latex of *Achras sapota*, a mixture of cis- and trans-polyisoprene plus polysaccharides. (M. S. Alger, *Polymer Science Dictionary*, Elsevier, New York, 1989.)

Table 1.3 Examples of polymers at room temperature by transition behavior

	Crystalline	Amorphous
Above T_g	Polyethylene	Natural rubber
Below T_g	Cellulose	Poly(methyl methacrylate)

regions, respectively. Examples of each region are shown in Table 1.2. These regions will be discussed further in Chapters 8 and 9.

Polymers may thus be partly crystalline or totally amorphous. Simultaneously, the amorphous portions may be above or below the glass transition temperature, creating four subclasses of materials. Table 1.3 gives a common example of each. While polyethylene and natural rubber need no further introduction, common names for processed cellulose are rayon and cellophane. Cotton is nearly pure cellulose, and wood pulp for paper is 80–90% cellulose. A well-known trade name for poly(methyl methacrylate) is Plexiglas®.

1.4 POLYMER SYNTHESIS AND STRUCTURE

1.4.1 Chain Polymerization

Polymers may be synthesized by two major kinetic schemes, chain and stepwise polymerization. The most important of the chain polymerization methods is called free radical polymerization.

1.4.1.1 Free Radical Polymerization

The synthesis of poly(ethyl acrylate) will be used as an example of free radical polymerization. Benzoyl peroxide is a common initiator. Free radical polymerization has three major kinetic steps—initiation, propagation, and termination.

1.4.1.2 *Initiation*

On heating, benzoyl peroxide decomposes to give two free radicals:

Benzoyl peroxide   Free radical, R·   (1.8)

In this reaction, the electrons in the oxygen–oxygen bond are unpaired and become the active site. With R representing a generalized organic chemical group, the free radical can be written R · . (It should be pointed out that hydrogen peroxide undergoes the same reaction on a wound, giving a burning sensation as the free radicals "kill the germs.")

The initiation step usually includes the addition of the first monomer molecule:

Free radical   Ethyl acrylate   Growing chain   (1.9)

In this reaction, the free radical attacks the monomer and adds to it. The double bond is broken open, and the free radical reappears at the far end.

1.4.1.3 *Propagation*

After initiation reactions (1.8) and (1.9), many monomer molecules are added rapidly, perhaps in a fraction of a second:

(1.10)

On the addition of each monomer, the free radical moves to the end of the chain.

1.4.1.4 Termination

In the termination reaction, two growing chain ends react with each other. Termination is either by combination,

$$
2R-CH_2-\underset{\underset{O=C-O-C_2H_5}{|}}{\overset{\overset{H}{|}}{C}}\cdot \quad \longrightarrow \quad R-CH_2-\underset{\underset{\underset{O-C_2H_5}{|}}{O=C}}{\overset{\overset{H}{|}}{C}}-\underset{\underset{O=C-O-C_2H_5}{}}{\overset{\overset{H}{|}}{C}}-CH_2-R
$$

$$(1.11)$$

where R now represents a long-chain portion, or by disproportionation, where a hydrogen is transferred from one chain to the other. This latter results in two final chains. While the normal mode of addition is a head-to-tail reaction (1.10), this termination step is normally head-to-head.

As a homopolymer, poly(ethyl acrylate) is widely used as an elastomer or adhesive, being a polymer with a low T_g, $-22°C$. As a copolymer with other acrylics it is used as a latex paint.

1.4.1.5 Structure and Nomenclature

The principal method of polymerizing monomers by the chain kinetic scheme involves the opening of double bonds to form a linear molecule. In a reacting mixture, monomer, fully reacted polymer, and only a small amount of rapidly reacting species are present. Once the polymer terminates, it is "dead" and cannot react further by the synthesis scheme outlined previously.

Polymers are named by rules laid out by the IUPAC Nomenclature Committee (8, 9). For many simple polymers, the source-based name utilizes the monomer name prefixed by "poly." If the monomer name has two or more words, parentheses are placed around the monomer name. Thus, in the above, the monomer ethyl acrylate is polymerized to make poly(ethyl acrylate).

Table 1.4 provides a selected list of common chain polymer structures and names along with comments as to how the polymers are used. The "vinyl" monomers are characterized by the general structure $CH_2 = CHR$, where R represents any side group. One of the best-known vinyl polymers is poly(vinyl chloride), where R is $-Cl$.

The diene monomer has the general structure $CH_2 = CR - CH = CH_2$, where on polymerization one of the double bonds forms the chain bonds, and the other goes to the central position. The vinylidenes have two groups on one carbon. Table 1.4 also lists some common copolymers, which are formed by reacting two or more monomers together.

Table 1.4 Selected chain polymer structures and nomenclature

Structure	Name	Where Used
$+CH_2-CH+_n$ $\quad\quad\quad\vert$ $\quad\quad\quad R$	"Vinyl" class	
R = — H	Polyethylene	Plastic
R = — CH_3	Polypropylene	Rope
R = —⟨benzene ring⟩	Polystyrene	Drinking cups
R = — Cl	Poly(vinyl chloride)	"Vinyl"
$\quad\quad\quad\quad O$ $\quad\quad\quad\quad\Vert$ R = —O—C—CH_3	Poly(vinyl acetate)	Latex paints
R = — OH	Poly(vinyl alcohol)	Fiber
$\quad\quad\quad X$ $\quad\quad\quad\vert$ $+CH_2-C+_n$ $\quad\quad\quad\vert$ $\quad\quad O=C-O-R$	X = — H, acrylics X = — CH_3, methacrylics	
X = — H, R = — C_2H_5	Poly(ethyl acrylate)	Latex paints
X = — CH_3, R = — CH_3	Poly(methyl methacrylate)	Plexiglas
X = — CH_3, R = C_2H_5	Poly(ethyl methacrylate)	Adhesives
$\quad\quad\quad H$ $\quad\quad\quad\vert$ $+CH_2-C+_n$ $\quad\quad\quad\vert$ $\quad\quad\quad C\equiv N$	Polyacrylonitrile[a]	Orlon®
$+CH_2-C=CH-CH_2+_n$ $\quad\quad\quad\vert$ $\quad\quad\quad R$	"Diene" class	
R = — H	Polybutadiene	Tires
R = — CH_3	Polyisoprene	Natural rubber
R = — Cl	Polychloroprene	Neoprene
$+CX_2-CR_2+_n$	Vinylidenes	
X = — H, R = — F	Poly(vinylidene fluoride)	Plastic
X = — F, R = — F	Polytetrafluoroethylene	Teflon®
X = — H, R = — CH_3	Polyisobutene[b]	Elastomer

Common Copolymers

EPDM	Ethylene–propylene–diene–monomer	Elastomer
SBR	Styrene–butadiene–rubber	Tire rubber
	Poly(styrene-*stat*-butadiene)[c]	
NBR	Acrylonitrile–butadiene–rubber	Elastomer
	Poly(acrylonitrile-*stat*-butadiene)	
ABS	Acrylonitrile–butadiene–styrene[d]	Plastic

[a]Polyacrylonitrile is technically a number of the acrylic class because it forms acrylic acid on hydrolysis.
[b]Also called polyisobutylene. The 2% copolymer with isoprene, after vulcanization, is called butyl rubber.
[c]The term –*stat*– means statistical, as explained in Chapter 2.
[d]ABS is actually a blend or graft of two random copolymers, poly(acrylonitrile-*stat*-butadiene), and poly(acrylonitrile-*stat*-styrene).

1.4.2 Step Polymerization

1.4.2.1 *A Polyester Condensation Reaction*
The second important kinetic scheme is step polymerization. As an example of a step polymerization, the synthesis of a polyester is given.

The general reaction to form esters starts with an acid and an alcohol:

$$CH_3-CH_2OH + CH_3-\overset{\displaystyle O}{\overset{\|}{C}}-OH \rightarrow CH_3-CH_2-O-\overset{\displaystyle O}{\overset{\|}{C}}-CH_3 + H_2O$$

Ethyl alcohol   Acetic acid   Ethyl acetate   Water

$$(1.12)$$

where the ester group is $-O-\overset{\displaystyle O}{\overset{\|}{C}}-$, and water is eliminated.

The above chemicals cannot form a polyester, because they have only one functional group each. When the two reactants each have bifunctionality, a linear polymer is formed.

$$nHO-CH_2-CH_2-OH + nHO-\overset{\displaystyle O}{\overset{\|}{C}}\!\!-\!\!\bigcirc\!\!-\!\!\overset{\displaystyle O}{\overset{\|}{C}}-OH \rightarrow$$

Ethylene glycol   Terephthalic acid

$$H\!\!\left[\!O-CH_2-CH_2-O-\overset{\displaystyle O}{\overset{\|}{C}}\!\!-\!\!\bigcirc\!\!-\!\!\overset{\displaystyle O}{\overset{\|}{C}}\!\right]_n\!\!OH + (2n-1)H_2O \quad (1.13)$$

Poly(ethylene terephthalate)

In the stepwise reaction scheme, monomers, dimers, trimers, and so on may all react together. All that is required is that the appropriate functional groups meet in space. Thus the molecular weight slowly climbs as the small molecule water is eliminated.

Poly(ethylene terephthalate) is widely known as the fiber Dacron®. It is highly crystalline, with a melting temperature of about $+265°C$.

Another well-known series of polymers made by step polymerization reactions is the polyamides, known widely as the nylons. In fact, there are two series of nylons. In the first series, the monomer has an amine at one end of the molecule and a carboxyl at the other. For example,

$$nH_2N-CH_2-CH_2-CH_2-\overset{\displaystyle O}{\overset{\|}{C}}-OH \rightarrow$$

$$H\!\!\left(\!\!\overset{\displaystyle H}{\overset{\|}{N}}\!-CH_2-CH_2-CH_2-\overset{\displaystyle O}{\overset{\|}{C}}\!\right)_n\!\!OH + (n-1)H_2O \quad (1.14)$$

which is known as nylon 4. The number 4 indicates the number of carbon atoms in the mer.

In the second series, a dicarboxylic acid is reacted with a diamine:

$$
nH_2N(CH_2)_4NH_2 + nH-O-\overset{\overset{\displaystyle O}{\|}}{C}(CH_2)_6\overset{\overset{\displaystyle O}{\|}}{C}-OH \longrightarrow
$$

$$
H\left[\overset{\overset{\displaystyle H}{|}}{N}-(CH_2)_4-\overset{\overset{\displaystyle H}{|}}{N}-\overset{\overset{\displaystyle O}{\|}}{C}-(CH_2)_6\overset{\overset{\displaystyle O}{\|}}{C}\right]_n OH + (2n-1)H_2O \quad (1.15)
$$

which is named nylon 48. Note that the amine carbon number is written first, and the acid carbon number second. For reaction purposes, acyl chlorides are

Table 1.5 Selected stepwise structures and nomenclature

Structure	Name	Where Known
$X = -\overset{\overset{\displaystyle O}{\|}}{C}-O-$	Polyester	
$+CH_2-CH_2-O-\overset{\overset{\displaystyle O}{\|}}{C}-\!\!\bigcirc\!\!-\overset{\overset{\displaystyle O}{\|}}{C}-O\!\!\rightarrow_n$	Poly(ethylene terephthalate)	Dacron®
$X = -\overset{\overset{\displaystyle H}{|}}{N}-\overset{\overset{\displaystyle O}{\|}}{C}-$	Polyamide	
$+N(CH_2\!\!\rightarrow_6\overset{\overset{\displaystyle H}{|}}{N}-\overset{\overset{\displaystyle O}{\|}}{C}+CH_2\!\!\rightarrow_8\overset{\overset{\displaystyle O}{\|}}{C}\!\!\rightarrow_n$	Poly(hexamethylene sebacamide)	Nylon 610[a]
$+\overset{\overset{\displaystyle H}{|}}{N}(CH_2\!\!\rightarrow_5\overset{\overset{\displaystyle O}{\|}}{C}\!\!\rightarrow_n$	Polycaprolactam	Nylon 6
$X = -\overset{\overset{\displaystyle H}{|}}{N}-\overset{\overset{\displaystyle O}{\|}}{C}-O-$	Urethane[b]	
$+CH_2\!\!\rightarrow_4 O\!\!\rightarrow_n$	Polytetrahydrofuran	
$+O(CH_2\!\!\rightarrow_4]_n \overset{\overset{\displaystyle H}{|}}{N}-\overset{\overset{\displaystyle O}{\|}}{C}\!\!\rightarrow_m$	One possible polyurethane	Spandex®
$+O-\underset{\underset{\displaystyle CH_3}{|}}{\overset{\overset{\displaystyle CH_3}{|}}{Si}}\!\!\rightarrow_n$	Poly(dimethyl siloxane)	Silicone rubber

Table 1.5 (*Continued*)

Structure	Name	Where known

Cellulose — Cotton

Cellulose triacetate — Arnel®

Epoxy resins — Epon®

[a]The "6" refers to the number of carbons in the diamine portion, and the "10" to the number of carbons in the diacid.

[b]The urethane group usually links polyethers such as $-CH_2-CH_2-O-$ or polyesters together. The polyethers or polyester moieties may have molecular weights of 600–4000 g/mol.

frequently substituted for the carboxyl groups. An excellent demonstration experiment is described by Morgan and Kwolek (10).

1.4.2.2 Stepwise Nomenclature and Structures

Table 1.5 names some of the more important stepwise polymers. The polyesters have already been mentioned. The nylons are known technically as polyamides. There are two important subseries of nylons, where amine and the carboxylic acid are on different monomer molecules (thus requiring both monomers to make the polymer) or one each on the ends of the same monomer molecule. These are numbered by the number of carbons present in the monomer species. It must be mentioned that the proteins are also polyamides.

Other classes of polymers mentioned in Table 1.5 include the polyurethanes, widely used as elastomers; the silicones, also elastomeric; and the cellulosics, used in fibers and plastics. Cellulose is a natural product.

1.5   CROSS-LINKING, PLASTICIZERS, AND FILLERS

The above provides a brief introduction to simple homopolymers, as made pure. Only a few of these are finally sold as "pure" polymers, such as polystyrene drinking cups and polyethylene films. Much more often, polymers are sold with various additives. That the student may better recognize the polymers, the most important additives are briefly discussed.

On heating, linear polymers flow and are termed thermoplastics. To prevent flow, polymers are sometimes cross-linked (●):

$$(1.16)$$

The cross-linking of rubber is sometimes called vulcanization. Cross-linking bonds the chains together to form a network. The resulting product is called a thermoset, because it does not flow on heating.

Plasticizers are small molecules added to soften a polymer by lowering its glass transition temperature or reducing its crystallinity or melting temperature. The most widely plasticized polymer is poly(vinyl chloride). The distinctive odor of new "vinyl" shower curtains is caused by the plasticizer, for example.

Fillers may be of two types, reinforcing and nonreinforcing. Common reinforcing fillers are the silicas and carbon blacks. The latter are most widely used in automotive tires to improve wear characteristics such as abrasion resistance. Nonreinforcing fillers, such as calcium carbonate, may provide color or opacity or may merely lower the price of the final product.

1.6   THE MACROMOLECULAR HYPOTHESIS

Up until the year 1920, scientists lumped several kinds of finely divided matter into the category of colloids. In that year, Herman Staudinger (11) enunciated the *macromolecular hypothesis*, which states that certain kinds of colloids actually consisted of very long-chained molecules. These came to be called polymers, because many (but not all) were composed of the same repeating unit, or mer. The macromolecular hypothesis is the origin of modern polymer science, leading to our current understanding of how and why such materials as plastics and rubber have the properties they do.

At this time, of course, the field of colloids is flourishing as well. A colloid is defined as a finely divided state of matter in dispersion, each particle being small enough to exhibit Brownian motion. Examples include sulfur sols, soap micelles, and certain forms of fine dusts. Polymer molecules in solution are in the size range of many colloids and also exhibit Brownian motion, so the cause of the confusion among early scientists is understandable.

1.7 HISTORICAL DEVELOPMENT OF INDUSTRIAL POLYMERS

Like most other technological developments, polymers were first used on an empirical basis, with only very incomplete understanding of the relationships between structure and properties. The first polymers used were natural products that date back to antiquity. The first chemical derivative of a natural polymer manufactured was cellulose nitrate, which was popular around 1885 for stiff collars and cuffs as celluloid (11).

The first truly synthetic polymer was a densely cross-linked material based on the reaction of phenol and formaldehyde. The product, called Bakelite, was manufactured from 1910 onward for applications ranging from electrical appliances to phonograph records (12, 13). However, very little was known about the actual chemical structure of the polymer until after 1920. In that year, Herman Staudinger enunciated the macromolecular hypothesis (see Section 1.6).

Bakelite was a thermoset; that is, it did not flow after the synthesis was complete. The first synthetic thermoplastics, that is, materials that could flow on heating, were poly(vinyl chloride), poly(styrene–stat–butadiene), polystyrene, and nylon 66 (see Table 1.6) (14). Other breakthrough polymers have included the very high modulus aromatic polyamides, known as Kevlar™, and a host of high-temperature polymers.

Table 1.6 Commercialization dates of selected synthetic polymers (14)

Year	Polymer	Producer
1909	Poly(phenol-*co*-formaldehyde)	General Bakelite Corporation
1927	Poly(vinyl chloride)	B. F. Goodrich
1929	Poly(styrene–*stat*–butadiene)	I. G. Farben
1930	Polystyrene	I. G. Farben/Dow
1936	Poly(methyl methacrylate)	Rohm and Haas
1936	Nylon 66	DuPont
1939	Polyethylene	ICI
1943	Poly(dimethyl siloxane)	Dow Corning
1954	Poly(ethylene terephthalate)	ICI
1960	Poly(*p*-phenylene terephthalamide)[a]	DuPont
1982	Polyetherimide	GEC

[a]Kevlar, see Chapter 7.

1.8   MOLECULAR ENGINEERING

The above discussion shows that polymer science is an admixture of pure and applied science. The structure, molecular weight, and shape of the polymer molecule are all closely tied to the physical and mechanical properties of the final material.

This book emphasizes physical polymer science, the science of the interrelationships between polymer structure and properties. Although much of the material (except the polymer syntheses) is developed in greater detail in the remaining chapters, the intent of this chapter is to provide an overview of the subject and a simple recognition of polymers as encountered in everyday life. In addition to the books in the General Reading section, a listing of handbooks and encyclopedias is given at the end of this chapter.

REFERENCES

1. L. Mandelkern and G. M. Stack, *Macromolecules*, **17**, 87(1984).
2. P. J. Flory, *Principles of Polymer Chemistry*, Cornell University Press, Ithaca, NY, 1953.
3. P. J. Flory, *J. Am. Chem. Soc.*, **67**, 2048 (1945).
4. L. E. Nielsen, *Mechanical Properties of Polymers*, Reinhold, New York, 1962, pp. 115–118.
5. P. E. Slade, Jr., Ed., *Polymer Molecular Weights*, Parts I and II, Dekker, New York, 1975.
6. T. C. Ward, *J. Chem. Ed.*, **58**, 867(1981).
7. L. H. Sperling et al., *J. Chem. Ed.*, **62**, 780, 1030(1985).
8. IUPAC Nomenclature of Regular Single-Stranded Polymers, *Pure Appl. Chem.*, **48**, 373(1976).
9. A Classification of Linear Single-Strand Polymers, *Pure Appl. Chem.*, **61**, 243(1989).
10. P. W. Morgan and S. L. Kwolek, *J. Chem. Ed.*, **36**, 182, 530(1959).
11. H. Staudinger, *Die Hochmolecular Organischen Verbindung*, Springer-Verlag, Berlin, 1932 (reprinted 1960).
12. H. Morawitz, *Polymers: The Origins and Growth of a Science*, Wiley-Interscience, New York, 1985.
13. L. H. Sperling, *Polymer News*, **12**, 332(1987).
14. L. A. Utracki, *Polymer Alloys and Blends*, Hanser, New York, 1990.

GENERAL READING

J. J. Aklonis and W. J. MacKnight, *Introduction to Polymer Viscoelasticity*, Wiley, New York, 1983.

H. R. Allcock and F. W. Lampe, *Contemporary Polymer Chemistry*, 2nd ed., Prentice-Hall, Englewood Cliffs, NJ, 1990.

F. W. Billmeyer, Jr., *Textbook of Polymer Science*, 3rd ed., Wiley, New York, 1984.

U. Eisele, *Introduction to Polymer Physics*, Springer-Verlag, Berlin, 1990.

H. G. Elias, *Macromolecules*, Vols. 1 and 2, Plenum, New York, 1984.

H. G. Elias et al., *Polym. News*, **9**, 101(1983).

H. Feuerberg, *J. Polym. Sci. Polym. Lett. Ed.*, **22**, 413(1984).

C. Hall, *Polymer Materials: An Introduction for Technologists and Scientists*, Wiley, New York, 1981.

P. C. Heimenz, *Polymer Chemistry: The Basic Concepts*, Dekker, New York, 1984.

N. Lozac'h, *Pure Appl. Chem.*, **55**(9), 1463(1983).

L. Mandelkern, *An Introduction to Macromolecules*, 2nd ed., Springer-Verlag, New York, 1983.

P. Munk, *Introduction to Macromolecular Science*, Wiley-Interscience, New York, 1989.

I. I. Perepechko, *Introduction to Polymer Physics*, Mir, Moscow, 1981.

F. Rodriguez, *Principles of Polymer Systems*, 3rd ed., Hemisphere, New York, 1989.

R. B. Seymour and C. E. Carraher, *Polymer Chemistry: An Introduction*, 2nd ed., Dekker, New York, 1988.

A. Tager, *Physical Chemistry of Polymers*, 2nd ed., Mir, Moscow, 1978.

HANDBOOKS AND ENCYCLOPEDIAS

1. H. Saechtling, *International Plastics Handbook*, 2nd ed. English translation, Hanser-Macmillan, New York, 1987.

2. *Engineered Materials Handbook*, *Volume 2*, *Engineering Plastics*, ASM International, Metals Park, OH, 1988.

3. J. Brandrup and E. H. Immergut, Eds., *Polymer Handbook*, 3rd ed., Wiley-Interscience, New York, 1990.

4. J. I. Kroschwitz, Ed., *Encyclopedia of Polymer Science and Engineering*, 2nd ed., Wiley, New York, 1985–1990.

5. *Modern Plastics Encyclopedia*, McGraw-Hill, New York, 1990.

6. G. Allen, Ed., *Comprehensive Polymer Science*, Pergamon, Oxford, 1989.

7. *Compendium of Macromolecular Nomenclature*, IUPAC, CRC Press, Boca Raton, FL, 1991.

STUDY PROBLEMS

1. Polymers are obviously different from small molecules. How does polyethylene differ from oil, grease, and wax, all of these materials being essentially $-CH_2-$?

2. Write chemical structures for polyethylene, polypropylene, poly(vinyl chloride), polystyrene, and nylon 66.

3. Name the following polymers:

$$+CH_2-\underset{\underset{O=C-O-CH_3}{|}}{\overset{\overset{H}{|}}{C}}\!\!\underset{n}{}$$

(a)

$$+CH_2-\underset{\underset{O=C-O-C_2H_5}{|}}{\overset{\overset{CH_3}{|}}{C}}\!\!\underset{n}{}$$

(b)

$$+CH_2-\underset{\underset{O-C-CH_3}{\underset{\|}{|}}}{\overset{\overset{H}{|}}{C}}\!\!\underset{n}{}$$
$$O$$

(c)

$$+CH_2-CF_2\!\!\underset{n}{}$$

(d)

4. What molecular characteristics are required for good mechanical properties? Distinguish between amorphous and crystalline polymers.

5. Show the synthesis of nylon 610 from the monomers.

6. Name some commercial polymer materials by chemical name which are: (a) amorphous, cross-linked, and above T_g; (b) crystalline at use temperatures.

7. Take any 10 books off a shelf and note the last page number. What are the number-average and weight-average number of pages of these books? Why is the weight-average number of pages greater than the number-average? What is the polydispersity index? Can it ever be unity?

8. Draw a log modulus–temperature plot for an amorphous polymer. Where do the five regions of viscoelasticity fit? To which regions do the following belong at room temperature: chewing gum, rubber bands, Plexiglas?

9. Define the terms: Young's modulus, ultimate tensile strength, chain entanglements, and glass–rubber transition.

10. A cube 1 cm on a side is made up of one giant polyethylene molecule, having a density of 1.0 g/cm^3. (a) What is the molecular weight of this molecule? (b) Assuming an all trans conformation, what is the contour length of the chain (length of the chain stretched out)?

APPENDIX 1.1 NAMES FOR POLYMERS

The IUPAC Macromolecular Nomenclature Commission has developed a systematic nomenclature for polymers (9). The Commission recognized, however, that a number of common polymers have semisystematic or trivial names that are well established by usage. For the reader's convenience, the recommended trivial name (or the source-based name) of the polymer is given under the polymer structure, and then the structure-based name is given. For example, the trivial name, polystyrene, is a source-based name, literally "the polymer made from styrene." The structure-based name, poly(1-phenylethylene), is useful both in addressing people who may not be familiar with the structure of polystyrene and in cases where the polymer is not well known. This book uses a source-based nomenclature, unless otherwise specified.

$\{CH_2CH_2\}_n$

polyethylene
poly(methylene)

$\{CHCH_2\}_n$
$\quad\ |$
$\quad CH_3$

polypropylene
poly(1-methylethylene)

$$\{CH_2\!-\!\underset{\underset{CH_3}{|}}{\overset{\overset{CH_3}{|}}{C}}\}_n$$

polyisobutylene
poly(1,1-dimethylethylene)

$\{CHCH_2\}_n$
$\quad\ |$
$\quad OH$

poly(vinyl alcohol)
poly(1-hydroxyethylene)

$\{CHCH_2\}_n$
$\quad\ |$
$\quad Cl$

poly(vinyl chloride)
poly(1-chloroethylene)

$\{CH\!=\!CHCH_2CH_2\}_n$

polybutadiene*
poly(1-butenylene)

$\{C\!=\!CHCH_2CH_2\}_n$
$\ \ |$
$\ CH_3$

polyisoprene**
poly(1-methyl-1-butenylene)

$\{CHCH_2\}_n$
$\quad\ |$

polystyrene
poly(1-phenylethylene)

$\{CHCH_2\}_n$
$\quad\ |$
$\quad CN$

polyacrylonitrile
poly(1-cyanoethylene)

$\{CHCH_2\}_n$
$\quad\ |$
$\quad OOCCH_3$

poly(vinyl acetate)
poly(1-acetoxyethylene)

$\{\underset{\underset{F}{|}}{\overset{\overset{F}{|}}{C}}CH_2\}_n$

poly(vinylidene fluoride)
poly(1,1-difluoroethylene)

---

*Polybutadiene is usually written $\{CH_2CH\!=\!CHCH_2\}$, that is, with the double bond in the center. The structure-based name is given.

** Polyisoprene is usually written $\{CH_2\underset{}{\overset{\overset{CH_3}{|}}{C}}\!=\!CHCH_2\}_n$.

$$+CF_2CF_2\!+_n$$

poly(tetrafluoroethylene)
poly(difluoromethylene)

$$+CHCH_2\!+_n$$
$$\quad\ |$$
$$\quad COOCH_3$$

poly(methyl acrylate)
poly[1-(methoxycarbonyl)ethylene]

$$+OCH_2\!+_n$$

polyformaldehyde
poly(oxymethylene)

$$+NH(CH_2)_6NHCO(CH_2)_4CO\!+_n$$
nylon 66*
poly(hexamethylene adipamide)
poly(iminohexamethyleneiminoadipoyl)

$$+OCH_2CH_2OOC\!-\!\!\bigcirc\!\!-CO\!+_n$$

poly(ethylene terephthalate)
poly(oxyethyleneoxyterephthaloyl)

poly(vinyl butyral)
poly[(2-propyl-1,3-dioxane-4,
6-diyl)methylene]

$$\begin{array}{c} CH_3 \\ | \\ +C\!-\!CH_2\!+_n \\ | \\ COOCH_3 \end{array}$$

poly(methyl methacrylate)
poly[1-(methoxycarbonyl)-
1-methylethylene]

$$+O\!-\!\!\bigcirc\!\!+_n$$

poly(phenylene oxide)
poly(oxy-1,4-phenylene)

$$+OCH_2CH_2\!+_n$$

poly(ethylene oxide)
poly(oxyethylene)

$$+NHCO(CH_2)_5\!+_n$$

nylon 6†
poly(ε-caprolactam)
poly[imino(1-oxohexamethylene)]

---

*Common name. Other ways this is written include nylon 6,6, 66-nylon, and 6,6-nylon.
†Common name.

2

CHAIN STRUCTURE
AND CONFIGURATION

In the teaching of physical polymer science, a natural progression of material begins with chain structure, proceeds through morphology, and leads on to physical and mechanical behavior. To a significant measure, one step determines the properties of the next (1). Polymer chains have three basic properties:

1. The molecular weight and molecular weight distribution of the molecules. These properties are discussed in Chapter 3.

2. The conformation of the chains in space. The term conformation refers to the different arrangements of atoms and substituents of the polymer chain brought about by rotations about single bonds. Examples of different polymer conformations include the fully extended planar zigzag, helical, folded chain, and random coils. Some conformations of a random coil might be

$$\text{(2.1)}$$

The methods of determining polymer chain conformation are discussed in Chapters 3 and 5.

3. The configuration of the chain. The term configuration refers to the organization of the atoms along the chain. Some authors prefer the term "microstructure" rather than configuration. Configurational isomerism in-

volves the different arrangements of the atoms and substituents in a chain, which can be interconverted only by the breakage and reformation of primary chemical bonds. The configuration of polymer chains constitutes the principal subject of this chapter.

2.1 EXAMPLES OF CONFIGURATIONS AND CONFORMATIONS

2.1.1 Head-to-Head and Head-to-Tail Configurations

Before proceeding with the development of theory and instruments, a simple but important example of chain configuration is given. This involves the difference between head-to-head and head-to-tail placement of the monomeric units, or mers, during polymerization. The head-to-tail structure of polystyrene may be written

$$\sim CH_2 - CH - CH_2 - CH \sim \qquad (2.2)$$

and its head-to-head structure may be written

$$\sim CH_2 - CH - CH - CH_2 \sim \qquad (2.3)$$

The thermodynamically and spatially preferred structure is usually the head-to-tail configuration, although most addition polymers contain a small percentage of head-to-head placements. If the synthesis is deliberately arranged so that the head-to-head configuration is obtained, the properties of the polymer are far different. Using polyisobutylene as an example, Malanga and Vogl (2) showed that the melting temperature of the head-to-head configuration was 187°C, whereas the head-to-tail configuration could only be crystallized under stress, and then with a melting temperature of 5°C. The head-to-head and head-to-tail configurations cannot be interchanged without breaking primary chemical bonds.

2.1.2 Trans–Gauche Conformations

The trans–gauche conformations of a polymer chain can be interchanged by simple rotation about the single bond linking the moieties. The trans and gauche states are defined as follows. Consider a sequence of carbon–carbon single bonds delineated $\ldots i - 1, i, i + 1, \ldots$. The rotational angle of bond

i is defined as the angle between two planes, the first plane defined by bonds $i - 1$ and i and the second by bonds i and $i + 1$. The planar zigzag conformation, where the angle between the two planes is zero, is called the trans state, t. When the angle between the two planes is $\pm 120°$, the gauche plus, g^+, and gauche minus, g^-, states are defined:

$$ \tag{2.4} $$

$$t\ t\ t \qquad\qquad g^+ t g^-$$

The bond symbol ➤ means pointed out of the paper toward the reader. The trans and gauche positions are located 120° apart on an imaginary cone of rotation, where the preferred positioning avoids groups on the neighboring chain carbon atoms, the trans being the more extended conformation. The trans and gauche conformations are discussed further in Section 2.8.

2.2 THEORY AND INSTRUMENTS

Koenig (3) defines the microstructure of a polymer in terms of its conformation and configuration. The term conformation has taken on two separate meanings: (a) the long-range shape of the entire chain, which is discussed in Chapter 5, and (b) the several possibilities of rotating atoms or short segments of chain relative to one another, to be discussed later. The term configuration includes its composition, sequence distribution, steric configuration, geometric and substitutional isomerism, and so on and is the major concern of this chapter.

The several aspects of polymer chain microstructure have been studied by both chemical and physical methods. Koenig (3) describes several of these methods, which are summarized in Tables 2.1 and 2.2.

2.2.1 Chemical Methods of Determining Microstructure

The most basic method of characterizing any material uses elemental analysis (Table 2.1). Elemental analysis helps identify unknowns, confirms new syntheses, and yields information on the purity of the polymer.

Functional group analysis relates to those reactions that polymers undergo, either intentionally or accidently. Selective degradation refers to those chemical reactions that a polymer undergoes which cut particular bonds. These may be main chain or side chain. Similarly, cyclization reactions and cooperative reactions enable particular sequences to be identified. It must be emphasized that all these methods of characterization are widely used

Table 2.1 Chemical methods of determining polymer chain microstructure (3)

Method	Application	Reference
Elemental analysis	Gross composition of polymers and copolymers, yielding the percent composition of each element; C, H, N, O, S, and so on.	(a)
Functional group analysis	Reaction of a specific group with a known reagent. Acids, bases, and oxidizing and reducing agents are common. Example: titration of carboxyl groups.	(b, c)
Selective degradation	Selective scissions of particular bonds, frequently by oxidation or hydrolysis. Example: ozonalysis of polymers containing double bonds.	(d)
Cyclization reactions	Sequence analysis through formation of lactones, lactams, imides, α-tetralenes, and endone rings.	(e)
Cooperative reactions	Sequence analysis using reactions of one group with a neighboring group.	(f)

References: (a) F. E. Critchfield and D. P. Johnson, *Anal. Chem.*, **33**, 1834(1961). (b) S. Siggia, *Quantitative Organic Analysis via Functional Groups*, 3rd ed., Wiley, New York, 1963. (c) N. Bikales, *Characterization of Polymers*, Encyclopedia of Polymer Science and Technology, Wiley-Interscience, New York, 1971, p. 91. (d) R. Hill, J. R. Lewis, and J. Simonsen, *Trans. Faraday Soc.*, **35**, 1073(1939). (e) M. Tanaka, F. Nishimura, and T. Shono, *Anal. Chim. Acta*, **74**, 119(1975). (f) J. J. Gonzales and P. C. Hammer, *Polym. Lett.* **14**, 645(1976).

throughout the field of chemistry for big and little molecules alike. This last statement holds also for the physical methods.

2.2.2 General Physical Methods

The more important physical methods of characterizing the microstructure of a polymer are summarized in Table 2.2. Nuclear magnetic resonance and infrared and Raman spectroscopy are considered in the following sections (4).

Ultraviolet and visible light spectroscopy makes use of the quantized nature of the electronic structure of molecules. One example that is commonly observed by eye is the yellow color of polymeric materials that have been slightly degraded by heat or oxidation. Frequently, this is due to the appearance of conjugated double bonds (5). For example, the 10-polyene conjugated structure absorbs light at 473 nm, in the blue region.

Table 2.2   Physical methods of determining polymer chain microstructure (3)

Method	Application	Reference
Nuclear magnetic resonance	Determination of steric configuration in homopolymers; composition of copolymers, including proteins; chemical functionality, including oxidation products; determination of structural and geometric and substitutional isomerism, conformation, and copolymer microstructure.	(a–c)
Infrared and Raman spectroscopy (considered together)	Molecular identification: determination of chemical functionality; chain and sequence length; quantitative analysis; stereochemical configuration; chain conformation.	(d, e)
Ultraviolet and visible light spectroscopy	Sequence length; conformation and spatial analysis	(f)
Mass Spectroscopy	Polymer degradation mechanisms; order and randomness of block copolymers.	(g)
Electron spectroscopy (ESCA)	Microstructure of polymers, particularly surfaces.	(h)
X-ray and electron diffraction (considered together)	Identification of repeat unit in crystalline polymers; inter- and intramolecular spacings; chain conformation and configuration.	(i)

References: (a) F. Bovey, *High Resolution NMR of Macromolecules*, Academic Press, New York, 1972. (b) C. C. McDonald, W. D. Phillips, and J. D. Glickson, *J. Am. Chem. Soc.*, **93**, 235(1971). (c) J. C. Randall, *J. Polym. Sci. Polym. Phys. Ed.*, **13**, 889(1975). (d) J. Haslam, H. A. Willis, and M. Squirrell, *Identification and Analysis of Plastics*, 2nd ed., Ileffe, London, 1972. (e) J. L. Koenig, *Appl. Spectrosc. Rev.*, **4**, 233(1971). (f) Y. C. Wang and M. A. Winnik, *Macromolecules*, **23**, 4731(1990). (g) A. K. Lee and R. D. Sedgwick, *J. Polym. Sci. Polym. Chem. Ed.*, **16**, 685(1978). (h) D. T. Clark and W. J. Feast, *J. Macromol. Sci.*, **C12**, 191(1975). (i) G. Natta, *Makromol. Chem.*, **35**, 94(1960).

Mass spectroscopy makes use of polymer degradation, particular masses emerging being identified. Electron spectroscopy for chemical applications (ESCA) is a relatively new method useful for surface analysis of polymers (see Section 11.1).

X-ray (6) and electron diffraction methods are most useful for determining the structure of polymers in the crystalline state and are discussed in Chapter 6. These methods do, however, provide a wealth of information relative to the inter- and intramolecular spacings, which can be interpreted in terms of conformations and configurations.

2.2.3   Infrared and Raman Spectroscopic Characterization

The total energy of a molecule, big or small, consists of contributions from the rotational, vibrational, electronic and electromagnetic spin energies. These states define the temperature of the system. Specific energies may be increased or decreased by interaction with electromagnetic radiation of a specified wavelength. In the following discussion, it is important to remember that all such interactions are quantized; that is, only specific energy levels are permitted.

Infrared spectra are obtained by passing infrared radiation through the sample of interest and observing the wavelength of absorption peaks. These peaks are caused by the absorption of the electromagnetic radiation and its conversion into specific molecular motions, such as $C - H$ stretching.

The older, conventional instruments are known as dispersive spectrometers, where the infrared radiation is divided into frequency elements by the use of a monochromator and slit system. Although these instruments are still in use today, the recent introduction of Fourier transform infrared (FT-IR) spectrometers has revitalized the field (4). The FT-IR system is based on the Michelson interferometer. The total spectral information is contained in an interferogram from a single scan of a movable mirror. There are no slits, and the amount of infrared energy falling on the detector is greatly enhanced. Together with the use of modern computer techniques, an entirely new breed of instrument has been created.

Raman spectra (7) are obtained by a variation of a light-scattering technique whereby visible light is passed into the sample. In addition to light of the same wavelength being scattered, there is an inelastic component. The physical cause relates to the light's exchanging energy with the molecule. This inelastic scattering causes light of slightly longer or shorter wavelengths to be scattered. As above, there is an increase or decrease in a specific molecular motion.

Raman and infrared spectroscopy are complementary because they are governed by different selection rules (4, 7, 8). In order for Raman scattering to occur, the electric field of the light must induce a dipole moment by changing the polarizability of the molecule. By contrast, infrared requires an intrinsic dipole moment to exist, which must change with molecular vibration.

The fields have advanced way beyond the simple determination of spectra and correlating particular bands with particular chemical groups. Today, specific motions are calculated. As an example, see Figure 2.1 (4). Here, two conformational displacements of isotactic polystyrene[†] are shown—one near 550 cm^{-1} in the infrared spectrum, and one near 225 cm^{-1} in the Raman spectrum. These motions illustrate a degree of coupling between the ring and backbone vibrations.

[†]The term isotactic refers to a specific configuration to be defined later.

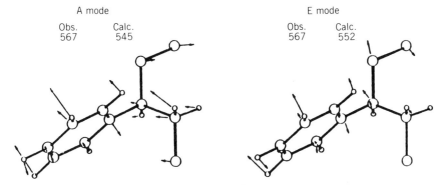

Figure 2.1 Motions associated with the 567 cm^{-1} infrared peak. Quite different motions are associated with the Raman peak at 225 cm^{-1}, not shown. Isotactic polystyrene.

2.2.4 Nuclear Magnetic Resonance Methods

Although x-ray (6), Raman spectroscopy (7), and infrared methods (8) are at the disposal of the polymer scientist for stereochemical configuration analyses, by far the most powerful method is nuclear magnetic resonance (NMR). Briefly, when the spin quantum number of a nucleus is $\frac{1}{2}$ or greater, it possesses a magnetic moment. A proton has a spin of $\frac{1}{2}$ and is widely used in NMR studies. When placed in a magnetic field H_0, it can occupy either of two energy levels, which corresponds to its magnetic moment, μ, being aligned with or against the field. The energy differences in the two orientations are given by Bovey et al. (9):

$$\Delta E = h\nu = 2\mu H_0 \tag{2.5}$$

The quantity ΔE indicates the energy that must be absorbed to raise the nuclei in the lower state up to the higher level and is emitted in the reverse process. The separation of energy levels is proportional to the magnetic field strength. In a field of 9400 gauss, the resonant frequency for protons is about 40 MHz. For field strengths of the order of 10,000 gauss and up, the frequency, ν, is in the microwave region.

In a molecule containing many atoms, the field on any one of these is altered by the presence of the others:

$$\Delta E = 2\mu(H_0 + H_L) \tag{2.6}$$

where H_L is the local field with a strength of 5–10 gauss. It is these changes that are important in NMR characterization.

Other nuclei besides hydrogen (^{1}H) that have a spin of $\frac{1}{2}$ or greater, and are used in NMR studies include deuterium, (^{2}H), fluorine. (^{19}F), carbon-13 (^{13}C), nitrogen-14 and -15 (^{14}N and ^{15}N), and phosphorus-31 (^{31}P). Much higher resolutions are often possible with these nuclei, allowing exact sequences of structures to be determined along the chain.

A new technique for ^{13}C NMR is the so-called magic angle method, which uses oriented specimens spun around an axis at $\theta = 54.7°$ to reduce line broadening due to anisotropic contributions. This particular angle arises because the broadening component is proportional to the quantity $3\cos^2\theta - 1$, where θ is the angle between the line connecting the nuclei and the direction of the magnetic field in isotropic compositions. At $54.7°$, this quantity is zero.

While the fundamental unit for determining shifts in NMR peaks is the change in frequency in hertz, an important practical scale is based on the position of the tetramethylsilane peak, leading to the τ scale (see Figure 2.6), for example. Scales based on shifts of parts per million, ppm, are also common (see Figure 2.8).

2.3 STEREOCHEMISTRY OF REPEATING UNITS

2.3.1 Chiral Centers

Early in the history of organic compounds, two substances were sometimes found that appeared to be chemically identical except that they rotated plane-polarized light equally, but in opposite directions. With the development of the tetrahedral carbon bond model, it gradually became clear that the two isomers were mirror images of each other. As an illustration, the two possible spatial configurations of the compound

$$
\begin{array}{c}
H \\
| \\
Br - C - Cl \\
| \\
CH_3
\end{array}
\qquad (2.7)
$$

form the mirror images shown in Figure 2.2 (10). The cause of the optical activity is the asymmetric carbon in the center, known as a chiral center. The two different compounds are known as enantiomorphs, or enantiomers. The important point is that the two mirror images are nonsuperimposable, and the two compounds are really different.

2.3.2 Tacticity in Polymers

The polymerization of a monosubstituted ethylene, such as a vinyl compound, leads to polymers in which every other carbon atom is a chiral center. This is

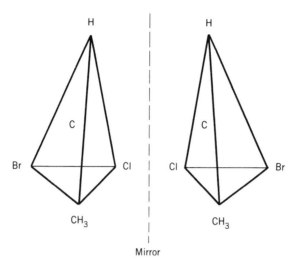

Figure 2.2   Two optical isomers of the same compound as mirror images (10).

often marked with an asterisk for emphasis:

$$CH_2 {=} \overset{\overset{\displaystyle H}{|}}{\underset{\underset{\displaystyle R}{|}}{C}} \rightarrow {-}CH_2 {-} \overset{\overset{\displaystyle H}{|}}{\underset{\underset{\displaystyle R}{|}}{C^*}} {-} \qquad (2.8)$$

Such carbon atoms are referred to as pseudochiral centers in long-chain polymers, because the polymers do not in fact exhibit optical activity (11). The reason for the lack of optical activity can be seen through a closer examination of the substituents on such a pseudochiral center:

$$\text{\wwww} \overset{\overset{\displaystyle H}{|}}{\underset{\underset{\displaystyle R}{|}}{C^*}} \text{\wwww} \qquad (2.9)$$

The two chain segments are indicated by ⌇⌇ and ⌇⌇⌇ and in general will be of unequal length. The first few atoms of the two chain segments attached to C* are responsible for the optical activity, not those farther away. These near atoms are seen to be the same, and hence the polymer is optically inactive.

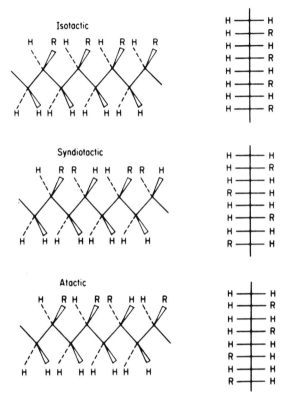

Figure 2.3   Three different configurations of a monosubstituted polyethylene,

$$+CH_2 - CHR +_n .$$

The dotted and triangular lines represent bonds to substitutents below and above the plane of the carbon – carbon backbone chain, respectively (11).

The two mirror image configurations remain distinguishable, however. The different possible spatial arrangements are called the tacticity of the polymer. If the R groups on successive pseudochiral carbons all have the same configuration, the polymer is called isotactic (see Figure 2.3) (11). When the pseudochiral centers alternate in configuration from one repeating unit to the next, the polymer is called syndiotactic. If the pseudochiral centers do not have any particular order, but in fact are statistical arrangements, the polymer is said to be atactic.

Figure 2.3 portrays the general three-dimensional structure of the polymer (11). Using Fischer–Hirshfelder or similar models, the actual differences between isotactic and syndiotactic poly(vinyl chloride) can be illustrated (see Figure 2.4). A two-dimensional analogue can be made using what are known as Fisher projections. In these projections, the R groups are placed either up

Figure 2.4 Isotactic and syndiotactic structures of poly(vinyl chloride). Allyn and Bacon Molecular Model Set for Organic Chemistry.

or down. All up (or all down) indicates the isotactic structure:

$$
\begin{array}{cccccccc}
H & H & H & H & H & H & H & H \\
| & | & | & | & | & | & | & | \\
-C & -C & -C & -C & -C & -C & -C & -C- \\
| & | & | & | & | & | & | & | \\
H & R & H & R & H & R & H & R
\end{array}
\tag{2.10}
$$

Alternating up and down indicates syndiotactic:

$$
\begin{array}{cccccccc}
H & H & H & R & H & H & H & R \\
| & | & | & | & | & | & | & | \\
-C & -C & -C & -C & -C & -C & -C & -C- \\
| & | & | & | & | & | & | & | \\
H & R & H & H & H & R & H & H
\end{array}
\tag{2.11}
$$

and random up and down indicates atactic:

$$
\begin{array}{cccccccc}
H & H & H & H & H & R & H & H \\
| & | & | & | & | & | & | & | \\
-C & -C & -C & -C & -C & -C & -C & -C- \\
| & | & | & | & | & | & | & | \\
H & R & H & R & H & H & H & R
\end{array}
\tag{2.12}
$$

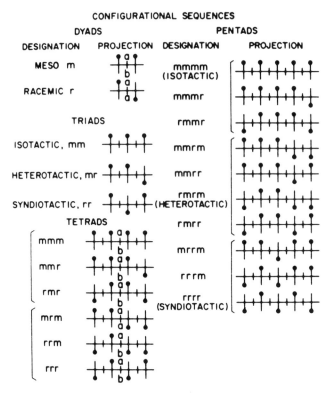

Figure 2.5 Configurational sequences in monosubstituted ethylenes projected in two dimensions (12).

In specifying the tacticity of the polymer, the prefixes *it* and *st* are placed before the name or structure to indicate isotactic and syndiotactic structures, respectively. For example, *it*-polystyrene means that the polystyrene is isotactic. Such polymers are known as stereoregular polymers. The absence of these terms denotes the corresponding atactic structure.

The structures shown in equations (2.10)–(2.12) result in profoundly different physical and mechanical behavior. The isotactic and syndiotactic structures are both crystallizable, because of their regularity along the chain. However, their unit cells and melting temperatures are not the same. Atactic polymers, on the other hand, are usually completely amorphous unless the side group is so small or so polar as to permit some crystallinity.[†]

2.3.3 Meso- and Racemic Placements

The Fisher projection in equation (2.10) shows that the placement of the groups corresponds to a meso- (same) or *m* placement of a pair of consecu-

[†]Two atactic polymers that crystallize are poly(vinyl chloride) and poly(vinyl alcohol).

tive pseudochiral centers. The syndiotactic structure in equation (2.11) corresponds to a racemic (opposite) or r placement of the corresponding pair of pseudochiral centers. It must be emphasized that the m or r notation refers to the configuration of one pseudochiral center relative to its neighbor. Several possible configurational sequences are illustrated in Figure 2.5 (12). Each of these, and even more complicated combinations, can be distinguished through NMR studies, as described later.

2.3.4 Proton Spectra by NMR

The 40-MHz ^1H spectrum of two samples of poly(methyl methacrylate) are illustrated in Figure 2.6 (13). The sample marked (a) was prepared via free radical polymerization methods (see Chapter 1). The sample marked (b) was synthesized by a then new method, anionic polymerization. The anionic polymerization method was thought to make samples predominantly isotactic, whereas free radical methods resulted in atactic polymers.

The peaks in Figure 2.6 were assigned as follows (13). The large peak at the left of both (a) and (b) is that of the chloroform solvent. The methyl ester group appears at 6.40 τ in both spectra and is unchanged by the chain configuration. At 8.78 τ, 8.95 τ, and 9.09 τ are three α-methyl peaks whose

Figure 2.6 The 40-MHz NMR proton spectra of poly(methyl methacrylate) in chloroform. (a) Atactic polymer prepared using a free radical initiator. (b) Isotactic polymer prepared using n-butyllithium initiator by anionic polymerization (13).

relative heights vary greatly with the method of synthesis. Note that the peak at 8.78 τ is much larger in (b) than in (a), and the peak at 9.09 τ is the more prominent in (a). (The τ values and the τ scale refer to a system in which the tetramethylsilane peak is assigned the arbitrary value of $+10.000$ ppm by definition and is shown on the extreme right in Figure 2.6. These τ values, measured for thousands of organic compounds in carbon tetrachloride solution, are widely used for comparison and identification.)

The peak at 8.78 τ was assigned to the configuration wherein the α-methyl groups of the monomer residues are flanked on both sides by mers of the same configuration—that is, all m-placement. The most prominent peak in the free radical polymerized polymers, at 9.09 τ, is attributed to α-methyl groups of central monomer units in syndiotactic, r-placement configuration. The peak at 8.95 τ is assigned to α-methyl groups in heterotactic configurations. On noting that free radical polymerizations, especially those at low temperatures, tend to be predominantly syndiotactic, the assignment becomes clear. In these early materials, however, the structures were not all one configuration, especially the anionically prepared polymer.

Returning to Figure 2.5, the relative frequencies of the several possibilities have certain necessary relationships (see Table 2.3) (14). For example, considering the triad relationships,

$$mm + mr + rr = 1 \qquad (2.13)$$

if the polymer is entirely isotactic, the terms mr and rr are both zero. These

Table 2.3   Algebraic relations among sequence frequencies (14)

Dyad	$(m) + (r) = 1$
Triad	$(mm) + (mr) + (rr) = 1$
Dyad–triad	$(m) = (mm) + \frac{1}{2}(mr)$
	$(r) = (rr) + \frac{1}{2}(mr)$
Triad–tetrad	$(mm) = (mmm) + \frac{1}{2}(mmr)$
	$(mr) = (mmr) + 2(rmr) + (mrr) + 2(mrm)$
	$(rr) = (rrr) + \frac{1}{2}(mrr)$
Tetrad–tetrad	sum = 1
	$(mmr) + 2(rmr) = 2(mrm) + (mrr)$
Pentad–pentad	sum = 1
	$(mmmr) + 2(rmmr) = (mmrm) + (mmrr)$
	$(mrrr) + 2(mrrm) = (rrmr) + (rrmm)$
Tetrad–pentad	$(mmm) = (mmmm) + \frac{1}{2}(mmmr)$
	$(mmr) = (mmmr) + 2(rmmr) = (mmrm) + (mmrr)$
	$(rmr) = \frac{1}{2}(mrmr) + \frac{1}{2}(rmrr)$
	$(mrm) = \frac{1}{2}(mrmr) + \frac{1}{2}(mmrm)$
	$(rrm) = 2(mrrm) + (mrrr) = (mmrr) + (rmrr)$
	$(rrr) = (rrrr) + \frac{1}{2}(mrrr)$

algebraic relationships provide a quantitative basis for determining the probability of certain sequences occurring.

There are, of course, other types of stereoregular polymers. Some of these are briefly described in Appendix 2.1.

2.4 REPEATING UNIT ISOMERISM

2.4.1 Optical Isomerism

There is one important class of polymers that do exhibit strong optical activity, as opposed to the above tactic structures. These are the polymers in which the chiral center is surrounded by different atoms or groups, and a true local center of asymmetry exists.

An example is poly(propylene oxide),

$$\left[\begin{array}{c} H \\ | \\ -C^* - CH_2 - O - \\ | \\ CH_3 \end{array}\right]_n \tag{2.14}$$

where the chiral center is surrounded by $-H$, $-CH_3$, $-CH_2-$, and $-O-$.

2.4.2 Geometric Isomerism

The most important examples in this class are the cis and trans isomerism about double bonds. Take polybutadiene as an example,

$$\left(\begin{array}{c} \begin{array}{cc} CH_2 & CH_2 \\ \diagdown & \diagup \\ C=C \\ \diagup & \diagdown \\ H & H \end{array} \end{array}\right)_n \quad \left(\begin{array}{c} \begin{array}{cc} CH_2 & H \\ \diagdown & \diagup \\ C=C \\ \diagup & \diagdown \\ H & CH_2 \end{array} \end{array}\right)_n \tag{2.15}$$

$$\text{cis-polybutadiene} \qquad\qquad \text{trans-polybutadiene}$$

The cis–trans isomerism arises because rotation about the double bond is impossible without disrupting the structure. Thus the formula on the left of equation (2.15) is written cis-polybutadiene. The reader should note that the cis–trans isomerism is entirely different from the trans–gauche structures written in equation (2.4).

The cis and trans formulas are both crystallizable when appearing in pure form, but with different melting temperatures. If a mixture of cis and trans isomers occurs, crystallization may be suppressed, similar to the atactic polymers.

2.4.3 Substitutional Isomerism

In the synthesis of diene type polymers, yet another type of isomerism may occur, that of 1,2 versus 1,4 addition:

$$
\begin{array}{cc}
\underset{\substack{\text{1,2-addition}}}{\overset{\displaystyle \begin{array}{c} H \quad H \\ | \quad | \\ -C-C- \\ | \quad | \\ H \quad C-H \\ \| \\ CH_2 \end{array}}{}} &
\underset{\substack{\text{1,4-addition}}}{\overset{\displaystyle \begin{array}{c} H \quad H \quad H \quad H \\ | \quad | \quad | \quad | \\ -C-C=C-C- \\ | \qquad\qquad | \\ H \qquad\qquad H \end{array}}{}}
\end{array}
\qquad (2.16)
$$

In the case of 1,2 addition, polymerization is similar to that of vinyl structures. Note that if the diene is substituted, as in isoprene,

$$
\begin{array}{c}
H \qquad H \quad H \\
| \qquad\; | \quad | \\
C=C-C=C \\
| \qquad | \qquad | \\
H \quad CH_3 \quad H
\end{array}
\qquad (2.17)
$$

1,2, 1,4, and 3,4 polymerizations are each distinguished. Of course, the 1,4 polymerizations also exhibit the cis–trans isomerism simultaneously. All these may appear together in various percentages in a given preparation.

2.4.4 Infrared and Raman Spectroscopic Characterization

The various 1,4 cis- and trans-, as well as the 1,2-structures of polybutadiene, polyisoprene, and the other diene polymers have been characterized by both the Raman (15) and infrared (16) methods. Figure 2.7 (16) illustrates the infrared spectra of several polybutadienes. Note that while some of the bands are common to all the samples, others are different. On the basis of comparison with known molecules including model systems, wet chemistry, and other methods, each band has been assigned to particular segment motions (12). Trans and gauche conformations can be identified by analysis of infrared and Raman spectra. Table 2.4 (17) illustrates the findings for poly(ethylene oxide).

This sort of information serves important analytical purposes as well as being useful in research. For example, the emulsion[†] polybutadiene in Figure

---

[†]Emulsion polymerization is a specialized form of free radical polymerization widely used to make both plastics and elastomers. The particles of polymer formed are submicroscopic. Many emulsions are used directly to make latex paints. See Section 4.5.

Figure 2.7   Infrared spectra of polybutadienes of various configurations (16).

2.7 contains some of each of the possible configurations. By examining the area under each curve, quantitative analyses can be made. Since the physical and mechanical properties of polymers composed of the various ratios of the possible isomers differ, careful determination of the configuration of each preparation is made daily in many chemical industries.

Much addition information can be obtained from infrared and Raman spectra. When a polymer is in the crystalline state, the chains are aligned. If the polymer is melted, the chain conformation becomes disordered. The spectra change accordingly. Many new frequencies appear arising from the new conformations in the melt, and some of the frequencies characteristic of the crystalline state disappear.

For example, in the solid state one structure predominates for poly(ethylene oxide), the *tgt* conformation. The O — C bond is trans, the C — C bond is gauche, and the C — O bond is trans. As illustrated in Table 2.4 (17), all possible conformations exist in the molten state—*tgt*, *tgg*, *ggg*, *ttt*, *ttg*, and

Table 2.4   IR and Raman spectra of molten poly(ethylene oxide) (17)

IR Frequency (cm^{-1})	Raman Frequency (cm^{-1})	Assignment	Form	Models (Tentative)
1485 sh		CH_2 scissor	*t*	*ttt, ttg, gtg*
1460 m	1470 s	CH_2 scissor	*g*	*tgt, tgg, ggg*
	1448 sh	CH_2 scissor	*g*	*tgg, ggg*
1352 m	1352 m	CH_2 wag	*g*	*ggg*
1326 (m)	1326 w	CH_2 wag	*t*	*ttt, ttg, gtg*
1296 m	1292 m	CH_2 twist	*g, t*	All
	1283 s	CH_2 twist	*t*	*ttt, ttg, gtg*
1249 m		CH_2 twist	*g*	*tgg, ggg*
	1239 m	CH_2 twist	*t*	
1140 sh	1134 s	$C-O, C-C$	*g, t*	All
1107 (s)		$C-O, C-C, CH_2$ rock	*g, t*	All
	1052 m(P)	$C-O, C-C$	*t*	*tgg*
1038 (m)		$C-O, C-C, CH_2$ rock	*t*	*ttt*
992 w		$C-O, C-C, CH_2$ rock	*t*	*ttt, ttg*
945 (m)		$C-O, C-C, CH_2$ rock	*g*	*tgt*
~ 915	919 sh	CH_2 rock, $C-O, C-C$	*g*	*tgg, ggg*
886	884 mw(P)	CH_2 rock	*g*	*ggg*
855 (m)	—	CH_2 rock	*g*	*tgg, ggg*
	834 m(P)	CH_2 rock	*t*	*ttt, ttg, gtg*
~ 810 (sh)	807 m(P)	CH_2 rock	*g*	*tgg*
	556 w			
	524 w			
	261 (P)			

gtg. The intensities of the Raman line at 807 cm^{-1} indicate that the *tgg* isomer predominates.

2.5   COMMON TYPES OF COPOLYMERS

In the above, polymers made from only one kind of monomeric unit, or mer, were considered. Many kinds of polymers contain two kinds of mers. These can be combined in various ways to obtain interesting and often highly useful materials. Some of the basic copolymer nomenclature is presented in Table 2.5 (18, 19). If three mers—*A*, *B*, and *C*—are considered, some of the possible copolymers are also named in Table 2.5. The connectives in copolymer nomenclature will be defined below.

Table 2.5   Some copolymer terminology (18, 19)

Type	Connective	Example
Short Sequences		
Unspecified	*–co–*	poly(*A–co–B*)
Statistical	*–stat–*	poly(*A–stat–B*)
Random	*–ran–*	poly(*A–ran–B*)
Alternating	*–alt–*	poly(*A–alt–B*)
Periodic	*–per–*	poly(*A–per–B–per–C*)
Long Sequences		
Block	*–block–*	poly *A–block*–poly *B*
Graft	*–graft–*	poly *A–graft*–poly *B*
Star	*–star–*	*star*–poly *A*
Blend	*–blend–*	poly *A–blend*–poly *B*
Starblock	‚ *–star–* · · · *–block–*	*star*–poly *A–block*–poly *B*
Networks		
Cross-linked	*–cross–*	*cross*–poly *A*
Interpenetrating	*–inter–*	*cross*–poly *A–inter–cross*–poly *B*
Conterminously	*–cross–*	poly *A–cross*–poly *B*

2.5.1   Unspecified Copolymers

An unspecified sequence arrangement of different monomeric units in a polymer is represented by

$$\text{poly}(\,A–co–B\,) \qquad (2.18)$$

Thus an unspecified copolymer of styrene and methyl methacrylate is named

$$\text{poly}[\text{styrene}–co–(\text{methyl methacrylate})] \qquad (2.19)$$

In the older literature, *–co–* was used to indicate a random copolymer, where the mers were added in random order, or perhaps addition preference was dictated by thermodynamic or spatial considerations. These are now distinguished from one another.

2.5.2   Statistical Copolymers

Statistical copolymers are copolymers in which the sequential distribution of the monomeric units obeys known statistical laws. The term *–stat–* embraces a large proportion of those copolymers that are prepared by simultaneous polymerization of two or more monomers in admixture. Thus the term *–stat–* is now preferred over *–co–* for most usage.

The arrangement of mers in a statistical copolymer of A and B might appear as follows:

$$\ldots -A-A-B-A-B-B-B-A-B-A-A-B-A- \ldots \qquad (2.20)$$

The statistical arrangement of mers and A and B is indicated by

$$\text{poly}(A-stat-B) \qquad (2.21)$$

See Table 2.5.

2.5.3 Random Copolymers

A random copolymer is a statistical copolymer in which the probability of finding a given monomeric unit at any given site in the chain is independent of the nature of the neighboring units at that position. Stated mathematically, the probability of finding a sequence $\ldots ABC \ldots$ of monomeric units $A, B, C, \ldots, P(\ldots ABC \ldots)$ is

$$P(\ldots ABC \ldots) = P(A) \cdot P(B) \cdot P(C) \ldots = \prod_i P(i); \qquad i = A, B, C \ldots$$

$$(2.22)$$

where $P(A), P(B), P(C)$, and so on are the unconditional probabilities of the occurrence of the various monomeric units.

2.5.4 Alternating Copolymers

In the above, various degrees of randomness were assumed. An alternating copolymer is just the opposite, comprising two species of monomeric units distributed in alternating sequence:

$$\ldots -A-B-A-B-A-B-A-B-A-B- \ldots \qquad (2.23)$$

Alternating copolymerization is caused either by A or B being unable to add itself, or the rate of addition of the other monomer being much faster than the addition of itself. An important example of an alternating copolymer is

$$\text{poly}[\text{styrene}-alt-(\text{maleic anhydride})] \qquad (2.24)$$

2.5.5 Periodic Copolymers

The alternating copolymer is the simplest case of a periodic copolymer. For three mers,

$$\ldots A-B-C-A-B-C-A-B-C- \ldots \qquad (2.25)$$

the structure is indicated by

$$\text{poly}(\,A\text{--}per\text{--}B\text{--}per\text{--}C\,) \qquad\qquad (2.26)$$

2.6 QUANTITATIVE DETERMINATION OF MER DISTRIBUTION

The above definitions of statistical and random copolymers are idealized. In reality, significant nonrandomness may exist. Since the physical and mechanical behavior of polymers sometimes depends critically on the exact order or lack of order in the copolymer structure, this demands special attention. Randall and Hsieh (20) studied the ^{13}C NMR spectrum of a series of copolymers of ethylene and 1-hexene,

$$CH_2 = CH - CH_2 - CH_2 - CH_2 - CH_3$$

(see Figure 2.8) (20). These data were analyzed in the form of sequence distributions, where E and H represent the two mers, respectively. The resulting triad distributions from this copolymer and another are shown in

Figure 2.8   The 50.3-MHz ^{13}C NMR spectrum for an 83 / 17 ethylene / 1-hexene copolymer. The temperature was 125°C, and the concentration was 15% by weight in 1,2,4-trichlorobenzene (20).

Table 2.6   Triad distributions in two ethylene / 1-hexene copolymers (20)

	83/17 Copolymer	97/3 Copolymer
(EHE)	0.098	0.031
(EHH)	0.053	0.000
(HHH)	0.022	0.000
(HEH)	0.043	0.000
(HEE)	0.164	0.061
(EEE)	0.620	0.908

Table 2.6 (20). Since both of these polymers are rich in ethylene, it is not unexpected that the triad sequence EEE predominates. More interesting are the other triad concentrations, which describe the statistical arrangements of the two mers along the chain.

With the information given in Table 2.6, a "run number" may be calculated. A run number, first introduced by Harwood and Ritchey (21), is defined as the average number of like mer sequences or "runs" occurring in a copolymer per 100 mers. This is calculated as follows:[†]

$$(H) = (HHH) + (EHH) + (EHE) \tag{2.27}$$

$$(E) = (EEE) + (HEE) + (HEH) \tag{2.28}$$

$$\text{Run number} = \left(\tfrac{1}{2}\right)(HE)$$

$$= (EHE) + \tfrac{1}{2}(EHH)$$

$$= (HEH) + \tfrac{1}{2}(HEE) \tag{2.29}$$

The average sequence lengths can then be calculated as follows:

$$\text{Average "}E\text{" sequence length} = (E)/\text{run number} \tag{2.30}$$

$$\text{Average "}H\text{" sequence length} = (H)/\text{run number} \tag{2.31}$$

2.7 MULTICOMPONENT POLYMERS

The above statistical, random, and alternating copolymers describe sequence lengths of one, two, three, or at most several mers. This section treats cases where whole polymer chains are linked together to form still larger polymer structures (10). These structures have been variously named "polymer alloys,"

[†]Note that the number of runs of *both* kinds of mers is twice that calculated here.

or "interpolymers," but the term "multicomponent polymers" is used here to describe this general class of materials.

2.7.1 Block Copolymers

A block copolymer contains a linear arrangement of blocks, a block being defined as a portion of a polymer molecule in which the monomeric units have at least one constitutional or configurational feature absent from the adjacent portions. A block copolymer of A and B may be written

$$\ldots -A-A-A-A-A-A-B-B-B-B-B-B-B-B-B-B- \ldots \quad (2.32)$$

Note that the blocks are linked end on end. Since the individual blocks are usually long enough to be considered polymers in their own right, the polymer is named (18)

$$\text{poly}A-block-\text{poly}B \quad (2.33)$$

An especially important block copolymer is the triblock copolymer of styrene and butadiene (10),

$$\text{polystyrene}-block-\text{polybutadiene}-block-\text{polystyrene} \quad (2.34)$$

In the older literature, $-b-$ was used for $-block-$, and $-g-$ was used for $-graft-$ (below). Only the first poly was indicated. Structure (2.34) was then written

$$\text{poly}(\text{styrene}-b-\text{butadiene}-b-\text{styrene}) \quad (2.35)$$

Table 2.7 defines a number of terms used in block copolymer terminology, as well as other structures described later (19). These structures are also illustrated in Figure 2.9 (see page 47) (19).

2.7.2 Graft Copolymers

A graft copolymer comprises a backbone species and a side-chain species. The side chains comprise units of mer that differ from those comprising the backbone chain. If the two mers are the same, the polymer is said to be branched. The name of a graft copolymer of A and B is written (18)

$$\text{poly}A-graft-\text{poly}B \quad (2.36)$$

Although many of the block copolymers reported in the literature are actually highly blocked, some of the most important "graft copolymers" described in the literature have been shown to be only partly grafted, with much homopolymer being present. To some extent, then, the term graft

Table 2.7 Specialized nomenclature terms (19)

Link—A covalent chemical bond between two monomeric units, or between two chains.

Chain—A linear polymer formed by covalent linking of monomeric units.

Backbone—Used in graft copolymer nomenclature to describe the chain onto which the graft is formed.

Side chain—The grafted chain in a graft copolymer.

Cross-link—A structure bonding two or more chains together.

Network—A three-dimensional polymer structure, where (ideally) all the chains are connected through cross-links.

Multicomponent polymer, multipolymer, and multicomponent molecule—General terms describing intimate solutions, blends, or bonded combinations of two or more polymers.

Copolymer—Polymers that are derived from more than one species of monomer.

Block—A portion of a polymer molecule in which the monomeric units have at least one constitutional or configurational feature absent from the adjacent portions.

Block copolymer—A combination of two or more chains of constitutionally or configurationally different features linked in a linear fashion.

Graft copolymer—A combination of two or more chains of constitutionally or configurationally different features, one of which serves as a backbone main chain, and at least one of which is bonded at some point(s) along the backbone and constitutes a side chain.

Polymer blend—An intimate combination of two or more polymer chains of constitutionally or configurationally different features, which are not bonded to each other.

Conterminous—At both ends or at points along the chain.

Conterminously linked copolymer—A polymer chain that is linked at both ends to the same or to constitutionally or configurationally different chain or chains; a polymer cross-linked by a second species of polymer.

Interpenetrating polymer network—An intimate combination of two polymers both in network form, at least one of which is synthesized and/or cross-linked in the immediate presence of the other.

Semi-interpenetrating polymer network [a]—A combination of two polymers, one cross-linked and one linear, at least one of which was synthesized and/or cross-linked in the immediate presence of the other.

Star polymer—Three or more chains linked at one end through a central moiety.

Star block copolymer—Three or more chains of different constitutional or configurational features linked at one end through a central moiety.

---

[a] Also called a pseudo-interpenetrating polymer network. See D. Klempner, K. C. Frisch, and H. L. Frisch, *J. Elastoplastics*, **5**, 196(1973).

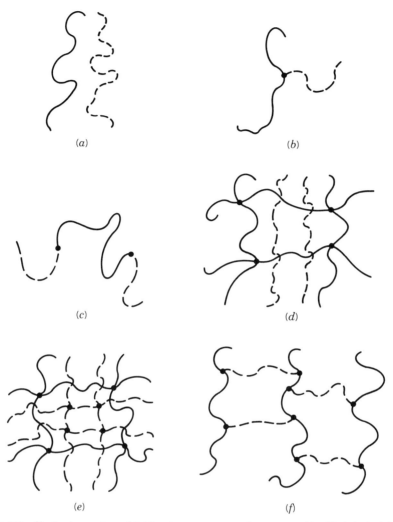

Figure 2.9 Six basic modes of linking two or more polymers are identified (19). (*a*) A polymer blend, constituted by a mixture or mutual solution of two or more polymers, not chemically bonded together. (*b*) A graft copolymer, constituted by a backbone of polymer I with covalently bonded side chains of polymer II. (*c*) A block copolymer, constituted by linking two polymers end on end by covalent bonds. (*d*) A semi-interpenetrating polymer network constituted by an entangled combination of two polymers, one of which is cross-linked, that are not bonded to each other. (*e*) An interpenetrating polymer network, abbreviated IPN, is an entangled combination of two cross-linked polymers that are not bonded to each other. (*f*) A conterminously linked polymer, constituted by having the polymer II species linked, at both ends, onto polymer I. The ends may be grafted to different chains or the same chain. The total product is a network composed of two different polymers.

copolymer may also mean, "polymer B synthesized in the immediate presence of polymer A." Only by a reading of the context can the two meanings be distinguished.

2.7.3 Conterminously Grafted Copolymers

The polymers of Section 2.7.2 are soluble, at least in the ideal case. A conterminously grafted copolymer has polymer B grafted at both ends, or at various points along the structure to polymer A, and hence it is a network and not soluble. [See structure (f) in Figure 2.9.]

2.7.4 Interpenetrating Polymer Networks

This is an intimate combination of two polymers in network form. At least one of the polymers is polymerized and/or cross-linked in the immediate presence of the other (22). While ideally the polymers should interpenetrate on the molecular level, actual interpenetration may be limited owing to phase separation. (Phase separation in polymer blends, grafts, blocks, and interpenetrating polymer networks is the more usual case and is discussed in Chapter 4.)

2.7.5 Other Polymer – Polymer Combinations

According to new proposed nomenclature, a polymer blend would be accorded the connective –blend–. Many of these blends are prepared by highly sophisticated methods and are actually on a parallel with blocks, grafts, and interpenetrating polymer networks.

Block copolymers may also be arranged in various star arrangements. In this case, polymer A radiates from a central point, with a number of arms to be specified. Then polymer B is attached to the end of each arm.

2.7.6 Separation and Identification of Multicomponent Polymers

The methods of separation and identification of multicomponent polymers are far different from the methods described previously for the statistical type of polymer. First, only the blends are separable by extraction techniques. The remainder are bound together by either chemical bonds or interpenetration. The interpenetrating polymer networks and the conterminously grafted polymers are insoluble in all simple solvents and do not flow on heating. The graft and block copolymers, on the other hand, do dissolve and flow on heating above T_f and/or T_g.

Most, but not all, of the multicomponent polymer combinations exhibit some type of phase separation, which is discussed in Chapter 4. Where the polymers are stainable and observable under the electron microscope, characteristic morphologies are often manifest. The principal polymers that are stainable include the diene types and those containing ester groups. For those combinations exhibiting phase separation, two characteristic glass temperatures are also usually observed.

2.8 CONFORMATIONAL STATES IN POLYMERS

This chapter would not be complete without a further discussion of the various conformational states in polymers (23–29). The rotational potential energy diagram (Figure 2.10) (23) indicates three stable positions or conformations—the trans, the gauche plus, and the gauche minus.

The barriers separating the three conformational states have heights several times the thermal energy, kT, which means that the lifetime in a given state will be much longer than the vibration periods within the well. The sequence of bond conformations at a given instant defines the rotational isomeric state of the chain.

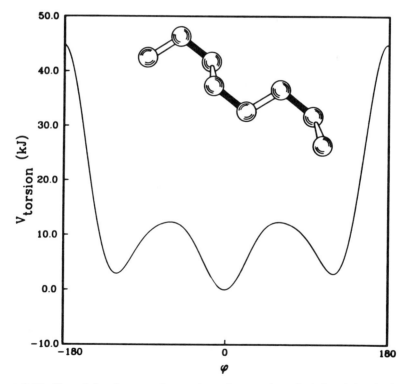

Figure 2.10 The rotational energy diagram for carbon–carbon single bonds in a hydrocarbon polymer such as polyethylene. Illustrated are the energy wells of the trans, gauche plus, and the gauche minus positions (23).

Helfand and co-workers investigated the various transitions among the conformational states by means of computer simulations (25, 26) and by applications of a kinetic theory (27–29). This analysis yielded the details of the long periods of motion near the bottom of the conformational wells, and the occasional transition to a different well. The activation energy for the transition was found to be approximately equal to the barrier height between the two states, as is required.

One surprising finding was that the transitions frequently occur in pairs, cooperatively. Immediately following the transition of one bond, a strong increase in the transition rate of its second-neighbor bonds was found. The intermediate bond usually remained unchanged. Thus the transitions might be

$$g^{\pm}tt \rightleftharpoons ttg^{\pm} \tag{2.37}$$

$$ttt \rightleftharpoons g^{\pm}tg^{\mp} \tag{2.38}$$

The significance of this observation arises from the geometric properties of the two transitions. In both cases, the first two and last two bonds translate relative to each other in opposite directions. Except for the central bond, the final state of each bond is parallel to its initial state. The cooperative pair transitions of equations (2.37) and (2.38) greatly reduce the motion of the long tail chains attached to the rotating segment, and hence the frictional resistance that the tails would present to the transition (23).

The rate of the transitions is given by two factors. First is the Arrhenius factor $\exp(-E_{act}/kT)$, where E_{act} represents the free energy of activation. The Arrhenius factor yields the probability of being near the saddle point joining the two energy wells in question. This is multiplied by a factor reflecting the frequency of saddle traversal. The "reaction" coordinate moves along the path of steepest descent from the saddle point. Helfand (23) points out that the two cooperative bond changes must take place in a coherent, sequential fashion to minimize the effect of the activation energy barrier.

The trans–gauche transitions underlie the diffusional motions of de Gennes (Section 5.4), the Shatzki transition (Section 6.4.1), and the glass transition itself.

2.9 PHOTOPHYSICS OF POLYMERS

Photophysics is the science of the absorption, transfer, localization, and emission of electromagnetic energy, with no chemical reactions occurring. By contrast, photochemistry deals with those processes by which light interacts with matter so as to induce chemical reactions (30). The portion of the electromagnetic spectrum of interest to photophysics includes both the ultraviolet and the visible wavelength ranges. In many of the experiments performed, light is absorbed in the ultraviolet range, and a fluorescence is measured in the visible range. Often, two molecular moieties must be in proper juxtaposition for the phenomenon of interest to be measured.

The first step, of course, is the absorption of electromagnetic energy, transforming it into excited molecular states,

$$A + h\nu = A^* \tag{2.39}$$

where A is the molecule to be excited, A^* represents the excited state, and $h\nu$ represents the electromagnetic energy absorbed.

Next most important is energy migration, either along the chain or among the chains. This allows the energy to reach the sites of interest. Such energy migration mimics that observed in the ordered chlorophyll regions of green plant chloroplasts, that is, the antenna chlorophyll pigments (30–32). These light-gathering antennas are composed of chlorophylls, carotenoids, and special pigment-containing proteins. These large organic molecules, some of them natural polymers, harvest light energy by absorbing a photon of light and storing the absorbed energy temporarily in the form of an electron in an excited singlet energy state. The energy migrates throughout the system of antennas within about 100 psec, being transmitted to the reaction center protein (32). Hence, in polymer photophysics, this phenomenon is termed the "antenna effect."

2.9.1 Quenching Phenomena

In situations where bimolecular encounters dominate, typical for polymers, such encounters may lead to an electronic relaxation of the system, termed quenching. In general, such collisions may be written

$$A^* + B = A + B^* \tag{2.40}$$

where the excited molecule A^* encounters another molecule, B. Most often, the bimolecular interaction is between an excited molecule in the singlet state and a quencher molecule in the ground state.

The possible bimolecular quenching process includes (33) (a) chemical reaction, (b) enhancement of nonradiative decay, (c) electronic energy transfer, or (d) complex formation.

Chemical reactions involve cross-linking, degradation, and rearrangement. Enhancement of nonradiative decay is not discussed. Electronic energy transfer involves exothermic processes, where part of the energy is absorbed as heat, and part is emitted via fluorescence or phosphorescence from the donor molecule.

Polarized energy is absorbed in fluorescence depolarization. This phenomenon is also known as luminescence anisotropy (31). If the chain portions are moving at about the same rate as the reemission, the energy is partly depolarized. The extent of depolarization is related to the various motions and their relative rates. Important are the inherent degree of anisotropy of the fluorescent chromophore, the degree of energy migration, and rotation of the chromophore during its excited state lifetime. From steady-state and transient emission anisotropic measurements, rotational relaxation times can be deduced.

Complex formation between two species is very important in photophysics. Two terms need definition—exciplex and excimer. An exciplex is an *excited state complex* between two different kinds of molecules, one being initially excited and the other in the ground state. A complex between an excited molecule and a ground-state molecule of the same species is called an excimer, being derived from the phrase *excited dimer* (30). Excimer formation is emphasized in this discussion. In excimer formation, excited-state complexes are usually formed between two aromatic structures. Resonance interactions lead to a weak intermolecular force, which binds the two species together, involving π bonds. Such excimers exhibit strong fluoresence characterized by being red-shifted with respect to the uncomplexed fluorescence.

Features of excimer fluorescence important in polymer characterization include intensity of radiation, intensity changes, decay rates, extent of frequency shifting of the fluorescence, and depolarization effects.

2.9.2 Excimer Formation

The formation of an excimer from an excited-state moiety A* and a ground-state moiety A may be illustrated as

$$A^* + A = (AA)^* \qquad (2.41)$$

The excimer, $(AA)^*$, decomposes due to a variety of interactions, the most

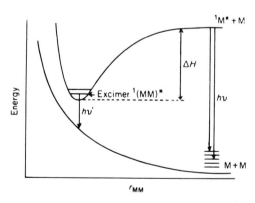

Figure 2.11 An energy-well diagram for excimer formation, illustrating the effects as a function of the distance, r_{MM}, between the two moieties (33).

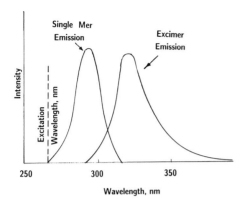

Figure 2.12   Schematic description of the excitation and fluorescence phenomena.

Figure 2.13   Geometry of the naphthalene excimer (30).

important one being the emission of fluorescence:

$$(AA)^* = 2A + h\nu_E \qquad (2.42)$$

where the emitted frequency, ν_E, is lower than the input frequency, the remaining energy being required to separate the two moieties.

The stability of excimers can be examined with the aid of Figure 2.11 (33). Highly stable excimers lie in the bottom of the energy well. Figure 2.12 illustrates the relationships among the initial light frequency, single mer emission, and excimer emission. The fluoresced light is more red-shifted in the more stable excimers, because it takes more energy to break them apart. There are three parameters determining excimer stability. With reference to the relative positions of the two aromatic groups in Figure 2.13 (30), these are the angle that the two planes make with each other, their distance apart, and their lateral displacement.

2.9.3   Experimental Studies

2.9.3.1   Microstructure of Polystyrene

One of the first polymers studied was polystyrene (Figure 2.14) (34). The single mer emission is at about 290 nm, with the band at 335 nm being

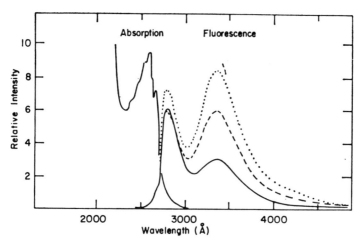

Figure 2.14   Room-temperature absorption and fluorescence spectra of atactic polystyrene in 1,2-dichloroethane (1×10^{-3} M). The three fluorescence curves denote nitrogen solution (dotted line), aerated solution (broken line), and oxygenated solution (solid line) (34).

attributed to excimer emission. In dilute solutions, excimer formation is largely intramolecular rather than intermolecular, the excimers arising from adjacent phenyl groups. This is because the chains are far separated from one another. In the bulk state, excimers may involve neighboring chains as well. Excimer formation in atactic polystyrene requires the *tt* conformation of the meso-isomer, or the *tg⁻* or *g⁻t* isomers of the racemic isomer.

David et al. (35) found that the ratio of excimer emission to single mer emission, I_D/I_M, increased from 10 to 100 times with increasing degree of tacticity; see Table 2.8. Since isotactic polystyrene exists in a 3-1 helix in the crystalline state with ring spacings of the order of 3 Å, this arrangement provides more excimer sites than in the atactic configuration (35).

There is generally an increase in the ratio I_D/I_M as the molecular weight of the polymer increases. This has been taken as a confirmation of the energy migration along polymer chains.

Table 2.8   Ratio of excimer to normal fluoresence intensity (I_D / I_M) in polystyrene at 77°C (35)

Polystyrene Type	I_D/I_M
Atactic, unoriented	1.43
Atactic, oriented ($F = 0.07$)	1.67
Atactic, oriented ($F = 0.1$)	2.08
Isotactic (25% crystalline)	10
Isotactic (35% crystalline)	100

Table 2.9   Relatively phenyl positions

Quantity	Polystyrene	Poly(p-t-butylstyrene)
Angle from parallel, degrees	6	22
Center of ring distance, Å	3.0	3.5
Lateral displacement, Å	0.2	1.3

2.9.3.2   Excimer Stability

While the adjacent phenyl rings in polystyrene form relatively stable excimers, putting substituent groups on the phenyls may alter the bonding
energy. Chakraborty et al. (36) substituted bulky t-butyl groups in the para
position,

$$\tag{2.43}$$

The t-butyl group forces the phenyl groups away from their ideal excimer
positions, resulting in a blue shift of the fluorescence to 320 nm. Thus
excimers in poly(p-t-butylstyrene) are less stable than those in unsubstituted
polystyrene. Analysis of the resulting blue shift, together with molecular
modeling, led to a comparison of the relative phenyl positions, shown in
Table 2.9.

2.9.3.3   Chain End Contact Probability

One of the more interesting polymer shape and motion questions in polymer
science relates to the probability of one end of a polymer chain being in
contact with the other, albeit transient in nature. Winnik and co-workers (37,
38), Cuniberti and Perico (39), and Char et al. (40) examined the conformation and motion of polystyrene, poly(dimethyl siloxane), and poly(ethylene
oxide) chains terminated by pyrene[†] groups. In order to obtain information of
an intramolecular nature, dilute solutions were used. It was of interest to
determine the cyclization rate constants, k_1 and k_{-1}; see Figure 2.15 (37).

 [†] pyrene structure, showing bonding position.

Figure 2.15 Illustration of chain motions involving temporary cyclization (37).

Intramolecular excimer formation and deactivation can be described by the double exponential kinetics scheme of Birks (41):

$$I_M(t) = A_1 e^{-\lambda_1 t} + A_2 e^{-\lambda_2 t} \tag{2.44}$$

$$I_E(t) = A_3 e^{-\lambda_1 t} - A_4 e^{-\lambda_2 t} \tag{2.45}$$

$$\lambda_{1,2} = \tfrac{1}{2}\left[(X + Y) \pm \left((X - Y)^2 + 4k_1 k_{-1} \right)^{1/2} \right] \tag{2.46}$$

$$A_2/A_1 = (\lambda_1 - X)/(X - \lambda_2) \tag{2.47}$$

$$X = K_M + k_1; \; Y = K_E + k_{-1} \tag{2.48}$$

where K_M and K_E are the rate constants for the unimolecular decay of the excited mer, M^*, and the excimer, E^*, respectively, and k_1 and k_{-1} are excimer formation and dissociation rate constants, respectively (42).

Table 2.10 illustrates typical kinetic data for chain end contact. In general, k_1 decreases with chain length, showing that as the molecular weight of the chain increases, it is more difficult for the ends to find one another. This has been modeled by Wilemski and Fixman (43), who showed that k_1 scales as the molecular weight of a chain to the $-\tfrac{3}{2}$ power in poor solvents. The constant k_{-1} was found to be independent of molecular weight. This means that although the chain ends find it increasingly difficult to find each other as the molecular weight increases, once having found each other, the kinetics of dissociation are not affected by the intervening chain. The decay constants, τ_1 and τ_2, in most instances were of the order of 10–100 nsec. for polystyrene of

Table 2.10   Chain end contact kinetics

Polymer	$k_1 \times 10^{-7} \, \text{s}^{-1}$	$k_{-1} \times 10^{-7} \, \text{s}^{-1}$	Ref.
py-polystyrene M = 9200 g/mol T = 71°C	0.50	3.07	(35)
py-poly(ethylene glycol) M = 1000 g/mol T = 20°C	1.67	0.66	(40)

9200 g/mol (a very low molecular weight), the probability of one end finding the other was calculated to be about 5×10^{-6} (37). Of course, as described in Chapter 3, these polymer chains form random coils.

In an extension of this type of study, Char et al. (40) studied the adsorption of py-poly(ethylene glycol) onto silica particles. They found that adsorption substantially reduced the excimer intensity ratio I_E/I_M, indicating a decrease in pyrene–pyrene contacts. This was attributed to the flattened conformation of the poly(ethylene glycol) chains adsorbed on the silica surface.

Excimer intensity can also be used to study the local environment of a polymer. For example, Frank and Gashgari (44) dispersed poly(2-vinyl naphthalene) in a range of other polymers. When the polymers were more miscible, I_D/I_M decreased, indicating a decrease in the availability of 2-vinyl naphthalene groups from neighboring molecules. This study is also of direct interest to polymer–polymer miscibility; see Section 4.3.7.

2.9.4 Photophysics Kinetics

Suppose a mer, M, is raised to the excited singlet state, designated $^1M_0^*$. The superscript 1 refers to multiplicity, the asterisk identifies the excited state, and the subscript 0 refers to the vibrational energy content of the state, zero. While there are many reactions that take place, the most important of these include (33) the following:

Phenomenon	Process	Rate
$M + h\nu \rightarrow {}^1M_0^*$	Absorption	I_a
$^1M_0^* \rightarrow M + h\nu$	Fluorescence	k_F
$M^* + Q \rightarrow (MQ)^*$	Quenching	$k_Q[^1M_0^*][Q]$

If Q is identical to M, the state $(MQ)^*$ is the excimer complex. Ignoring competing phenomena, under continuous illumination, the rate of formation of the excited state, $^1M_0^*$, can be equated to the rate of loss:

$$\frac{d[^1M_0^*]}{dt} = I_a - k_F[^1M_0^*] = 0 \qquad (2.49)$$

where I_a represents the illumination intensity.

A useful quantity in photophysics is the primary quantum yield, ϕ, defined as the rate of any process relative to the rate of photon absorption. The quantum yield for the fluorescence phenomenon can be written

$$\phi_F = k_F[^1M_0^*]/I_a \qquad (2.50)$$

Under equilibrium conditions, the ratio of fluorescence quantum yields ϕ_D

and ϕ_M of excimer and single mer, respectively, is

$$\frac{\phi_D}{\phi_M} = \frac{k_{FD}k_{DM}[M]}{k_F(k_{MD} + k_D)} \tag{2.51}$$

where k_{FD} is the constant for the fluorescent decomposition of $[MM]^*$, and k_{DM} and k_{MD} are the constants for the formation and decomposition of the excimer. These equations lay the foundation for understanding the structure and dynamics of polymer chains in both the solution and bulk states.

2.10 CONFIGURATION AND CONFORMATION

Through the development of instrumental techniques such as infrared, Raman spectroscopy, nuclear magnetic resonance, x-ray diffraction, fluorescence, and other methods, the organization of the individual atoms along the chain has gradually become clear. In many cases, two or more isomeric forms may be simultaneously present. The configurational properties of a polymer determine if it is crystallizable and, if so, its melting temperature.

If two or more monomeric units are used to make one polymer, a copolymer is formed. The statistical, random, alternating, and periodic copolymers show the relationship of two mers on an individual basis. The block, graft, conterminously grafted, and interpenetrating polymer network copolymers comprise large portions of chain or chains containing only one mer. It must be pointed out that each of these may be subject to being composed of the various tacticities and so forth that describe the configurational properties.

By contrast, the conformational properties of a polymer are determined by rotations about single bonds. The overall shape and size of the chain and many of its motions are determined by its conformation. Both conformation and configuration contribute, albeit in different ways, to the behavior of the polymer.

REFERENCES

1. F. W. Harris, *J. Chem. Ed.*, **58**(11), 836(1981).
2. M. Malanga and O. Vogl, *Polym. Eng. Sci.*, **23**, 597(1983).
3. J. L. Koenig, *Chemical Microstructure of Polymer Chains*, Wiley-Interscience, New York, 1980, Chaps. 6–8.
4. P. C. Painter and M. M. Coleman, in *Static and Dynamic Properties of the Polymeric Solid State*, R. A. Pethrick and R. W. Richards, Eds., D. Reidel, Boston, 1982.
5. A. Winston and P. Wichackeewa, *Macromolecules*, **6**, 200(1973).
6. G. Natta, *Makromol. Chem.*, **35**, 94(1960).
7. S. W. Cornell and J. L. Koenig, *J. Polym. Sci.*, *A*2, **7**, 1965(1969).

8. H. W. Siesler and K. Holland-Moritz, *Infrared and Raman Spectroscopy of Polymers*, Dekker, New York, 1980, Chap. 2.

9. F. A. Bovey, G. V. D. Tiers, and G. Filipovitch, *J. Polym. Sci.*, **38**, 73(1959).

10. J. A. Manson and L. H. Sperling, *Polymer Blends and Composites*, Plenum, New York, 1976, Chap. 1.

11. G. Odian, *Principles of Polymerization*, 2nd ed., Wiley-Interscience, New York, 1981, Chap. 8.

12. F. A. Bovey, Chap. 1 in *NMR and Macromolecules*, J. C. Randall, Jr., Ed., ACS Symposium Series No. 247, American Chemical Society, Washington, DC, 1984.

13. F. A. Bovey and G. V. D. Tiers, *J. Polym. Sci.*, **44**, 173(1960).

14. F. A. Bovey, *Polymer Conformation and Configuration*, Academic Press, New York, 1969, Chap. 1.

15. S. W. Cornell and J. L. Koenig, *Macromolecules*, **2**, 540, 564(1969).

16. J. L. Biner, *J. Polym. Sci.*, **A1**, 47(1963).

17. A. C. Angood and J. L. Koenig, *J. Polym. Sci.*, **A28**, 1787(1970).

18. W. Ring et al., *Pure and Appl. Chem.*, **57**, 1427(1985).

19. L. H. Sperling, *Source-Based Nomenclature for Polymer Blends, Interpenetrating Polymer Networks, and Related Polymers*, prepared for the Nomenclature Committee of the Polymer Chemistry Division of the American Chemical Society, Washington, DC, 1984.

20. J. C. Randall and E. T. Hsieh, in *NMR and Macromolecules*, J. C. Randall, Jr., Ed., American Chemical Society, Washington, DC, 1984.

21. H. J. Harwood and W. M. Ritchey, *Polym. Lett.*, **2**, 601(1964).

22. L. H. Sperling, *Interpenetrating Polymer Networks and Related Materials*, Plenum, New York, 1981.

23. E. Helfand, *Science*, **226**(4675), 647(1984).

24. W. H. Stockmeyer, *Pure Appl. Chem. Suppl. Macromol. Chem.*, **8**, 379(1973).

25. E. Helfand, Z. R. Wasserman, and T. A. Weber, *Macromolecules*, **13**, 526(1980).

26. T. A. Weber and E. Helfand, *J. Phys. Chem.*, **87**, 2881(1983).

27. E. Helfand, *J. Chem. Phys.* **54**, 4651(1971).

28. J. Skolnick and E. Helfand, *J. Chem. Phys.*, **72**, 5489(1980).

29. E. Helfand and J. Skolnick, *J. Chem. Phys.* **77**, 3275(1982).

30. J. Guillet, *Polymer Photophysics and Photochemistry*, Cambridge University Press, Cambridge, England, 1985.

31. C. E. Hoyle, in *Photophysics of Polymers*, C. E. Hoyle and J. M. Torkelson, Eds., ACS Symposium Series No. 358, ACS Books, Washington, DC, 1987.

32. J. R. Norris and M. Schiffer, *C&E News*, **68**(31), July 30, 22(1990).

33. D. Phillips, in *Polymer Photophysics*, D. Phillips, Ed., Chapman and Hall, London, 1985.

34. M. T. Vala, Jr., J. Haebig, and S. A. Rice, *J. Chem. Phys.*, **43**, 886(1965).

35. C. David, N. Putman-de Lavarielle, and G. Geuskens, *Eur. Polym. J.*, **10**, 617(1974).

36. D. K. Chakraborty, K. D. Heitzhaus, F. J. Hamilton, H. J. Harwook, and W. L. Mattice, *Polym. Prepr.*, **31**(2), 590(1990).

37. A. E. C. Redpath and M. A. Winnik, *J. Am. Chem. Soc.*, **104**, 5604(1982).

38. M. A. Winnik, A. E. C. Redpath, P. Svirskaya, and A. Mar, *Polymer*, **24**, 473(1983).

39. C. Cuniberti and A. Perico, *Eur. Polym. Jr.*, **13**, 369(1977).

40. K. Char, A. P. Gast, and C. W. Frank, *Langmuir*, **4**, 989(1988).

41. J. B. Birks, *Photophysics of Aromatic Molecules*, Wiley-Interscience, London, 1970.

42. S. E. Webber and C. H. Chang, *J. Luminescence*, **44**, 113(1989).

43. G. Wilemski and M. Fixman, *J. Chem. Phys.*, **60**, 866, 878(1974).

44. C. W. Frank and M. A. Gashgari, *Macromolecules*, **12**, 163(1979).

GENERAL READING

F. A. Bovey, *Polymer Conformation and Configuration*, Academic Press, Orlando, 1969.

F. A. Bovey, *High Resolution NMR of Macromolecules*, Academic Press, Orlando, 1972.

C. E. Hoyle and J. M. Torkelson, Eds., "Photophysics of Polymers," ACS Books, Washington, DC, 1987.

J. L. Koenig, *Chemical Microstructure of Polymer Chains*, Wiley-Interscience, New York, 1980.

R. W. Lenz, *Organic Chemistry of Synthetic High Polymers*, Interscience, New York, 1967.

G. Odian, *Principles of Polymerization*, 2nd ed., Wiley-Interscience, New York, 1981.

P. C. Painter, M. M. Coleman, and J. L. Koenig, *The Theory of Vibrational Spectroscopy and Its Application to Polymeric Materials*, Wiley-Interscience, New York, 1982.

E. M. Pearce, C. E. Wright, and B. B. Bordoloi, *Laboratory Experiments in Polymer Synthesis and Characterization*, The Pennsylvania State University EMMSE Project, 1982.

J. C. Randall, Jr., Ed., *NMR and Macromolecules*, ACS Symposium Series N. 247, American Chemical Society, Washington, DC, 1984.

H. W. Siesler and K. Holland-Mority, *Infrared and Raman Spectroscopy of Polymers*, Dekker, New York, 1980.

STUDY PROBLEMS

1. Write chemical structures for isotactic, syndiotactic, and atactic polystyrene.

2. (a) Write chemical structures for cis- and trans-polybutadiene and (b) the 1,2 and 3,4 structures for polyisoprene.

3. How do head-to-head and head-to-tail structures of poly(methyl methacrylate) differ?

4. Show the structures of random and alternating copolymers of vinyl chloride and ethyl acrylate.

5. Cis-polyisoprene has been totally hydrogenated. What new polymer is formed?

6. What are the two possible triblock copolymer structures of polybutadiene and cellulose?

7. Based on Table 2.6, calculate the run numbers and average sequence lengths for the two poly(ethylene–*stat*–1-hexene) copolymers. Do they indeed appear to be statistical copolymers?

8. A graft copolymer is formed with polybutadiene as the backbone and polystyrene as the side chains. What is the name of this material?

9. Compare and contrast infrared and Raman spectra with NMR techniques for their capability of characterizing (a) tacticity and (b) cis and trans double bonds in polymers.

10. Chemical nomenclature forms the alphabet of polymer science. (a) What is the chemical structure of *it*–poly(vinyl chloride)–*block*–cis-1,4-polyisoprene? (b) Poly(vinyl acetate) is totally hydrolyzed. What new polymer is formed? What polymer is formed if the hydrolysis is only partial?

11. In the accompanying structures, P_1 is poly(vinyl acetate), P_2 is poly(ethyl acrylate), and P_3 is polystyrene. What are the chemical names of these structures?

(a)   (b)   (c)

(d)   (e)

12. Your new assistant copolymerized styrene and *n*-butyl acrylate. "I made it in such a way as to produce an alternating copolymer," he said.

"No," you replied, "I think you really made a statistical copolymer."

At that instant, your boss walks in. "I'll bet this new synthesis really made a block copolymer," she volunteers. Using the concepts of photo-physics, devise an experiment to distinguish the three possibilities. How would the resulting data look as a function of composition? (Prepare appropriate plots.)

APPENDIX 2.1 ASSORTED ISOMERIC AND COPOLYMER MACROMOLECULES

In addition to the types of isomeric and copolymer structures illustrated in the text of the chapter, there are several other structures of which the student must be aware.

Proteins

Proteins have the general structure

$$\left(\begin{array}{c} O \quad H \quad H \\ \| \quad | \quad | \\ -C-C-N- \\ | \\ R \end{array}\right)_n \qquad (A2.1.1)$$

and hence are sometimes called nylon 2. There are 23 common types of amino acids (A1), each with a specific type of R group. In proteins, these amino acids follow very specific sequences, which frequently differ from species to species of plant or animal. (Note the various kinds of insulin, for example.) However, in the broad sense of the term, they are copolymers.

In addition to the copolymer structure, proteins are also optically active. All amino acids except glycine possess at least one asymmetric carbon atom, as illustrated in equation (A2.1.1). The language commonly used to characterize the structure is different, however. The Fischer convention of designating the active group as *D*- or *L*- is used (A1). All the natural proteins contain the *L* configuration. Woe to the person who feeds an earth-bound creature the *D* configuration! This difference, of course, has generated many a science fiction story.

The proteins also have specific spatial arrangements. This is partly aided by sulfur–sulfur cross-links linking cystine residues. On cooking, these bonds break or rearrange, which is called denaturing.

Figure A2.1.1   A comparison of the structures of cellulose and starch. Note that cellulose has two mers in the repeat unit, caused by β-1,4-glucoside linkage (11).

Cellulose and Starch

Both of these natural polymers are composed of glucose, a six-membered ring (see Chapter 1). Both are linked together at the 1,4 position, but differ in that cellulose has a β linkage and starch has the α linkage. The β linkage has the effect of alternating the structure of the glucosides in an up-down-up-down configuration, while the α linkage makes them all up-up-up-up (See Figure A2.1.1). The difference in physical properties reflects the altered chemistry. Cellulose is highly crystalline and nondigestible by humans. Starch is much less crystalline but highly digestible, as too many overweight people know. The difference in digestibility is caused by the lack of an enzyme (a protein!) to attack the β linkage. It must be pointed out that there are many structurally different polysaccharides in nature.

Ditactic Polymers

These structures are generated by polymerizing 1,2-disubstituted ethylenes having the general structure

$$\left(\begin{array}{cc} H & H \\ | & | \\ -C-C- \\ | & | \\ R' & R \end{array}\right)_n \qquad (A2.1.2)$$

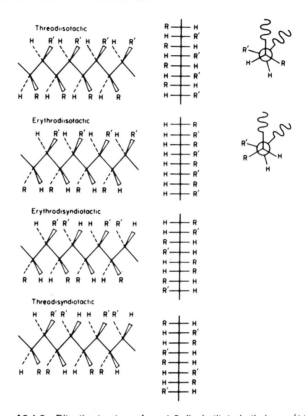

Figure A2.1.2 Ditactic structures from 1,2-disubstituted ethylenes (11).

Ditacticity occurs when the individual carbon atoms possess specific stereoisomerism. Four different stereoregular structures may be identified, as shown in Figure A2.1.2 (11).

REFERENCE

A1. H. K. Salzberg, in *Encyclopedia of Polymer Science and Technology*, Vol. 11, N. M. Bikales, Ed., Interscience, New York, 1969, p. 620.

3

MOLECULAR WEIGHTS
AND SIZES

3.1 POLYMER SIZE AND SHAPE

The general problem of the size and shape of polymer molecules stands at the very heart of polymer science and engineering. If the molecular weight[†] and molecular weight distribution are known along with a good understanding of the polymer chain conformation, many mechanical and rheological properties can be predicted. While the question of molecular weights pervades the entire area of the chemical arts, polymers have classically presented several special problems:

1. The molecular weights are very high, ranging from about 25,000 to 1,000,000 g/mol or higher.

2. The molecular weight of ordinary size molecules is fixed (e.g., benzene has a molecular weight of 78 g/mol regardless of its source). Most polymer molecular weights, on the other hand, vary greatly depending on the method of preparation. In addition, most polymers are polydisperse; that is, the sample contains more than one species.

3. The spatial arrangement of the polymer chain is called its "conformation." Conformations can be determined in dilute solutions by light-scattering, and in the bulk state by small-angle neutron scattering (SANS). Conformations can also be estimated theoretically, from the structure and molecular weight of the polymer.

4. Polymer chain conformations are functions of temperature, solvent, structure, crystallization, extension, and the presence of other polymers.

[†]Some authors prefer the term "relative molar mass," using units of daltons.

Most of the methods for determining the molecular weight and size of polymers (except for small-angle neutron scattering) depend on dissolving the polymer in an appropriate solvent and measuring the required properties in dilute solution. The molecular weights of gaseous molecules such as oxygen or methane can be determined via an understanding of the gas laws and the fact that the molecules are separated in space. The same is true for polymer solutions. Solution laws must be understood, and the polymer chains must be separated from each other in the solution. The following two sections develop the basic polymer solution thermodynamics for the purposes of determining molecular weights and sizes. In Chapter 4, these same concepts are developed further in the study of concentrated solutions, phase separation, and diffusion problems.

3.2   THE SOLUBILITY PARAMETER

One of the simplest notions in chemistry is that "like dissolves like." Qualitatively, "like" may be defined variously in terms of similar chemical groups or similar polarities.

Quantitatively, solubility of one component in another is governed by the familiar equation of the free energy of mixing,

$$\Delta G_M = \Delta H_M - T \Delta S_M \tag{3.1}$$

where ΔG_M is the change in Gibbs' free energy, T is the absolute temperature, and ΔS_M is the entropy of mixing. A negative value of ΔG_M indicates that the solution process will occur spontaneously. The term $T \Delta S_M$ is always positive because there is an increase in the entropy on mixing. Therefore the sign of ΔG_M depends on ΔH_M, the enthalpy of mixing.

Surprisingly, the heat of mixing is usually positive, opposing mixing. This is true for big and little molecules alike. Exceptions occur most frequently when the two species in question attract one another in some way, perhaps by having opposite polarities, being acid and base relative to one another, or through hydrogen bonding. However, positive heats of mixing are the more usual case for relatively nonpolar organic compounds. On a quantitative basis, Hildebrand and Scott (1) proposed that, for regular solutions,

$$\Delta H_M = V_M \left[\left(\frac{\Delta E_1}{V_1} \right)^{1/2} - \left(\frac{\Delta E_2}{V_2} \right)^{1/2} \right]^2 v_1 v_2 \tag{3.2}$$

where V_M represents the total volume of the mixture, ΔE represents the energy of vaporization to a gas at zero pressure (i.e., at infinite separation of the molecules), and V is the molar volume of the components, for both species 1 and 2. The quantity v represents the volume fraction of component 1 or 2 in the mixture. The quantity $\Delta E / V$ represents the energy of vaporization per cm^3. This term is sometimes called the cohesive energy density. By convention, component 1 is the solvent, and component 2 is the polymer.

The reader should note that according to equation (3.2), "like dissolves like" means that the two terms $\Delta E_1/V_1$ and $\Delta E_2/V_2$ have nearly the same numerical values. Equation (3.2) also yields only positive values of ΔH_M, a serious fault in the theory. However, since the majority of polymer solutions do have positive heats of mixing, the theory has found very considerable application.

The square root of the cohesive energy density is widely known as the solubility parameter,

$$\delta = (\Delta E/V)^{1/2} \tag{3.3}$$

Thus the heat of mixing of two substances is dependent on $(\delta_1 - \delta_2)^2$.

These relationships are meaningful only for positive heats of mixing; that is, when the heat of mixing term opposes solution. Since $(\delta_1 - \delta_2)^2$ cannot be negative, equations (3.2) and (3.3) break down for negative heats of mixing.

3.2.1 Solubility Parameter Tables

Tables 3.1 (2) and 3.2 (2) present the solubility parameters of common solvents and polymers, respectively. These tables provide a quantitative basis

Table 3.1 Solubility parameters of some common solvents

Solvent	$\delta(\text{cal}/\text{cm}^3)^{1/2}$	H-bonding[a] Group	Specific Gravity[b] 20°C (g/cm^3)
Acetone	9.9	m	0.7899
Benzene	9.2	p	0.87865
n-Butyl acetate	8.3	m	0.8825
Carbon tetrachloride	8.6	p	1.5940
Cyclohexane	8.2	p	0.7785
n-Decane	6.6	p	—
Dibutyl amine	8.1	s	—
Difluorodichloromethane	5.1	p	—
1,4-Dioxane	10.0	m	1.0337
Low odor mineral spirits	6.9	p	—
Methanol	14.5	s	0.7914
Toluene	8.9	p	0.8669
Turpentine	8.1	p	—
Water	23.4	s	0.99823
Xylene	8.8	p	0.8611

Source: J. Brandrup and E. H. Immergut, Eds., *Polymer Handbook*, 2nd ed., Wiley-Interscience, New York, 1975, Sec. IV.

[a]Hydrogen bonding is an important secondary parameter in predicting solubility. p, Poorly H-bonded; m, moderately H-bonded; and s, strongly H-bonded.

[b]J. Brandrup and E. H. Immergut, *Polymer Handbook*, 3rd ed., Wiley-Interscience, New York, 1989, Sec. III, p. 29.

Table 3.2   Solubility parameters and densities of common polymers (2)

Polymer	$\delta(cal/cm^3)^{1/2}$	Density(g/cm^3)
Polybutadiene	8.4[a]	1.01
Polyethylene	7.9	0.85 (amorphous)
Poly(methyl methacrylate)	9.45	1.188
Polytetrafluorethylene	6.2	2.00 amorphous, estimated
Polyisobutene	7.85	0.917
Polystyrene	9.10	1.06
Cellulose triacetate	13.60	1.28[b]
Cellulose tributyrate	—	1.16[b]
Nylon 66	13.6	1.24
Poly(ethylene oxide)	9.9	1.20
Poly(vinyl chloride)	9.6	1.39

[a]*Note:* $1(cal/cm^3)^{1/2} = 2.046 \times 10^3 (J/m^3)^{1/2}$.
[b]C. J. Malm, C. R. Fordyce, and H. A. Tanner, *Ind. Eng. Chem.*, **34**, 430 (1942).

for understanding why methanol or water does not dissolve polybutadiene or polystyrene. However, benzene and toluene are predicted to be good solvents for these polymers, which they are. While solubility of a polymer also depends on its molecular weight, the temperature, and so on, it is frequently found that polymers will dissolve in solvents having solubility parameters within about one unit of their own.

3.2.2   Experimental Determination

The solubility parameter of a new polymer may be determined by any of several means. If the polymer is cross-linked, the solubility parameter may be determined by swelling experiments (3). The best solvent is defined for the purposes of the experiment as the one with the closest solubility parameter. This solvent also swells the polymer the most. Several solvents of varying solubility parameter are selected, and the cross-linked polymer is swelled to equilibrium in each of them. The swelling coefficient, Q, is plotted against the various solvent's solubility parameter, the maximum defining the solubility parameter of the polymer. The theoretical extent of swelling is predicted by the Flory–Rehner theory on the basis of the cross-link density and the attractive forces between the solvent and the polymer (see Section 9.12).

The swelling coefficient, Q, is defined by

$$Q = \frac{m - m_0}{m_0} \times \frac{1}{\rho_s} \qquad (3.4)$$

where m is the weight of the swollen sample, m_0 is the dry weight, and ρ_s is

Figure 3.1 The swelling coefficient, Q, reaches a maximum when the solubility parameter of the solvent nearly matches that of the polymer, for several cross-linked systems: polyurethane (■), polystyrene (▲), and a polyurethane – polystyrene interpenetrating polymer networks (●) (5).

the density of the swelling agent (4, 5). Typical results are shown in Figure 3.1 (5). Here, the swelling behavior of a cross-linked polyurethane and a cross-linked polystyrene are shown, together with the 50/50 interpenetrating polymer network made from these two polymers. Both the homopolymers and the interpenetrating polymer network exhibit single peaks, albeit that the IPN peak is somewhat broader and appears in-between its two homopolymers. Polymer networks swollen to equilibrium are discussed in Section 9.12.

Alternatively, the solubility parameter may be determined by measuring the intrinsic viscosity of the polymer in these solvents, if the polymer is soluble in them. Then, the intrinsic viscosity is plotted against the solubility parameter of the several solvents. Since the chain conformation is most expanded in the best solvent [see equation (3.83)], the intrinsic viscosity will be highest for the best match in solubility parameter. Such an experiment is illustrated in Figure 3.2 (6) for polyisobutene and polystyrene. The results of such experiments are collected in Table 3.2.

Figure 3.2 Determination of the solubility parameter, using the intrinsic viscosity method (4), for polyisobutene (A) and polystyrene (B). The intrinsic viscosity, $[\eta]$, is a measure of the individual chain size. See Section 3.8.

Table 3.3 Group molar attraction constants at 25°C (according to Small; derived from measurement of heat of evaporation)

Group	G	Group	G	Group	G
$-CH_3$	214	Ring 5-membered	105–115	Br single	340
$-CH_2-$ single-bonded	133	Ring 6-membered	95–105	I single	425
$-CH<$	28	Conjugation	20–30	CF_2 n-fluorocarbons only	150
$>C<$	–93	H (variable)	80–100	CF_3	274
$CH_2=$	190	O ethers	70	S sulfides	225
$-CH=$ double-bonded	111	CO ketones	275	SH thiols	315
$>C=$	19	COO esters	310	ONO nitrates	~440
$-CH=C<$	285	CN	410	NO_2 (aliphatic nitro-compounds)	~440
$-C\equiv C-$	222	Cl (mean)	260	PO_4 (organic phosphates)	~500
Phenyl	735	Cl single	270	Si (in silicones)	–38
Phenylene (o,m,p)	658	Cl twinned as in $>CCl_2$	260		
Naphthyl	1146	Cl triple as in $-CCl_3$	250		

Source: P. A. Small, J. Appl. Chem., 3, 71(1953).

3.2.3 Theoretical Calculation: An Example

Values of the solubility parameter may be calculated from a knowledge of the chemical structure of any compound, polymer or otherwise. Use is made of the group molar attraction constants, **G**, for each group,

$$\delta = \frac{\rho \Sigma \mathbf{G}}{M} \tag{3.5}$$

where ρ represents the density and M is the molecular weight. For a polymer, M is mer molecular weight.

Group molar attraction constants have been calculated by Small (7) and Hoy (8). Table 3.3 (7) presents a wide range of values of **G** for chemical groups.

For example, the solubility parameter of polystyrene may be estimated from Table 3.3. The structure is

$$-CH_2-CH-$$

which contains $-CH_2-$ with a **G** value of 133, a $-CH-$ with **G** equal to 28, and a phenyl group with **G** equal to 735. The density of polystyrene is 1.05 g/cm^3, and the mer molecular weight is 104 g/mol. Then equation (3.5) gives

$$\delta = \frac{1.05}{104}(133 + 28 + 735) \tag{3.6}$$

$$\delta = 9.05 \; (cal/cm^3)^{1/2} \tag{3.7}$$

Table 3.2 gives a value of 9.1 $(cal/cm^3)^{1/2}$ for polystyrene.

3.3 THERMODYNAMICS OF MIXING

3.3.1 Types of Solutions

3.3.1.1 The Ideal Solution

In the previous section, the solubility of a polymer in a given solvent was examined on the basis of their respective solubility parameters, which was governed by the heats of mixing. The entropy of mixing was entirely ignored.

In an ideal solution, the circumstances are reversed, and the heat of mixing is zero, by definition. Raoult's law is obeyed.

$$p_1 = p_1^\circ n_1 \tag{3.8}$$

where p_1 is the partial vapor pressure, n_1 is the mole fraction of component 1, and p_1° is the vapor pressure of the pure component.

The free energy of mixing is given as the sum of the free energies of dilution per molecule, incompressibility of the mixture implicitly assumed:

$$\Delta G_M = N_1 \, \Delta G_1 + N_2 \, \Delta G_2 \qquad (3.9)$$

$$\Delta G_M = kT[N_1 \, \ln(p_1/p_1^\circ) + N_2 \, \ln(p_2/p_2^\circ)] \qquad (3.10)$$

where N_1 and N_2 are the numbers of molecules of the 1 and 2 species, respectively. Then, from equation (3.8), for small molecules,

$$\Delta G_M = kT(N_1 \, \ln n_1 + N_2 \, \ln n_2) \qquad (3.11)$$

Since $\Delta H_M = 0$,

$$\Delta S_M = -k(N_1 \, \ln n_1 + N_2 \, \ln n_2) \qquad (3.12)$$

Since the entropy of mixing is always positive, and the heat of mixing is zero for an ideal solution, mixing in all proportions always occurs spontaneously.

3.3.1.2 Statistical Thermodynamics of Mixing

Equations (3.8)–(3.12) present the classically derived entropy of mixing. More generally, equation (3.12) can be derived through the application of statistical thermodynamics. According to statistical thermodynamics, the entropy of mixing is determined by counting the number of possible arrangements in space that the molecules may assume, Ω. The entropy of mixing is given by Boltzmann's relation,

$$\Delta S_M = k \ln \Omega \qquad (3.13)$$

For small molecules of about the same size, this is given by the total number of ways of arranging the N_1 identical molecules of the solute on a lattice comprising $N_0 = N_1 + N_2$ cells (see Figure 3.3). The total number of such arrangements is given by $\Omega = N_0!/N_1!N_2!$.

Applying Stirling's approximation,

$$\ln N! = N \ln N - N \qquad (3.14)$$

the entropy of mixing is given by

$$\Delta S_M = k[(N_1 + N_2)\ln(N_1 + N_2) - N_1 \ln N_1 - N_2 \ln N_2] \qquad (3.15)$$

which rearranges to give equation (3.12).

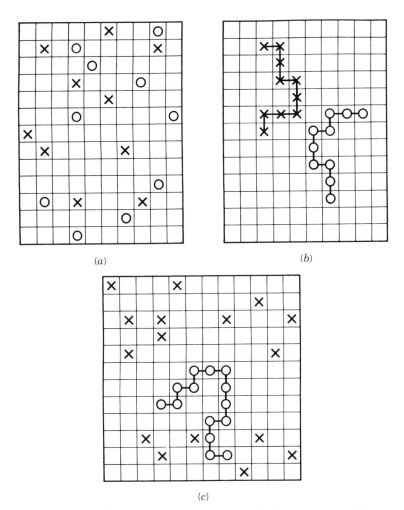

Figure 3.3 Illustration of two types of molecules on quasilattice structures. (*a*) Two types of small molecules; (*b*) a blend of two types of polymer molecules; (*c*) a polymer dissolved in a solvent. The entropy of mixing decreases from (*a*) to (*c*) to (*b*) because the number of different ways of arranging the molecules in space decreases. Note that the mers of the polymer chains are constrained to remain in juxtaposition with their neighbors.

When the polymer has x chain segments (mers), the entropy of mixing is given by

$$\Delta S_M = -k(N_1 \ln v_1 + N_2 \ln v_2) \tag{3.16}$$

Equation (3.16) yields the total entropy. The entropy per unit volume is given by

$$\frac{\Delta S_M}{kV} = -\sum_i N_i \ln v_i \Big/ \sum_i V_i \tag{3.16a}$$

In equation (3.16) v_1 and v_2 represent the volume fractions of solvent and polymer, respectively, and

$$v_1 = \frac{N_1}{N_1 + xN_2} \tag{3.17}$$

$$v_2 = \frac{xN_2}{N_1 + xN_2} \tag{3.18}$$

For polymer–polymer solutions,

$$v_1 = \frac{x_1 N_1}{x_1 N_1 + x_2 N_2} \tag{3.19}$$

$$v_2 = \frac{x_2 N_2}{x_1 N_1 + x_2 N_2} \tag{3.20}$$

It must be pointed out that ΔS_M is the combinatorial entropy computed by considering the possible arrangements of the molecules on the lattice in Figure 3.3. Furthermore, it is observed even qualitatively that the number of ways that the system can be rearranged in space is reduced when one or both species exist as long chains.

3.3.1.3   Other Types of Solutions

The ideal solution has a zero heat of mixing. Other types of solutions include the athermal solutions, in which ΔH_M is still zero, but the entropy of mixing is not given by equation (3.16). A regular solution is defined as one in which ΔS_M has the ideal value but ΔH_M is finite. In an irregular solution, both ΔH_M and ΔS_M deviate from ideal values. As will be shown later, a principal cause of nonideality arises from changes in volume nonadditivity on mixing.

3.3.2   Dilute Solutions

The Flory–Huggins theory (9–13) introduces the quantity χ_1 to represent the heat of mixing,

$$\chi_1 = \Delta H_M / kTN_1 v_2 \tag{3.21}$$

which combined with equation (3.16) leads to the free energy of mixing in statistical thermodynamic terms[†]:

$$\Delta G_M = kT(N_1 \ln v_1 + N_2 \ln v_2 + \chi_1 N_1 v_2) \tag{3.22}$$

[†]The convention assumed is that a positive ΔG_M leads to phase separation, and a negative value leads to molecular solutions.

Again, incompressibility is assumed, meaning that the free volume in the system is constant. Equation (3.22) provides a starting point for many equations of interest. The partial molar free energy of mixing may be written, after multiplying by Avogadro's number,

$$\overline{\Delta G_1} = RT\left[\ln(1 - v_2) + (1 - 1/x)v_2 + \chi_1 v_2^2\right] \qquad (3.23)$$

Since the osmotic pressure is given by

$$\pi = -\frac{\overline{\Delta G_1}}{v_1} \qquad (3.24)$$

then

$$\pi = -\frac{RT}{V_1}\left[\ln(1 - v_2) + (1 - 1/x)v_2 + \chi_1 v_2^2\right] \qquad (3.25)$$

where V_1 represents the molar volume of the solvent.

The second virial coefficient, see equation (3.41), may be given a theoretical interpretation. On expanding equation (3.25), the theoretical value for the second virial coefficient can be separated out,

$$A_2 = \frac{\bar{v}_2^2}{N_A V_1}\left(\frac{1}{2} - \chi_1\right) \qquad (3.26)$$

where $\bar{v}_2$ is the specific volume of the polymer. The Flory θ-temperature can now be seen as the point where $\chi_1 = \frac{1}{2}$.

At the temperature where $\chi_1 = \frac{1}{2}$, A_2 is zero. Another form of equation (3.26) is

$$A_2 = Nu/2M^2 \qquad (3.27)$$

where N represents Avogadro's number, and u is the intermolecular excluded volume between molecular pairs. When the quantity u is zero, A_2 is zero, and the chains mix without recognizing each other's presence. Also, since the forces are balanced out, the chains cannot distinguish between solvent molecules and segments of the other polymer chains. Thus the simplified conditions of mixing existing at the Flory θ-temperature become apparent.

Another equation interrelating thermodynamic terms may be written (14)

$$\chi_1 = \beta_1 + \frac{V_1}{RT}(\delta_1 - \delta_2)^2 \qquad (3.28)$$

Thus χ_1 is related to the solubility parameters. While β_1, sometimes called the lattice constant of entropic origin, is often assigned a value of 0.35 in the older literature, a value of zero yields better correlations with other quantities in most cases. The quantity α equals r/r_0, the expansion of the polymer chain over that of the theta condition state.

The quantity α may be expressed in terms of the Flory θ-temperature (15),

$$\alpha^5 - \alpha^3 = 2C_m\psi_1\left(1 - \frac{\theta}{T}\right)M^{1/2} \tag{3.29}$$

where ψ_1 is an entropic factor and C_m is a constant. Thus, at $T = \theta$, $\alpha = 1$. For high molecular weights, equation (3.29) shows that the chains expand as $M^{0.1}$ power. Since the Mark–Houwink equation depends on $M^{0.5}$ in the θ condition, the quantity a is seen to vary theoretically from 0.5 to 0.8; see Section 3.8.

While the above relations must be used with caution, they provide a clear theoretical interpretation and interrelationships for the equations used in molecular weight determination. As will be seen later, they can also express quantities useful for phase separation problems. The theory expressed in equations (3.21)–(3.29) has been reviewed many times (16) and derived by Flory (13).

3.3.3   Values for the Flory–Huggins χ_1 Parameter

The Flory–Huggins χ_1 parameter has been one of the most widely used quantities, characterizing a variety of polymer–solvent and polymer–polymer interactions. It is a unitless number. Sometimes the Flory–Huggins parameter is written $\chi_{1,2}$, and sometimes just plain χ. While the original theory proposed that χ_1 be concentration independent, many polymer–solvent systems exhibit increases of χ_1 with polymer concentration (17). In that case, the analytical representation of the experimentally found concentration dependence can be written as a power series,

$$\chi_1 = \chi^0 + \chi^1 v_2 + \chi^2 v_2^2 \tag{3.30}$$

where χ^0, χ^1, and χ^2 are determined experimentally.

However, for many simple calculations, it is valuable to have a single number parameter. Typical values of χ_1 are illustrated in Table 3.4, where values were selected at low concentrations of polymer. If the value of χ_1 is below 0.5, the polymer should be soluble if amorphous and linear. When χ_1 equals 0.5, as in the case of the polystyrene–cyclohexane system at 34°C in Table 3.4, then the Flory θ conditions exist; see Section 4.1. If the polymer is crystalline, as in the case of polyethylene, it must be heated to near its melting temperature, so that the total free energy of melting plus dissolving is negative. For very many nonpolar polymer–solvent systems, χ_1 is in the range of 0.3–0.4.

Table 3.4 Flory–Huggins χ_1 values

Polymer	Solvent	T, °C	χ_1
Polystyrene	Toluene	25	0.37
Polystyrene	Cyclohexane	34	0.50
Polyisoprene	Benzene	25	0.40
Cellulose nitrate	Acetone	20	0.14
Cellulose nitrate	n-Propylacetate	20	−0.38
Poly(ethylene oxide)	Benzene	70	0.19
Poly(dimethyl siloxane)	Toluene	20	0.45
Polyethylene	n-Heptane	109	0.29
Poly(butadiene- stat-styrene)	Toluene	25	0.39

Source: J. Brandrup and E. H. Immergut, Eds., *Polymer Handbook*, 3rd ed., Wiley, New York, 1989, Sec. VII, pp. 176–178.

3.3.4 A Worked Example

What is the free energy of mixing polystyrene of molecular weight 10,000 g/mol with cyclohexane at 34°C, to make up a 10% solution by volume?

The starting point is equation (3.22),

$$\Delta G_M = kT(N_1 \ln v_1 + N_2 \ln v_2 + \chi_1 N_1 v_2)$$

The basis usually taken is 1 cm^3. The number of molecules of each species needs to be calculated for $v_1 = 0.90$ and $v_2 = 0.10$. Densities are shown in Tables 3.1 and 3.2. For cyclohexane, C_6H_{12}, the molecular weight is 84 g/mol and the molar volume is (84 g/mol)/(0.7785 g/cm^3) $\cong$ 108 cm^3/mol. For 0.90 cm^3, there are 0.0083 mol, or, using Avogadro's number, 5.02×10^{21} molecules, N_1.

The density of polystyrene is 1.06 g/cm^3, yielding a molar volume of 9.43×10^3 cm^3/mol for a molecular weight of 10,000 g/mol, and for 0.10 cm^3, 1.06×10^{-5} mol of polystyrene, or 6.38×10^{19} molecules, N_2. The value of χ_1 is 0.50, taken from Table 3.4. The free energy of mixing is given by

$$\Delta G_M = 1.38 \times 10^{-23} \text{ J/K} \times 307 \text{ K}$$
$$\times (5.02 \times 10^{21} \ln 0.90 + 6.38 \times 10^{19} \ln 0.10$$
$$+ 0.50 \times 5.02 \times 10^{21} \times 0.10)$$
$$\Delta G_M = -1.24 \text{ J}$$

for each cm^3 of solution. The value of ΔG_M is small and negative, because of the entropy of mixing term.

3.4 MOLECULAR WEIGHT AVERAGES

There are four molecular weight averages in common use; the number-average molecular weight, M_n; the weight-average molecular weight, M_w; the z-average molecular weight, M_z; and the viscosity-average molecular weight, M_v. These are defined below in terms of the numbers of molecules N_i having molecular weights M_i; or w_i, the weight of species with molecular weights M_i.

$$M_n = \frac{\Sigma_i N_i M_i}{\Sigma_i N_i} = \frac{\Sigma_i w_i}{\Sigma_i (w_i / M_i)} \tag{3.31}$$

$$M_w = \frac{\Sigma_i N_i M_i^2}{\Sigma N_i M_i} = \frac{\Sigma_i w_i M_i}{\Sigma w_i} \tag{3.32}$$

$$M_z = \frac{\Sigma_i N_i M_i^3}{\Sigma N_i M_i^2} = \frac{\Sigma_i w_i M_i^2}{\Sigma w_i M_i} \tag{3.33}$$

$$M_v = \left[\frac{\Sigma_i N_i M_i^{1+a}}{\Sigma_i N_i M_i} \right]^{1/a} \tag{3.34}$$

For a random molecular weight distribution, such as produced by many free radical or condensation syntheses, $M_n : M_w : M_z = 1:2:3$. This is illustrated

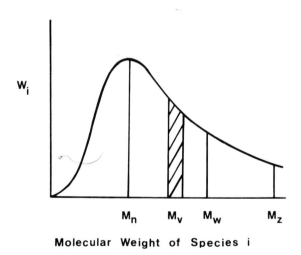

Figure 3.4 Schematic of a simple molecular weight distribution, showing the various averages.

in Figure 3.4. The quantity a in equation (3.34) varies from 0.5 to 0.8 (see Section 3.8.3).

An absolute method of measuring the molecular weight is one that depends solely on theoretical considerations, counting molecules and their weight directly. The relative methods require calibration based on an absolute method and include intrinsic viscosity and gel permeation chromatography (GPC).

Absolute methods of determining the number average molecular weight include osmometry and other colligative methods, and end-group analysis. Light-scattering yields an absolute weight-average molecular weight. The problem is compounded for light-scattering and SANS studies, because these techniques measure not only M_w but also the z-average radius of gyration (R_g^z). A knowledge of the molecular weight distribution is then required to obtain R_g^w. Only then can values of R_g be studied in relation to M_w.

3.5 DETERMINATION OF THE NUMBER-AVERAGE MOLECULAR WEIGHT

The number-average molecular weight involves a count of the number of molecules of each species, $N_i M_i$, summed over i, divided by the total number of molecules; see equation (3.31). It is the simple average that most people think about. For many simple single-peaked distributions, M_n is near the peak (see Figure 3.4). There are two important groups of methods for determining M_n.

3.5.1 End-Group Analyses

The first group of methods involves end-group analyses. Many types of syntheses leave a special group on one or both ends of the molecule, such as hydroxyl and carboxyl. These can be titrated or analyzed instrumentally by such methods as infrared. For molecular weights above about 25,000 g/mol, however, the method becomes insensitive because the end groups are present in too low a concentration.

3.5.2 Colligative Properties

The second group of methods makes use of the colligative properties of solutions. Colligative properties depend on the number of molecules in a solution, and not their chemical constitution (18). The colligative properties include boiling point elevation, melting point depression, vapor pressure lowering, and osmotic pressure. The basic equations for the first two may be

Table 3.5   A comparison of the colligative solution properties of a 1% polymer solution with $M = 20,000$ g / mol (19)

Property	Value
Vapor pressure lowering	4×10^{-3} mm Hg
Boiling point elevation	1.3×10^{-3} °C
Freezing point depression	2.5×10^{-3} °C
Osmotic pressure	15 cm solvent

written (19)

$$\lim_{c \to 0} \frac{\Delta T_b}{c} = \frac{RT^2}{\rho \, \Delta H_v} \left(\frac{1}{M_n} \right) \tag{3.35}$$

$$\lim_{c \to 0} \frac{\Delta T_f}{c} = -\frac{RT^2}{\rho \, \Delta H_f} \left(\frac{1}{M_n} \right) \tag{3.36}$$

where ΔT_b and ΔT_f are the boiling point elevation and freezing point depression, respectively; ρ is the solvent density; ΔH_v and ΔH_f are the latent heats of vaporization and fusion per gram of solvent, and c is the solute concentration in grams per cubic centimeter.

If the vapor pressure of the solute is small, and the solvent follows Raoult's vapor pressure law,

$$\frac{P^{\circ}_1 - P_1}{P^{\circ}_1} = X_2 \tag{3.37}$$

where $P^{\circ 1}$ is the vapor pressure of the pure solvent, P_1 is that of the solution, and X_2 is the mole fraction of the solute.

The osmotic pressure π depends on the molecular weight as follows:

$$\lim_{c \to 0} \frac{\pi}{c} = \frac{RT}{M_n} \tag{3.38}$$

Typical values for the colligative properties for a polymer having a molecular weight of 20,000 g/mol are shown in Table 3.5 (19). Only osmotic pressure is large enough for fruitful studies at this molecular weight or higher.

3.5.3   Osmotic Pressure

3.5.3.1   Thermodynamic Basis

Polymer solutions exhibit osmotic pressures because the chemical potentials of the pure solvent and the solvent in the solution are unequal. Because of this inequality, there is a net flow of solvent, through a connecting membrane, from the pure solvent side to the solution side. When sufficient pressure is built up on the solution side of the membrane, so that the two sides have the same activity, equilibrium will be restored (20).

While the above gives an exact thermodynamic interpretation of the phenomenon, a consideration in terms of the number of solute molecules per unit volume is useful for practical calculations. An analogy exists between equation (3.38) and the ideal gas law:

$$PV = nRT \tag{3.39}$$

where n is in moles, as usual. The quantity n/V is equal to c/M, yielding

$$P = \frac{c}{M}RT \tag{3.40}$$

Setting the gas pressure equal to the osmotic pressure, $P = \pi$, and rearranging, we obtain equation (3.38).

3.5.3.2 Instrumentation

A typical static osmometer design is shown in Figure 3.5. Typical membranes are made from regenerated cellulose or other microporous materials. Such an instrument usually requires 24 h to reach equilibrium.

There are now several types of automatic osmometers that operate with essentially zero flow and that reach equilibrium very rapidly, usually within minutes. Osmotic equilibrium depends on an equal and opposite pressure being developed. Table 3.6 (see page 83) (20) summarizes several commercial designs. The critical part of their design relates to the method of automatic adjustment of the osmotic pressure of the solution side so that the activity of the two sides is equal. Since several concentrations usually need to be run, the time required to determine a molecular weight by osmometry has been reduced from a week to a few hours by these automatic instruments.

3.5.3.3 Experimental Treatment of Data

The basis for determining the molecular weight by osmometry has been given in equation (3.38). At finite concentrations, interactions between the solvent and the solute result in the virial coefficients A_2, A_3, and so on. The full equation may be written

$$\frac{\pi}{c} = RT\left(\frac{1}{M_n} + A_2 c + A_3 c^2 + \cdots\right) \tag{3.41}$$

Interactions between one polymer molecule and the solvent result in the

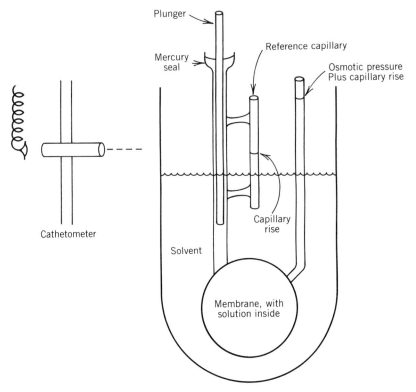

Figure 3.5 Design of the Stabin osmometer. The cell holds 10 cm^3 of solution. The plunger is used to preset the pressure near that expected, to reduce the time to equilibrium. The difference in heights of the solution and reference capillaries is read via a cathetometer. The temperature must be maintained constant, $\pm 0.01°C$; otherwise the solution capillary meniscus may rise and fall like in a thermometer tube.

second virial coefficient, A_2. Multiple polymer–solvent interactions produce higher virial coefficients, A_3, A_4, and so on. A further interpretation of the virial coefficients is considered along with solution theory (Section 3.3). Equation (3.41) is illustrated schematically in Figure 3.6. For medium molecular weights, the slope is substantially linear below about 1% solute concentration. Of course, $A_1 = 1/M_n$.

Since A_2 and A_3 are approximately related by $A_3 = gM_n A_2^2$, equation (3.41) may be rewritten (after separating out M_n), with $\Gamma_2 = M_n A_2$:

$$\frac{\pi}{c} = \left(\frac{\pi}{c}\right)_{c=0}(1 + \Gamma_2 c + g\Gamma_2^2 c^2) \tag{3.42}$$

The quantity g, which depends on the polymer–solvent interactions, often equals about 0.25 for polymer solutions (21–23). Therefore equation (3.42)

Table 3.6 Automatic osmometers for rapid molecular weights (20)

Instrument	Pressure Adjustment Mechanism
Mechrolab (Hewlett-Packard)	Solvent reservoir raised or lowered by servomechanism. Instantaneous flow observed by motion of a bubble.
Shell (Hallikainen Instruments and Dohrman Instrument Co.)	Volume change in solution compartment monitored; servomechanism adjusts pressure on a diaphragm.
Metabs	Strain gauges detect pressure differences.
Knauer	High-grade porous steel backing; membrane deflects, motion detected by an electronic displacement device and converted into a signal.
Wescan	Strain gauge on diaphragm.

may be treated as a perfect square. Taking the square root, we have

$$\left(\frac{\pi}{c}\right)^{1/2} = \left(\frac{\pi}{c}\right)^{1/2}_{c=0}\left(1 + \frac{\Gamma_2 c}{2}\right) \tag{3.43}$$

Thus a plot of $(\pi/c)^{1/2}$ versus c will tend to be more linear than π/c versus c, which is particularly helpful for determining higher molecular weights or working in more concentrated solutions.

The quantity A_2 depends on both the temperature and the solvent, for a given polymer. A unique and much desired state arises when A_2 equals zero. The temperature at which this condition holds is called the Flory θ-temperature (23). In this state, π/c is independent of concentration, so that only one concentration need be studied to determine M_n. Since it is also the state where an infinite molecular weight polymer just precipitates (see Section 4.1), considerable care must be taken to keep the polymer in solution.

Figure 3.7 illustrates the determination of the Flory θ-temperature for cellulose tricaproate dissolved in dimethylformamide (24). Table 3.7 presents a selected list of polymers and their theta solvents (25).

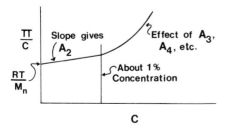

Figure 3.6 Schematic illustration of the dependence of osmotic pressure on concentration.

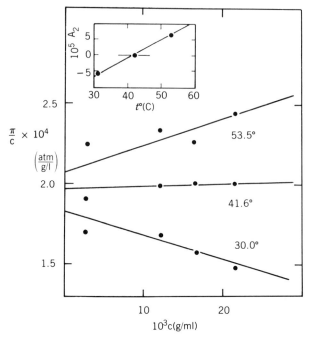

Figure 3.7 The osmotic pressure data for cellulose tricaproate in dimethylformamide at three temperatures. The Flory θ-temperature was determined to be 41 ± 1°C (24).

Table 3.7 Polymers and their θ-solvents (25)

Polymer	Solvent	Temperature (°C)
Cis-polybutadiene	*n*-Heptane	−1
Polyethylene	Biphenyl	125
Poly(*n*-butyl acrylate)	Benzene/methanol 52/48	25
Polystyrene	Cyclohexane	34
Poly(oxytetramethylene)	Chlorobenzene	25
Cellulose tricaproate	Dimethylformamide	41

Source: J. Brandrup and E. H. Immergut, Eds., *Polymer Handbook*, 2nd ed., Wiley-Interscience, New York, 1975.

It must be emphasized that molecular weights determined by any of the colligative properties, and osmometry in particular, are absolute molecular weights; that is, the values are determined by theory and not by prior calibration.

3.5.3.4 *Example Calculation Using Osmometry*

Osmotic pressure is conveniently measured in centimeters of solvent:

$$\pi(\text{cm solvent}) \times \text{g/cm}^3 \text{ (solvent density)} = \pi(\text{g/cm}^2)$$

For such calculations

$$R = 8.48 \times 10^4 \frac{\text{g} \cdot \text{cm}}{\text{mol} \cdot \text{K}} \qquad T \text{ in K (kelvin)}$$

At a concentration of 2×10^{-3} g/cm^3, a polymer exhibited an osmotic pressure of 0.30 cm in a solvent of 1.0 g/cm^3 density at 30°C. If A_2 is zero, what is its molecular weight?

$$\frac{\pi}{c} = \frac{RT}{M_n} \qquad M_n = 1.6 \times 10^5 \frac{\text{g}}{\text{mol}}$$

The practical limit of osmometry is about 500,000 g/mol, because the pressures become too small. Light-scattering, to be considered in the following, becomes more sensitive at higher molecular weights.

3.6 WEIGHT-AVERAGE MOLECULAR WEIGHTS AND RADII OF GYRATION

The principal method of determining the weight-average molecular weight is light-scattering, although small-angle neutron scattering is now becoming important, especially in the bulk state, and x-ray scattering is also sometimes employed. For generality, all three methods are considered.

First, a few general terms need to be defined. Radiation from an object is said to be reflected when the object is much larger than the wavelength of the radiation. The radiation is said to be scattered when the object begins to approach the wavelength of the radiation in size, down to atomic dimensions. Common examples of light-scattering include the blue of the sky and the rainbow. The latter, involving scattering from spheres of about 50 μm in diameter, is connected to other types of scattering through the Mie (26) equations.

Scattering from a single electron isolated in space provides a beginning. As the electron is approached by an electromagnetic wave, the electron absorbs the energy and begins to oscillate (see Figure 3.8). Now an accelerating electric charge is itself a radiator, and the energy is re-radiated, but in all directions. This re-radiation of energy is called scattering.

For more complex systems, there are two general cases. First, if the atoms, molecules, or particles are organized into a regular array, the radiation will be diffracted; that is, scattering will be observed only at special angles. This arises because at all other angles there is total destructive interference between the scattered radiation arising from different parts of the array. An example is the diffraction of x-rays by a crystal.

If the atoms, molecules, or particles are not organized into a regular array, scattering will be observed at all angles. In general, the angular variation of the scattering intensity provides a measure of the size of the structures.

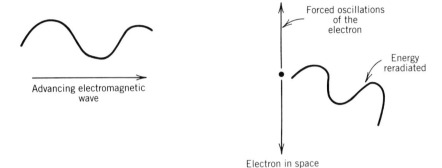

Figure 3.8 The effect of an electromagnetic wave on a free electron. The forced oscillations of the electron involve accelerations and decelerations, which cause the electromagnetic energy to be re-radiated.

The phenomenon of light-scattering is caused by fluctuations in the refractive index of the medium on the molecular or supermolecular scale. For example, the blue-of-the-sky scattering mentioned previously is caused by the random presence of gas molecules in what is otherwise a vacuum. The blue of large bodies of water is caused by slight fluctuations in the spacing of the water molecules. In both cases, the scattered intensity varies as the wavelength to the inverse fourth power, which causes the characteristic blue color.

The application of light-scattering to polymer molecular weights follows a trail begun in part by Smoluchowski (27, 28) and Einstein (29), who considered the fluctuations of refractive index in liquids. The basic equations used in light-scattering of polymer solutions today were developed by Debye (30) and Zimm (31, 32). They replaced fluctuations in the refractive index of the solvent itself by the changes caused by the polymer molecules. The final result relates the observed light-scattering intensity to the osmotic pressure, π, of the polymer as follows:

$$\frac{Hc}{R(\theta)} = \frac{1}{RT}\left(\frac{\partial \pi}{\partial c}\right)_T \tag{3.44}$$

Where $R(\theta)$ is called Rayleigh's ratio and is equal to $I_\theta w^2/I_0 V_s$ where I_θ represents the light intensity observed at angle θ scattered from a volume V_s, if the distance from the source is w and the intensity of the incident light is I_0. The optical constant H is given by

$$H = \frac{2\pi^2 n_0^2 \left(\dfrac{dn}{dc}\right)^2}{N_A \lambda^4} \tag{3.45}$$

where n_0 represents the refractive index at wavelength λ and N_A is Avogadro's number. The quantity dn/dc must be determined for each polymer–solvent pair. Then H is a constant for a particular polymer and solvent, but determined theoretically rather than empirically.

3.6.1 Scattering Theory and Formulations

The basic theory of light-scattering applied to polymer solutions dates from the works of Debye (30), who formulated the absolute molecular weight determination in terms of an optical constant H, as shown in equation (3.44). The corresponding theory for x-rays was developed by Guinier and Fournet (33). The theory for small-angle neutron scattering was derived by Kirste, Ballard, and Ibel (34). Several reviews have been written (35–38).

The basic equation for the molecular weight and size for all three modes of scattering can be written

$$\frac{Hc}{R(\theta) - R(\text{solvent})} = \frac{1}{M_w P(\theta)} + 2A_2 c \qquad (3.46)$$

This equation corrects $R(\theta)$ for the Einstein–Smoluchowski solvent scattering. The factor of 2 appearing in the second term, RHS of equation (3.46) arises from first multiplying equation (3.41) by c, then differentiating as shown in equation (3.44). If the particles are very small compared to the wavelength of the radiation, $R(\theta)$ reduces to $3\tau/16\pi$, where τ is the turbidity in Beer's law[†]:

$$I = I_0 e^{-\tau x} \qquad (3.47)$$

where x is the sample thickness. Under these conditions, $P(\theta)$, the scattering form factor, equals unity. The calibration and use of light-scattering instrumentation is discussed in Appendix 3.1 for the case of small molecules, where $P(\theta) = 1$.

Incidentally, the inverse fourth power of the wavelength shown in equation (3.45) can be quantified,

$$\tau = \frac{\tilde{K}}{\lambda^{-4}} \qquad (3.48)$$

where $\tilde{K}$ is a constant that can be calculated from the above relationships for any particular system. As a general phenomenon for particles much smaller than the wavelength (such as gas molecules), equation (3.48) then quantifies the blue of the sky.

[†]Use of τ in equation (3.46) leads to an optical constant of $32\pi^3 n_0^2 (dn/dc)^2 / 3N_A \lambda^4$.

For particles (or molecules) larger than about 0.05 times the wavelength, $P(\theta)$ differs from unity. The quantity $P(\theta)$ is called the single chain form factor, which describes the angular scattering arising from the conformation of an individual chain. This molecular structure factor becomes independent of particle shape as θ approaches zero. The region of very small angles, $K^2 R_g^2 < 1$, is known as the Guinier region. In the Guinier region, $P(\theta)$ becomes a measure of the radius of gyration, R_g.

For a random coil (30), $P(\theta)$ in equation (3.46) is expressed by

$$P(\theta) = \frac{2}{R_g^4 K^4}\left\{R_g^2 K^2 - \left[1 - \exp\left(-R_g^2 K^2\right)\right]\right\} \tag{3.49}$$

Equation (3.49) may be given in the expanded form as

$$P(\theta) = 1 - \frac{K^2 R_g^2}{3} + \cdots \tag{3.50}$$

where

$$K = \frac{4\pi}{\lambda}\sin(\theta/2) \tag{3.51}$$

The quantity λ represents the wavelength of the radiation, and θ is the angle of scatter. The quantity K, sometimes written q or Q, is variously called the wave vector or the range of momentum transfer, especially for neutron scattering.

According to Zimm (31, 32), the key equations at the limit of zero angle and zero concentration, respectively, relating the light-scattering intensity to the weight-average molecular weight M_w and the z-average radius of gyration R_g, may be written

$$\left(H\frac{c}{R(\theta)}\right)_{\theta=0} = \frac{1}{M_w} + 2A_2 c + \cdots \tag{3.52}$$

$$\left(H\frac{c}{R(\theta)}\right)_{c=0} = \frac{1}{M_w}\left[1 + \frac{1}{3}\left(\frac{4\pi}{\lambda'}\right)^2 R_g^2 \sin^2\frac{\theta}{2} + \cdots\right] \tag{3.53}$$

where λ' is the wavelength of the light in solution (λ_0/n_0). To construct a Zimm plot, equations (3.52) and (3.53) are added together. To make a more aesthetic plot, the concentration term is usually multiplied by an arbitrary factor. (See Section 3.6.3.)

Thus three quantities of interest can be determined from the same experiment: the weight-average molecular weight, the z-average radius of

Table 3.8   Convenient light-scattering units

$$H = \frac{\text{mol} \cdot \text{cm}^2}{\text{g}^2} \qquad\qquad A_2 = \frac{\text{mol} \cdot \text{cm}^3}{\text{g}^2}$$

$$c = \text{g}/\text{cm}^3 \qquad\qquad R(\theta) = \text{cm}^{-1}$$

$$\theta = \text{degrees}$$

$$(\sin^2 \theta/2 \text{ for } 90° \text{ is } 0.500, \text{ unitless})$$

$$M_w = \text{g}/\text{mol}$$

gyration, and the second virial coefficient. Note that the numeral 2, which appears in front of A_2 in equations (3.46) and (3.52), arises from the partial differentiation given in equation (3.44). A useful practical equation for the determination of R_g from a plot of $H[c/R(\theta)]$ versus $\sin^2 \theta/2$ is

$$R_g^2 = \frac{3(\lambda')^2(\text{initial slope})}{16\pi^2(\text{intercept})} \tag{3.54}$$

Convenient light-scattering units are shown in Table 3.8.

The quantity H depends on the kind of radiation. For light-scattering, the formulation has been given in equation (3.45).

For x-ray scattering,

$$H = N_a i_e \left(\frac{\partial \rho_e}{\partial c}\right)^2 = \frac{N_a i_e}{e^2}(\rho_{e_s} - \rho_{e_p})^2 \tag{3.55}$$

and for neutron scattering,

$$H = \frac{N_a}{M_p^2}\left[a_s\left(\frac{V_p}{V_s}\right) - a_p\right]^2 \tag{3.56}$$

where M_p is the mer[†] molecular weight, and a_p and a_s are the coherent neutron scattering lengths of the polymer mer units and solvent, respectively. The quantity ρ is the density of the polymer (37). The quantity i_e is the Thomson scattering factor for a single electron, and ρ_e is the electron density of the solution; subscripts p and s represent polymer and solvent, respectively. Of course, the solvent can also be polymeric, and frequently the

---

[†]The term "mer," derived from poly*mer*, means the individual monomeric unit molecular entity. Some texts call this entity the monomeric unit.

"solvent" differs from the polymer by having hydrogen or deuterium (H or D) atoms where the polymer has D or H atoms, especially for neutron scattering.

3.6.2 The Appropriate Angular Range

To determine the radius of gyration, R_g, the basic mathematical relationship must hold,

$$K^2 R_g^2 < 1 \tag{3.57}$$

The corresponding physical requirement indicates that there must be only partial destructive interference between two waves striking the same particle, so that the waves should not be out of phase by more than 180° (see Figure 3.9). Referring to Figure 3.9, the distance A, B, C is seen to be, in general, different from the distance X, Y, Z, where the line A–X is perpendicular to the direction of radiation flux before scattering, and the line C–Z is perpendicular to the scattered flux (35–39).

Before scattering, the radiation has a certain degree of coherency; that is, the waves tend to be in phase. (Ordinary light is coherent over the size of the object; laser light is coherent over larger distances.) On scattering from points B and Y at an angle 2θ, the waves become out of phase. If the angle of scatter is large enough, the waves are out of phase by 180°; that is, one wave lags behind the other by $\frac{1}{2}\lambda$, and the radiation intensity observed at angle 2θ will be at the minimum. At still higher angles, the intensity increases again until a 360° phase difference is attained. The scattering must be observed for angles smaller than $KR_g = 1$ [see equation (3.57) for radii of gyration to be determined]. The partial destructive interference between the waves causes the intensity to vary according to angle (see Figure 3.10). Until 180° phase difference is attained, scattering intensity decreases with increasing angle.

For ordinary-size polymer molecules, R_g is of the order of 100–200 Å. For light-scattering θ is usually measured in the range of 45–135°, because

Figure 3.9 Schematic of the scattering phenomenon. When the waves are out of phase, the intensity of scattered light is reduced.

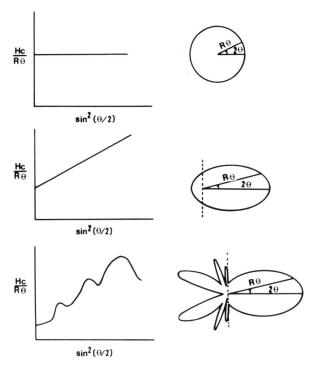

Figure 3.10 Scattering intensity envelopes for small, medium, and large particles. The Guinier region contains both the small and medium ranges, but the medium range is far more useful for scattering experiments.

$\lambda \simeq 5000$ Å. For x-rays ($\lambda \simeq 1$–2 Å) and thermal neutrons ($\lambda \simeq 5$ Å), scattering measurements are usually made below 1°.

The requirement of small angles for x-ray and neutron scattering has led to the construction of huge instruments. For example, x-ray scattering instruments may be 10 m long. The newest scattering instruments in the world, however, are the three giant small-angle neutron scattering instruments, located at the Institute Laue-Langevin, Grenoble, France (80 m), Jülich, West Germany (40 m), and Oak Ridge National Laboratory, Tennessee (30 m) (40). The workings of the Oak Ridge instruments are illustrated in the frontispiece for the 30-m instrument and in Figure 3.11 for the 10-m instrument. These instruments employ thermal or cool neutrons to increase their wavelengths.

Before proceeding further, a word should be said about the term "radius of gyration." The quantity R_g^2 is defined as the mean square distance away from the center of gravity, $R_g^2 = (1/N)\Sigma_{i=1}^N r_i^2$, for N scattering points of distance r_i. It is sometimes helpful to view R_g as a mechanical term wherein the radius of gyration of a body is the radius of a thin ring that has the same

Figure 3.11   Schematic diagram of the 10-m small-angle neutron scattering facility at Oak Ridge, TN. The neutron beam used by the instrument is transported from the beam room to the SANS facility level by Bragg reflection from pyrolitic graphite crystals scattering at 90°. Courtesy of Dr. G. D. Wignall.

mass and same moment of inertia as the body when centered at the same axis (38). For a random coil, it is related to the end-to-end distance, r, by

$$R_g^2 = r^2/6 \tag{3.58}$$

where r is the distance between the ends of the chain. Different relationships hold for spheres, rods, and coils. These differences are expressed quantitatively in terms of the quantity $P(\theta)$ (39):

Sphere $$P(\theta) = \left[\frac{3}{x^3}(\sin x - x \cos x)\right]^2 \qquad x = \frac{ksD}{2} \tag{3.59}$$

Rod $$P(\theta) = \frac{1}{x}\int_0^{2x} \frac{\sin w}{w}\, dw - \left(\frac{\sin x}{x}\right)^2 \qquad x = \frac{ksL}{2} \tag{3.60}$$

Coil $$P(\theta) = \frac{2}{x^2}\left[e^{-x} - (1 - x)\right] \qquad x = \frac{k^2 s^2 r^2}{6} \tag{3.61}$$

where D = diameter of sphere, L = length of rod, $ks = K$ and r = root

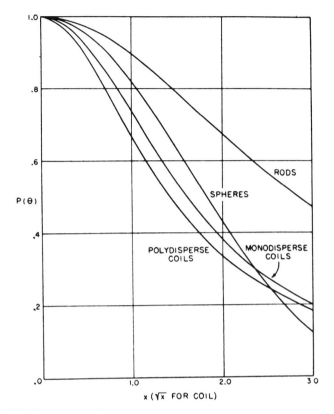

Figure 3.12 Particle scattering factors plotted against the size factor x. The size determined depends on the model of the particle shape (39).

mean square of the distance between ends of the random coil. Rearranged, equation (3.61) yields equation (3.49), discussed previously.

The various relations for $P(\theta)$ are plotted versus x in Figure 3.12 (39). While the present text is interested primarily in random coils, scattering phenomena are widely used to determine quantities related to many shapes.

3.6.3 The Zimm Plot

Equations (3.52) and (3.53) show the function $H[c/R(\theta)]$ in the limit of $\theta = 0$ and $c = 0$, respectively. Three important pieces of information can be extracted from this experiment: the weight-average molecular weight, z-average radius of gyration, and the second virial coefficient. A simple but laborious method of plotting the data is shown in Figure 3.13. Here, (a) shows a plot of $H[c/R(\theta)]$ versus concentration for the several angles,

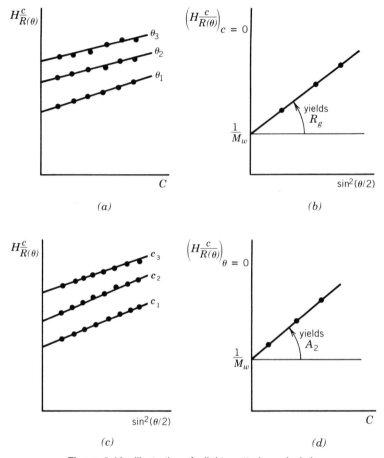

Figure 3.13 Illustration of a light-scattering calculation.

extrapolated to $c = 0$. Part (b) is a replot of the intercepts from (a) versus $\sin^2(\theta/2)$, which yields M_w and R_g. In (c), the same data as (a) are replotted, this time against $\sin^2(\theta/2)$ first. Then, in (d), the intercepts of (c) are plotted against concentration, yielding M_w and A_2. The intercepts from (b) and (d) must be equal, since they both equal $1/M_w$. This last provides an internal test of the data.

A most powerful advance was the introduction of the Zimm plot (31, 32), which enabled the radius of gyration, the molecular weight, and the second virial coefficient to be calculated from a single master figure, by plotting $H[c/R(\theta)]$ versus a function of both angle and concentration (see Figure 3.14) (41). In a Zimm plot, the concentration is multiplied by an arbitrary constant, in this case 1×10^3, which is, of course, divided out of the final answer.

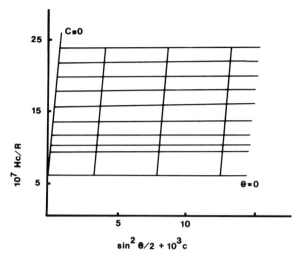

Figure 3.14 A Zimm plot for cellulose tricaproate in dimethylformamide. Note that A_2 is zero, indicating that at the temperature of measurement, 41°C, this is a Flory θ-system (33). $\lambda = 5460$ Å; refractive index of dimethyl formamide is 1.43 (41).

3.6.4 Polymer Chain Dimensions

For random coils obeying Gaussian statistics, the end-to-end distance squared depends on the molecular weight,

$$r^2 = CM \tag{3.62}$$

where C is a function of the chain molecular structure. Equation (3.62) has the same form as the relation equating the total distance traveled for a particle undergoing Brownian motion as a function of time, except that in Brownian motion, the distance traveled under each impact is variable, where mer bond lengths are fixed. In addition, the Brownian particle may turn at any angle, whereas carbon–carbon bonds are fixed. Perhaps a more fundamental difference is that the path of a Brownian particle may cross itself, whereas a polymer chain may not. All three of these effects tend to increase the constant C but do not change the fundamental relationship expressed in equation (3.62).

Light-scattering studies such as illustrated in Figure 3.14 showed that for high enough molecular weight, polymer chains were indeed random coils. During the 1940s and 1950s, scientists theorized that the conformation of polymer chains in an amorphous bulk polymer were similar to that which existed under Flory θ-conditions, but there was only limited direct experimental evidence to support this theory.

The problem was that nobody could figure out a good way to measure the molecular dimensions in the bulk state, although on theoretical grounds (see

Chapter 5) they were thought to be similar to the dilute solution results. After about 1960, polymer scientists turned their attention to other problems, and there the results stood until the development of small-angle neutron scattering (SANS) in the early 1970s.

3.6.5   Scattering Data

One problem that had to be solved, and was, relates to the fact that all these scattering techniques yield the z-average radius of gyration and the weight-average molecular weight. To correct the data for proper comparison, the molecular weight distribution needs to be known, or very sharp molecular weight distributions need to be made. The preferred solution has been to work with nearly monodisperse polymer samples, such as prepared by anionic polymerization. In other cases, the ratio of M_z to M_w is known or estimated from polymerization kinetics (see Section 3.4). To properly estimate the molecular weight distribution, the number-average molecular weight is required, bringing into play osmometry or gel permeation chromatography (see Sections 3.5.3 and 3.9). Generally, the z-average R_g values are corrected back to the weight-average, and the weight-averages of both quantities are reported.

When SANS instrumentation became available in the early 1970s, polymer scientists sought to reexamine chain conformation behavior, this time directly in the bulk amorphous state. The earliest studies were on amorphous poly(methyl methacrylate) (42, 43) and polystyrene (44–46). These and several subsequent papers (47–55) indeed confirmed that values of $(R_g^2/M_w)^{1/2}$ [see equation (3.62)] were substantially the same in the bulk as in Flory θ-solvents. The results of the SANS experiments are substantially the same as those obtained in θ-solvents (35).

Some light-scattering results for polystyrene are shown in Table 3.9 (56); as a function of temperature for molecular weights of the order of 10^5, radii

Table 3.9   Dilute solution behavior of polystyrene
[$M_w = 7.65 \times 10^5$ g / mol in cyclohexane (56)]

T (°C)	$A_2 \times 10^4$ (ml · mol/g^2)	$\langle R_g^2 \rangle \times 10^{12}$ (cm^2)	$[\eta]$ (dl/g)
60.1	0.93$_3$	8.96	1.15
50.1	0.67$_7$	8.39	1.04
45.4	0.55$_4$	7.98	0.985
40.2	0.31$_5$	7.56	0.916
37.3	0.18$_7$	7.25	0.858
35.2	0.08$_2$	6.77	0.817
34.2	-0.02_5	6.51	0.793
32.2	-0.18_5	5.85	0.749

of gyration of about 100 Å are obtained. This is in the true colloidal range. (It must be remembered that before Staudinger's macromolecular hypothesis in 1920, polymers and colloids were classed together.) Molecular dimensions by SANS are discussed further in Section 5.2.2.1.

The quantity $(R_g^2/M_w)^{1/2}$ is, of course, a measure of chain stiffness. For example, polycarbonate, with $(R_g^2/M_w)^{1/2} = 0.457$, is stiffer than polystyrene, which has a value of 0.275. The importance of these quantities lies in their relation to physical and mechanical behavior. Both melt and solution viscosities depend directly on the radius of gyration of the polymer and on the chain's capability of being deformed. The theory of the random coil, strongly supported by these measurements, is used in rubber elasticity theory (Chapter 9) and many mechanical and relaxation calculations.

3.7 MOLECULAR WEIGHTS OF COMMON POLYMERS

3.7.1 Molecular Weight of Commercial Polymers

The molecular weight of polymers used in commerce varies from about 30,000 to over 1,000,000 g/mol. Sometimes conflicting requirements include the use of high enough molecular weights to obtain good physical properties, and low enough molecular weights to permit reasonable processing conditions, such as melt viscosity.

Poly(vinyl chloride) Commercial poly(vinyl chloride) "vinyl" polymers range from 60,000 to about 90,000 g/mol. The above restrictions hold in this case.

Poly(methyl methacrylate) Those polymers that are used in such products as Plexiglas® have high molecular weights with broad distributions. A viscous syrup containing low-molecular-weight polymer and monomer is poured into a mold and allowed to polymerize. Late in the polymerization, the phenomenon known as autoacceleration takes place, where the molecular weight increases dramatically owing to a suppression of the termination step. This high molecular weight produced at the end may be over 1×10^6 g/mol, contributing strength and toughness to the final sheet. This material is used, however, without further processing.

Cellulose This natural polymer occurs with extremely high molecular weights, sometimes in the several million range, and with molecular weight distributions of M_w/M_n in the range of 10–50. For commercial applications such as rayon, the polymer is deliberately degraded down to the 50,000–80,000 g/mol range to increase processibility. The better products often utilize the higher end of this range.

3.7.2 Molecular Weight Distributions

If the termination reaction in chain polymerization is by disproportionation, then the polydispersity index, M_w/M_n, is 2. Termination by combination yields a polydispersity index of 1.5. Stepwise polymerizations, such as polyester formation, yield a value of 2. Anionic polymerizations yield surprisingly narrow distribution, with values sometimes less than 1.05.

Proteins are almost the only source of truly monodisperse polymers. Nature makes all these molecules exactly alike. Polymers like cellulose have very broad distributions, as mentioned previously.

Of course, polymerization need not be ideal in its kinetics. Branching may occur, which broadens the molecular weight distribution. There may even be two or more peaks in the molecular weight distribution. A powerful method for directly observing the shape of the distribution curve is gel permeation chromatography (see Section 3.9).

The various molecular weight distributions have been modeled. Two of the most important are the Schultz (57) distribution and the Poisson (58) distribution. Taking x to be the degree of polymerization and w_x as the differential distribution by weight, the Schultz distribution is

$$w_x = \frac{a}{x_n \Gamma(a + 1)} \left(\frac{ax}{x_n} \right)^a \exp\left(-\frac{ax}{x_n} \right) \tag{3.63}$$

where $x_w/x_n = (a + 1)/a$, and $\Gamma(a + 1)$ is the gamma function of $a + 1$. When $a = 1$ and x is large, the polydispersity index is near 2.

The Poisson distribution assumes that all the chains are initiated simultaneously. Growth continues at approximately the same rate in each chain, until the monomer runs out. An analogy would be a horse race, where all the horses start at the bell and finish at nearly the same time. The result is a narrow molecular weight distribution. (For anionic polymerizations following these statistics, the ends of the chains remain "living," even though the monomer has run out. Addition of a second monomer then yields well-defined block copolymers; see Section 4.3.9.)

The Poisson distribution is given by (58)

$$(P_r)_x = \frac{e^{1-x_n}(x_n - 1)^{x-1}}{(x - 1)!} \tag{3.64}$$

where $(P_r)_x$ is the mole fraction of x-mer. Then

$$\frac{x_w}{x_n} = 1 + \frac{x_n - 1}{x_n^2} \tag{3.65}$$

For reasonable values of x_n, the polydispersity index is nearly unity. This distribution is realized for carefully prepared anionic polymerizations.

For stepwise polymerizations, the kinetics are considered in terms of the extent of reaction, p, defined as the fraction of the functional groups reacted at time t. If only bifunctional reactants are present, then the degree of

Table 3.10   Typical molecular weight distributions

Method	Polydispersity	Stereospecificity
Natural Proteins	1.0	Perfect
Anionic Polymerization	1.02–1.5	None
Chain Polymerization	1.5–3	None
Step Polymerization	2.0–4	None
Ziegler-Natta	2–40	High
Cationic	Broad	None

polymerization may be deduced by considering the total number of structural units present, N_0, and the total number of molecules, N:

$$x_n = \frac{N_0}{N} = \frac{1}{1-p} \qquad (3.66)$$

and

$$x_w = \frac{1+p}{1-p} \qquad (3.67)$$

An important difference between stepwise reactions and chain polymerization is that the former type of molecules can keep on reacting. As p approaches unity, both x_w and x_n increase. Their ratio, however, approaches 2, called the most probable distribution.

The molecular weight distributions ordinarily obtained by the more important polymerization methods are shown in Table 3.10. Natural proteins, nature's best, are both monodisperse and 100% stereospecific. The Ziegler–Natta polymerization route, while highly stereospecific, has a broad molecular weight distribution. In general, the polydispersity index can be determined from an analysis of the kinetics of the reaction; in practice, various phenomena cause the products to be much broader in molecular weight distributions (59).

3.7.3   Gelation and Network Formation

If the functionality of the monomer is 2, as in the case of vinyl groups for stepwise polymerization,

$$C=C \rightarrow -C-C-$$

or monomers containing one carboxyl group and one hydroxyl group for stepwise polymerization,

$$HO-R-COOH \rightarrow -O-R-COO- + H_2O$$

linear polymers are formed. If some trifunctional, tetrafunctional, or higher monomers are incorporated, the polymer will be either branched or cross-linked.

An example of a trifunctional monomer is glycerol, with three hydroxyl groups,

$$
\begin{array}{l}
CH_2-OH \\
| \\
CH-OH \\
| \\
CH_2-OH
\end{array}
$$

Divinyl benzene, with two vinyl groups, is a common cross-linker for chain polymerizations,

An important question is: When in a polymerization will gelation occur? Gelation is defined as when a single molecule, connected by ordinary covalent bonds, extends throughout the polymerization vessel. (Most of the material in the vessel need not be part of the molecule, however.) Alternately, gelation may be conceived as the point where a three-dimensional network is formed. From a physical point of view, the viscosity of the reacting mass goes to infinity at the gelation point.

According to Flory and Stockmayer (60), the critical extent of reaction, P_c, at the gel point is given by

$$
P_c = \frac{1}{(f-1)^{1/2}} \tag{3.68}
$$

where f represents the functionality of the branch units—that is, of the monomer with functionality greater than 2. This simple equation has been modified many times for particular stoichiometries and mixtures of monomers. Gelation is considered further in Section 9.1 and in Section 9.14.

3.8 INTRINSIC VISCOSITY

Both the colligative and the scattering methods result in absolute molecular weights; that is, the molecular weight can be calculated directly from first principles based on theory. Frequently, these methods are slow, and some-

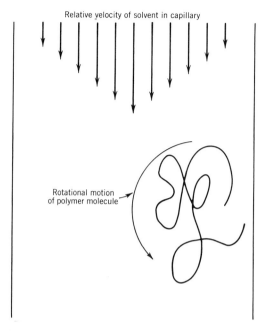

Figure 3.15 The effect of shear rates on polymer chain rotation. Hydrodynamic work is converted into heat, resulting in an increased solution viscosity.

times expensive. In order to handle large numbers of samples, especially on a routine basis, rapid, inexpensive methods are required. This need is fulfilled by intrinsic viscosity and by gel permeation chromatography. The latter is discussed in the next section.

Intrinsic viscosity measurements are carried out in dilute solution and result in the viscosity-average molecular weight [see Figure 3.4 and equation (3.34)]. Consider such a dilute solution flowing down a capillary tube (Figure 3.15). The flow rate, and hence the shear rate, is different depending on the distance from the edge of the capillary. The polymer molecule, although small, is of finite size and "sees" a different shear rate in different parts of its coil. This change in shear rate results in an increase in the frictional drag and rotational forces on the molecule, yielding the mechanism of viscosity increase by the polymer in the solution.

3.8.1 Definition of Terms

Several terms need defining. The solvent viscosity is η_0, usually expressed in poises, Stokes, or, more recently, Pascal seconds. The viscosity of the polymer solution is η. The relative viscosity is the ratio of the two,

$$\eta_{\text{rel}} = \eta / \eta_0 \tag{3.69}$$

Of course, the relative viscosity is a quantity larger than unity. The specific viscosity is the relative viscosity minus one:

$$\eta_{sp} = \eta_{rel} - 1 \tag{3.70}$$

Usually, η_{sp} is a quantity between 0.2 and 0.6 for the best results.

The specific viscosity, divided by the concentration and extrapolated to zero concentration, yields the intrinsic viscosity:

$$\left[\frac{\eta_{sp}}{c}\right]_{c=0} = [\eta] \tag{3.71}$$

For dilute solutions, where the relative viscosity is just over unity, the following algebraic expansion is useful:

$$\ln \eta_{rel} = \ln(\eta_{sp} + 1) \cong \eta_{sp} - \eta_{sp}^2/2 + \cdots \tag{3.72}$$

Then, dividing $\ln \eta_{rel}$ by c and extrapolating to zero concentration also yields the intrinsic viscosity:

$$\left[\frac{\ln(\eta_{rel})}{c}\right]_{c=0} = [\eta] \tag{3.73}$$

Note that the logarithm of η_{rel} is divided by c in equation (3.73), not η_{rel} itself. Also, note that the intrinsic viscosity is written with η enclosed in brackets. This is not to be confused with the plain η, which is used to indicate solution or melt viscosities.

Two sets of units are in use for $[\eta]$. The "American" units are $100 \text{ cm}^3/\text{g}$, whereas the "European" units are cm^3/g. Of course, this results in a factor of 100 difference in the numerical result. Lately, the European units are becoming preferred.

3.8.2 The Equivalent Sphere Model

Consider a coiled polymer molecule as being impenetrable to solvent in the first approximation (23). A hydrodynamic sphere of equivalent radius R_e will be used to approximate the coil dimensions (see Figure 3.16). In shear flow, it exhibits a frictional coefficient of f_0. Then according to Stokes law,

$$f_0 = 6\pi\eta_0 R_e \tag{3.74}$$

where R_e remains quantitatively undefined.

The classical Einstein viscosity relationship for spheres may be written

$$\frac{\eta - \eta_0}{\eta_0} = \eta_{sp} = 2.5\left(\frac{n_2}{V}\right)V_e \tag{3.75}$$

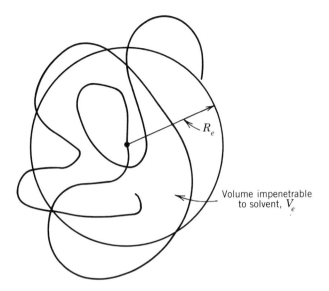

Figure 3.16 The equivalent sphere model.

where n_2/V is the number of molecules per unit volume. Of course, $V_e = (4\pi/3)R_e^3$. The quantity $n_2 V_e$ is the volume fraction of equivalent spheres, yielding the familiar result that the viscosity of an assembly of spheres is independent of the size of the spheres, depending only on their volume fraction.

Writing

$$\frac{n_2}{V} = \frac{cN}{M} \tag{3.76}$$

where c is the concentration and N is Avogadro's number,

$$\left[\frac{\eta_{sp}}{c}\right]_{c=0} = [\eta] = 2.5\frac{NV_e}{M} \tag{3.77}$$

Note that

$$\frac{V_e}{M} = \frac{4\pi}{3}\frac{R_e^3}{M} = \frac{4\pi}{3}\left(\frac{R_e^2}{M}\right)^{3/2}M^{1/2} \tag{3.78}$$

and

$$R_e = R_{e0}\alpha \tag{3.79}$$

where α is the expansion of the coil in a good solvent over that of a Flory θ-solvent.

The quantity R_{e0}^2/M is roughly constant. The same constant appears in Brownian motion statistics, where time takes the place of the molecular weight. This expresses the distance traveled by the chain in a random walk as a function of molecular weight. According to Flory (23), the expansion of the coil increases with molecular weight for high molecular weights as $M^{0.1}$, yielding

$$[\eta] = 2.5\frac{4\pi}{3}N\left(\frac{R_{e0}^2}{M}\right)^{3/2}M^{1/2}\alpha^3 \tag{3.80}$$

3.8.3 The Mark–Houwink Relationship

In 1938 Mark and Houwink arrived at an empirical relationship between the molecular weight and the intrinsic viscosity (61):

$$[\eta] = KM_V^a \tag{3.81}$$

where **K** and a are constants for a particular polymer–solvent pair at a particular temperature. This equation is in wide use today, being one of the most important relationships in polymer science. Values of **K** and a for selected polymers are given in Table 3.11 (62). It must be pointed out that since viscosity-average molecular weights are difficult to obtain directly, the weight-average molecular weights of sharp fractions or narrow molecular weight distributions are usually substituted to determine **K** and a.

Table 3.11   Selected intrinsic viscosity–molecular weight relationships, $[\eta] = KM_v^a$ (62)

Polymer	Solvent	$T(°C)$	$K \times 10^{3*}$	a
Cis-polybutadiene	Benzene	30	33.7	0.715
it-Polypropylene	1-Chloronaphthalene	139	21.5	0.67
Poly(ethyl acrylate)	Acetone	25	51	0.59
Poly(methyl methacrylate)	Acetone	20	5.5	0.73
Poly(vinyl acetate)	Benzene	30	22	0.65
Polystyrene	Butanone	25	39	0.58
Polystyrene	Cyclohexane (θ-solvent)	34.5	84.6	0.50
Polytetrahydrofuran	Toluene	28	25.1	0.78
Polytetrahydrofuran	Ethyl acetate hexane (θ-solvent)	31.8	206	0.49
Cellulose trinitrate	Acetone	25	6.93	0.91

Source: J. Brandrup and E. H. Immergut, Eds., *Polymer Handbook*, 2nd ed., Wiley, New York, 1975, Sec. IV.
*European units, concentrations in g/ml.

According to equation (3.80) the value of a is predicted to vary from 0.5 for a Flory θ-solvent to about 0.8 in a thermodynamically good solvent. This corresponds to α increasing from a zero dependence on the molecular weight to a 0.1 power dependence. More generally, it should be pointed out that a equals zero for hard spheres, about unity for semicoils, and 2 for rigid rods.

The quantity $\mathbf{K}$ is often given in terms of the universal constant Φ,

$$\mathbf{K} = \Phi \left(\frac{\overline{r_0^2}}{M} \right)^{3/2} \tag{3.82}$$

where $\overline{r_0^2}$ represents the mean square end-to-end distance of the unperturbed coil. If the number-average molecular weights are used, then Φ equals 2.5×10^{21} dl/mol $\cdot$ cm^3 (56). A theoretical value of 3.6×10^{21} dl/mol $\cdot$ cm^3 can be calculated from a study of the chain frictional coefficients (23).[†] For many theoretical purposes, it is convenient to express the Mark–Houwink equation in the form

$$[\eta] = \Phi \left(\frac{\overline{r_0^2}}{M} \right)^{3/2} M^{1/2}\alpha^3 = \mathbf{K} M^{1/2}\alpha^3 \tag{3.83}$$

If the intrinsic viscosity is determined in both a Flory θ-solvent and a "good" solvent, the expansion of the coil may be estimated. Noting equation (3.79),

$$[\eta]/[\eta]_\theta = \alpha^3 \tag{3.84}$$

Values of α vary from unity in Flory θ-solvents to about 2 or 3, increasing with molecular weight.

3.8.4 Intrinsic Viscosity Experiments

In most experiments, dilute solutions of about 1% polymer are made up. The quantity η_{rel} should be about 1.6 for the highest concentration used. The most frequent instrument used is the Ubbelhode viscometer, which equalizes the pressure above and below the capillary.

Several concentrations are run and plotted according to Figure 3.17. Two practical points must be noted:

1. Both lines must extrapolate to the same intercept at zero concentration.

2. The sum of the slopes of the two curves is related through the Huggins (63) equation,

$$\frac{\eta_{\text{sp}}}{c} = [\eta] + k'[\eta]^2 c \tag{3.85}$$

[†]A widely used older value of Φ is 2.1×10^{21} dl/mol $\cdot$ cm^3.

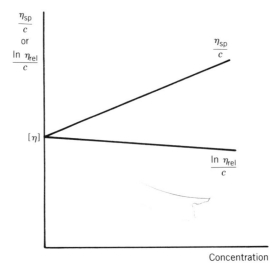

Figure 3.17 Schematic of a plot of η_{sp}/c and $\ln \eta_{rel}/c$ versus c, and extrapolation to zero concentration to determine $[\eta]$.

and the Kraemer (64) equation,

$$\frac{\ln \eta_{rel}}{c} = [\eta] - k''[\eta]^2 c \tag{3.86}$$

Algebraically,

$$k' + k'' = 0.5 \tag{3.87}$$

If either of these requirements is not met, molecular aggregation, ionic effects, or other problems may be indicated. For many polymer–solvent systems, k' is about 0.35, and k'' is about 0.15, although significant variation is possible.

Some intrinsic viscosities are shown in Table 3.11. Note that as A_2 goes to zero (the Flory θ-temperature is at 34.5°C), both R_g and $[\eta]$ decrease.

The molecular weight is usually determined through light-scattering, as indicated previously. In order to determine the constants **K** and a in the Mark–Houwink equation, a double logarithmic plot of molecular weight versus intrinsic viscosity is prepared (see Figure 3.18) (56). Results of this type of experiment were used in compiling Table 3.11.

3.8.5 Calculations Involving Intrinsic Viscosity

Suppose you are interested in a fast, approximate molecular weight of a polystyrene sample. You dissolve 0.10 g of the polymer in 100 ml of butanone

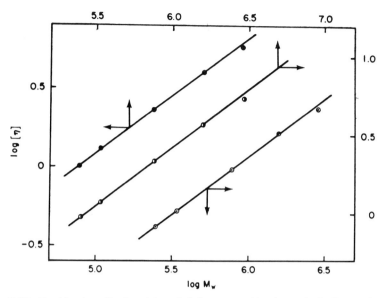

Figure 3.18 Double logarithmic plots of $[\eta]$ versus M_w for anionically synthesized polystyrenes, which were then fractionated leading to values of M_w / M_n of less than 1.06. Filled circles in benzene, half-filled circles in toluene, and open circles in dichloroethylene, all at 30°C (56). The arrows indicate the axes to be used.

and measure the flow times at 25°C in an Ubbelhode capillary viscometer. The results are

Pure butanone	110 sec
0.10% Polystyrene solution	140 sec

Starting with equation (3.33),

$$\eta_{rel} = \eta / \eta_0 = 140/110 = 1.273$$
$$\eta_{sp} = \eta_{rel} - 1 = 0.273$$
$$\eta_{sp}/c = 0.273/0.001 \text{ g/ml} = 2.73 \times 10^2 \text{ ml/g}$$

As an approximation, we assume the concentration is near zero, and $[\eta] = 2.73 \times 10^2$ ml/g, equation (3.71). This obviates the extrapolation in Figure 3.17, required for more accurate results. Using the Mark–Houwink relation, equation (3.81) and Table 3.11,

$$[\eta] = \mathbf{K} M_V^a$$
$$2.73 \times 10^2 = 39 \times 10^{-3} M_V^{0.58}$$
$$M_V = 4.10 \times 10^6 \text{ g/mol}$$

Note that the units of **K** are irregular, depending on the value of *a*. Usually, the units of **K** are omitted from tables.

3.9 GEL PERMEATION CHROMATOGRAPHY

Gel permeation chromatography (GPC) makes use of the size exclusion principle. The size of the molecule, defined by its hydrodynamic radius, can or cannot enter small pores in a bed of cross-linked polymer (65). The smaller molecules can diffuse into the pores (see Figure 3.19) and are delayed. The larger molecules pass by and continue in the solvent phase.

3.9.1 The Experimental Method

The instrumentation commonly used in GPC work is illustrated in Figure 3.20 (66). The flowing stream of solvent is called the mobile phase, which flows through the columns (67). The stationary phase consists of the small, porous particles. The sample is injected into the mobile phase and enters the columns, while the mobile phase flows at a specified rate controlled by the solvent delivery system. The length of time that a particular fraction remains

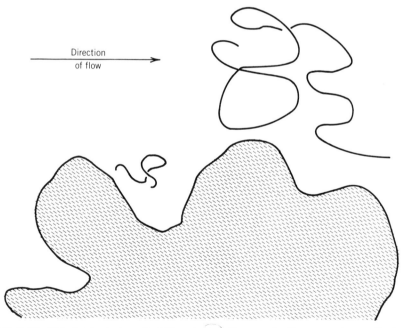

Figure 3.19 The size exclusion effect. The short chain can enter the pore, whereas the long chain passes by.

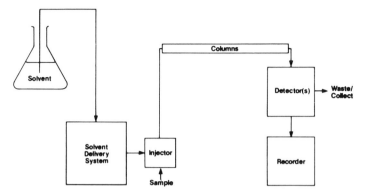

Figure 3.20   An illustration of the modules that make up GPC instrumentation (66).

in the columns is called the retention time. As the mobile phase passes the porous particles, the separation between the smaller and larger molecules becomes greater (see Figure 3.21) (65).

Gel permeation chromatography columns are some 30–60 cm in length. The main column is usually divided in tubes, each packed with a different pore size gel. Alternately, a mixed pore size column may be employed.

Gel Permeation Chromatography

Figure 3.21   Illustration of the GPC experiment (65). The sample is injected into a solvent, which flows into a porous packed bed. The larger molecules flow straight through, whereas the small ones are temporarily held up.

After passing through the columns, the solution goes to the detector. The larger molecules enter the detector first. The most common types of detectors include differential refractometers and ultraviolet or visible absorption methods. The detector monitors the concentration of the solute molecules in the solvent, with the objective being to follow the changes in concentration continuously. The molecular weight is then determined by the retention time of the particular fraction. Infrared absorption systems, particularly FT-IR detection systems, are also employed and are especially useful for determining the complete spectrum of the polymer on line for analytical purposes.

Data are collected as the height of the chromatogram versus retention counts (Figure 3.22) (66). The resultant data are then analyzed, usually by computer, to yield the various molecular weight averages (see Table 3.12). Thus the entire molecular weight distribution is determined in one experiment.

Most of the column materials in use today are cross-linked polystyrene (65). These polystyrene particles are heavily cross-linked, but they are made in such a way that small pores of controlled sizes exist within. These particles are fairly rigid and do not swell in the solvent.

PACKING:	STYRAGEL
SOLVENT:	THF
FLOW RATE:	2.0 ml/min
DETECTOR:	RI: 8X

Figure 3.22 A typical GPC chromatogram. The largest molecules appear at the lowest retention counts. Typical data for polystyrene (66).

w_i M_i N_i $w_i M_i$

Table 3.12   Calculation of molecular weight averages from GPC chromatograms (66)

1	2	3	4	5
Retention (Counts)	Height (mm)	Chain Length or Mol Wt	Col 2/Col 3	Col 2 × Col 3
30	1.0	340K	0.0000029	340,000
31	17	162K	0.000105	2,754,000
32	82	77K	0.001065	6,314,000
33	194	35K	0.005543	6,790,000
34	180	19K	0.009474	3,420,000
35	90	12K	0.007500	1,080,000
36	41	7.8K	0.005256	319,800
37	26	5.2K	0.00500	135,200
38	13.5	3.6K	0.003750	48,600
39	8.5	2.0K	0.004250	17,000
40	6	1.3K	0.004615	7,800
41	2.5	820	0.003049	2,050
42	0.5	510	0.000980	255
	662		0.050616	24,288,705

$\bar{A}_n = \Sigma\,\text{Col 2}/\Sigma\,\text{Col 4}$;
$\bar{M}_N = \Sigma\,\text{Col 2}/\Sigma\,\text{Col 4}$;
$\bar{M}_N = 662/0.05062 = 13,000$;
$\bar{A}_w = \Sigma\,\text{Col 5}/\Sigma\,\text{Col 2}$;
$\bar{M}_w = \Sigma\,\text{Col 5}/\Sigma\,\text{Col 2}$;
$\bar{M}_w = 24,288,705/662 = 36,600$.

$PDI = \dfrac{36600 \; M_w}{13000 \; M_n}$

It must be emphasized that the motion in and out of the gel particles by the polymer is strictly governed by the size of the chains and Brownian motion. The columns are calibrated with known molecular weight polymers, usually polystyrene. Today, substantially monodisperse polystyrenes synthesized via anionic polymerization methods, $M_w/M_n \cong 1.05$, are available as standards. These have been calibrated using light-scattering or osmometry. Other polymer standards available include poly(methyl methacrylate), poly(ethylene glycol), cis-polyisoprene, poly(ethylene oxide), poly(ethylene glycol), and polyethylene (68).

Important factors to consider for individual instruments are efficiency and selectivity; see Figure 3.23 (69). In general, the peaks will always be slightly broader than the actual molecular weight distribution, because of the statistical breadth of retention times for individual molecules.

Often, the molecular weights reported are actually those of the equivalent polystyrene radius of gyration. Since the relationship between the radius of gyration and the molecular weight varies from polymer to polymer and from solvent to solvent, the reported values may differ significantly from the absolute molecular weight, unless significant precautions are made. Nevertheless, the power of the method, yielding the complete molecular weight distribution in a matter of minutes, is great enough that GPC has become

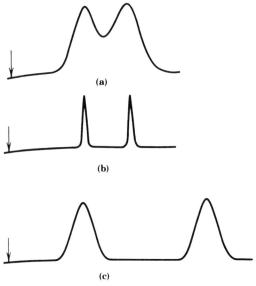

Figure 3.23 Resolution and efficiency versus selectivity due to column characteristics. (a) Poor resolution. (b) Good resolution (column efficiency). (c) Good resolution (column selectivity).

one of the most popular instruments in the polymer laboratories of the world. Besides simple molecular weights and distributions, GPC finds wide use as an analytical tool to determine the extent of block and graft copolymer formation, branching, degradation, and detection of impurities.

3.9.2 The Universal Calibration

Beginning with the Mark–Houwink relationship, equation (3.81), it is easy to show that the average molecular size is given by

$$[\eta]M = \Phi\left(\overline{r_0^2}\right)^{3/2}\alpha^3 \tag{3.88}$$

where the right-hand side is proportional to the polymer's hydrodynamic volume (70). A new aspect of GPC calibration arises from the recognition that a polymer's hydrodynamic volume might form the basis for molecular weight determination. Since GPC depends on the hydrodynamic volume per se rather than its molecular weight per se, a new calibration method is suggested. This is the "universal calibration," which calls for a plot of $[\eta]M$ versus elution volume.

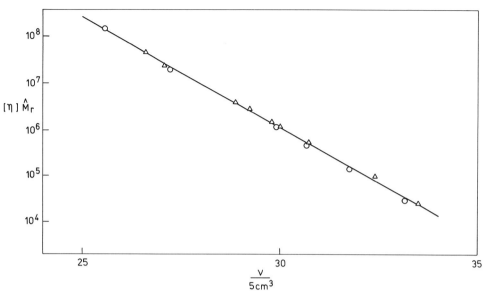

Figure 3.24 The universal calibration curves for polystyrene and poly(vinyl acetate) (71). The number 5 in the *x*-axis units means that the scale is in siphon "counts" of 5 cm^3, so that the *x*-ordinate 30 corresponds to an elution volume of 150 cm^3. (R. Dietz, private communication, November 1984.) $\hat{M}_r$ is the "peak" GPC molecular weight, usually the unknown. $\hat{M}_r$ values are close to the geometric mean of M_n and M_w.

Figure 3.24 (71) illustrates the universal calibration procedure for poly(vinyl acetate) and polystyrene. Note that the two sets of data lie on the same straight line. The universal calibration is valid for a range of topologies and chemical compositions. However, it cannot be used for highly branched materials or polyelectrolytes, which have different or varying hydrodynamic volume relationships. The universal calibration procedure is especially useful for estimating the molecular weight of new polymers, since the intrinsic viscosity is usually easy to obtain. The procedure also tends to correct for differences in the hydrodynamic relationships when several polymers are compared, and only one of them (such as polystyrene) is used as the calibration material.

3.10   SOLUTION THERMODYNAMICS AND MOLECULAR WEIGHTS

A knowledge of solution thermodynamics is critically important in determining the suitability of a solvent for a molecular weight determination. Once having decided on a suitable solvent, solution thermodynamics provides a

Table 3.13 Summary of molecular weight methods and results

	Absolute	Relative	M_n	M_w	Other
Gel Permeation Chromatography		X	X	X	Molecular Weight Distribution
Scattering—					
Light, Neutrons	X			X	R_g^z, A_2
Osmometry	X		X		A_2
Intrinsic Viscosity [η]		X			M_v

Source: D. A. Thomas, unpublished observations.

basis for determining how an extrapolation to zero concentration is to be carried out.

Below the Flory θ-temperature, polymer solutions may phase-separate. The higher the molecular weight is, the higher the upper critical solution temperature. At infinite molecular weight, the Flory θ-temperature is reached. Thus the Flory θ-temperature is defined by several different criteria:

1. It is the temperature where A_2 is zero for dilute solutions, and $\chi_1 = \frac{1}{2}$.
2. It is the temperature where the radius of gyration approximates that of the bulk polymer (see Chapter 5).
3. It is the temperature at which an infinite molecular weight fraction would just precipitate.

The molecular weight and polydispersity of polymers remain among the most important properties that are measured. The methods are divided into absolute methods, which determine the molecular weight from first principles, and relative methods, which depend on prior calibration. The latter are usually selected because they are fast and inexpensive. Values obtained from the several methods are summarized in Table 3.13.

While polymer molecular weights vary from about 20,000 to over 1,000,000 g/mol for linear polymers, many polymers used in commerce have molecular weights around 10^5 g/mol and polydispersity indices of about 2. This is governed by polymerization kinetics and by the balance between good physical properties and processibility.

It must be noted that sometimes the molecular weight distribution can be important in ways that are not obvious. For example, the low-molecular-weight component behaves substantially as a plasticizer, weakening the material rather than strengthening it. The high-molecular-weight tail adds much to the melt viscosity, since the melt viscosity of high-molecular-weight polymers depends on M_w to the 3.4 power (see Chapter 10). Thus most industries try to minimize the polydispersity index of their polymers.

REFERENCES

1. J. Hildebrand and R. Scott, *The Solubility of Nonelectrolytes*, 3rd ed., Reinhold, New York, 1949.

2. J. Brandrup and E. H. Immergut, Eds., *Polymer Handbook*, 2nd ed., Wiley-Interscience, New York, 1975, Sect. IV.

3. G. M. Bristow and W. F. Watson, *Trans. Faraday Soc.*, **54**, 1731, 1742(1958).

4. H. Mark and A. V. Tobolsky, *Physical Chemistry of High Polymer Systems*, 2nd ed., Interscience, New York, 1950.

5. S. K. Kim and S. C. Kim, *Polym. Bull.*, **23**, 141(1990).

6. D. Mangaraj, S. K. Bhatnagar, and S. B. Rath, *Makromol. Chem.* **67**, 75(1963).

7. P. A. Small, *J. Appl. Chem.*, **3**, 71(1953).

8. K. L. Hoy, *J. Paint Technol.*, **46**, 76(1970).

9. P. J. Flory, *J. Chem. Phys.*, **10**, 51(1942).

10. M. L. Huggins, *Ann. N.Y. Acad. Sci.*, **42**, 1(1942).

11. M. L. Huggins, *J. Phys. Chem.*, **46**, 151(1942).

12. M. L. Huggins, *J. Am. Chem. Soc.*, **64**, 1712(1942).

13. P. J. Flory, *Principles of Polymer Chemistry*, Cornell University Press, Ithaca, NY, 1953.

14. G. M. Bristow and W. F. Watson, *Trans. Faraday Soc.*, **54**, 1731(1958).

15. P. J. Flory and W. R. Krigbaum, *J. Chem. Phys.*, **18**, 1086(1950).

16. F. W. Billmeyer, Jr., *Textbook of Polymer Science*, 3rd ed., Wiley-Interscience, New York, 1984, Chap. 7.

17. F. Gundert and B. A. Wolf, in J. Brandrup and E. H. Immergut, Eds., *Polymer Handbook*, 3rd ed., Wiley, New York, 1989, Sect. VII, p. 173.

18. C. A. Glover, Chap. 4 in *Polymer Molecular Weights*, Part I, P. E. Slade, Jr., Ed., Dekker, New York, 1975.

19. F. W. Billmeyer, Jr., *Textbook of Polymer Sciences*, Interscience, New York, 1962.

20. R. D. Ulrich, Chap. 2 in *Polymer Molecular Weights*, Part I, P. E. Slade, Jr., Ed., Dekker, New York, 1975.

21. P. J. Flory and W. R. Krigbaum, *J. Chem. Phys.*, **18**, 1086(1950).

22. W. H. Stockmayer and E. F. Casassa, *J. Chem. Phys.*, **20**, 1560(1952).

23. P. J. Flory, *Principles of Polymer Chemistry*, Cornell University Press, Ithaca, NY, 1953, pp. 280, 606.

24. W. R. Krigbaum and L. H. Sperling, *J. Phys. Chem.*, **64**, 99(1960).

25. J. Brandrup and E. H. Immergut, Eds., *Polymer Handbook*, 2nd ed., Wiley-Interscience, New York, 1975.

26. G. Mie, *Ann. Phys.*, **25**, 377(1908).

27. M. Smoluchowski, *Ann. Phys.*, **25**, 205(1908).

28. M. Smoluchowski, *Philos. Mag.*, **23**, 165(1912).

29. A. Einstein, *Ann. Phys.*, **33**, 1275(1910).

30. P. Debye, *J. Phys. Coll. Chem.*, **51**, 18(1947).

31. B. H. Zimm, *J. Chem. Phys.*, **16**, 1093(1948).

32. B. H. Zimm, *J. Chem. Phys.*, **16**, 1099(1948).

33. A. Guinier and G. Fournet, *Small-Angle Scattering of X-Rays*, translation by C. B. Walker, Wiley, New York, 1955.

34. R. G. Kirste, W. A. Kruse, and K. Ibel, *Polymer*, **16**, 120(1975).

35. L. H. Sperling, *Polym. Eng. Sci.*, **24**, 1(1984).

36. K. A. Stacy, *Light-Scattering in Physical Chemistry*, Academic Press, Orlando, 1956.

37. G. Oster, *Chem. Rev.*, **43**, 319(1948).

38. M. Bender, *J. Chem. Ed.*, **29**, 15(1952).

39. P. Doty and J. T. Edsall, *Advances in Protein Chemistry VI*, Academic Press, Orlando, 1951, pp. 35–121.

40. J. J. Rush, *Current Status of Neutron-Scattering Research and Facilities in the United States*, National Academy Press, Washington, DC, 1984.

41. W. R. Krigbaum and L. H. Sperling, *J. Phys. Chem.*, **64**, 99(1960).

42. R. G. Kirste, W. A. Kruse, and J. Schelten, *Makromol. Chem.*, **162**, 299(1973).

43. J. Schelten, W. A. Kruse, and R. G. Kirste, *Kolloid Z. Z. Polym.*, **251**, 919(1973).

44. J. P. Cotton, B. Farnoux, G. Jannink, J. Mons, and C. Picot, *C. R. Acad. Sci. (Paris)*, **C275**, 175(1972).

45. H. Benoit, D. Decker, J. S. Higgins, C. Picot, J. P. Cotton, B. Farnoux, G. Jannink, and R. Ober, *Nature*, **245**, 13(1973).

46. D. G. H. Ballard, J. Schelten, and G. D. Wignall, *Eur. Polym. J.*, **9**, 965(1973).

47. J. P. Cotton, D. Decker, H. Benoit, B. Farnoux, J. S. Higgins, G. Jannink, R. Ober, C. Picot, and J. des Cloiseaux, *Macromolecules*, **7**, 863(1974).

48. G. D. Wignall, D. G. Ballard, and J. Schelten, *Eur. Polym. J.*, **10**, 861(1974).

49. J. Schelten, D. G. H. Ballard, G. Wignall, G. Longman, and W. Schmatz, *Polymer*, **17**, 751(1976).

50. G. Lieser, E. W. Fischer, and K. Ibel, *J. Polym. Sci. Polym. Lett. Ed.*, **13**, 39(1975).

51. R. G. Kirste and B. R. Lehnen, *Makromol. Chem.*, **177**, 1137(1976).

52. G. Allen, *Proc. R. Soc. London Ser. A*, **351**, 381(1976).

53. P. Herchenroeder, M. Dettenmaier, E. W. Fischer, M. Stamm, J. Hass, H. Reimann, B. Tieke, G. Wegner, and E. L. Zichny, *Europhys. Conf. Abstr.*, C.A. **89**, 198133Z(1978).

54. R. G. Kirste, W. A. Kruse, and K. Ibel, *Polymer*, **16**, 120(1975).

55. J. Schelten, G. D. Wignall, and D. G. H. Ballard, *Polymer*, **15**, 682(1974).

56. A. Yamamoto, M. Fujii, G. Tanaka, and H. Yamakawa, *Polym. J.*, **2**, 799(1971).

57. G. V. Schultz, *Z. Phys. Chem.*, **B43**, 25(1939).

58. See M. Swarc, *Polymerization and Polycondensation Processes*, p. 96, *Adv. Chem. Ser.* 34, American Chemical Society, Washington, DC, 1962.

59. G. C. Odian, *Principles of Polymerization*, 2nd ed., Wiley, New York, 1981.

60. W. H. Stockmayer, *J. Chem. Phys.*, **11**, 45(1943).

61. H. Mark, *Der feste Korper*, Hirzel, Leipzig, 1938, p. 103.

62. J. Brandrup and E. H. Immergut, Eds., *Polymer Handbook*, 2nd ed., Wiley, New York, 1975, Sec. IV.

63. M. L. Huggins, *J. Am. Chem. Soc.*, **64**, 2716(1942).

64. E. O. Kraemer, *Ind. Eng. Chem.*, **30**, 1200(1938).

65. W. W. Yau, J. J. Kirkland, and D. D. Bly, *Modern Size-Exclusion Liquid Chromatography*, Wiley-Interscience, New York, 1979.

66. Waters Associates Liquid Chromatography School, Manual, *LC Short Course*, Waters Associates, Morristown, NJ, 1983.

67. F. M. Rabel, *J. Chromatogr. Sci.*, **18**, 394(1980).

68. D. Campbell and J. R. White, *Polymer Characterization: Physical Techniques*, Chapman and Hall, London, 1989.

69. J. F. Johnson, in *Encyclopedia of Polymer Science and Engineering*, 2nd ed., Vol. 3, J. I. Kroschwitz, Ed., Wiley, New York, 1985.

70. T. C. Ward, Jr., *Chem. Ed.*, **58**, 867(1981).

71. C. M. L. Atkinson and R. Dietz, *Eur. Polym. J.*, **15**, 21(1979).

GENERAL READING

S. T. Balke, *Quantitative Column Liquid Chromatography*, Elsevier, New York, 1984.

F. W. Billmeyer, Jr., *Textbook of Polymer Science*, 3rd ed., Interscience, New York, 1984.

R. U. Bonnar, M. Dimbat, and F. H. Stross, *Number-Average Molecular Weights*, Interscience, New York, 1958.

E. A. Collins, J. Bares, and F. W. Billmeyer, Jr., *Experiments in Polymer Science*, Wiley, New York, 1973.

P. J. Flory, *Principles of Polymer Chemistry*, Cornell University Press, Ithaca, NY, 1953.

M. Kerker, *The Scattering of Light and Other Electromagnetic Radiation*, Academic Press, Orlando, 1969.

L. Mandelkern, *An Introduction to Macromolecules*, 2nd ed., Springer-Verlag, New York, 1983.

P. E. Slade, Jr., Ed., *Polymer Molecular Weights*, Parts I and II, Dekker, New York, 1975.

K. A. Stacy, *Light-Scattering in Physical Chemistry*, Academic Press, Orlando, 1956.

H. Tompa, *Polymer Solutions*, Academic Press, Orlando, 1956.

W. W. Yau, J. J. Kirkland, and D. D. Bly, *Modern Size-Exclusion Chromatography*, Wiley-Interscience, New York, 1979.

STUDY PROBLEMS

1. Calculate the solubility parameter of polyisobutene from Table 3.3. How does this value compare with that shown in Figure 3.1?

2. What is the free energy of mixing 1 mol of polystyrene, $M = 2 \times 10^5$ g/mol, with 1×10^4 liters of toluene, at 298 K?

3. What is the calculated heat of mixing of 1000 g of polystyrene with 1000 g of polybutadiene? If the molecular weight of both polymers is on the order of 1×10^5 g/mol, would you anticipate that two polymers are mutually miscible at 150°C? What are the experimental results in this case?

4. Calculate the entropy of mixing of the red and black checkers on an ordinary checkerboard. Assuming an ideal solution, what is the free energy of mixing? After "polymerizing" the checkers, what is the new entropy and free energy of mixing of the blend?

5. Why does a polymer always dissolve in an ideal solvent, but may or may not dissolve in a regular solution?

6. A 5-g sample of a polyester having one carboxyl group per molecule is to be titrated by sodium hydroxide solutions to determine its number-average molecular weight. How much 0.01 molar solution is required if the polymer has a molecular weight of approximately 1000 g/mol? 10,000 g/mol? 100,000 g/mol? Discuss the practicality of the experiment.

7. Calculate ΔH_v and ΔH_f from Table 3.5, assuming $T_b = 150°C$, $T_f = 10°C$, and $\rho = 1.0$ g/cm³.

8. What is the molecular weight of the cellulose tricaproate sample in Figure 3.7? Note that $R = 0.08205$ atm · liter/mol · K for this problem.

9. Derive the units for A_2 in both the cgs and SI unit systems.

10. In Figure 3.14, a Zimm plot of cellulose tricaproate in dimethylformamide is given. What is the molecular weight and radius of gyration of this material? $\lambda = 5461$ Å, $n(DMF) = 1.429$.

11. A sample of polystyrene has a radius of gyration of approximately 150 Å. Calculate the maximum usable angle to determine R_g for (a) light-scattering, (b) small-angle neutron scattering, and (c) small-angle x-ray scattering.

12. Values of **K** and a are listed in Table 3.11 for polystyrene in cyclohexane, a θ-solvent at 34.5°C. How does the calculated value of r_0^2/M [see equation (3.82)] compare with the corresponding values from Table 3.9? (Don't forget to convert r_0 into R_g!)

13. The intrinsic viscosity of a sample of poly(methyl methacrylate) in acetone at 20°C was found to be 6.7 ml/g. What is its viscosity-average molecular weight?

14. Prove the relation $k' + k'' = 0.5$, equation (3.87).

15. Estimate the Mark–Houwink constants from the data in Figure 3.18.

16. The chain expansion quantity α varies with both R_g (by definition) and $[\eta]$. Using Table 3.9, show the relationships between α, R_g, and $[\eta]$. Can α and $[\eta]$ be related theoretically?

17. What is the z-average molecular weight of the sample data shown in Table 3.12?

18. Given the following data, what is the number-average molecular weight and the second virial coefficient? ($T = 25°C$; density $= 1.0$ g/cm^3.)

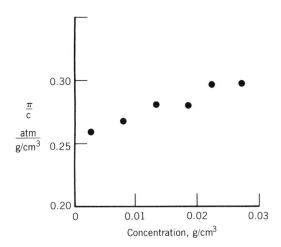

19. "But I know its a block copolymer!" your technician exclaimed. "Poly$A$ is a living polymer, which adds poly$B$ at its end!" You are not so sure. The synthesis might have taken another route, ending in a blend of the two polymers. The starting molecular weight of poly$A$ is known to be 100,000 g/mol, and poly$B$ is estimated to be around 15,000 g/mol. Devise a GPC experiment for your technician, which will distinguish (a) the 100% block copolymer from (b) a 50% block, from (c) a simple blend. What would the resulting GPC chromatograms look like?

20. A new polymer has $[\eta] = 5.5$ cm^3/g, and an elution volume of 160 cm^3. Based on the method of Figure 3.24, what is its molecular weight?

21. What are the solubility parameters of the polyurethane and polystyrene in Figure 3.2?

22. We tend to think of molecules as being of finite size. The polymer networks used in Figure 3.2 are clearly the size of the sample, while the molecules used in Figure 3.1 are clearly finite and soluble. In what philosophical sense are (or are not) the materials making up Figure 3.2 molecules?

23. Two syntheses of the same polymer are made, but with different molecular weights, M_A and M_B, with intrinsic viscosities $[\eta]_A$ and $[\eta]_B$. Derive a general equation to express the intrinsic viscosity of various blends of the two syntheses as a function of composition.

APPENDIX 3.1 CALIBRATION AND APPLICATION OF LIGHT-SCATTERING INSTRUMENTATION FOR THE CASE WHERE $P(\theta) = 1$

A cylindrical cell with flat portions where the beam enters and leaves is presumed (Figure A3.1.1). A turbid calibrating liquid such as a soap solution or a Ludox® dispersion is made up. Ideally, the light intensity after passing through the solution, I, should be about $0.7I_0$. Beers law gives the turbidity, τ,

$$I = I_0 e^{-\tau x} \tag{A3.1.1}$$

where x is the path length of the light. The receiving photomultiplier tube is then turned to 90°, and the scattered light intensity i recorded. The quantity i is related to the turbidity through the relation

$$\tau = ki \tag{A3.1.2}$$

where k is a proportionality constant.

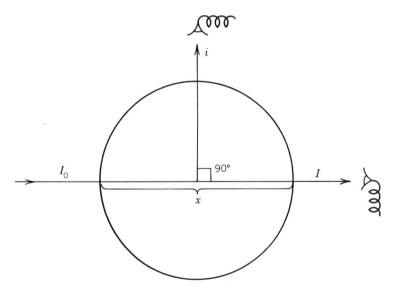

Figure A3.1.1 Light-scattering from a cylindrical cell.

The polymer solution of interest is then placed in the cell, and the scattering intensity i determined at 90°. The turbidity of the solution is determined through equation (A3.1.2). The turbidity is determined for several concentrations and extrapolated to zero concentration. For this simple experiment,

$$H\frac{c}{\tau} = \frac{1}{M_w} + 2A_2c \qquad\qquad (A3.1.3)$$

In reality, a secondary standard is also calibrated, so that the electronic sensitivity of the instrument may be varied widely and τ still calculated easily.

4

CONCENTRATED SOLUTIONS AND PHASE SEPARATION BEHAVIOR

Polymer solutions are important for a variety of purposes. In Chapter 3, dilute polymer solutions were required for the determination of molecular weights and sizes.

It must be emphasized that a wide variety of engineering applications exist for polymer solutions (see Table 4.1). These range from maintaining constant viscosity in motor oil to the production of paints, varnishes, and glues.

Not every polymer will dissolve in every solvent, however. When one attempts to dissolve a polymer in solvents selected at random, many, perhaps most, will not work. The experimenter rapidly discovers that the higher the molecular weight of the polymer, the more difficult it is to select a good solvent. Polymer–polymer mutual solutions are even more difficult to attain.

Attempts to understand polymer–solvent and polymer–polymer mutual solution behavior led to many new theoretical developments, particularly in thermodynamics. To an increasing extent, the polymer scientist is now able to predict phase relationships involving polymers. The development of polymer solution and phase separation behavior is the subject of this chapter.

4.1 PHASE SEPARATION AND FRACTIONATION

According to thermodynamic principles, the condition for equilibrium between two phases requires that the partial molar free energy of each component be equal in each phase. This condition requires that the first and second derivatives of $\overline{\Delta G_1}$ in equation (3.23) with respect to v_2 be zero. The critical concentration at which phase separation occurs may be written

$$v_{2c} = \frac{1}{1 + x^{0.5}} \tag{4.1}$$

Table 4.1 Selected industrial applications of polymer solutions and precipitates

Polymer	Solvent	Effect	Application
Sodium carboxymethyl cellulose	Soapy water	Selective precipitation onto clothing fibers	Prevents oils from redepositing on clothing during detergent washing: an antiredeposition agent
Diblock copolymers	Motor oil	Colloidal suspensions dissolve at high temperatures, raising viscosity	Multiviscosity (constant viscosity) motor oil, Example: 10W-40
Poly(ethylene oxide) $M = 10^6$ g/mol	Water	Reduces turbulent flow	Heat exchange systems, reduces pumping costs
Proteins	Wine	Gels on reacting with tannin	Clarification of wines, removes colloidal matter
Polystyene, various	Triglyceride oils	Viscosity control, phase-separates during oil polymerization	Oil-based house paints, makes coatings harder, tougher
Polyurethanes, various	Esters, alcohols, various	Solvent vehicle evaporates, leaving polymer film for glues, solvent enters mating surfaces	Varnishes, shellac, and glues (adhesives)
Poly(vinyl chloride)	Dibutyl phthalate	Plasticizes polymer	Lower polymer T_g, soften polymer, makes "vinyl"
Polystyrene	Poly(2,6-dimethyl-1,4-phenyleneoxide)	Mutual solution; toughens polystyrene	Impact-resistant objects, such as appliances
Poly(methyl methacrylate)	Poly(vinylidine fluoride)	Increases PMMA oil and solvent resistance	Automotive applications, parts that might contact gasoline

Figure 4.1 Phase diagrams for polystyrene fractions in cyclohexane. Circles and solid lines, experimental. Theoretical curves are shown for two of the fractions. The viscosity-average molecular weights are PSA, 43,600; PSB, 89,000; PSC, 250,000; PSD, 1,270,000 g / mol (1).

For large x, the right-hand side of equation (4.1) reduces to $1/x^{0.5}$. For $x = 10^4$, v_{2c} equals about 0.01, a very dilute solution. The critical value of the Flory-Huggins polymer-solvent interaction parameter, χ_1, is given by

$$\chi_{1c} = \frac{\left(1 + x^{1/2}\right)^2}{2x} \cong \frac{1}{2} + \frac{1}{x^{1/2}} \tag{4.2}$$

which suggests further that as x approaches infinity, χ_{1c} approaches $\frac{1}{2}$.

Experimental data are given in Figure 4.1 (1) for the system polystyrene in cyclohexane. The Flory θ-temperature for this system was already given as 34.5°C.

The critical temperature is the highest temperature of phase separation. The equation for the critical temperature is given by

$$\frac{1}{T_c} = \frac{1}{\theta}\left[1 + \frac{1}{\psi_1}\left(\frac{1}{x^{1/2}} + \frac{1}{2x}\right)\right] \tag{4.3}$$

Thus a plot of $1/T_c$ versus $1/x^{0.5} + 1/2x$ should yield the Flory θ-temperature at $x =$ infinity (see Figure 4.2) (1).

If a range of molecular species exists, which is true for all synthetic polymers, then the lower molecular weight species will tend to remain in

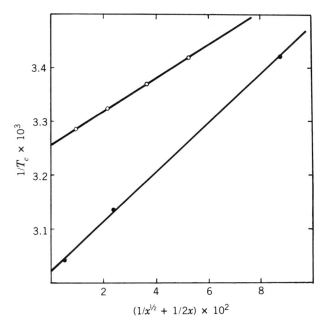

Figure 4.2 Dependence of the critical temperature on the number of segments per polymer chain. Polystyrene in cylcohexane, open circles; polyisobutene in diisobutyl ketone, solid circles (1).

solution at a temperature where the higher molecular weights phase-separate. Actually, there is always a partition of molecular weights between the more concentrated and more dilute phases. The fractionation becomes more efficient at very low concentrations.

While the above describes the phase behavior of a polymer in a poor solvent as the temperature is changed, quite often a polymer is placed in a solvent–nonsolvent mixture. As nonsolvent is added to the solution, the polymer may phase-separate. The phase diagram at constant temperature is shown in Figure 4.3. With maturation time, both mechanisms produce a continuous concentrated polymer phase and a dilute polymer phase, the concentrations being connected by tie lines.

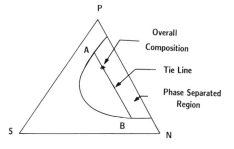

Figure 4.3 Schematic polymer – solvent – nonsolvent phase diagram, illustrating concentrated polymer phase A and dilute polymer phase B.

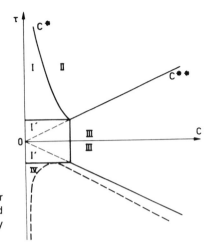

Figure 4.4 Phase diagram of a typical polymer solution. The quantity τ represents the reduced temperature, $(T - \theta) / \theta$, where θ is the Flory θ-temperature (2).

4.2 REGIONS OF THE POLYMER–SOLVENT PHASE DIAGRAM

A polymer dissolves in two stages. First, solvent molecules diffuse into the polymer, swelling it to a gel state. Then the gel gradually disintegrates, and the molecules diffuse into the solvent-rich regions. In this discussion, linear amorphous polymers are assumed. Cross-linked polymers may reach the gel state, but they do not dissolve.

The concentration of the final solution, of course, depends on the relative proportions of polymer and solvent. In Chapter 3, the solutions were assumed to be dilute, generally below 1% concentration, because this is required to obtain molecular weights. However, many solutions are used in the 10–50% concentration range. More concentrated systems are better described as plasticized polymers.

Daoud and Jannink (2) divided polymer–solvent space into four regions (see Figure 4.4). Each of these regions exhibits distinct molecular characteristics, as derived by scaling concepts.

Scaling concepts in polymer physics were brought to their modern state of importance by de Gennes (3). They consist of a series of proportionalities, showing the algebraic relationships between many quantities, particularly between macroscopic and microscopic variables, molecular dimensions, and thermodynamics.

Region I in Figure 4.4 is the dilute region. According to the scaling laws, region I is limited by the lines where τ is proportional to R_g^{-1} and the concentration line c^*. The quantity c^* is also proportional to R_g^{-1}.[†]

[†]Some authors prefer to use the more formal language here of $(\overline{R_g^2})^{-1/2}$, which reads the mean square radius of gyration to the inverse one-half power. Where any confusion might occur, this text has this meaning in the statistical terminology.

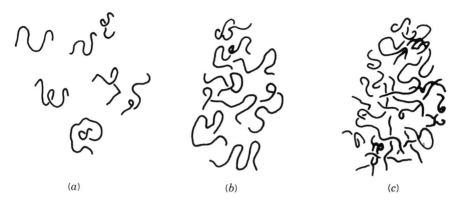

(a)   *(b)*   *(c)*

Figure 4.5 Relationships of polymer chains in solution at different concentration regions. (a) Region I, where $c < c^*$. (b) The transition region, where $c = c^*$. (c) Region II, where $c > c^*$. Note overlap of chain portion in space in region II.

Region II is the semidilute region. It is bound by the lines c^* and c^{**}. The line c^* is now seen as the critical concentration at which the chains begin to overlap (see Figure 4.5). The concentration c^{**} is proportional to τ.

Region III consists of the semidilute and concentrated θ-region, which has not yet been studied theoretically.

Region IV occurs in the negative τ range. It is bound by the coexistence phase diagram curve (dashed line). This dashed line is the equivalent of the phase separation lines shown in Figure 4.1.

Region I′ represents the Flory θ-region of dilute solutions, where the concentration is low, and $T = \theta$.

An interesting quantity useful in region II is the screening length, ξ, first introduced by Edwards (4). Roughly speaking, the screening length measures the average distance between nearest chain contacts (5) in semidilute solutions; the scaling law

$$\xi \propto c^{-3/4} \tag{4.4}$$

applies, where ξ is independent of the molecular weight. At c^*, ξ can be identified with R_g. If the polymer is cross-linked, ξ is a measure of the net size, which rapidly decreases as the cross-link concentration increases. Thus the quantity ξ measures the distance between chain portions, either intramolecularly, when the chains are far separated, or both inter- and intramolecularly in more concentrated systems.

4.3   POLYMER–POLYMER PHASE SEPARATION

When two polymers are mixed, the most frequent result is a system that exhibits almost total phase separation. Qualitatively, this can be explained in terms of the reduced combinatorial entropy of mixing of two types of polymer

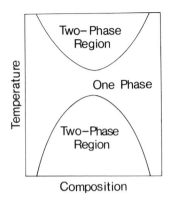

Figure 4.6 Phase diagram for a polymer blend illustrating an upper critical solution temperature, UCST (apex of lower curve), and a lower critical solution temperature, LCST (apex of upper curve). Because of the low entropy of mixing, high-molecular-weight polymer blends exhibit the LCST phenomenon. If the chains are short, an UCST may be seen.

chains, as illustrated in Figure 3.3*b*. In this case, neither type of chain can interchange its segments, because of covalent bonding.

Prior to the 1970s, the scientific literature on polymer blends was dominated by the idea that polymer–polymer miscibility would always be the rare exception (6–9). This was based on numerous experiments, and the theoretical work of Scott (9). Thus the usual endothermic heat of mixing and the very small combinatorial entropy of mixing made it seem unlikely to realize the necessary negative free energy of mixing.

When two polymers do mutually dissolve, they are generally found to phase-separate at some higher temperature rather than at some lower temperature. This is called a lower critical solution temperature (LCST) (see Figure 4.6, upper portion). This unusual result can be interpreted by considering the unusual features of the mixing process. At the critical point, the heat of mixing must balance the entropy of mixing times the absolute temperature. The latter is known to be unusually small (see above), so that the free energy of mixing, according to the Flory–Huggins theory, is the difference between two small quantities.

The Flory–Huggins theory does not permit volume change on mixing, and it ignores the equation of state properties of the pure components. It also does not properly consider the enormous size differences between polymer and solvent.

In response to the developing field of polymer blends, two new theories of polymer mixing were developed. The first was the Flory equation of state theory (10, 11), and the second was Sanchez's lattice fluid theory (12–15). These theories were expressed in terms of the reduced temperature, $\tilde{T} = T/T^*$, reduced pressure, $\tilde{P} = P/P^*$, reduced volume, $\tilde{V} = V/V^*$, and reduced density, $\tilde{\rho} = 1/\tilde{V}$. The starred quantities represent characteristic values for particular polymers, often referred to as the "hard core" or "close-packed" values, that is, with no free volume.

The new theories pointed out that LCST behavior is characteristic of exothermic mixing and negative excess entropy (16). This last is caused by densification of the polymers on mixing. The entropy of volume change, which ordinarily is relatively small compared to other quantities, comes to the fore in polymer–polymer blends (15).

4.3.1 Phase Diagrams

Phase separation and dissolution are controlled by three variables: temperature, pressure, and concentration. While simultaneous variation of all three yields a three-dimensional figure of significant complexity, the effect of systematic variation of two of these quantities, holding the third constant, leads to the main stream of polymer research. The schematic phase diagrams in Figure 4.7, while drawn specifically for polymer–polymer miscibility analysis, are quite general. The solid line is called the binodal and the dashed line the spinodal (17). As defined below, these two lines demarcate regions of different kinetics of phase separation. All the phase diagrams in Figure 4.7 are for liquid–liquid phase separation. If one of the components crystallizes, quite different phase diagrams emerge (18). While the figures are drawn symmetrically, the shapes and positions of the various curves depend on the ratio of the two polymer molecular weights.

Figure 4.7a shows the LCST behavior of a typical polymer blend. Again, as in Figure 4.6, as the temperature is raised through the binodal line, phase

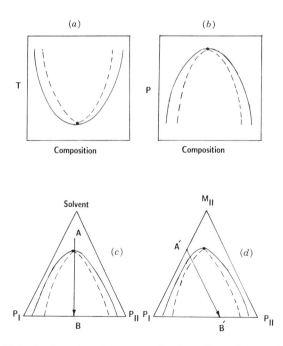

Figure 4.7 Methods of causing phase separation in multiconstituent polymer materials.

separation will occur. Figure 4.7b shows the corresponding behavior as a function of pressure. While not widely investigated, early results show that increasing pressure increases miscibility (19). Thus the pressure coefficient of an LCST is positive. This result agrees with the finding that miscible polymer pairs undergo densification on mixing.

Figure 4.7c illustrates the phase diagram of a solvent and two polymers, P_I and P_{II}. As the solvent is removed, the mixture becomes less miscible and undergoes phase separation on going from point A to point B.

The case of a mutual solution of a polymer, P_I, and a monomer M_{II} is illustrated in Figure 4.7d. As monomer II polymerizes to form polymer II, the composition follows the trajectory A' to B'. Here, phase separation occurs during polymerization. While this case is the most important industrially, encompassing such materials as high-impact polystyrene and ABS resins, it is the least understood. The reasons lie in the complexity of the thermodynamics of a system undergoing two simultaneous changes: (a) removal of the monomer solvent (as in Figure 4.7c) and (b) replacement of the solvent by polymer II. Grafting during the polymerization, if present, constitutes another complication. Generally, grafted polymers are more miscible than the corresponding blends.

4.3.2   Thermodynamics of Phase Separation

The basic thermodynamic relationships developed in Section 3.2 hold for polymer blends. The basic equation for mixing of blends reads

$$\frac{\Delta G_M}{kT} = \frac{V}{V_r} v_1 v_2 \chi_{12} \left(1 - \frac{2}{z}\right) + N_c [v_1 \ln v_1 + v_2 \ln v_2] \qquad (4.5)$$

where V is the volume of the sample (usually taken as 1 cm^3), V_r is the volume of one cell (see Figure 3.3), z is the lattice coordination number (z is usually between 6 and 12), and N_c is the number of molecules in 1 cm^3. Then, the term V/V_r is a count of the number of cells in 1 cm^3, the first term on the right being the heat of mixing term, ΔH_M. This may be positive or negative, depending on the sign of χ_{12}. The second term on the right is the statistical entropy of mixing term, ΔS_M. Since v_1 and v_2 are always fractions, this second term is always negative. Of course, the classical representation of equation (4.5) is

$$\Delta G_M = \Delta H_M - T \Delta S_M \qquad (4.6)$$

The statistical thermodynamic approach provides a molecular basis for calculating the properties of interest.

4.3.3   An Example Calculation

What is the molecular weight limit for mixing polystyrene and polybutadiene at 25°C and obtaining one phase? For simplicity, we assume $v_1 = v_2 = 0.5$, unit density, and a mer molecular weight of 100 g/mol for both polymers. This last relieves the requirement somewhat of having exactly one mer in each cell; see Figure 3.3. The quantity $\Delta G_M/kT$ equals zero at the critical point. The basic unknown is reached through N_c.

The quantity χ_{12} is found through Table 3.2 and equation (3.28):

$$\chi_{12} = \frac{V_1}{RT}(\delta_1 - \delta_2)^2 = \frac{(100 \text{ cm}^3/\text{mol})(9.1 - 8.4)^2 \text{ cal}/\text{cm}^3}{(1.987 \text{ cal}/\text{mol. K}) \times 298 \text{ K}}$$

$$\chi_{12} = +0.083$$

Assuming $z = 6$, the first term in equation (4.5) may be evaluated:

$$\frac{V}{V_r} = \frac{1 \text{ cm}^3 \times 6.023 \times 10^{23} \text{ molecules}/\text{mol}}{(100 \text{ g}/\text{mol}) \times 1 \text{ cm}^3/\text{g}}$$

$$V/V_r = 6.023 \times 10^{21} \text{ mers (or cells)}/\text{cm}^3$$

leading to

$$\frac{V}{V_r} v_1 v_2 \chi_{12}\left(1 - \frac{2}{z}\right) = 8.3 \times 10^{19}$$

The second term on the right of equation (4.5) is

$$N_c[v_1 \ln v_1 + v_2 \ln v_2] = -0.69 N_c$$

Then, from equation (4.5),

$$0 = 8.3 \times 10^{19} - 0.69 N_c$$

$N_c = 1.2 \times 10^{20}$ chains/cm³, which occupy 6.023×10^{21} cells:

$$\frac{6.023 \times 10^{21} \text{ cells(mers)}/\text{cm}^3}{1.2 \times 10^{20} \text{ chains}/\text{cm}^3} = 50 \text{ mers}/\text{chain}$$

Since each "mer" weighs 100 g/mol, each chain is 5000 g/mol.

4.3.4 Equation of State Theories

However, specific new theories, called the equation of state theories, have been worked out in an effort to understand the special conditions of mixing long chains (20–25). At equilibrium, an equation of state is a constitutive equation that relates the thermodynamic variables of pressure, volume, and temperature. A simple example is the ideal gas equation, $PV = nRT$. The van der Waals equation provides a fundamental correction for molecular volume and attractive forces. Equations of state may also include mechanical terms. Thus rubber elasticity phenomena are also described by equations of state; see Chapter 9.

A serious deficiency in the classical theory, as exemplified by equation (4.5), is the assumption of incompressibility. This deficiency can easily be remedied by the addition of free volume in the form of "holes" to the system. These holes will be about the size of a mer and occupy one lattice site. Imagine that a multicomponent mixture is mixed with N_0 holes of volume fraction v_0. Then the entropy of mixing is

$$\frac{\Delta S_M}{k} = -N_0 \ln v_0 - \sum_i N_i \ln v_i \qquad (4.7)$$

which follows from equation (3.15). This allows for compressibility, since the number of holes may be varied. Noting that the fractional free volume is given by $1 - \tilde{\rho}$, the entropy of mixing vacant sites with the molecules in equation of state terminology is given by

$$\Delta S_M = -\frac{k}{\rho} \left[(1 - \tilde{\rho}) \ln(1 - \tilde{\rho}) + \tilde{\rho} \ln \tilde{\rho} \right] \qquad (4.8)$$

where $\tilde{\rho}$ is less than unity. When all the sites are occupied, $\tilde{\rho} = 1$, and the RHS is zero. In reality, the cell size also increases with temperature, to accommodate increased vibrational amplitudes.

The Gibbs free energy of mixing is given by

$$\Delta G_M = \tilde{\rho}\epsilon^* - T\Delta S_M \qquad (4.9)$$

where the quantity ϵ^* is a van der Waals type of energy of interaction.

Note that $\tilde{\rho} = \tilde{\rho}(P, T)$; $\tilde{\rho} \to 1$ as $T \to 0$; $\tilde{\rho} \to 1$ as $P \to \infty$.

By taking $\partial \Delta G_M / \partial \tilde{\rho} = 0$, the equation of state via the lattice fluid theory (22) is obtained

$$\tilde{\rho}^2 + \tilde{P} + \tilde{T}\left[\ln(1 - \tilde{\rho}) + (1 - 1/r)\tilde{\rho}\right] = 0$$

where r is the number of sites (mers, if each occupies a site) in the chains, and

$$\frac{1}{r} = \frac{v_1}{r_1} + \frac{v_2}{r_2}$$ (4.11)

For high polymers, r goes substantially to infinity, yielding a general equation of state for both homopolymers and miscible polymer blends,

$$\tilde{\rho}^2 + \tilde{P} + \tilde{T}\left[\ln(1 - \tilde{\rho}) + \tilde{\rho}\right] = 0$$ (4.12)

The corresponding equation of state derived by Flory (20) is

$$\tilde{\rho}^2 + \tilde{P} - \tilde{T}\tilde{\rho}\left(1 - \tilde{\rho}^{1/3}\right)^{-1} = 0$$ (4.13)

Workers in the field prefer to state the equations in terms of density relations, because for condensed systems, density is easier to measure than volume. Again, $\tilde{\rho} = \rho/\rho^* = V^*/V$, where V^* represents the close-packed volume per unit mass.

Data for several polymers of interest are shown in Table 4.2 (22). This information may be used to determine miscibility criteria. Basically, T^* values must be similar for miscibility. If $T_1^* > T_2^*$, then it is desirable to have $P_1^* > P_2^*$. Polymers must have similar coefficients of expansion for miscibility; that is, the ratio of the densities must remain similar as the temperature is changed.

From such information, it may be found that the system poly(2,6-dimethyl phenylene oxide)–*blend*–polystyrene is miscible, but that poly(dimethyl siloxane), which is so different from the others, should be immiscible with all of them.

The basis for phase separation in polymer blends must be considered further here. For the Flory–Huggins theory of incompressible polymer mix-

Table 4.2 Equation of state parameters for some common polymers (22)

Polymer	T^* (K)	P^* (bars)	ρ^* (g/cm^3)
Poly(dimethylsiloxane), PDMS	476	3020	1.104
Poly(vinyl acetate), PVAc	590	5090	1.283
Poly(n-butyl methacrylate), PnBMA	627	4310	1.125
Polyisobutylene, PIB	643	3540	0.974
Polyethylene (low density), LDPE	673	4250	0.887
Poly(methyl methacrylate), PMMA	696	5030	1.269
Poly(cyclohexyl methacrylate), PcHMA	697	4260	1.178
Polystyrene (atactic), PS	735	3570	1.105
Poly(2,6-dimethylphenylene oxide), PPO	739	5170	1.161
Poly(o-methylstyrene), PoMS	768	3780	1.079

tures (see Section 3.3), the small entropy of mixing is dominated by the heat of mixing, leading to the conclusion that the heat of mixing must be zero or negative to induce miscibility. When the system is compressible, the reverse may be true. "Compressible" in the current sense means that the theory includes terms for volume changes.

When two compressible polymers are mixed together at sufficiently high temperature, negative heats of mixing cause a negative volume change. Since reducing the volume of the system reduces the number of holes, the entropy of mixing is negative. At high enough temperatures, the unfavorable entropy change associated with the densification of the mixture becomes prohibitive, that is, $T \Delta S > \Delta H$, and the mixture phase-separates. The two-phase system has a larger volume as the holes are reintroduced, and hence a larger positive contribution to the entropy. This scenario leads to lower critical solution temperatures. Significantly differing coefficients of expansion contribute to phase separation. In virtually all polymer–polymer systems exhibiting critical phenomena, both ΔH and $T \Delta S$ are relatively small quantities. Thus relatively modest changes in either the enthalpy or the entropy alter the phase diagram significantly.

4.3.5 Kinetics of Phase Separation

There are two major mechanisms by which two components of a mutual solution can phase-separate: nucleation and growth, and spinodal decomposition; see Figure 4.8 (25). Nucleation and growth are associated with metastability, implying the existence of an energy barrier and the occurrence of large composition fluctuations. Domains of a minimum size, the so-called critical nuclei, are a necessary condition. Nucleation and growth (NG) are the usual mechanisms of phase separation of salts from supersaturated aqueous solutions, for example. Spinodal decomposition (SD), on the other hand, refers to phase separation under conditions in which the energy barrier is negligible, so even small fluctuations in composition grow.

The two kinetic modes are compared in Table 4.3. The most interesting differences are the growth mechanisms. As illustrated in Figure 4.8, the nucleation and growth mechanism results in domain size increasing with time. The domains tend to be spheroidal in nature. For spinodal decomposition, interconnected cylinders tend to form. Their initial growth mechanism involves an exchange of mass across the boundary, purer phases forming with time. Thus the amplitude of the waves increases, but not necessarily the wavelength. The domains are of about the same size as the original wavelength of the wavelike fluctuation during the early stages of phase separation. On annealing, it must be pointed out, both kinds of domains (NG and SD) may coalesce, forming larger and larger spheroidal structures.

The best characterized system has been polystyrene blended with poly(vinyl methyl ether). Several studies showed that this system is miscible in all proportions below about 80°C or so, depending on the molecular weight. Nishi et al. (26) prepared a number of mutual solutions at low temperatures.

Figure 4.8 A schematic of concentration profiles during phase separation (25), for (a) nucleation and growth and (b) spinodal decomposition.

After equilibration, they rapidly raised the temperature, then held the temperature constant again, making observations in the microscope. Nucleation and growth could be seen as tiny spheres, while spinodal decomposition looked like tiny overlapping worms. The more unusual of these two results, that of spinodal decomposition, has been modeled by Reich (27); see Figure 4.9. The results are shown in Figure 4.10 (see page 137) (26). Experiments

Table 4.3   Mechanisms of phase separation

1. Nucleation and growth
   (a) Initial fragment of more stable phase forms.
   (b) Two contributions to free energy: (i) work spent in forming the surface and (ii) work gained in forming the interior.
   (c) Concentration in immediate vicinity of nucleus is reduced; hence diffusion into this region is *downhill*. (Diffusion coefficient is positive.)
   (d) Droplet size increases by growth initially.
   (e) Requires activation energy.
2. Spinodal decomposition
   (a) Initial small-amplitude composition fluctuations.
   (b) Amplitude of wavelike composition fluctuations increases with time.
   (c) Diffusion is *uphill* from the low concentration region into the domain. (Diffusion coefficient is negative.)
   (d) Unstable process: no activation energy required.
   (e) Phases tend to be interconnected.

90×90μm

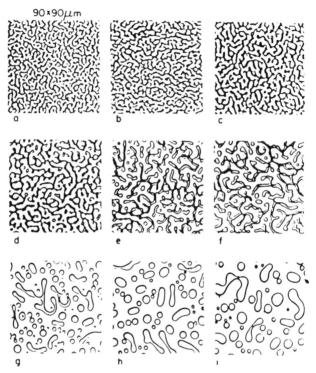

Figure 4.9 Coarsening during the latter stages of spinodal phase separation (27), polystyrene-*blend*-poly(vinyl methyl ether).

like these delineate the regions of nucleation and growth from those of spinodal decomposition in phase-separating polymer blends.

The kinetics of phase separation proper have been described by Cahn and Hilliard (28, 29). For the early stages of decomposition, the diffusion equation is of the following form:

$$\frac{\partial \phi}{\partial t} = M\left[\left(\frac{\partial^2 G}{\partial \phi^2}\right)\nabla^2 \phi - 2K \nabla^4 \phi\right] \tag{4.14}$$

where G is the free energy, ϕ is the composition, M is the mobility coefficient, and K is the energy gradient coefficient arising from contributions of composition gradients to the free energy. Solution of this equation leads to theories of the wavelength of the system and the rate at which the domains will form. The spinodal line is defined by the locus of temperatures in the phase diagram, where $\partial^2 G / \partial \phi^2 = 0$. The critical point is given where the third derivative is equal to zero. It is interesting to note that the Cahn–Hilliard theory was originally derived to obtain an understanding of the phase-separation kinetics in certain inorganic glasses.

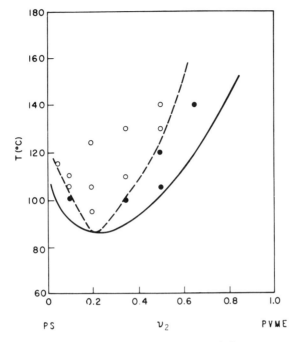

Figure 4.10 Phase separation by spinodal decomposition (○) and nucleation and growth (●), as observed under a microscope (26).

4.3.6 Miscibility in Statistical Copolymer Blends

As stated previously, most homopolymer blends are immiscible due to the negative entropy of mixing and negative heats of mixing. Sometimes, however, miscibility can be achieved with the introduction of comonomers. These may be, in general,

$$\text{poly}(\text{A}-stat-\text{B})-blend-\text{poly}(\text{C}-stat-\text{D}) \qquad (4.15)$$

or, more frequently,

$$\text{poly}(\text{A}-stat-\text{B})-blend-\text{poly}(\text{A}-stat-\text{C}) \qquad (4.16)$$

The phenomenon makes use of the fact that miscibility in copolymer–copolymer blends may be enhanced by the repulsion of dissimilar mers on the *same* chain: A hates B more than it hates C. An example is poly(vinyl chloride)–*blend*–poly(ethylene–*stat*–vinyl acetate), where poly(vinyl chloride) is immiscible with either polyethylene or poly(vinyl acetate).

Karasz and MacKnight (30) approached the problem through mean field thermodynamic considerations, arguing that negative net interactions are necessary to induce miscibility. The general equation for the free energy was adopted from equation (4.5):

$$\frac{\Delta G_M}{RT} = \frac{v_1}{n_1}\ln v_1 + \frac{v_2}{n_2}\ln v_2 + v_1 v_2 \chi_{\text{blend}} \tag{4.17}$$

where v_1 and v_2 are volume fractions, n_1 and n_2 are degrees of polymerization, and χ_{blend} is a dimensionless interaction parameter defined as

$$\chi_{\text{blend}} = \sum_{i,j} c_{ij}\chi_{ij} \tag{4.17a}$$

where the coefficients c_{ij} are functions of the copolymer compositions, with $0 \le c_{ij} \le 1$. For $A_n/(B_x C_{1-x})_{n'}$ blends,

$$\chi_{\text{blend}} = x\chi_{AB} + (1-x)\chi_{AC} - x(1-x)\chi_{BC} \tag{4.18}$$

"Windows of miscibility" result when $\Delta G_M < 0$. This is illustrated for the system poly(2,6-phenylene oxide)–*blend*–poly(*o*-chlorostyrene–*stat*–*p*-chlorostyrene) in Figure 4.11 (31). The experimental determination of miscibility involved a DSC characterization of each sample for one or two glass transition temperatures. As mentioned previously, poly(2,6-phenylene oxide) is miscible with polystyrene in all proportions; indeed, this forms the basis for an important impact-resistant plastic, Noryl®. However, miscibility is limited with the halogenated styrenes.

Miscibility for $(A_x B_{1-x})_n/(C_y D_{1-y})_{n'}$ may even be brought about if all the χ_{ij} values are positive, as illustrated in Figure 4.12 (30). The ellipse shown cannot extend to the corners of the diagram because these areas correspond to immiscible blends of the respective homopolymers.

The history of polymer blending has taken several interesting turns. Before about 1974, virtually all polymer blends known were immiscible. Polymer scientists thought that thermodynamics was against such mixing, particularly that the very small entropies of mixing would always dominate the circumstances. Kinetically, diffusion was thought to be so slow that even if a miscible polymer pair could be found, inordinate amounts of time would elapse before mixing would occur.

With the development of the equation of state theories, however, phase diagrams began to be constructed. To everyone's surprise, virtually all these systems exhibited LCST behavior. (Earlier, many scientists were trying to *raise* the temperature to bring about miscibility.) It was discovered that for very small domains, equilibrium mixing could be achieved in a matter of minutes. The concept of specific interactions was developed, where hydrogen bonds, polar attractions, and so on between the polymers each played a role in developing miscibility. Today, very many miscible polymer pairs are known. While the science of polymer blending was qualitative in nature in the 1950s

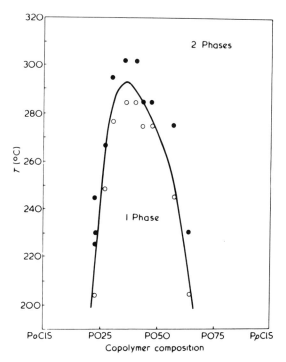

Figure 4.11 Lower critical solution temperature of 40 / 60 PO-PPO blends versus polystyrene copolymer composition: P_oClS, poly(ortho-chlorostyrene); P_pClS, poly(para-chlorostyrene) (31). The system goes from one phase to two as the temperature is raised. Open circles indicate one phased region, and filled circles indicte two phased region.

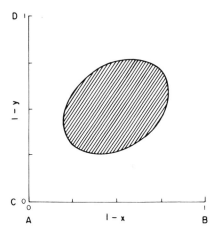

Figure 4.12 Theoretical miscibility domain (shaded region) in copolymer blends, assuming infinite degree of polymerization. The Flory interaction parameters are χ_{AB}, χ_{AC}, χ_{AD}, χ_{BC}, χ_{BD}, and χ_{CD} are 0.3, 0.1, 0.2, 0.2, 0.1, and 0.4, respectively (30).

and 1960s, and primarily concerned with toughness and impact resistance (still important!), major advances in the field have now turned toward thermodynamics and kinetics.

From an engineering point of view, the need is to control the miscibility and the phase domain size. Very small sizes (200 Å) serve to make good damping compositions, while domains of the order of 1000 Å make better impact-resistant materials. The extent of mixing, particularly at the interface between the domains, has now been found to be critical in determining many properties. All these are subjects of current research (17, 21, 32–37).

4.3.7 Polymer Blend Characterization

There has been increasing interest in mutually miscible polymer blends (38–43). Rather than trying to find polymer–polymer systems that are "like," the emphasis has been on finding systems that are complementary, attracting one another by hydrogen bonds or polarity. The first important polymer–polymer phase diagram was worked out by McMaster (38) on poly(styrene–*stat*–acrylonitrile)–*blend*–poly(methyl methacrylate), who determined the phase diagram by cloud point determination. He found a LCST behavior. Other miscible systems include poly(methyl methacrylate)–*blend*–poly(vinylidene chloride) (39) and polystyrene–*blend*–poly(2,6-dimethyl–1,4-phenylene oxide) (40). These materials are characterized by a negative heat of mixing.

Polymer–polymer interactions in the bulk state have been studied by a number of experimental tools; see Table 4.4. For phase-separated compositions, dynamic mechanical spectroscopy (Chapter 8) measures shifts in the glass transition temperature, which provides a measure of the extent of molecular mixing in partly mixed systems. Electron microscopy is used to determine domain size and shape (see Section 4.5.6).

For measurements on miscible systems, a variety of instruments are used. These include small-angle neutron scattering (see Section 5.2.2), which provides a measure of the second virial coefficient thermodynamic interaction between the two polymers in miscible blends, inverse gas chromatography, and solvent vapor sorption; the last two measure the Flory interaction parameter between the two polymers. (*Note*: The subscript 1 stands for the solvent; subscripts 2 and 3 stand for the two polymers.) NMR and ESR instrumentation each measure different relaxation times for segments of one component in contact with segments of the other component. FT-IR provides a measure of the specific interactions occurring between polymer segments. For example, the carbonyl vibration band of polycaprolactone shifts to lower frequencies on blending with poly(vinyl chloride), providing a measure of the strength of interaction between the carbonyl and the chlorine. Heats of mixing of two polymers, of course, provide a sum of all such energies of interaction. Differential scanning calorimetry measures the changes in heat capacity between the individual components and the blend.

Table 4.4   Methods of characterizing polymer blends

Method	Parameter Determined	Reference
Inverse gas chromatography	χ_{23}	(a)
Differential scanning calorimetry	ΔC_p	(b)
Dynamic mechanical spectroscopy	ΔT_g	(c)
NMR	τ	(d)
ESR	τ	(e)
Solvent vapor sorption	χ_{23}	(f)
Heat of mixing	ΔH_M	(g)
Light-, x-ray, and neutron scattering	A_2	(h)
Photophysics	I_D/I_M	(i)
FT-IR	$\Delta \nu$	(j)
Electron microscopy	Domain size and shape	(k)

References: (a) M. Galin and L. Maslinka, *Eur. Polym. J.*, **23**, 923(1987). (b) A. Natansohn, *J. Polym. Sci. Polym. Lett. Ed.*, **23**, 305(1985). (c) L. H. Sperling, J. J. Fay, C. J. Murphy, and D. A. Thomas, *Macromol. Chem. Macromol. Symp.*, **38** 99(1990). (d) T. K. Kwei, T. Nishi, and R. F. Roberts, *Macromolecules*, **7**, 667(1974). (e) Y. Shimada and H. Kashiwabara, *Macromolecules*, **21**, 3454(1988). (f) S. Saeki, S. Tsubotani, H. Kominami, M. Tsubokawa, and T. Yamaguchi, *J. Polym. Sci. Polym. Phys. Ed.*, **24**, 325(1986). (g) R. E. Bernstein, D. C. Wahrmund, J. W. Barlow, and D. R. Paul, *Polym. Eng. Sci.*, **18**, 1220(1978). (h) R. J. Roe and W. C. Sin, *Macromolecules*, **13**, 1221(1980). (i) C. W. Frank and M. A. Gashgari, *Macromolecules*, **12**, 163(1979). (j) P. Cousin and R. E. Prud'homme, in *Multicomponent Polymer Materials*, D. R. Paul and L. H. Sperling, Eds., American Chemical Society, Washington, DC, 1986. (k) A. A. Donatelli, L. H. Sperling, and D. A. Thomas, *Macromolecules*, **9**, 671(1976).

Another method involves excimer fluorescence as a molecular probe; see Section 2.9. The question may be raised as to whether polymer blends will become more miscible if the differences in their solubility parameters are reduced. Excimer fluorescence provides some evidence; see Figure 4.13 (44). Here, 0.2 wt.% of poly(2-vinyl naphthalene), P2VN, is dispersed in a series of poly(alkyl methacrylates). These include the following, which are identified in Figure 4.13 by acronym: methyl, PMMA; ethyl, PEMA; *n*-propyl, PnPMA; isopropyl, PiPMA; *n*-butyl, PnBMA; isobutyl, PiBMA; *sec*-butyl, PsBMA; *tert*-butyl, PtBMA; phenyl, PPhMA; isobornyl, PiBoMA; benzyl, PBzMA; and cyclohexyl, PCMA. Two other host polymers were polystyrene, PS, and poly(vinyl acetate), PVAc.

Figure 4.13 shows the average values of the excimer to monomer ratio, I_D/I_M, plotted against the difference in host and guest solubility parameters. There are two phenomena involved. The intramolecular site concentration depends on the population of suitable local conformation states of the chain, that is, rotational dyads. However, the concentration of intermolecular sites is sensitive solely on clustering of the guest P2VN, or the chains bending back on themselves. In a thermodynamically good host medium, there will be extensive mixing, causing the local concentration of aromatic rings to drop. In

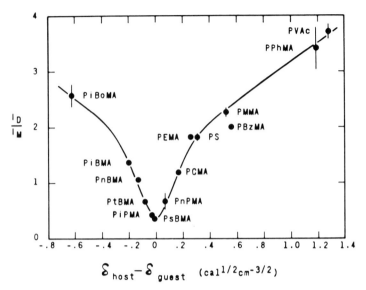

Figure 4.13   Poly(2-vinyl naphthalene) becomes quite miscible with acrylic polymers when the differences between their solubility parameters become small (44).

addition, chain expansion reduces the likelihood of chain back-bending. Both effects lead to a minimum in I_D/I_M for $\Delta\delta = 0$. It must be noted that in all cases $\delta_{host} - \delta_{guest}$ is less than unity, and at the extremes studied I_D/I_M has not yet leveled out, which would be a sign of more or less complete phase separation.

4.3.8   Graft Copolymers and IPNs

Most polymer blends, grafts, blocks, and IPNs exhibit phase separation, as discussed previously. It must be emphasized that their wide application in commerce arises largely because of the synergistic properties exhibited by these materials. Applications have included impact-resistant plastics, thermoplastic elastomers, coatings, and adhesives.

The morphology of some graft copolymers and IPNs based on SBR[†] and polystyrene is illustrated in Figure 4.14 (41). Here the various morphologies, and hence their subsequent physical and mechanical properties, are controlled by cross-linking and/or mixing. The graft copolymer in the upper left of Figure 4.14 is called high-impact polystyrene (HIPS). The SBR phase, stained dark, toughens the otherwise brittle polystyrene. Important features of the rubber phase include its low glass transition temperature, small domain size, and extent of grafting to the polystyrene phase. These features

---

[†]Styrene–butadiene rubber, poly(butadiene–*stat*–styrene).

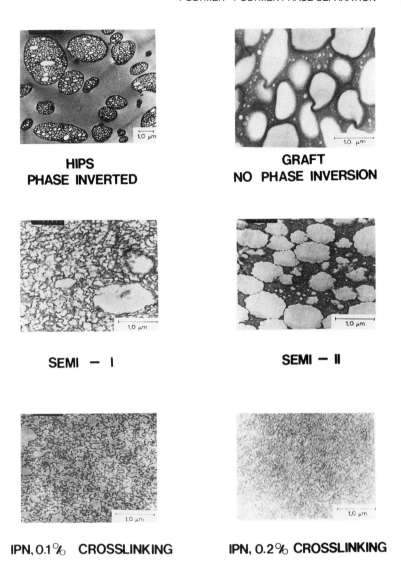

HIPS
PHASE INVERTED

GRAFT
NO PHASE INVERSION

SEMI — I

SEMI — II

IPN, 0.1% CROSSLINKING

IPN, 0.2% CROSSLINKING

Figure 4.14 Phase morphology of graft copolymers and IPNs of SBR rubber and polystyrene by transmission electron microscopy. Upper two figures, graft copolymers; middle two, semi-IPNs; bottom two, full IPNs. Diene portion stained dark with OsO_4 (41).

will be discussed further in Chapter 11. It must be pointed out, however, that actual extraction studies show that most of the polystyrene is not grafted. Only that portion in contact with polybutadiene is significantly bonded. The upper right structure is the graft copolymer made without proper stirring. Note that the rubber phase is continuous. As a consequence, the material is much softer than the HIPS composition.

The bottom four structures are IPNs or semi-IPNs, as indicated. Note that all six compositions have essentially the same chemical composition, except for cross-linking. Higher cross-linking makes the domains smaller.

4.3.9 Block Copolymers

In the above, the limited solubility of one polymer in another was ascribed to the unusually low entropy of mixing. In a block copolymer, one chain portion is attached to another, end on end. It is interesting to note the miscibility characteristics and morphology of these materials.

First of all, the presence of the chemical bond between the blocks definitely improves the mutual miscibility[†] of the two polymers (45–47). Thus block copolymers are miscible to higher block molecular weight than their corresponding blends.

If block copolymers phase-separate, how big are the domains?[‡] If the bonds holding the blocks together are maintained, then the domains must be small enough that one block resides in one phase while the other block is in the neighboring phase. The junction, or bond between the blocks, tends to be in the interface between the two blocks (49, 50).

The morphology of the phases changes from spheres to cylinders to alternating lamellae depending on the relative length of the two blocks. Spheres containing the short blocks are formed within the continuous phase of the longer block (see Figure 4.15). Alternating lamellae form when the blocks are about the same length. Cylinders are formed for intermediate cases as illustrated in Figure 4.16 (51).

The actual size of the three types of domains was calculated by Meier (49, 50):

$$\text{Spheres} \qquad R = 1.33\alpha \bar{K} M^{1/2} \qquad\qquad (4.19)$$

$$\text{Cylinders} \qquad R = 1.0\alpha \bar{K} M^{1/2} \qquad\qquad (4.20)$$

$$\text{Lamellae} \qquad R = 1.4\alpha \bar{K} M^{1/2} \qquad\qquad (4.21)$$

where $\bar{K}$ is the experimental constant relating the unperturbed root-mean-square end-to-end distance to the molecular weight, $\overline{r_0^2}/M$, and R is a characteristic dimension, the radius of the spheres and cylinders, and the half-thickness of the corresponding lamellae. Typically, the minor component forms spheres up to 25% concentration, cylinders between 25 and 40%, and alternating lamellae in midrange. For some common polymers, values of $\bar{K}$

[†] The term "miscibility" is preferred by Olabisi et al. (42) and (45–47) Paul (48) and others rather than "solubility" because the very large size of the polymer molecules has sometimes created the appearance of slight demixing, when they were in fact mixed on a molecular level according to all thermodynamic criteria. An older term for miscibility is "compatibility." Use of that term is now recommended in terms of satisfactory engineering properties.

[‡] A domain is a discrete region of space occupied by one phase and surrounded by another.

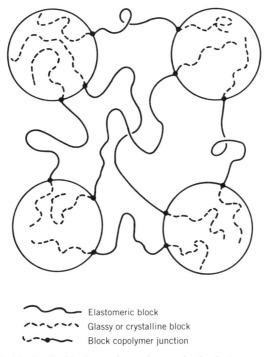

～～～   Elastomeric block
＾～＾～＾   Glassy or crystalline block
＼～＿●～   Block copolymer junction

Figure 4.15   Idealized triblock copolymer thermoplastic elastomer morphology.

Figure 4.16   Polystyrene – *block* – polybutadiene – *block* – polystyrene (40% butadiene), showing the existence of the cylindrical structures. Polymer cast from toluene solution, and the polybutadiene portion stained with OsO_4. Two cuts through the film are (*a*) normal and (*b*) parallel. Apparent spheres in (*a*) are end-on cuts (51).

for R in Ångstroms are

Polystyrene	670×10^{-3}
Polyisobutylene	700×10^{-3}
Poly(dimethyl siloxane)	880×10^{-3}
Poly(methyl methacrylate)	565×10^{-3}

Values of α are slightly larger than unity because the chains are slightly strained at the common junction point between the two phases. For practical calculations, a value of 1.2 might be assumed, since α varies between 1.0 and 1.5 for most cases of interest.

It must be pointed out that many block copolymers are synthesized in the form of triblock or multiblock copolymers. As an example, the triblock copolymer polystyrene–*block*–polybutadiene–*block*–polystyrene is widely used as a rubbery shoe sole material. In this case, the long, rubbery center block forms the continuous phase whereas the short, glassy polystyrene blocks form submicroscopic spheres (see Figure 4.15). The hard domains constitute a type of physical cross-link, holding the whole mass together. On heating above the glass transition of the hard phase, the material softens and flows. Other examples are the so-called segmented polyurethanes, which form the elastic thread in clothing, particularly undergarments, a multiblock material.

4.4   DIFFUSION AND PERMEABILITY IN POLYMERS

Permeation is the rate at which a gas or vapor passes through a polymer. The mechanism by which permeation takes place involves three steps: (a) absorption of the permeating species into the polymer; (b) diffusion of the permeating species through the polymer, traveling, on average, along the concentration gradient; and (c) desorption of the permeating species from the polymer surface and evaporation or removal by other mechanisms (52).

Factors affecting permeability include the solubility and diffusivity of the penetrant into the polymer, polymer packing and side-group complexity, polarity, crystallinity, orientation, fillers, humidity, and plasticization. For example, polymers with high crystallinity usually are less permeable because their ordered structure has fewer holes through which gases may pass.

The concept of holes, or free volume, in polymers has already been introduced in relation to phase diagrams. It must be emphasized that holes in materials are required for all types of molecular motion beyond simple vibrational and rotational states. One must ask the question: When a molecule moves from position A to position B, into what does it move, and what does it leave behind? The answer is that it moves into a hole. The hole and the molecule are transposed, so the hole is where the molecule was before the action started. The general concept of free volume is developed in Chapter 8, in relation to the glass transition.

4.4.1 Swelling Phenomena

Consider a polymer in contact with a solvent. Diffusion takes place in both directions, the polymer into the solvent and vice versa. However, the rate of diffusion of the solvent, being a small molecule, is much faster. Hence, for a time, the polymer really acts as the solvent.

If the polymer is glassy, the solvent lowers the T_g by a plasticizing action. Polymer molecular motion increases. Diffusion rates above T_g are far higher than below T_g. Thus diffusion may depend on the concentration of the diffusing species (53, 54).

4.4.2 Fick's Laws

Perhaps one of the most interesting cases of the polymer behaving as the solvent has to do with permeability of water and gases. Often polymers, in the form of films, are used as barriers to keep out water and air. In the case of food wrappers, it is often desired to keep in water but keep out oxygen.

The general case of diffusion in materials is given by Fick's laws (55). His first law governs the steady-state diffusion circumstance:

$$J = -\tilde{D}\frac{\partial c}{\partial x} \qquad (4.22)$$

Simple steady-state diffusion through a film is modeled in Figure 4.17. The flux J gives the quantity of permeant passing through a unit cross-section of membrane per unit time. Thus equation (4.22) leads to first-order transport kinetics, where the quantity transported at any instant depends on the concentration on the high concentration side to the first power.

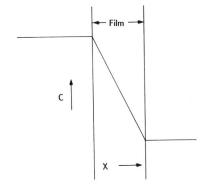

Figure 4.17 Illustration of permeability through a film under steady-state conditions, showing the decrease in concentration, c.

Fick's second law controls the unsteady state:

$$\frac{\partial c}{\partial t} = \frac{\partial}{\partial x}\left[\tilde{D}\frac{\partial c}{\partial x}\right] \tag{4.23}$$

The quantity J represents the net flux of diffusing material across unit area of a reference plane and has the units of $mol \cdot cm^{-2} \cdot sec^{-1}$; c is the vapor concentration, x represents the distance diffused in time t, and $\tilde{D}$ is the diffusion coefficient.

The permeability coefficient, $\tilde{P}$, is defined as the volume of vapor passing per unit time through unit area of polymer having unit thickness, with a unit pressure difference across the sample. The solubility coefficient, $\tilde{S}$, determines the concentration. For the simplest case,

$$\tilde{P} = \tilde{D} \cdot \tilde{S} \tag{4.24}$$

which expresses the permeability in terms of solubility and diffusivity.

The temperature dependence of the solubility obeys the Clausius–Clapeyron equation written in the form (56)

$$\Delta H_s = -R\frac{d \ln \tilde{S}}{d(1/T)} \tag{4.25}$$

A study of vapor solubility as a function of temperature allows the heat of solution ΔH_s to be evaluated. Since the size of the diffusing molecule per se is important, it turns out that the logarithm of the diffusion coefficient depends inversely on the molar volume (57, 58).

Of course, the permeability coefficients depend on the temperature according to the Arrhenius equation,

$$\tilde{P} = \tilde{P}_0 e^{-\Delta E/RT} \tag{4.26}$$

where ΔE is the activation energy for permeation (59).

Similar relations hold for solubility and diffusivity.

4.4.3 Permeability Units

While there are many dimensions and units found in the literature for permeability, the preferred current dimensions are (60)

$$\tilde{P} = \frac{(\text{quantity of permeant}) \times (\text{film thickness})}{(\text{area}) \times (\text{time}) \times (\text{pressure drop across the film})} \tag{4.27}$$

Table 4.5   Permeability in polymers at 25°C (60)

Polymer	Permeant	$\tilde{P} \times 10^{13}$	$\tilde{P}_0 \times 10^7$	ΔE
Polyethylene	O_2	2.2	4.62	42.7
(LDPE)	N_2	0.73	329	49.9
	CO_2	9.5	62	38.9
	H_2O	68	48.8	33.5
cis-1,4-polybutadiene	N_2	14.4	0.078	21.3
	He	24.5	0.0855	20.3
Poly(ethyl methacrylate)	O_2	0.889	2.1	36.4
	CO_2	3.79	0.435	28.9
Poly(ethylene	O_2	0.0444	0.227	37.7
terephthalate)[a]	CO_2	0.227	0.00021	27.0
(amorphous)	CH_4	0.0070	0.9232	24.7
Poly(vinylidene	N_2	0.00070	900	70.3
chloride)[b]	O_2	0.00383	825	66.6
	CO_2	0.0218	24.8	51.5
	H_2O	7.0	863	46.1
Cellulose	H_2O	18900	—	—

[a] Major component of plastic soft drink bottles.
[b] Major component of Saran®.
Units: $\tilde{P}$, $\tilde{P}_0$ = cm³ (273 K; 1 × 10⁵ Pa) × (cm/cm²) × sec × Pa.
   ΔE = kJ/mol.

The preferred SI unit of the permeability coefficient is (60)

$$\text{Units of } \tilde{P} = \frac{\text{cm}^3(273.15 \text{ K}, 1.013 \times 10^5 \text{ Pa}) \times \text{cm}}{\text{cm}^2 \times \text{sec} \times \text{Pa}} \tag{4.28}$$

Thus permeability coefficients range from 10^{-11} to 10^{-16} cm³ × cm/cm² × sec × Pa for many polymers and permeants. The units of activation energy are kJ/mol.
   Similarly, the units of $\tilde{D}$ and $\tilde{S}$ are

$$\text{Units of } \tilde{D} = \text{cm}^2/\text{sec} \tag{4.29}$$

$$\text{Units of } \tilde{S} = \frac{\text{cm}^3(273.15 \text{ K}; 1.013 \times 10^5 \text{ Pa})}{\text{cm}^3 \times \text{Pa}} \tag{4.30}$$

4.4.4   Permeability Data

Typical permeability data are shown in Table 4.5 (60) for a range of elastomers, amorphous plastics, and semicrystalline plastics. In general, per-

Table 4.6   Kinetic diameters[a] of various penetrants (62), nm

Molecule	Diameter	Molecule	Diameter
He	0.26	C_2H_4	0.39
H_2	0.289	Xe	0.396
NO	0.317	C_3H_8	0.43
CO_2	0.33	n-C_4H_{10}	0.43
Ar	0.34	CF_2Cl_2	0.44
O_2	0.346	C_3H_6	0.45
N_2	0.364	CF_4	0.47
CO	0.376	i-C_4H_{10}	0.50
CH_4	0.38		

[a]Zeolite sieving diameters.

meability decreases from elastomers to amorphous plastics to semicrystalline plastics.

One of the more recent applications is the use of amorphous poly(ethylene terephthalate)* for soft drink bottles. The major requirements are to keep carbon dioxide and water in and to keep oxygen out. One must realize that these gases are continuously being transported across the plastic even if at a low rate, causing the soft drink eventually to go "flat." Thus these drinks have a "shelf life," after which they must be discarded if not sold.

4.4.5   Effect of Permeant Size

For elastomers, concerted movements of several adjacent chain segments take place to provide rapid transport. Such motions are restricted in glassy polymers (61). Also, free volume is much less in glasses than in plastics. The size of the permeant (Table 4.6) (62) is critical in determining its diffusion rate in polymers. Sizes range from 2 to 5 Å for many molecules. The larger the molecule, the smaller the diffusion rate; see Figure 4.18 (63). In Figure 4.18 diffusion is seen to be significantly lower in poly(vinyl chloride), a glassy polymer, than in natural rubber. Of course, polymers have a distribution of hole sizes (64). Note that the permeability of a given molecule depends on both its diffusion coefficient and its solubility; see equation (4.24). Solubility depends on the solubility parameter, as described in Section 3.2.

Other modern areas of application must be mentioned. First is the development of controlled drug delivery systems (65). For some applications, the drug is dissolved in suspension sized[†] polymer particles, which are ingested. Drug release may be controlled further by coating the particle with a polymer of low permeability.

---

*Small amounts of comonomer destroy crystallinity.
[†]Suspension particles are usually 100–1000 μm in diameter, see Section 4.5.

Figure 4.18 Rigid poly(vinyl chloride) has lower diffusion coefficients than natural rubber for a variety of gases and liquids (63).

The second area relates to oxygen diffusion through soft contact lenses (66). Soft contact lenses are made of poly(2-hydroxyethyl methacrylate) and its copolymers, in the form of cross-linked networks. These are swollen to thermodynamic equilibrium in water or saline solution. The hydroxyl group provides the hydrophilic characteristic and is also important for oxygen permeability. Oxygen permeability is important because of the physiological requirements of the eye. Thus the polymer is highly swollen with water and also serves as a semipermeable material.

Reverse osmosis membranes are important in making potable water from saline or alkaline waters. Here, large permeabilities of water and small permeabilities of salt ions are required. Cellulose membranes are used, because of their high permeability coefficient to water; see Table 4.5.

On the other hand, ion exchange membranes require large permeabilities for ions. This is obtained through the use of polymers with large bonded ion concentrations such as $— SO_4Na$ and $— NH_4Cl$, which allow the polymer to swell in water. In both contact lenses and ion exchange membranes, large permeation rates take place because the polymer is so swollen with water that diffusion substantially takes place in the aqueous medium.

Table 4.7 **Latexes and suspensions**

Property	Latex	Suspension
Size	0.04–5 μm	10–1000 μm
Shape	spheres	spheres to potato
Emulsifier	surfactant	polar polymer
Initiator	water soluble	oil soluble

4.5 LATEXES AND SUSPENSIONS

Polymers in water dispersion serve as coatings, adhesives, and the basis for many plastics and elastomers. The two most important classes are latexes (or latices) and suspensions. These differ by size and method of synthesis; see Table 4.7. While latex particles are normally small enough to be kept in dispersion by Brownian motion, most suspensions will "cream" on standing and hence must be stirred continuously.

Basically, a latex is synthesized by mixing monomer, surface active agent (surfactant), and an initiator (free radical source) into the water; see Figure 4.19 (67). Usually, the initiator is activated by heat. A widely used initiator is potassium persulfate. Surfactants are molecules that have a hydrophilic end and a hydrophobic end. These form micelles in the water, with the hydrophilic end facing outward, shielding the hydrophobic portion. The surfactant can be a soap or detergent, sodium lauryl sulfate being a typical

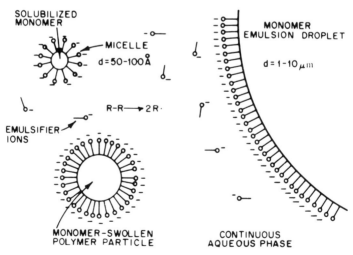

Figure 4.19 Illustration of the emulsion polymerization process during interval I (early stage). The monomer emulsion droplet is large and hence has a very small surface area compared to the soap micelles and monomer swollen particles (67).

example. The turbid or hazy appearance of soap in water is due to the presence of micelles. The surfactant dissolves the oil-soluble monomer in much the same way as the soap micelles dissolve skin oil in ordinary toiletry. Some of the surfactant remains water soluble, as does some of the monomer. Monomer diffuses from the large droplets into the water, and to the micelles. Some of it begins to polymerize on contact with the free radicals. After adding a few mers, the much less water-soluble free radical diffuses (randomly) into a micelle, where it begins to polymerize the monomer present there. It is terminated when a second free radical enters the micelle. The fully polymerized latex then is a submicroscopic sphere of polymer emulsified by a layer of surfactant (68). The kinetics of emulsion polymerization are treated in many organic polymer science textbooks.

The reasons why emulsion polymerization is so popular include the following: (a) heat removal through the aqueous phase during the polymerization is very efficient; (b) the final latex has a very low viscosity, approaching that of water; and (c) the polymer is easily recovered via coagulation and/or film formation.

By contrast, a suspension polymerization behaves kinetically much more like a bulk polymerization. Heat removal during polymerization is also important. Special advantages include ease of handling the filterable powder and dry blending of the powder with plasticizer. This last step involves diffusion of the plasticizer into the particles, aided by a large surface area. For such cases, the particles are made with a porous structure, which easily adsorbs plasticizer. The plasticizer then diffuses in when the polymer is heated above its glass transition temperature.

4.5.1 Natural Rubber Latex

Natural rubber latexes occur in many plants. The white sap of the common milkweed and dandelion is a rubber latex. The most important source of natural rubber latex is the Hevea brasiliensis tree, now grown in plantations in many tropical parts of the world.

The polymer is a highly linear cis-1,4-polyisoprene of about 5×10^5 g/mol. It owes its colloid stability to the presence of adsorbed proteins at the surface of the rubber particles. These adsorbed proteins are in the anionic state, so that the rubber particles carry negative charges at their surfaces (69).

4.5.2 Colloidal Stability and Film Formation

Latexes constitute a subgroup of colloid systems known as lyophobic sol. Sometimes they are called polymer colloids. The stability of these colloids is determined by the balance between attractive and repulsive forces affecting two particles as they approach one another. Stability is conferred on these latexes by electrostatic forces, which arise because of the counterion clouds surrounding the particles. Other forces of an enthalpic or entropic nature arise when the lyophilic molecules on the surfaces of the latexes interact on close approach. These can be overcome by evaporation of the water, heating, freezing, or by chemically modifying the surfactant, such as by acidification.

If the polymer in the latex is above its glass transition temperature, it may form a film on evaporation of the water. A simple example is the drying of a latex paint film on a wall. As the water evaporates, coalescence surface tension forces proceed from the presence of water menisci of very small radii of curvature. These menisci develop between the particles as the last traces of water are removed. The forces that these menisci generate drive the particles together. Interdiffusion of the polymer chains takes place, forming coherent films.

4.6 MULTICOMPONENT AND MULTIPHASED MATERIALS

Solutions may be dilute or concentrated up to substantially 100% solute. A question arises in considering concentrated polymer solutions: When do we stop considering the polymer dissolved in the solvent, and start considering the solvent dissolved in the polymer? It must be pointed out that many polymer solutions containing more than about 25% polymer may no longer be pourable but behave like plasticized solid materials. A key issue for the "semidilute" systems is the overlap of the polymer chains, at which point the properties shift to much higher viscosities and/or solid behavior profiles. Another critical problem is diffusion of the polymer through the solvent and the diffusion of the solvent through the polymer.

Of course, a multicomponent solution is one that contains more than one component, while a multiphase system is one in which the various components have phase-separated. Polymer blends usually exhibit phase separation, principally because the entropy of mixing and the enthalpy of mixing are both negative. Phase-separated polymer–polymer systems exhibit quite complex morphologies. Many polymer blends, blocks, and grafts possess great toughness and/or impact resistance, a subject of Chapter 11. In this case, the size, shape, and interfacial characteristics of the domains are all important.

REFERENCES

1. A. Shultz and P. J. Flory, *J. Am. Chem. Soc.*, **74**, 4760(1952).
2. M. Daoud and G. Jannink, *J. Phys. Paris*, **37**, 973(1976).
3. P. G. de Gennes, *Scaling Concepts in Polymer Physics*, Cornell University Press, Ithaca, NY, 1979.
4. S. F. Edwards, *Proc. Phys. Soc.*, **88**, 265(1966).
5. S. Candau, J. Bastide, and M. Delsanti, Polymer networks, *Adv. Polym. Sci.*, **44**, 27(1982).

6. A. Dobry and F. Boyer-Kawenoki, *J. Polym. Sci.*, **2**, 90(1947).

7. L. Bohn, *Rubber Chem. Tech.*, **41**, 495(1968).

8. S. Krause, *J. Macromol. Sci. Rev. Macromol. Chem.*, **C7**, 251(1972).

9. R. L. Scott, *J. Chem. Phys.*, **17**, 279(1949).

10. P. J. Flory, R. A. Orwall, and A. Vrij, *J. Am. Chem. Soc.*, **86**, 3515(1964).

11. P. J. Flory, *Discuss. Faraday Soc.*, **49**, 7(1970).

12. I. C. Sanchez and R. H. Lacombe, *J. Phys. Chem.*, **80**, 2352(1976).

13. I. C. Sanchez and R. H. Lacombe, *J. Polym. Sci. Polym. Lett. Ed.*, **15**, 71(1977).

14. R. H. Lacombe and I. C. Sanchez, *J. Phys. Chem.*, **80**, 2568(1976).

15. I. C. Sanchez, Chap. 3 in *Polymer Blends*, Vol. I, D. R. Paul and S. Newman, Eds., Academic Press, Orlando, 1978.

16. L. D. Taylor and L. D. Cerankowski, *J. Polym. Sci. Polym. Chem. Ed.*, **13**, 2551(1975).

17. O. Olabisi, L. M. Robeson, and M. T. Shaw, *Polymer–Polymer Miscibility*, Academic Press, Orlando, 1979.

18. D. R. Paul, in *Multicomponent Polymer Materials*, Advances in Chemistry Series No. 211, ACS Books, Washington, DC, 1986.

19. D. S. Lee and S. C. Kim, *Macromolecules*, **17**, 268(1984).

20. P. J. Flory, *J. Am. Chem. Soc.*, **87**, 1833(1965).

21. L. A. Utracki, *Polymer Alloys and Blends*, Hanser, New York, 1990.

22. (a) I. C. Sanchez, Chap. 3 in *Polymer Blends*, Vol. I, D. R. Paul and S. Newman, Eds., Academic Press, Orlando, 1978; (b) I. C. Sanchez, *Encyclopedia of Physical Science and Technology*, R. A. Meyers, Ed., Vol. 11, Academic Press, Orlando, 1987, p. 1.

23. P. J. Flory, R. A. Orwoll, and A. Vrij, *J. Am. Chem. Soc.*, **86**, 3515(1964).

24. D. Patterson, *Polym. Eng. Sci.*, **22**, 64(1982).

25. J. H. An and L. H. Sperling, in *Cross-Linked Polymers*: *Chemistry, Properties and Applications*, R. A. Dickie S. S. Labana and R. A. Baues, Eds., American Chemical Society, Washington, DC, 1988.

26. T. Nishi, T. T. Wang, and T. K. Kwei, *Macromolecules*, **8**, 227(1975).

27. S. Reich, *Phys. Lett.*, **114A**, 90(1986).

28. J. W. Cahn, *J. Chem. Phys.*, **422**, 93(1963).

29. J. W. Cahn and J. E. Hilliard, *J. Chem. Phys.*, **28**, 258(1958).

30. F. E. Karasz and W. J. MacKnight, Chap. 4 in *Multicomponent Polymer Materials*, D. R. Paul and L. H. Sperling, Eds., Advances in Chemistry Series No. 211, ACS Books, Washington, DC, 1986.

31. P. Alexandrovich, F. E. Karasz, and W. J. MacKnight, *Polymer*, **17**, 1023(1977).

32. L. A. Utracki and R. A. Weiss, Eds., *Multiphase Polymers*: *Blends and Ionomers*, ACS Symposium Series No. 395, ACS Books, Washington, DC, 1989.

33. D. R. Paul and L. H. Sperling, Eds., *Multicomponent Polymer Materials*, Advances in Chemistry Series No. 211, ACS Books, Washington, DC, 1986.

34. B. M. Culbertson, Ed., *Multiphase Macromolecular Systems*, Plenum, New York, 1989.

35. C. R. Riew, Ed., *Rubber-Toughened Plastics*, Advances in Chemistry Series No. 222, ACS Books, Washington, DC, 1989.

36. D. Klempner and K. C. Frisch, Eds., *Advances in Interpenetrating Polymer Networks*, Vol. II, Technomic, Lancaster, PA, 1990.

37. T. Saegusa, T. Higashimura, and A. Abe, Eds., *Frontiers of Macromolecular Science*, Blackwell, Oxford, England, 1989.

38. L. P. McMaster, in *Copolymers, Polyblends, and Composites*, N. A. J. Platzer, Ed., Advances in Chemistry Series No. 142, American Chemical Society, Washington, DC, 1975.

39. J. S. Noland, N. N. C. Hsu, R. Saxon, and J. M. Schmitt, Chap. 2 in *Multicomponent Polymer Systems*, N. A. J. Platzer, Ed., Advances in Chemistry Series No. 99, American Chemical Society, Washington, DC, 1971.

40. J. Stoelting, F. E. Karasz, and W. J. MacKnight, *Polym. Eng. Sci.*, **10**, 133(1970).

41. A. A. Donatelli, L. H. Sperling, and D. A. Thomas, *Macromolecules*, **9**, 671(1976).

42. O. Olabisi, L. M. Robeson, and M. T. Shaw, *Polymer–Polymer Miscibility*, Academic, Orlando, 1979.

43. L. H. Sperling and D. R. Paul, *Multicomponent Polymer Materials*, Advances in Chemistry Series, No. 211, American Chemical Society, Washington DC, 1986.

44. C. W. Frank and M. A. Gashgari, *Macromolecules*, **12**, 163(1979).

45. S. Krause, *J. Polym. Sci.*, *A-2*, **7**, 249(1969).

46. S. Krause, *Macromolecules*, **3**, 84(1970).

47. S. Krause, in *Colloidal and Morphological Behavior of Block and Graft Copolymers*, G. E. Molair, Ed., Plenum, New York, 1971.

48. D. R. Paul, *Polym. Mater. Sci. Eng. Prepr.*, **50**, 1(1984).

49. D. J. Meier, *J. Polym. Sci.*, **26C** 81(1969).

50. D. J. Meier, *Polym. Preprints*, **11**, 400(1970).

51. M. Matsuo, *Jpn. Plastics*, **2**, 6(1968).

52. W. A. Combellick, in *Encyclopedia of Polymer Science and Engineering*, Vol. 2, Wiley, New York, 1985, p. 176.

53. J. Crank and G. S. Park, *Diffusion in Polymers*, Academic Press, Orlando, 1968.

54. C. E. Rogers and D. Machin, *Crit. Rev. Macromol. Sci.*, **1**, 245(1972).

55. A. Fick, *Ann. Phys.* (*Leipzig*), **170**, 59(1855).

56. P. Meares, *Polymers: Structure and Bulk Properties*, Van Nostrand, New York, 1965, Chap. 12.

57. G. van Amerongen, *J. Appl. Phys.*, **17**, 972(1946).

58. G. J. van Amerongen, *J. Polym. Sci.*, **5**, 307(1950).

59. V. T. Stannett, W. J. Koros, D. R. Paul, H. K. Lonsdale, and R. W. Baker, *Adv. Polym. Sci.*, **32**, 69(1979).

60. S. Pauly, in *Polymer Handbook*, 3rd ed., J. Brandrup and E. H. Immergut, Eds., Wiley-Interscience, New York, 1989, Sec. VI, p. 435.

61. W. J. Koros and M. W. Hellums, in *Encyclopedia of Polymer Science and Engineering*, 2nd ed., Supplement Volume, Wiley, New York, 1989, p. 724.

62. D. W. Breck, *Zeolite Molecular Sieves*, Wiley, New York, 1974, Chap. 8.

63. R. T. Chern, W. J. Koros, H. B. Hopfenberg, and V. T. Stannett, in *Materials Science of Synthetic Membranes*, D. R. Lloyd, Ed., ACS Symposium Series No. 269, ACS Books, Washington, DC, 1985.

64. S. Arizzi and U. W. Suter, *Polym. Mater. Sci. Eng. Prepr.*, **61**, 481(1989).

65. C. T. Reinhart, R. W. Korsmeyer, and N. A. Peppas, *Int. J. Pharm. Tech. Prod. Mfr.*, **2**(2), 9(1981).

66. N. A. Peppas and W. H. Yang, *Contact Introcular Lens Med. J.*, **7**, 300(1981).

67. J. W. Vanderhoff, in *Vinyl Polymerization*, Vol. 1, Part II, Chap. 1, G. E. Ham, Ed., Marcel Dekker, New York, 1969.

68. G. W. Poehlein, in *Encyclopedia of Polymer Science and Engineering*, 2nd ed., Vol. 6, Wiley, New York, 1986.

69. D. C. Blackley, in *Encyclopedia of Polymer Science and Engineering*, 2nd ed., Vol. 8, J. I. Kroschwitz, Ed., Wiley, New York, 1987.

GENERAL READING

F. W. Billmeyer, Jr., *Textbook of Polymer Science*, 3rd ed., Wiley-Interscience, New York, 1984.

J. Comyn, Ed., *Polymer Permeability*, Elsevier, London, 1985.

J. Crank and G. S. Park, *Diffusion in Polymers*, Academic Press, Orlando, 1968.

P. J. Flory, *Principles of Polymer Chemistry*, Cornell University Press, Ithaca, NY, 1953.

W. J. Koros, Ed., *Barrier Polymers and Structures*, ACS Symposium Series 423, ACS Books, Washington, DC, 1990.

J. A. Manson and L. H. Sperling, *Polymer Blends and Composites*, Plenum, New York, 1976.

J. E. Mark, A. Eisenberg, W. W. Graessley, L. Mandelkern, and J. L. Koenig, *Physical Properties of Polymers*, American Chemical Society, Washington, DC, 1984.

P. Meares, Ed., *Membrane Separation Processes*, Elsevier, Amsterdam, 1976.

O. Olabisi, L. M. Robeson, and M. T. Shaw, *Polymer–Polymer Miscibility*, Academic Press, Orlando, 1979.

D. R. Paul and S. Newman, Eds., *Polymer Blends*, Vols. 1 and 2, Academic Press, Orlando, 1978.

L. H. Sperling, *Interpenetrating Polymer Networks and Related Materials*, Plenum, New York, 1981.

STUDY PROBLEMS

1. Why do most polymer blends exhibit lower critical solution temperatures, rather than upper critical solution temperatures?

2. A sample of polystyrene–*block*–polybutadiene has block molecular weights of 20,000 g/mol and 80,000 g/mol, respectively. What is the diameter of the polystyrene domains, assuming spheres? How many blocks are in one of these domains? How many of these domains are there per cm^3?

3. A child's polybutadiene balloon, filled with helium, is accidently released into the air. What factors control its rate of rise into the atmosphere and its eventual fall back to earth? Assuming ground level is at 25°C and 760 mm Hg pressure, plot the projected rise and fall of the balloon, making any realistic assumptions required. How long will it stay in the air?

4. What is the analytical expression for χ_{blend} for the general system of two statistical copolymers, $(A_x B_{1-x})_n/(C_y D_{1-y})_{n'}$?

5. Based on equation-of-state concepts, how does the density of polystyrene vary (at 25°C) with hydrostatic pressure?

6. Polystyrene and polybutadiene are glassy and rubbery, respectively. They are substantially immiscible over their entire composition range. Show how both soft (low modulus) and stiff (high modulus) materials may be made from a 50/50 composition.

7. Obtain a large plastic soft drink bottle and measure the thickness of its walls. Assuming it is made of poly(ethylene terephthalate), what is the estimated length of time for the level of carbonation to decrease by a factor of 2? (Afterward, make sure the bottle is recycled!) Why isn't low-density polyethylene used for this purpose? (Pressure in fresh soda bottles is around 10 Atm.)

8. Obtain a sample of latex paint. What is its composition? Coat it on a piece of wood or cardboard and observe as it dries. Touch or rub it periodically with a paper towel. Does the film become strong immediately? Later? Why?

9. Figure 4.13 illustrates the effects of miscibility on excimer fluorescence. (a) If we assume that I_D/I_M arises solely from intermolecular interactions, what value is prediced for total phase separation? (b) If the phase-separated value of I_D/I_M is assumed to be 10, and the system is completely mixed at $\Delta\delta = 0$, what fraction of the excimer fluorescence arises from intramolecular contacts?

5

THE AMORPHOUS STATE

The bulk state, sometimes called the condensed or solid state, includes both amorphous and crystalline polymers. As opposed to polymer solutions, generally there is no solvent present. This state comprises polymers as ordinarily observed, such as plastics, elastomers, fibers, adhesives, and coatings.

While amorphous polymers do not contain any crystalline regions, "crystalline" polymers generally are only semicrystalline, containing appreciable amounts of amorphous material. When a crystalline polymer is melted, the melt is amorphous. In treating the kinetics and thermodynamics of crystallization, the transformation from the amorphous state to the crystalline state and back again is constantly being considered. The subjects of amorphous and crystalline polymers are treated in the next two chapters. This will be followed by a discussion of liquid crystalline polymers, Chapter 7. Although polymers in the bulk state may contain plasticizers, fillers, and other components, this chapter emphasizes the polymer molecular organization itself.

A few definitions are in order. Depending on temperature and structure, amorphous polymers exhibit widely different physical and mechanical behavior patterns. At low temperatures, amorphous polymers are glassy, hard, and brittle. As the temperature is raised, they go through the glass–rubber transition. The glass transition temperature (T_g) is defined as the temperature at which the polymer softens because of the onset of long-range coordinated molecular motion. This is the subject of Chapter 8.

Above T_g, cross-linked amorphous polymers exhibit rubber elasticity. An example is styrene–butadiene rubber (SBR), widely used in materials ranging from rubber bands to automotive tires. Rubber elasticity is treated in Chapter 9. Linear amorphous polymers flow above T_g.

Polymers that cannot crystallize usually have some irregularity in their structure. Examples include the atactic vinyl polymers and statistical copolymers.

5.1 THE AMORPHOUS POLYMER STATE

5.1.1 Solids and Liquids

An amorphous polymer does not exhibit a crystalline x-ray diffraction pattern, and it does not have a first-order melting transition. If the structure of crystalline polymers is taken to be regular or ordered, then by difference, the structure of amorphous polymers contains greater or lesser amounts of disorder.

The older literature often referred to the amorphous state as a liquid state. Water is a noncrystalline (amorphous) condensed substance and is surely a liquid. However, polymers such as polystyrene or poly(methyl methacrylate) at room temperature are glassy, taking months or years for significant creep or flow. By contrast, skyscrapers are also undergoing creep (or flow), becoming measurably shorter as the years pass, as the steel girders creep (or flow). Today, amorphous polymers in the glassy state are better called amorphous solids.

Above the glass transition temperature, if the polymer is amorphous and linear, it will flow, albeit the viscosity may be very high. Such materials are liquids in the modern sense of the term. It should be noted that the glass transition itself is named after the softening of ordinary glass, an amorphous inorganic polymer.

If the polymer is crystalline, it is clearly a solid. Above the melting temperature, it becomes amorphous. Since the melting temperature is always above the glass transition temperature, if the polymer is linear, it becomes a liquid.

If the polymer is cross-linked, then it never flows, and never really becomes a liquid, but remains a soft amorphous solid. An example is a rubber band.

5.1.2 Possible Residual Order in Amorphous Polymers?

As a point of focus, the evidence for and against partial order in amorphous polymers is presented. On the simplest level, the structure of bulk amorphous polymers has been likened to a pot of spaghetti, where the spaghetti strands weave randomly in and out among each other. The model would be better if the strands of spaghetti were much longer, because by ratio of length to diameter, spaghetti more resembles wax chain lengths than it does high polymers.

The spaghetti model provides an entry into the question of residual order in amorphous polymers. An examination of relative positions of adjacent strands shows that they have short regions where they appear to lie more or less parallel. One group of experiments finds that oligomeric polymers also exhibit similar parallel regions (1, 2). Accordingly, the chains appear to lie parallel for short runs because of space-filling requirements, permitting a higher density (3). This point, the subject of much debate, is discussed further later.

Questions of interest to amorphous state studies include the design of critical experiments concerning the shape of the polymer chain, the estimation of type and extent of order or disorder, and the development of models suitable for physical and mechanical applications. It must be emphasized that our knowledge of the amorphous state remains very incomplete, and that this and other areas of polymer science are the subjects of intensive research at this time. Pechhold and Grossmann (4) capture the spirit of the times exactly:

> Our current knowledge about the level of order in amorphous polymers should stimulate further development of competing molecular models, by making their suppositions more precise in order to provide a bridge between their microscopic structure description and the understanding of macroscopic properties, thereby predicting effects which might be proved experimentally.

The subject of structure in amorphous polymers has been entensively reviewed (5–10) and has been the subject of two recent symposia (11, 12).

5.2 EXPERIMENTAL EVIDENCE REGARDING AMORPHOUS POLYMERS

The experimental methods used to characterize amorphous polymers may be divided into those that measure relatively short-range interactions (nonrandom versus random chain positions) (13), below about 20 Å, and those that measure longer-range interactions. In the following paragraphs, the role of these several techniques will be explored. The information obtainable from these methods is summarized in Table 5.1.

5.2.1 Short-Range Interactions in Amorphous Polymers

Methods that measure short-range interactions can be divided into two groups: those that measure the orientation or correlation of monomer residues along the *axial* direction of a chain, and those that measure the order between chains, in the *radial* direction. Figure 5.1 illustrates the two types of measurements.

One of the most powerful experimental methods of determining short-range order in polymers utilizes birefringence (5a). Birefringence measures order in the axial direction. The birefringence of a sample is defined by

$$\Delta n = n_1 - n_2 \tag{5.1}$$

Table 5.1 Selected studies of the amorphous state

Methods	Information Obtainable	Principal Findings	Reference
A. Short-Range Interactions			
Stress–optical coefficient	Orientation of segments in isolated chain	Orientation limited to 5–10 Å	(a)
Depolarized light-scattering	Segmental orientation correlation	2–3 $-CH_2-$ units along chain correlated	(b)
Magnetic birefringence	Segmental orientation correlation	Orientation correlations very small	(b)
Raman scattering	Trans and gauche populations	Little or no modification in chain conformation initiated by intermolecular forces	(b)
NMR relaxation	Relaxation times	Small fluctuating bundles in the melt	(c)
Small-angle X-ray scattering	Density variations	Amorphous polymers highly homogeneous; thermal fluctuations predominate	(d)
B. Long-Range Interactions			
Small-angle neutron scattering	Conformation of single chains	Radius of gyration the same in melt as in θ-solvents	(e, f)
Electron microscopy	Surface inhomogeneities	Nodular structures of 50–200 Å in diameter	(g, h)
Electron diffraction and wide-angle x-ray diffraction	Amorphous halos	Bundles of radial dimension = 25 Å and axial dimension = 50 Å, but order may extend to only one or two adjacent chains	(i, j)
C. General			
Enthalpy relaxation	Deviations from equilibrium state	Changes not related to formation of structure	(k)
Density	Packing of chains	Density in the amorphous state is about 0.9 times the density in the crystalline state	(l, m)

References: (a) R. S. Stein and S. D. Hong, *J. Macromol. Sci. Phys.*, **B12**(11), 125(1976). (b) E. W. Fischer, G. R. Strobl, M. Dettenmaier, M. Stamm, and N. Steidle, *Faraday Discuss. Chem. Soc.*, **68**, 26(1979). (c) W. L. F. Golz and H. G. Zachmann, *Makromol. Chem.*, **176**, 2721(1975). (d) D. R. Uhlmann, *Faraday Discuss. Chem. Soc.*, **68**, 87(1979). (e) H. Benoit, *J. Macromol. Sci. Phys.*, **B12**(1), 27(1976). (f) G. D. Wignall, D. G. H. Ballard, and J. Schelten, *J. Macromol. Sci. Phys.*, **B12**(1), 75(1976). (g) G. S. Y. Yeh, *Crit. Rev. Macromol. Sci.*, **1**, 173(1972). (h) R. Lam and P. H. Geil, *J. Macromol. Sci. Phys.*, **B20**(1), 37(1981). (i) Yu. K. Ovchinnikov, G. S. Markova, and V. A. Kargin, *Vipokomol. Soyed. AII*(2), 329(1969). (j) R. Lovell, G. R. Mitchell, and A. H. Windle, *Faraday Discuss. Chem. Soc.*, **68**, 46(1979). (k) S. E. B. Petrie, *J. Macromol. Sci. Phys.*, **B12**(2), 225(1976). (l) R. E. Robertson, *J. Phys. Chem.*, **69**, 1575(1965). (m) R. F. Boyer, *J. Macromol. Sci. Phys.*, **B12**, 253 (1976).

Figure 5.1   Schematic diagram illustrating the axial and radial correlation directions.

where n_1 and n_2 are the refractive indices for light polarized in two directions 90° apart. If a polymer sample is stretched, n_1 and n_2 are taken as the refractive indices for light polarized parallel and perpendicular to the stretching direction.

The anisotropy of refractive index of the stretched polymer can be demonstrated by placing a thin film between crossed polaroids. The field of view is dark before stretching, but vivid colors develop as orientation is imposed. For stretching at 45° to the polarization directions, the fraction of light transmitted is given by (5a)

$$\mathbf{T} = \sin^2\left(\frac{\pi d \, \Delta n}{\lambda_0}\right) \tag{5.2}$$

where d represents the thickness and λ_0 represents the wavelength of light in vacuum.

By measuring the transmitted light quantitatively, the birefringence is obtained. The birefringence is related to the orientation of molecular units such as mers, crystals, or even chemical bonds by

$$\Delta n = \frac{2}{9}\pi \frac{(\bar{n}^2 + 2)^2}{\bar{n}} \sum_i (b_1 - b_2)_i f_i \tag{5.3}$$

where f_i is an orientation function of such units given by

$$f_i = \frac{3\cos^2\theta_i - 1}{2} \tag{5.4}$$

where θ_i is the angle that the symmetry axis of the unit makes with respect to the stretching direction, $\bar{n}$ is the average refractive index, and b_1 and b_2 are the polarizabilities along and perpendicular to the axes of such units.

Equation (5.4) contains two important solutions for fibers and films:

$$\theta = 0°, \text{ perfect orientation}$$
$$\theta = 54°, \text{ zero orientation}$$

Many commercial fibers such as nylon or rayon will have θ equal to about 5°.

The stress–optical coefficient (SOC) is a measure of the change in birefringence on stretching a sample under a stress σ (14),

$$\text{SOC} = \frac{\Delta n}{\sigma} \tag{5.5}$$

If the polymer is assumed to obey rubbery elasticity relations (see Chapter 9), then

$$\text{SOC} = \frac{\Delta n}{\sigma} = \frac{2\pi}{45kT} \frac{(\bar{n}^2 + 2)^2}{\bar{n}} (b_1 - b_2) \tag{5.6}$$

The change in birefringence that occurs when an amorphous polymer is deformed yields important information concerning the state of order in the amorphous solid. It should be emphasized that the theory expressed in equation (5.6) involves the orientation of segments within a single isolated chain, with no correlation in orientation of segments on different chains. From an experimental point of view, it has been found that the strain–optical coefficient (STOC) is independent of the extension (15) but that the SOC is not.

The anisotropy of a segment is given by $b_1 - b_2$. Experiments carried out by Stein and Hong (14) on this quantity as a function of swelling and extension show no appreciable changes, leading to the conclusion that the order within a chain (axial correlation) does not change beyond a range of 5–10 Å, comparable with the range of ordering found for low-molecular-weight liquids.

Depolarized light-scattering (DPS) is a related technique whereby the intensity of scattered light is measured when the sample is irradiated by visible light. During this experiment, the sample is held between crossed Nicols. Studies on DPS on *n*-alkane liquids (13) reveal that there is a critical chain length of 8–9 carbons, below which there is no order in the melt. For longer chains, only 2–3 — CH_2 — units in one chain are correlated with regard to their orientation, indicating an extremely weak orientational correlation.

Other electromagnetic radiation interactions with polymers useful for the study of short-range interactions in polymers include:

1. *Rayleigh scattering*: elastically scattered light, usually measured as a function of scattering angle.
2. *Brillouin scattering*: in essence a Doppler effect, which yields small frequency shifts.
3. *Raman scattering*: an inelastic process with a shift in wavelength due to chemical absorption or emission.

Results of measurements utilizing SOC, DPS, and other short-range experimental methods such as magnetic birefringence, Raman scattering, Brillouin scattering, NMR relaxation, and small-angle x-ray scattering are summarized in Table 5.1. The basic conclusion is that intramolecular orientation is little affected by the presence of other chains in the bulk amorphous state. The extent of order indicated by these techniques is limited to at most a few tens of angstroms, approximately that which was found in ordinary low-molecular-weight liquids.

5.2.2 Long-Range Interactions in Amorphous Polymers

5.2.2.1 Small-Angle Neutron Scattering

The long-range interactions are more interesting from a polymer conformation and structure point of view. The most powerful of the methods now available is small-angle neutron scattering (SANS). For these experiments, the wave nature of neutrons is utilized. Applied to polymers, SANS techniques can be used to determine the actual chain radius of gyration in the bulk state (16–22).

The basic theory of SANS follows the development of light-scattering (see Section 3.6.1). For small-angle neutron scattering, the weight-average molecular weight, M_w, and the *z*-average radius of gyration, R_g, may be determined (16, 17):

$$\frac{Hc}{R(\theta) - R(\text{solvent})} = \frac{1}{M_w P(\theta)} + 2A_2 c \qquad (5.7)$$

where $R(\theta)$ is the scattering intensity known as the "Rayleigh ratio,"

$$R(\theta) = \frac{I_\theta \omega^2}{I_0 V_s} \tag{5.8}$$

where ω represents the sample-detector distance, V_s is the scattering volume, and I_θ/I_0 is the ratio of scattered radiation intensity to the initial intensity (18–20). The quantity $P(\theta)$ is the scattering form factor, identical to the form factor used in light-scattering formulations [see equation (3.49)]. The formulation for $P(\theta)$, originally derived by Peter Debye (24a), forms one of the mainspring relationships between physical measurements in both the dilute solution and solid states and in the interpretation of the data. For very small particles or molecules, $P(\theta)$ equals unity. In both equation (5.7) (explicit) and equation (5.8) (implicit), the scattering intensity of the solvent or background must be subtracted.

In SANS experiments of the type of interest here, a deuterated polymer is dissolved in an ordinary hydrogen-bearing polymer of the same type (or vice versa). The calculations are simplified if the two polymers have the same molecular weight. The background to be subtracted originates from the scattering of the protonated species, and the coherent scattering of interest originates from the dissolved deuterated species. The quantity H in equation (5.7) was already defined in equation (3.56) for neutron scattering. SANS instrumentation has evolved through several generations, as delineated in Table 5.2.

As currently used (23, 24), the coherent intensity in a SANS experiment is described by the cross section, $d\Sigma/d\Omega$, which is the probability that a neutron will be scattered in a solid angle, Ω, per unit volume of the sample. This cross section, which is normally used to express the neutron scattering power of a sample, is identical with the quantity R defined in equation (5.8).

Table 5.2 Evolution of SANS instrumentation

Method	Location	Comments
Long flight path	(a) ILL, Grenoble (b) Oak Ridge (c) Jülich	Inverse square distance law means long experimental times
Long wavelength	NIST	Neutrons cooled via liquid He or H_2
Time-of-flight[a] (TOF)	Los Alamos Nat. Lab	"White" neutrons, liquid H_2, pulsed source

[a]Pulsed neutrons are separated via TOF according to wavelength. TOF is to long flight path as FT-IR is to IR.

Then it is convenient to express equation (5.7) as

$$\frac{C_N}{d\Sigma/d\Omega} = \frac{1}{M_w P(\theta)} \tag{5.9}$$

where C_N, the analog of H, may be expressed

$$C_N = \frac{(a_H - a_D)^2 N_a \rho (1 - X) X}{M_p^2} \tag{5.10}$$

The quantities a_H and a_D are the scattering length of a normal protonated and deuterated (labeled) structural unit (mer), and X is the mole fraction of labeled chains. Thus C_N contains the concentration term as well as the "optical" constants. The quantities a_H and a_D are calculated by adding up the scattering lengths of each atom in the mer (see Table 5.3). In the case of high dilution, the quantity $(1 - X)X$ reduces to the concentration c, as in equation (5.7).

After rearranging, equation (5.7) becomes

$$\left[\frac{d\Sigma}{d\Omega}\right]^{-1} = \frac{1}{C_N M_w}\left(1 + \frac{K^2 R_g^2}{3} + \cdots\right) \tag{5.11}$$

Thus the mean square radius of gyration, R_g^2, and the polymer molecular weight, M_w, may be obtained from the ratio of the slope to the intercept and the intercept, respectively, of a plot of $[d\Sigma/d\Omega]^{-1}$ versus K^2, see Section 3.6.1, especially equation (3.54). If $A_2 = 0$, this result is satisfactory. For finite A_2 values, a second extrapolation to zero concentration is required (see below).

A problem in neutron scattering and light-scattering alike stems from the fact that R_g is a z-average quantity, whereas the molecular weight is a weight-average quantity. The preferred solution has been to work with nearly

Table 5.3   Scattering lengths of elements (18, 19)

Element	Coherent Scattering Length[a] $b \times 10^{12}$ cm
Carbon, ^{12}C	0.665
Oxygen, ^{16}O	0.580
Hydrogen, ^{1}H	−0.374
Deuterium, ^{2}H	0.667
Fluorine, ^{19}F	0.560
Sulfur, ^{32}S	0.280

[a]Here $a = \Sigma_i b_i$.

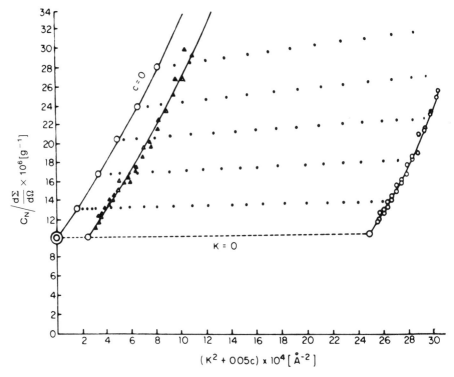

Figure 5.2 Small-angle neutron scattering of polyprotostyrene dissolved in polyde-uterostyrene. A Zimm plot with extrapolations to both zero angle and zero concentration. Note that the second virial coefficient is zero, because polystyrene is essentially dissolved in polystyrene (25).

monodisperse polymer samples, such as prepared by anionic polymerization. If the molecular weight distribution is known, an approximate correction can be made.

Typical data for polyprotostyrene dissolved in polydeuterostyrene are shown in Figure 5.2 (25). Use is made of the Zimm plot, which allows simultaneous plotting of both concentration and angular functions for a more compact representation of the data (26).

From data such as presented in Figure 5.2, both R_g and M_w may be calculated. The results are tabulated in Table 5.3. Also shown in Table 5.3 are corresponding data obtained by light-scattering in Flory θ-solvents, where the conformation of the chain is unperturbed because the free energies of solvent–polymer and polymer–polymer interactions are all the same (see Section 3.3).

Values of $(R_g^2/M)^{1/2}$ are shown in Table 5.4 because this quantity is independent of the molecular weight when the chain is unperturbed (27), being a constant characteristic of each polymer. An examination of Table 5.4

Table 5.4 Molecular dimensions in bulk polymer samples (18)

Polymer	State of Bulk	SANS Bulk	Light-Scattering θ-Solvent	SAX	Reference
		$\left(R_g^2/M_w\right)^{1/2} \dfrac{\text{Å} \cdot \text{mol}^{1/2}}{\text{g}^{1/2}}$			
Polystyrene	Glass	0.275	0.275	0.27 (i)	(a)
Polystyrene	Glass	0.28	0.275	—	(b)
Polyethylene	Melt	0.46	0.45	—	(c)
Polyethylene	Melt	0.45	0.45	—	(d)
Poly(methyl methacrylate)	Glass	0.31	0.30	—	(e)
Poly(ethylene oxide)	Melt	0.45	—	—	(f)
Poly(vinyl chloride)	Glass	0.30	0.37	—	(g)
Polycarbonate	Glass	0.457	—	—	(h)

References: (a) J. P. Cotton, D. Decker, H. Benoit, B. Farnoux, J. Higgins, G. Jannink, R. Ober, C. Picot, and J. desCloizeaux, *Macromolecules*, **7** 863(1974). (b) G. D. Wignall, D. G. Ballard, and J. Schelten, *Eur. Polym. J.*, **10**, 861(1974). (c) J. Schelten, D. G. H. Ballard, G. Wignall, G. Longman, and W. Schmatz, *Polymer*, **17**, 751(1976). (d) G. Lieser, E. W. Fischer, and K. Ibel, *J. Polym. Sci. Polym. Lett. Ed.*, **13**, 39(1975). (e) R. G. Kirste, W. A. Kruse, and K. Ibel, *Polymer*, **16**, 120(1975). (f) G. Allen, *Proc. R. Soc. Lond., Ser. A*, **351**, 381(1976). (g) P. Herchenroeder and M. Dettenmaier, unpublished (1977). (h) D. G. H. Ballard, A. N. Burgess, P. Cheshire, E. W. Janke, A. Nevin, and J. Schelten, *Polymer*, **22**, 1353(1981). (i) H. Hayashi, F. Hamada, and A. Nakajima, *Macromolecules*, **9**, 543(1976).

reveals that the values in θ-solvents and in the bulk state are identical within experimental error. This important finding confirms earlier theories (27) that these two quantities ought to be equal, since under these conditions the polymer chain theoretically is unable to distinguish between a solvent molecular and a polymer segment with which it may be in contact.[†] Since it was believed that polymer chains in dilute solution were random coils, this finding provided powerful evidence that random coils also existed in the bulk amorphous state.

The reader should note the similarities between $R_g = KM^{1/2}$ and the Brownian motion relationship, $X = k't^{1/2}$, where X is the average distance traversed.

5.2.2.2 Electron and X-Ray Diffraction

Under various conditions, crystalline substances diffract x-rays and electrons to give spots or rings. According to Bragg's law, these can be interpreted as

---

[†]This finding was predicted in 1953 by P. J. Flory, 20 years before it was confirmed experimentally.

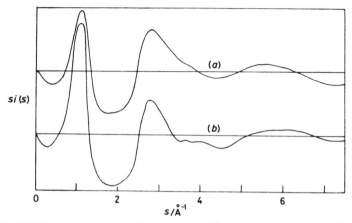

Figure 5.3 WAXS data on polytetrafluorethylene: (a) experimental data and (b) theory. Model is based on a disordered helix arranged with fivefold packing in a 24-Å diameter cylinder.

interplanar spacings. Amorphous materials, including ordinary liquids, also diffract x-rays and electrons, but the diffraction is much more diffuse, sometimes called halos. For low-molecular-weight liquids, the diffuse halos have long been interpreted to mean that the nearest-neighbor spacings are slightly irregular and that after two or three molecular spacings all sense of order is lost. The situation is complicated in the case of polymers because of the presence of long chains. Questions to be resolved center about whether or not chains lie parallel for some distance, and if so, to what extent (28–31).

X-ray diffraction studies are frequently called wide-angle x-ray scattering, or WAXS. Typical data are illustrated in Figure 5.3 for polytetrafluorethylene (30). The scattering maxima indicate chain spacing distances. The reduced intensity data are plotted as a function of angle,[†] $s = 4\pi \sin \theta / \lambda$, which is sometimes called inverse space, because the dimensions are Å^{-1}. The diffracted intensity is plotted in the y axis multiplied by the quantity s to permit the features to be more evenly weighted. Lovell et al. (30) fitted the experimental data with various theoretical models, also illustrated in Figure 5.3.

In analyzing WAXS data, the two different molecular directions must be borne in mind: (a) conformational orientation in the axial direction, which is a measure of how ordered or straight a given chain might be, and (b) organization in the radial direction, which is a direct measure of intermolecular order. WAXS measures both parameters. Lovell et al. (30) concluded from their study that the axial direction of molten polyethylene could be described by a chain with three rotational states, 0^0 and $\pm 120^0$, with an

[†]Variously, K is used by SANS experimenters for the angular function.

Table 5.5   Interchain spacing in selected amorphous polymers

Polymer	Spacing, Å	Reference
Polyethylene	5.5	(a)
Silicone rubber	9.0	(a)
Polystyrene	10.0	(b)
Polycarbonate	4.8	(c)

References: (a) Y. K. Ovchinnikov, G. S. Markova, and V. A. Kargin, *Vysokomol. Soyed.*, **A11**, 329(1969). (b) A. Bjornhaug, O. Ellefsen, and B. A. Tonnesen, *J. Polym. Sci.*, **12**, 621(1954). (c) A. Siegmann and P. H. Geil, *J. Macromol. Sci. (Phys.)*, **4**(2), 239(1970).

average trans sequence length of 3–4 backbone bonds. The best radial packing model consisted of flexible chains arranged in a random manner (see Figure 5.3*a* curve *b*).

Polytetrafluoroethylene, on the other hand, was found to have more or less straight chains in the axial direction for distances of at least 24 Å (30). Many other studies have also shown that this polymer has extraordinarily stiff chains, because of its extensive substitution. In the radial direction, a model of parallel straight-chain segments was strongly supported by the WAXS data, although the exact nature of the packing and the extent of chain disorder are still the subject of current research. Poly(methyl methacrylate) and polystyrene were found to have a level of order intermediate between polyethylene and polytetrafluoroethylene.

The first interchain spacing of typical amorphous polymers is shown in Table 5.5. The greater interchain spacing of polystyrene and silicone rubbers is in part caused by bulky side groups compared with polyethylene.

The interpretation of diffraction data on amorphous polymers is currently a subject of debate. Ovchinnikov et al. (28, 31) interpreted their electron diffraction data to show considerable order in the bulk amorphous state, even for polyethylene. Miller and co-workers (32, 33) found that spacings increase with the size of the side groups, supporting the idea of local order in amorphous polymers. Fischer et al. (29), on the other hand, found that little or no order fits their data best. Schubach et al. (34) take an intermediate position, finding that they were able to characterize first- and second-neighbor spacings for polystyrene and polycarbonate, but no further.

5.2.2.3   General Properties

Two of the most important general properties of the amorphous polymers are the density and the excess free energy due to nonattainment of equilibrium. The latter shows mostly smooth changes on relaxation and annealing (35) and is not suggestive of any particular order. Changes in enthalpy on relaxation and annealing are touched on in Chapter 8.

However, many polymer scientists have been highly concerned with the density of polymers (3, 29). For many common polymers, the density of the amorphous phase is approximately 0.85–0.95 that of the crystalline phase (3,9). Returning to the spaghetti model, some scientists think that the polymer chains have to be organized more or less parallel over short distances, or the experimental densities cannot be attained. Others (29) have pointed out that different statistical methods of calculation lead, in fact, to satisfactory agreement between the experimental densities and a more random arrangement of the chains.

Using computer simulation of polymer molecular packing, Weber and Helfand (35a) studied the relative alignment of polyethylene chains expected from certain models. They calculated the angle between pairs of chords of chains (from the center of one bond to the center of the next), which showed a small but clear tendency toward alignment between closely situated molecular segments, and registered well-developed first and second density peaks. However, no long-range order was observed.

Table 5.1 summarized the several experiments designed to obtain information about the organization of polymer chains in the bulk amorphous state,

Table 5.6   Major order – disorder arguments in amorphous polymers

Order	Disorder
Conceptual difficulties in dense packing without order (a, b)	Rubber elasticity of polymer networks (e, f)
Appearance of nodules (b, c)	Absence of anomalous thermodynamic dilution effects (g)
Amorphous halos intensifying on equatorial plane during extension (d)	Radii of gyration the same in bulk as in θ-solvents (h)
Nonzero Mooney–Rivlin C_2 constants (a)	Fit of $P(\theta)$ for random coil model to scattering data (i)
Electron diffraction (l) lateral order to 15–20 Å	Rayleigh–Brillouin scattering, x-ray diffraction (j, k), stress–optical coefficient, etc. studies showing only modest (if any) short-range order (j, k, m)

References: (a) R. F. Boyer, *J. Macromol. Sci. Phys.*, **B12**(2), 253(1976). (b) G. S. Y. Yeh, *Crit. Rev. Macromol. Sci.*, **1**, 173(1972). (c) P. H. Geil, *Faraday Discuss. Chem. Soc.*, **68**, 141(1979); but see S. W. Lee, H. Miyaji, and P. H. Geil, *J. Macromol. Sci.*, *Phys.*, **B22**(3), 489(1983); and D. R. Uhlmann, *Faraday Discuss. Chem. Soc.*, **68**, 87(1979). (d) S. Krimm and A. V. Tobolsky, *Text. Res. J.*, **21**, 805(1951). (e) P. J. Flory, *J. Macromol. Sci. Phys.*, **B12**(1), 1(1976). (f) P. J. Flory, *Faraday Discuss. Chem. Soc.*, **68**, 14(1979). (g) P. J. Flory, *Principles of Polymer Chemistry*, Cornell University Press, Ithaca, NY, 1953. (h) J. S. Higgins and R. S. Stein, *J. Appl. Crystallog.* **11**, 346(1978). (i) H. Hayashi, F. Hamada, and A. Nakajima, *Macromolecules*, **9**, 543(1976). (j) E. W. Fischer, J. H. Wendorff, M. Dettenmaier, G. Leiser, and I. Voigt-Martin, *J. Macromol. Sci. Phys.*, **B12**(1), 41(1976). (k) D. R. Uhlmann, *Faraday Discuss. Chem. Soc.*, **68**, 87(1979). (l) Yu. K. Ovchinnikov, G. S. Markova, and V. A. Kargin, *Vysokomol. Soyed.*, **A11**(2), 329(1969); *Polym. Sci. USSR*, **11**, 369(1969). (m) R. S. Stein and S. O. Hong, *J. Macromol. Sci. Phys.*, **B12**(1), 125(1976). (n) R. E. Robertson, *J. Phys. Chem.*, **69**, 1575(1965).

both for short- and long-range order and for the general properties. Table 5.6 outlines some of the major order–disorder arguments. Some of these are discussed below. The next section is concerned with the development of molecular models that best fit the data and understanding obtained to date.

5.3 CONFORMATION OF THE POLYMER CHAIN

One of the great classic problems in polymer science has been the conformation of the polymer chain in space. The data in Table 5.4 show that the radius of gyration divided by the square root of the molecular weight is a constant for any given polymer in the Flory θ-state, or in the bulk state. However, the detailed arrangement in space must be determined by other experiments and, in particular, by modeling. The resulting models are important in deriving equations for viscosity, diffusion, rubbery elasticity, and mechanical behavior.

5.3.1 The Freely Jointed Chain

The simplest mathematical model of a polymer chain in space is the freely jointed chain. It has x links, each of length l, joined in a linear sequence with no restrictions on the angles between successive bonds (see Figure 5.4). By analogy with Brownian motion statistics, the root-mean-square end-to-end distance is given by (35–37)

$$\left(\overline{r_f^2}\right)^{1/2} = lx^{1/2} \tag{5.12}$$

where the subscript f indicates free rotation.

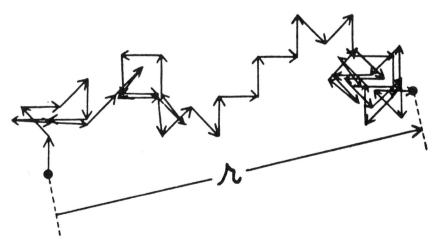

Figure 5.4 A vectorial representation of a freely jointed chain in two dimensions. A random walk of 50 steps (37).

A more general equation yielding the average end-to-end distance of a random coil, r_0, is given by

$$r_0^2 = l^2 x \frac{(1 - \cos\theta)(1 + \cos\phi)}{(1 + \cos\theta)(1 - \cos\phi)} \tag{5.13}$$

where θ is the bond angle between atoms, and ϕ is the conformation angle. This latter is as the angle of rotation for each bond, defined as the angle each bond makes from the plane delineated by the previous two bonds.

However, equation (5.13) still underestimates the end-to-end distance of the polymer chain, omitting such factors as excluded volume. This last arises from the fact that a chain cannot cross itself in space.

The placement of regular bond angles (109° between carbon atoms) expands the chain by a factor of $[(1 - \cos\theta)/(1 + \cos\theta)]^{1/2} = \sqrt{2}$, and the three major positions of successive placement obtained by rotation about the previous bond (two gauche and one trans positions) results in the chain being extended still further. Other short-range interactions include steric hindrances. Long-range interactions include excluded volume, which eliminates conformations in which two widely separated segments would occupy the same space. The total expansion is by a constant C, after squaring both sides of equation (5.12):

$$\overline{r_f^2} = Cl^2 x \tag{5.14}$$

The characteristic ratio $C = \overline{r_f^2}/l^2 x$ varies from about 5 to about 10, depending on the foliage present on the individual chains (see Table 5.7). The values of $l^2 x$ can be calculated by a direct consideration of the bond angles and energies of the various states and a consideration of longer-range interactions between portions of the chain (35b).

5.3.2 The Random Coil

The term "random coil" is often used to describe the unperturbed shape of the polymer chains in both dilute solutions and in the bulk amorphous state.

Table 5.7   Typical values of the characteristic ratio C

Polymer	C
Polyethylene	6.8
Polystyrene	9.85
it-Polyproplyene	5.5
Poly(ethylene oxide)	4.1
Nylon 66	5.9
Polybutadiene, 98% cis	4.75

Source: J. Brandrup and E. H. Immergut, Eds., Polymer Handbook, 2nd ed., Wiley-Interscience, New York, 1975.

In dilute solutions, the random coil dimensions are present under Flory θ-solvent conditions, where the polymer–solvent interactions and the excluded volume terms just cancel each other. In the bulk amorphous state, the mers are surrounded entirely by identical mers, and the sum of all the interactions is zero. Considering mer–mer contacts, the interaction between two distant mers on the same chain is the same as the interaction between two mers on different chains. The same is true for longer chain segments.

In the limit of high molecular weight, the end-to-end distance of the random coil divided by the square root of 6 yields the radius of gyration (Section 3.6.2). Since the x links are proportional to the molecular weight, these relations lead directly to the result that $R_g/M^{1/2}$ is constant.

Of course, there is a distribution in end-to-end distances for random coils, even of the same molecular weight. The distribution of end-to-end distances can be treated by Gaussian distribution functions (see Chapter 9). The most important result is that, for relaxed random coils, there is a well-defined maximum in the frequency of the end-to-end distances, this distance is designated as r_0.

Appendix 5.1 describes the historical development of the random coil.

5.3.3 Models of Polymer Chains in the Bulk Amorphous State

Ever since Hermann Staudinger developed the macromolecular hypothesis in the 1920s (36), polymer scientists have wondered about the spatial arrangement of polymer chains, both in dilute solution and in the bulk. The earliest models included both rods and bedspring-like coils. X-ray and mechanical studies led to the development of the random coil model. In this model, the polymer chains are permitted to wander about in a space-filling way as long as they do not pass through themselves or another chain (excluded-volume theory).

The development of the random coil model by Mark, and the many further developments by Flory (5–7, 37), led to a description of the conformation of chains in the bulk amorphous state. Neutron scattering studies found the conformation in the bulk to be close to that found in the θ-solvents, strengthening the random coil model. On the other hand, some workers suggested that the chains have various degrees of either local or long-range order (38–41).

Some of the better-developed models are described in Table 5.8. They range from the random coil model of Mark and Flory (37) to the highly organized meander model of Pechhold et al. (38). Several of the models have taken an intermediate position of suggesting some type of tighter than random coiling, or various extents of chain folding in the amorphous state (39–41). A collage of the most different models is illustrated in Figure 5.5.

The most important reasons why some polymer scientists are suggesting nonrandom chain conformations in the bulk state include the high amorphous/crystalline density ratio, and electron and x-ray diffraction studies, which suggest lateral order (see Table 5.6). Experiments that most favor the

Table 5.8   Major models of the amorphous polymer state

Principals	Description of Model	Reference
H. Mark and P. J. Flory	Random coil model; chains mutually penetrable and of the same dimension as in θ-solvents	(a, b)
B. Vollmert	Individual cell structure model, close-packed structure of individual chains	(c, d)
P. H. Lindenmeyer	Highly coiled or irregularly folded conformational model, limited chain interpenetration	(e)
T. G. F. Schoon	Pearl necklace model of spherical structural units	(f)
V. A. Kargin	Bundle model, aggregates of molecules exist in parallel alignment	(g)
W. Pechhold	Meander model, with defective bundle structure, with meander-like folds	(h, i)
G. S. Y. Yeh	Folded-chain fringed-micellar grain model. Contains two elements: grain (ordered) domain of quasi-parallel chains, and intergrain region of randomly packed chains	(j)
V. P. Privalko and Y. S. Lipatov	Conformation having folded structures with R_g equaling the unperturbed dimension	(j)
R. Hosemann	Paracrystalline model with disorder within the lamellae (see Fig. 6.38)	(l, m, n)
S. A. Arzhakov	Folded fibril model, with folded chains perpendicular to fibrillar axis	(o)

References: (a) P. J. Flory, *Principles of Polymer Chemistry*, Cornell University Press, Ithaca, NY, 1953. (b) P. J. Flory, *Faraday Discuss. Chem. Soc.*, **68**, 14(1979). (c) B. Vollmert, *Polymer Chemistry*, Springer-Verlag, Berlin, 1973, p. 552. (d) B. Vollmert and H. Stuty, in *Colloidal and Morphological Behavior of Block and Graft Copolymers*, G. E. Molau, Ed., Plenum, New York, 1970. (e) P. H. Lindenmeyer, *J. Macromol. Sci. Phys.*, **8**, 361(1973). (f) T. G. F. Schoon and G. Rieber, *Angew. Makromol. Chem.*, **15**, 263(1971). (g) Y. K. Ovchinnikov, G. S. Markova, and V. A. Kargin, *Polym. Sci. USSR* (Eng. Transl.), **11**, 369(1969); V. A. Kargin, A. I. Kitajgorodskij, and G. L. Slonimskii, *Kolloid-Zh.*, **19**, 131(1957). (h) W. Pechhold, M. E. T. Hauber, and E. Liska, *Kolloid Z. Z. Polym.*, **251**, 818(1973). (i) W. R. Pechhold and H. P. Grossmann, *Faraday Discuss. Chem. Soc.*, **68**, 58(1979). (j) G. S. Y. Yeh, *J. Macromol. Sci. Phys.*, **6**, 451(1972). (k) V. P. Privalko and Yu. S. Lipatov, *Makromol. Chem.*, **175**, 641(1974). (l) R. Hosemann, *J. Polym. Sci.*, **C20**, 1(1967). (m) R. Hosemann, *Colloid Polym. Sci.*, **260**, 864(1982). (n) R. Hosemann, *CRC Crit. Rev. Macromol. Sci.*, **1**, 351(1972). (o) S. A. Arzhakov, N. F. Bakeyev, and V. A. Kabanov, *Vysokomol. Soyed.*, **A15**(5), 1154(1973).

random coil model include small-angle neutron scattering and a host of short-range interaction experiments that suggest little or no order at the local level. Both the random coil proponents and the order-favoring proponents claim points in the area of rubber elasticity, which is examined further in Chapter 9.

The SANS experiments bear further development. As shown above, the radius of gyration (R_g) of the chains is the same in the bulk amorphous state

(a) P. J. Flory

(b) V. P. Privalko and
Y. S. Lipatov

(c) G. S. Y. Yeh

(d) W. Pechhold et al.

Figure 5.5 Models of the amorphous state in pictorial form. (a) Flory's random coil model; the (b) Privalko and Lipatov randomly folded chain conformations; (c) Yeh's folded-chain fringed-micellar model; and (d) Pechhold's meander model. Models increase in degree of order from (a) to (d). *References:* (a) P. J. Flory, *Principles of Polymer Chemistry*, Cornell University Press, Ithaca, NY, 1953. (b) V. P. Privalko and Y. S. Lipatov, *Makromol. Chem.*, **175**, 641(1972). (c) G. S. Y. Yeh, *J. Makoromol. Sci. Phys.*, **6**, 451(1972). (d) W. Pechhold, M. E. T. Hauber, and E. Liska, *Kolloid Z. Z. Polym.*, **251**, 818(1973). W. Pechhold, IUPAC Preprints, 789(1971).

as it is in θ-solvents. However, virtually the same values of R_g are also obtained in rapidly crystallized polymers (42–46), where significant order is known to exist. This finding at first appeared to support the possibility of short-range order of the type suggested by the appearance of x-ray halos. Two points need to be mentioned. (a) A more sensitive indication of random chains is the Debye scattering form factor for random coils [see equations (3.49) and (5.11)]. Plots of $P(\theta)$ versus $\sin^2(\theta/2)$ follow the experimental data over surprisingly long ranges of θ, including regions where the Guinier approximation, implicit in equation (5.11), no longer holds (47, 48). (b) The cases where R_g is the same in the melt as in the crystallized polymer appear to be in crystallization regime III, where chain folding is significantly reduced. (See Section 6.6.2.5.)

A major advantage of the random coil model, interestingly, is its simplicity. By not assuming any particular order, the random coil has become amenable to extensive mathematical development. Thus detailed theories have been developed including rubber elasticity (Chapter 9) and viscosity behavior (Section 3.8), which predict polymer behavior quite well. By difference, little or no analytical development of the other models has taken place, so few properties can be quantitatively predicted. Until such developments have taken place, their absence alone is a strong driving force for the use of the random coil model.

Some of the models may not be quite as far apart as first imagined, however. Privalko and Lipatov have pointed out some of the possible relationships between the random, Gaussian coil, and their own folded-chain model (41) (see Figure 5.5b). As a result of thermal motion, they suggest that both the size and location of regions of short-range order in amorphous polymers depend on the time of observation, assuming that the polymers are above T_g and in rapid motion. The instantaneous conformation of the polymer corresponds to a loosely folded chain. However, when the time of observation is long relative to the time required for molecular motion (see Section 5.4), the various chain conformations will be averaged out in time, yielding radii of gyration more like the unperturbed random coil. For polymers in the glassy state, a similar argument holds, because the very many different chains and their respective conformations replace the argument of a single chain varying its conformation with time.

Clearly, the issue of the conformation of polymer chains in the bulk amorphous state is not yet settled; indeed it remains an area of current research. The vast bulk of research to date strongly suggests that the random coil must be at least close to the truth for many polymers of interest. Points such as the extent of local order await further developments. Thus this book will expound the Mark–Flory theory of the random coil, except where specifically mentioned to the contrary.

5.4 MACROMOLECULAR DYNAMICS

Since the basic notions of chain motion in the bulk state are required to understand much of physical polymer science, a brief introduction is given here. Applications include chain crystallization (to be considered beginning in Chapter 6), the onset of motions in the glass transition region (Chapter 8), and the extension and relaxation of elastomers (Chapters 9 and 10).

Small molecules move primarily by translation. A simple case is of a gas molecule moving in space, following a straight line until hitting another molecule or a wall. In the liquid state, small molecules also move primarily by translation, although the path length is usually only of the order of molecular dimensions.

Polymer motion can take two forms: (a) the chain can change its overall conformation, as in relaxation after strain, or (b) it can move relative to its neighbors. Both motions can be considered in terms of self-diffusion. All such diffusion is a subcase of Brownian motion, being induced by random thermal processes. For center-of-mass diffusion, the center-of-mass distance diffused depends on the square root of time. For high enough temperatures, an Arrhenius temperature dependence is found.

While polymer chains can and do move "sideways" by simple translation, such motion is exceedingly slow for long, entangled chains. This is because the surrounding chains that block sideways diffusion are also long and entangled, and sideways diffusion can only occur by many cooperative motions. Thus polymer chain diffusion is more complex than small-molecule motion, and demands separate theoretical treatment.

5.4.1 The Rouse – Bueche Theory

The first molecular theories concerned with polymer chain motion were developed by Rouse (49) and Bueche (50), and modified by Peticolas (51). This theory begins with the notion that a polymer chain may be considered as a succession of equal submolecules, each long enough to obey the Gaussian distribution function; that is, they are random coils in their own right. These submolecules are replaced by a series of beads connected by springs with the proper Hooke's force constant; see Figure 5.6. The beads also act as universal joints. This model resembles a one-dimensional crystal.

In the development of rubber elasticity theory (Section 9.7.1), it will be shown that the restoring force, f, on a chain or chain portion large enough to be Gaussian, is given by

$$f = \frac{3kT\,\Delta x}{\overline{r^2}} \tag{5.15}$$

where Δx is the displacement and r is the end-to-end distance of the chain or Gaussian segment.

If the beads in Figure 5.6 are numbered $1, 2, 3, \ldots, z$, so that there are z springs and $z + 1$ beads, the restoring force on the ith bead may be written

$$f_i = \frac{-3kT}{a^2}(-x_{i-1} + 2x_i - x_{i+1}), \qquad 1 \le i \le z - 1 \tag{5.16}$$

where f_i represents the force on the ith bead in the x direction, x_i represents the amount by which the bead i has been displaced from its equilibrium position, and a is the end-to-end distance of the segment.

The segments move through a viscous medium (other polymer chains and segments), in which they are immersed. This viscous medium exerts a drag

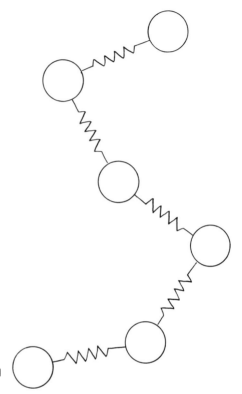

Figure 5.6 Rouse – Bueche bead and spring model of a polymer chain.

force on the system, damping out the motions. It is assumed that the force is proportional to the velocity of the beads, which is equivalent to assuming that the bead behaves exactly as if it were a macroscopic bead in a continuous viscous medium. The viscous force on the ith bead is given by

$$f_i = \rho \left(\frac{dx_i}{dt} \right) \tag{5.17}$$

where ρ is the friction factor.

Zimm (52) advanced the theory by introducing the concepts of Brownian motion and hydrodynamic shielding into the system. One advantage is that the unmeasurable friction factor is replaced by the macroscopic viscosity of the medium. This leads to a matrix algebra solution with relaxation times of

$$\tau_{p,i} = \frac{6\eta_0 M_i^2}{\pi^2 c R T M_w p^2} \tag{5.18}$$

where η_0 is the bulk-melt viscosity, p is a running index, and c is the polymer concentration.

Table 5.9   Rouse – Zimm relaxation times for polystyrene solutions (53)

C, g/cm³	log τ_1, obs	log τ_1, calc.
0.0072	− 2.89	− 2.87
0.0144	− 2.90	− 2.90
0.0286	− 2.50	− 2.50
0.0430	− 2.38	− 2.37
0.0568	− 2.01	− 2.02

Source: R. D. De Mallie et al., *J. Phys. Chem.*, **66**, 536(1962).

The Rouse–Bueche–Zimm theory was verified via the data of De Mallie et al. (53), using nearly monodisperse polystyrene of 267,000 g/mol weight-average molecular weight; see Table 5.9. The quantity τ_1 is the longest relaxation time. In Table 5.9, theory agrees with experiment, especially below 1% concentration. However, only poor agreement is obtained on studies of the bulk melt. The theory describes the relaxation of deformed polymer chains, leading to advances in creep and stress relaxation. While it does not speak about the center-of-mass diffusional motions of the polymer chains, the theory is important because it serves as a precursor to the de Gennes reptation theory, described next.

5.4.2   The de Gennes Reptation Theory[†]

While the Rouse–Bueche theory was highly successful in establishing the idea that chain motion was responsible for creep, relaxation, and viscosity, quantitative agreement with experiment was generally unsatisfactory. More recently, de Gennes (54) introduced his theory of reptation of polymer chains. His model consisted of a single polymeric chain, P, trapped inside a three-dimensional network, G, such as a polymeric gel. The gel itself may be reduced to a set of fixed obstacles—$O_1, O_2, \ldots, O_n \ldots$. His model is illustrated in Figure 5.7 (54). The chain P is not allowed to cross any of the obstacles; however, it may move in a snakelike fashion among them (55).

The snakelike motion is called reptation. The chain is assumed to have certain "defects," each with stored length, b (see Figure 5.8) (54). These defects migrate along the chain in a type of defect current. When the defects move, the chain progresses, as shown in Figure 5.9 (54). The velocity of the nth mer is related to the defect current J_n by

$$\frac{d\vec{r}_n}{dt} = bJ_n \tag{5.19}$$

where $\vec{r}_n$ represents the position vector of the nth mer.

[†]As this book was being readied for press, Dr. de Gennes was awarded the 1991 Nobel Prize in Physics for his work in liquid crystals and polymers. Several sections following discuss his contributions.

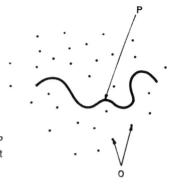

Figure 5.7 A model for reptation. The chain P moves among the fixed obstacles, O, but cannot cross any of them (54).

The reptation motion yields forward motion when a defect leaves the chain at the extremity. The end of the chain may assume various new orientations. In zoological terms, the head of the snake must decide which direction it will go through the bushes. De Gennes assumes that this choice is at random.

Using scaling concepts, de Gennes (54) found that the diffusion coefficient, D, of a chain in the gel depends on the molecular weight M as

$$D \propto M^{-2} \tag{5.20}$$

Numerical values of the diffusion coefficient in bulk systems range from 10^{-12} to 10^{-6} cm^2/sec. In data reviewed by Tirrell (56), polyethylene of 1×10^4 g/mol at 176°C has a value of D near 1×10^{-8} cm^2/sec. Polystyrene of 1×10^5 g/mol has a diffusion coefficient of about 1×10^{-12} cm^2/sec at

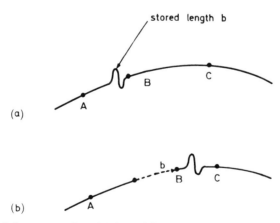

Figure 5.8 Reptation as a motion of defects. (*a*) The stored length *b* moves from A toward C along the chain. (*b*) When the defect crosses mer B, it is displaced by an amount *b* (54).

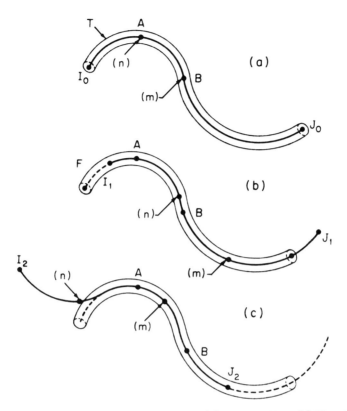

Figure 5.9 The chain is considered within a tube. (*a*) Initial position. (*b*) The chain has moved to the right by reptation. (*c*) The chain has moved to the left, the extremity *I* choosing another path, I_2J_2. A certain fraction of the chain, I_1J_2, remains trapped within the tube at this stage (54).   *Reptation = Snake*

175°C. The inverse second-power molecular weight relationship holds. The temperature dependence can be determined either through activation energies (E_a = 90 kJ/mol for polystyrene and 23 kJ/mol for polyethylene) or through the WLF equation (Chapter 8). Calculations using the diffusion coefficient are illustrated in Appendix 5.2.

The relaxation time, τ, depends on the molecular weight as

$$\tau \propto M^3 \tag{5.21}$$

Continuing these theoretical developments, Doi and Edwards (57) developed the relationship of the dynamics of reptating chains to mechanical properties. In brief, expressions for the rubbery plateau shear modulus, G_N^0, steady-state viscosity, η_0, and the steady-state recoverable compliance, J_e^0, were found to be related to the molecular weight as follows:

$$G_N^0 \propto M^0 \tag{5.22}$$

$$\eta_0 \propto M^3 \tag{5.23}$$

$$J_e^0 \propto M^0 \tag{5.24}$$

An important reason why the modulus and the compliance are independent of the molecular weight is that the number of entanglements (contacts between the reptating chain and the gel) are large for each chain and occur at roughly constant intervals. Experimentally, the viscosity is found to depend on the molecular weight to the 3.4 power (Chapter 10), more rapidly than predicted. The de Gennes theory of reptation as a mechanism for diffusion has also seen applications in the dissolution of polymers, termination by combination in free radical polymerizations, and polymer–polymer welding (56). Graessley (58) reviewed the theory of reptation recently.

5.4.3 Nonlinear Chains

The above discussion models the motion of *linear* chains in a tube. Physical entanglements define a tube of some 50 Å diameter. This permits the easy passage of defects but effectively prevents sideways motion of the chain. According to the reptation theory, the chains wiggle onward randomly from their ends. How do branched, star, and cyclic polymers diffuse?

Two possibilities exist for translational motion in branched polymers. First, one end may move forward, pulling the other end and the branch into the same tube. This process is strongly resisted by the chains as it requires a considerable decrease in entropy to cause a substantial portion of a branch to lie parallel to the main chain in an adjacent tube (59).

Instead, it is energetically cheaper for an entangled branched-chain polymer to renew its conformation by retracting a branch so that it retraces its path along the confining tube to the position of the center mer. Then, it may extend outward again, adopting a new conformation at random; see Figure 5.10 (60). The basic requirement is that the branch not loop around another chain in the process, or it must drag it along also.

(a)   (b)   (c)

Figure 5.10 The basic diffusion steps for a branched polymer. Note motion of mer C, which requires a fully retracted branch before it can take a step into a new topological environment (60).

De Gennes (61) calculated the probability P_1 of an arm of n-mers folding back on itself as

$$P_1 = \exp(-\alpha n / n_c) \qquad (5.25)$$

where n_c is the critical number of mers between physical entanglements and α is a constant. Note that in Section 2.9.3.3 the probability of a chain's two ends being in contact in dilute solution was measured using photophysical methods.

The result is that diffusion in branched-chain polymers is much slower than in linear chains. For rings, diffusion is even more sluggish, because the ring is forced to collapse into a quasilinear conformation in order to have center-of-mass motion. Since many commercial polymers are branched or star-shaped, the viscosity of the polymer is correspondingly increased.

5.4.4 Experimental Methods of Determining Diffusion Coefficients

Two general methods exist for determining the translational diffusion coefficient, D, in polymer melts: (a) by measuring the broadening of concentration gradients as a function of time (in such cases, two portions of polymer, which differ in some identifiable mode, are placed in juxtaposition and the two polymer portions are allowed to interdiffuse) and (b) by measuring the translation of molecules directly using local probes such as NMR.

Diffusion broadening has been the more widely reported. Small-angle neutron scattering (62), forward recoil spectrometry (63), radioactive labeling (64), and infrared absorption methods (65) have been employed, among others. A generalized scheme of sample preparation and analysis is shown in Figure 5.11 (65). Of course, pure labeled polymers may be used as well as the blend. The objective is to measure some characteristic change at the interface.

Bartels et al. (66) reacted H_2 or D_2 with polybutadiene to saturate the double bond, making the equivalent of polyethylene. Films with up to 25 alternating layers of HPB and DPB for the diffusion studies were prepared with layer thicknesses ranging from 3 to 15 μm, chosen to keep homogenization times in the order of a few hours. SANS measurements were made with the incident beam perpendicular to the film surface. The resulting diffusion coefficient on the linear polymer is compared to values obtained on three-armed stars of the same molecular weight in Table 5.10. The stars diffuse almost three orders of magnitude slower than linear polymer, illustrating the detriment that long side chains present to reptation.

Mixtures of deuterated and protonated polystyrene latexes have also been studied by SANS (67, 68). The main advantage is the relatively large surface area presented for interdiffusion, especially if the latex particles are small. Figure 5.12 (68) shows that the interdiffusion distance increases as the square root of time, theoretically predicted.

Figure 5.11   Schematic flowchart showing the stages involved in an interdiffusion experiment (60). The slices are scanned in an IR microdensitometer to obtain the broadened concentration profile, from which D is evaluated.

Of course, there are other methods for measuring interdiffusion in latex-based films. Most recently, Winnik et al. (69–71) employed photophysics technology; see Section 2.9. In this method, latexes are prepared in two different batches. In one batch, the chains contain a "donor" group, while in the other, an "acceptor" group is attached. When the two groups are close to one another, the excitation energy of the donor molecules may be transferred

Table 5.10   Diffusion coefficients of hydrogenated polybutadienes

Method	Shape	T, °C	M_w, g/mol	D, cm²/sec	Ref.
SANS	Linear	125	7.3×10^4	4.8×10^{-11}	(a)
Forward recoil Spectrometry	3-arm	125	7.5×10^4	2.4×10^{-14}	(b)
SANS	3-arm	165	7.5×10^4	1.4×10^{-13}	(c)

References: (a) C. R. Bartels, B. Crist, and W. W. Graessley, *Macromolecules,* **17,** 2702(1984). (b) B. Crist, P. F. Green, R. A. L. Jones, and E. J. Kramer, *Macromolecules,* **22,** 2857(1989). (c) C. R. Bartels, B. Crist, Jr., L. J. Fetters, and W. W. Graessley, *Macromolecules,* **19,** 785(1986).

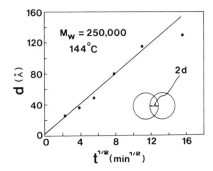

Figure 5.12 Interdiffusion depth, d, of polystyrene latex chains during film formation. The latex particle diameter was 544 Å (63).

by the resonance dipole–dipole interaction mechanism known as nonradiative energy transfer, if the emission spectrum of the "donor" overlaps the absorption spectrum of the "acceptor." The ratio of emission intensities to the acceptor intensities changes with interdiffusion, permitting the calculation of self-diffusion coefficients, and hence the interdiffusion depth. For poly(methyl methacrylate)-based latexes, self-diffusion coefficients of approximately 10^{-15} cm^2/sec were calculated, similar to the results obtained by Yoo et al. (68). The field has been reviewed by Morawetz (72), who pointed out that changes in the emission spectrum in nonradiative energy transfer between fluorescent labels has been used to characterize polymer miscibility, interpenetration of chain molecules in solution, micelle formation in graft copolymers, and other important effects.

5.5 CONCLUDING REMARKS

The amorphous state is defined as a condensed, noncrystalline state of matter. Many polymers are amorphous under ordinary use conditions, including polystyrene, poly(methyl methacrylate), and poly(vinyl acetate). Crystalline polymers such as polyethylene, polypropylene, and nylon become amorphous above their melting temperatures.

In the amorphous state, the position of one chain segment relative to its neighbors is relatively disordered. In the relaxed condition, the polymer chains making up the amorphous state form random coils. The chains are highly entangled with one another, with physical cross-links appearing at about every 600 backbone atoms.

While the amorphous polymer state is "liquid-like" in the classical sense, if the polymer is glassy, a better term would be "amorphous solid," since measurable flow takes years or centuries. Ordinary glass, an inorganic polymer, is such a glassy polymer. The chains are rigidly interlocked in glassy polymers, motion being restricted to vibrational modes.

Above the glass transition, the polymer may flow if it is not in network form. On a submicroscopic scale, the chains interdiffuse with one another with a reptating motion. Common values of the diffusion coefficient vary from 10^{-10} to 10^{-17} cm^2/sec, depending inversely on the square of the molecular weight.

REFERENCES

1. E. W. Fischer, G. R. Strobl, M. Dettenmaier, M. Stamm, and N. Steidle, *Faraday Discuss. Chem. Soc.*, **68**, 26(1979).
2. F. J. Balta-Calleja, K. D. Berling, H. Cackovic, R. Hosemann, and J. Loboda-Cackovic, *J. Macromol. Sci. Phys.*, **B12**, 383(1976).
3. R. E. Robertson, *J. Phys. Chem.*, **69**, 1575(1965).
4. W. R. Pechhold and H. P. Grossmann, *Faraday Discuss. Chem. Soc.*, **68**, 58(1979).
5. P. J. Flory, *J. Macromol. Sci. Phys.*, **B12**(1), 1(1976).
5a. R. S. Stein, *J. Chem. Ed.*, **50**, 748(1973).
6. P. J. Flory, *Faraday Discuss. Chem. Soc.*, **68**, 15(1979).
7. P. J. Flory, *Pure Appl. Chem. Macromol. Chem.*, **8**, 1(1972); reprinted in *Rubber Chem. Tech.*, **48**, 513(1975).
8. G. S. Y. Yeh, *Crit. Rev. Macromol. Sci.*, **1**, 173(1972).
9. R. F. Boyer, *J. Macromol. Sci. Phys.*, **B12**, 253(1976).
10. V. P. Privalko, Yu. S. Lipatov, and A. P. Lobodina, *J. Macromol. Sci. Phys.*, **B11**(4), 441(1975).
11. Symposium on "Physical Structure of the Amorphous State," *J. Macromol. Sci. Phys.*, **B12**, (1976).
12. "Organization of Macromolecules in the Condensed Phase," *Faraday Discuss. Chem. Soc.*, **68**, (1979).
13. E. W. Fischer, G. R. Strobl, M. Dettenmaier, M. Stamm, and N. Steidle, *Faraday Discuss. Chem. Soc.*, **68**, 26(1979).
14. R. S. Stein and S. D. Hong, *J. Macromol. Sci. Phys.*, **B12**(1), 125(1976).
15. G. M. Estes, R. W. Seymour, D. S. Huh, and S. L. Cooper, *Polym. Eng. Sci.*, **9**, 383(1969).
16. R. G. Kirste, W. A. Kruse, and K. Ibel, *Polymer*, **16**, 120(1975).
17. D. G. H. Ballard, A. N.. Burgess, P. Cheshire, E. W. Janke, A. Nevin, and J. Schelten, *Polymer*, **22**, 1353(1981).
18. J. S. Higgins and R. S. Stein, *J. Appl. Cryst.*, **11**, 346(1978).
19. A. Maconnachie and R. W. Richards, *Polymer*, **19**, 739(1978).
20. L. H. Sperling, *Polym. Eng. Sci.*, **24**, 1(1984).
21. H. Benoit, *J. Macromol. Sci. Phys.*, **B12**(1), 27(1976).
22. G. D. Wignall, D. G. H. Ballard, and J. Schelten, *J. Macromol. Sci. Phys.*, **B12**(1), 75(1976).
23. *National Center for Small-Angle Neutron Scattering Research User's Guide*, Oak Ridge National Laboratory, 1980. Solid State Divisions, Oak Ridge National Laboratory, Oak Ridge, TN 37830.
24. W. C. Koehler, R. W. Hendricks, H. R. Child, S. P. King, J. S. Lin, and G. D. Wignall, in *Proceedings of NATO Advanced Study Institute on Scattering Tech-*

niques Applied to Supramolecular and Nonequilibrium Systems, Vol. 73, S. H. Chen, B. Chu, and R. Nossal, Eds., Plenum, New York, 1981.

24a. P. Debye, *J. Phys. Coll. Chem.*, **51**, 18(1947). See also B. H. Zimm, R. S. Stein, and P. Duty, *Polym. Bull.*, **1**, 90(1945).

25. G. D. Wignall, D. G. H. Ballard, and J. Schelten, *Eur. Polym. J.*, **10**, 861(1974); reprinted in *J. Macromol. Sci. Phys.*, **B12**(1), 75(1976).

26. B. H. Zimm, *J. Chem. Phys.* **16**, 1098(1948).

27. P. J. Flory, *Principles of Polymer Chemistry*, Cornell University Press, Ithaca, NY, 1953.

28. Yu. K. Ovchinnikov, G. S. Markova, and V. A. Kargin, *Vysokol. Soyed.*, **A11**(2), 329(1969).

29. E. W. Fischer, J. H. Wendorff, M. Dettenmaier, G. Lieser, and I. Voigt-Martin, *J. Macromol. Sci. Phys.*, **B12**(1), 41(1976).

30. R. Lovell, G. R. Mitchell, and A. H. Windle, *Faraday Discuss. Chem. Soc.*, **68**, 46(1979).

31. Yu. K. Ovchinnikov, Ye. M. Antipov, and G. S. Markova, *Polymer Sci. U.S.S.R.*, **17**, 2081(1975).

32. R. L. Miller, R. F. Boyer, and J. Heijboer, *J. Polym. Sci. Polym. Phys. Ed.*, **22**, 2021(1984).

33. R. L. Miller and R. F. Boyer, *J. Polym. Sci. Polym. Phys. Ed.*, **22**, 2043(1984).

34. H. R. Schubach, E. Nagy, and B. Heise, *Coll. Polym. Sci. (Koll. Z.z. Polym.)*, **259**(8), 789(1981).

35. S. E. B. Petrie, *J. Macromol. Sci. Phys.*, **B12** 225(1976).

35a. T. A. Weber and E. Helfand, *J. Chem. Phys.*, **71**, 4760(1979).

35b. P. J. Flory, *Statistical Mechanics of Chain Molecules*, Interscience, New York, 1969.

36. H. Staudinger, *Die Hochmolekularen Organischen Verbindung*, Springer, Berlin, 1932.

37. P. J. Flory, *Principles of Polymer Chemistry*, Cornell University Press, Ithaca, NY, 1953.

38. W. Pechhold, M. E. T. Hauber, and E. Liska, *Kolloid Z.z. Polym.*, **251**, 818(1973).

39. B. Vollmert, *Polymer Chemistry*, Springer-Verlag, Berlin, 1973, p. 552.

40. P. H. Lindenmeyer, *J. Macromol. Sci. Phys.*, **8**, 361(1973).

41. V. P. Privalko and Yu. S. Lipatov, *Makromol. Chem.*, **175**, 641(1974).

42. J. Schelten, D. G. H. Ballard, G. Wignall, G. Longman, and W. Schmatz, *Polymer*, **17**, 751(1976).

43. J. Schelten, G. D. Wignall, D. G. H. Ballard, and G. W. Longman, *Polymer*, **18**, 1111(1977).

44. D. G. H. Ballard, P. Cheshire, G. W. Longman, and J. Schelten, *Polymer*, **19**, 379(1978).

45. E. W. Fischer, M. Stamm, M. Dettenmaier, and P. Herschenraeder, *Polym. Prepr. Am. Chem. Soc. Div. Polym. Chem.*, **20**(1), 219(1979).

46. J. M. Guenet, *Polymer*, **22**, 313(1981).

47. F. S. Bates, C. V. Berney, R. E. Cohen, and G. D. Wignall, *Polymer*, **24**, 519(1983).

48. A. M. Fernandez, J. M. Widmaier, G. D. Wignall, and L. H. Sperling, *Polymer*, **25**, 1718(1984).

49. P. E. Rouse, *J. Chem. Phys.*, **21**, 1272(1953).

50. F. Bueche, *J. Chem. Phys.*, **22**, 1570(1954).

51. W. L. Peticolas, *Rubber Chem. Tech.* **36**, 1422(1963).

52. B. H. Zimm, *J. Chem. Phys.*, **24**, 269(1956).

53. R. D. De Mallie, M. H. Birnboim, J. E. Fredrick, N. W. Tschoegl, and J. D. Ferry, *J. Phys. Chem.*, **66**, 536(1962).

54. P. G. de Gennes, *J. Chem. Phys.*, **55**, 572(1971).

55. P. G. de Gennes, *Phys. Today*, **36**(6), 33(1983).

56. M. Tirrell, *Rubber Chem. Tech.*, **57**, 523(1984).

57. M. Doi and S. F. Edwards, *J. Chem. Soc. Faraday Trans. 2*, **74**, 1789, 1802, 1818(1978); **75**, 38(1979).

58. W. W. Graessley, *Adv. Polym. Sci.*, **47** 67(1982).

59. J. Klein, in *Encyclopedia of Polymer Science and Engineering*, 2nd ed., Vol. 9, J. I. Kroschwitz, Ed., Wiley, New York, 1987.

60. J. Klein, *Macromolecules*, **19**, 105(1986).

61. P. G. de Gennes, *J. Phys. (Les Ulis, Fr.)*, **36**, 1199(1975).

62. C. R. Bartels, B. Crist, Jr., L. J. Fetters, and W. W. Graessley, *Macromolecules*, **19**, 785(1986).

63. B. Crist, P. F. Green, R. A. L. Jones, and E. J. Kramer, *Macromolecules*, **22**, 2857(1989).

64. F. Bueche, W. M. Cashin, and P. Debye, *J. Chem. Phys.*, **20**, 1956(1952).

65. J. Klein and B. J. Briscoe, *Proc. R. Soc. Lond. A*, **365**, 53(1979).

66. C. R. Bartels, B. Crist, and W. W. Graessley, *Macromolecules*, **17**, 2702(1984).

67. J. H. Jou and J. E. Anderson, *Macromolecules*, **20**, 1544 (1987).

68. J. N. Yoo, L. H. Sperling, C. J. Glinka, and A. Klein, *Macromolecules*, **23**, 3962(1990).

69. O. Pekan, M. A. Winnik, and M. D. Croucher, *Macromolecules*, **23**, 2673 (1990).

70. C. L. Zhao, W. C. Wang, Z. Hruska, and M. A. Winnik, *Macromolecules*, **23**, 4082(1990).

71. Y. C. Wang and M. A. Winnik, *Macromolecules*, **23**, 4731(1990).

72. H. Morawetz, *Science*, **240**, 172(1988).

GENERAL READING

M. Doi and S. F. Edwards, *The Theory of Polymer Dynamics*, Oxford University Press, New York, 1986.

P. G. de Gennes, *Scaling Concepts in Polymer Physics*, Cornell University Press, Ithaca, NY, 1979.

S. E. Keinath, R. E. Miller, and J. K. Reike, Eds., *Order in the Amorphous State of Polymers*, Plenum, New York, 1987.

V. P. Privalko, *Molecular Structure and Properties of Polymers*, Khimia, Leningrad, 1986.

STUDY PROBLEMS

1. Why is the radius of gyration of a polymer in the bulk state essentially the same as measured in a θ-solvent but not the same as in other solvents?

2. Estimate the radius of gyration and end-to-end of a polystyrene sample having $M_w = 1 \times 10^5$ g/mol.

3. In an actual kitchen experiment, one quart of cooked spaghetti was measured out level with cold water. Nine ounces of water were drained.
   (a) Assuming the spaghetti strands were polymer chains, what is the ratio of the specific volumes of the perfectly packed state to the actual disordered state? Assume a hexagonal close pack array. [*Hint:* Allow for water between perfectly aligned spaghetti strands.]
   (b) Calculate the average angle between strands, $\theta/2$, given by the ratio of the specific volumes (specific volume is the reciprocal of the density),[†]

$$\frac{v_c}{v_a} = \left(\frac{3}{2}\right)^3 \left\{ \left[\frac{1 - \cos^3(\theta/2)}{\sin^3(\theta/2)} + 1 \right]^2 \left(1 - \cos^3\left(\frac{\theta}{2}\right)\right) \right\}^{-1}$$

   (c) Interpret $\theta/2$ in terms of intermolecular orientation and the randomness of the "amorphous state."

4. Compare the Rouse–Bueche theory with the de Gennes theory. How do they model molecular motion?

5. A small-angle neutron scattering experiment was done on polystyrene containing 2 mol% of deuterated polystyrene. The following data were obtained for $\lambda = 4.75$ Å:[‡]

$\left[\dfrac{d\Sigma}{d\Omega}\right]^{-1}$ (cm)	$K^2 \times 10^4$, Å^{-2}
0.246	0.5
0.31	1.31
0.34	1.67
0.36	1.86
0.39	2.75
0.47	3.81
0.56	4.42
0.60	5.07

The constant C_N for this system was determined to be 7.95×10^{-5}

[†]R. E. Robertson, *J. Phys. Chem.*, **69**, 1575(1965).
[‡]A. M. Fernandez, J. M. Widmaier, G. D. Wignall, and L. H. Sperling, *Org. Coat. Appl. Polym. Sci. Proc.*, **48**(2), 327(1982).

mol/g · cm. What is the weight-average molecular weight and the z-average radius of gyration of the deuterated polystyrene?

6. With the advent of small-angle neutron scattering, molecular dimensions can now be determined in the bulk state. A polymer scientist determined the following data on a new deuterated polymer dissolved in a sample of (protonated) polymer:

$\left[\dfrac{d\Sigma}{d\Omega}\right]^{-1}$ (cm)	0.50	0.72	1.20
$K^2 \times 10^4$ (Å^{-2})	1.00	3.70	10.1

The constant C_N for this system was determined to be 10.0×10^{-5} mol/g · cm. What is the weight-average molecular weight and the z-average radius of gyration of the deuterated polymer? What third quantity is implicit in this experiment, and what is its probable numerical value?

7. What is the activation energy for the three-armed star's diffusion coefficient in Table 5.10, assuming an Arrhenius relationship? How do you interpret this result?

8. Calculate the first interchain radial spacing for polytetrafluoroethylene from the data given in Figure 5.3b. How are these data best interpreted?

APPENDIX 5.1 HISTORY OF THE RANDOM COIL MODEL FOR POLYMER CHAINS†

Introduction

Advances in science and engineering have never been completely uniform or followed an orderly pattern. The truth is that science advances by fits and jerks, with ideas propounded by individuals who see the world in a different light. Frequently, they face adversity when putting their ideas forward.

Before polymer science came to be, people had the concept of colloids. There were both inorganic and organic colloids, but they shared certain facts. They both were large compared to ordinary molecules, and both were of irregular sizes and shapes. While this concept "explained" certain simple experimental results, it left much to be desired in the way of understanding the properties of rubber and plastics, which were then considered to be colloids.

†L. H. Sperling, in *Pioneers in Polymer Science*, R. B. Seymour, Ed., Kluwer Academic Publishers, Dordrecht, Germany, 1989.

In 1920, Herman Staudinger formulated the macromolecular hypothesis: there was a special class of organic colloids of high viscosity, which were composed of long chains (A1, A2). This revolutionary idea was argued throughout important areas of chemistry (A3). One of the most important experiments was provided by Herman Mark, who showed that crystalline polymers that had cells of ordinary sizes had only a few mers in each cell, but that the mers were connected to those in the next cell. Eventually, the idea of long-chained molecules formed one of the most important cornerstones in the development of modern polymer science.

Early Ideas of Polymer Chain Shape

If one accepts the idea of long-chain macromolecules, the next obvious question relates to their conformation or shape in space. This was especially important since it was early thought that the physical and mechanical properties of the material were determined by the spatial arrangement of the long chains. Staudinger himself thought that most amorphous high polymers such as polystyrene were rod-shaped, and when in solution, the rods lay parallel to each other (A2, pp. 116–123).

Rubbery materials were different, however. According to early scientists (A4), elastomers were coils or spirals resembling bedsprings (see Figure

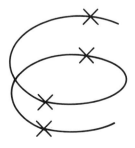

Figure A5.1.1 The spiral structure of natural rubber proposed to explain long-range elasticity. X indicates double bond locations.

A5.1.1). Staudinger himself described the idea as follows (A2):

> In order to clarify the elasticity of rubber, several investigators have stated that long molecules form spirals, and to be sure the spiral form of the molecules is promoted through the double bonds. By this arrangement, the secondary valences of the double bonds can be satisfied. The elasticity of rubber depends upon the extensibility of such spirals.

The Random Coil

According to H. Mark (A5), the story of the development of the random coil began with the x-ray work of Katz on natural rubber in 1925 (A6–A9). Katz studied the x-ray patterns of rubber both in the relaxed state and the extended or stretched state. In the stretched state, Katz found a characteristic fiber diagram, with many strong and clear diffraction spots, indicating a crystalline material. This contrasted with the diffuse halo found in the relaxed state, indicating that the chains were amorphous under that condition. The fiber periodicity of the elementary cell was found to be about 9 Å, which could only accommodate a few isoprene units. Since the question of how a long chain could fit into a small elementary cell is fundamental to the macromolecular hypothesis, Hauser and Mark repeated the "Katz effect" experiment and, on the basis of improved diagrams and x-ray techniques, established the exact size of the elementary cell (A10). Of course, the answer to the question of the cell size is that the cell actually accommodates the mer, or repeat unit, rather than the whole chain.

The "Katz effect" was particularly important because it established the first relationship between mechanical deformation and concomitant molecular events in polymers (A5). This led Mark and Valko (A11) to carry out stress–strain studies over a wide temperature range together with x-ray studies in order to analyze the phenomenon of rubber reinforcement. This paper contains the first clear statement that the contraction of rubber is not caused by an increase in energy but by the decrease in entropy on elongation.

This finding can be explained by assuming that the rubber chains are in the form of flexible coils (see Figure A5.1.2) (A12). These flexible coils have a high conformational entropy, but they lose their conformational entropy on being straightened out. The fully extended chain, which is rod-shaped, can have only one conformation, and its entropy is zero. This concept was extended to all elastic polymers by Meyer et al. (A13) in 1932. These ideas are developed quantitatively in Chapter 9. Although thermal motion and free chain rotation are required for rubber elasticity, the idea of the random coil was later adopted for glassy polymers such as polystyrene as well.

The main quantitative developments began in 1934 with the work of Guth and Mark (A14) and Kuhn (A15). Guth and Mark chose to study the entropic

Figure A5.1.2 Early drawing of the random coil. (*A*) Relaxed state. (*B*) Effect of deformation in the *Z* direction. After W. Kuhn (A12).

origin of the rubber elastic forces, whereas Kuhn was more interested in explaining the high viscosity of polymeric solutions. Using the concept of free rotation of the carbon–carbon bond, Guth and Mark developed the idea of the "random walk" or "random flight" of the polymer chain. This led to the familiar Gaussian statistics of today, and eventually to the famous relationship between the end-to-end distance of the chain and the square root of the molecular weight. Three stages in the development of the random coil model have been described by H. Mark (see Table A5.1.1) (A16). Like many great ideas, it apparently occurred to several people nearly simultaneously. It is also clear from Table A5.1.1 that Mark played a central role in the development of the random coil.

The random coil model has remained essentially the same until today (A17, A18), although many mathematical treatments have refined its exact definition. Its main values are twofold: by all experiments, it appears to be the best model for amorphous polymers, and it is the only model that has been extensively treated mathematically. It is interesting to note that by its very randomness, the random coil model is easier to understand quantitatively and analytically than models introducing modest amounts of order (A19).

Table A5.1.1 Early references on randomly linked flexible macromolecules (16)

Phase 1:	Hypothetic and accidental remarks:
	E. Wachlisch, *Z. S. Biol.*, **85**, 206(1927)
	H. Mark and E. Valko, *Kautschuk*, **6**, 210(1930)
	W. Haller, *Koll. Z. S.*, **56**, 257(1931)
Phase 2:	Elaborate *qualitative* interpretation:
	K. H. Meyer, G. von Susich, and E. Valko,
	Koll. Z. S., **59**, 208(1932)
Phase 3:	*Quantitative* mathematical treatment:
	E. Guth and H. Mark, *Monatschefte Chemie*, **65**, 93(1934)
	H. Mark, IX Congress for Chemistry, Madrid 1934,
	Vol. **4**, p. 197, (1934)
	W. Kuhn, *Koll. Z. S.*, **68**, 2(1934)
	H. Mark, *Der Feste Koerper*, 1937, p. 65

Source: H. Mark, private communication, May 2, 1983.

The random coil model has been supported by many experiments over the years. The most important of these has been light-scattering from dilute solutions and, more recently, small-angle neutron scattering from the bulk state (see earlier sections of Chapter 5). Both of these experiments support the famous relationships between the square root of the molecular weight and the end-to-end distance. The random coil model has been used to explain not only rubber elasticity and dilute solution viscosities but a host of other physical and mechanical phenomena, such as melt rheology, diffusion, and the equilibrium swelling of cross-linked polymers. Some important reviews include the works of Flory (A17), Treloar (A18), Staverman (A21), and Guth and Mark (A22).

REFERENCES

A1. H. Staudinger, *Ber. Dtsch. Chem. Ges.*, **53**, 1074(1920).

A2. H. Staudinger, *Die Hochmolekularen Organischen Verbindung*, Springer-Verlag, Berlin, 1932. Reprinted, 1960.

A3. G. A. Stahl, Ed., *"Polymer Science: Overview A Tribute to Herman F. Mark*, ACS Symposium No. 175, American Chemical Society, Washington, DC, 1981.

A4. F. Kirchhof, *Kautschuk*, **6**, 31(1930).

A5. H. Mark, unpublished, 1982.

A6. J. R. Katz, *Naturwissenschaften*, **13**, 410(1925).

A7. J. R. Katz, *Chem. Ztg.*, **19**, 353(1925).

A8. J. R. Katz, *Kolloid Z.*, **36**, 300(1925).

A9. J. R. Katz, *Kolloid Z.*, **37**, 19(1925).

A10. E. A. Hauser and H. Mark, *Koll. Chem. Beih.*, **22**, 63; **23**, 64(1929).

A11. H. Mark and E. Valko, *Kautschuk*, **6**, 210(1930).

A12. W. Kuhn, *Angew. Chem.*, **49**, 858(1936).

A13. K. H. Meyer, G. V. Susich, and E. Valko, *Koll. Z.*, **41**, 208(1932).

A14. E. Guth and H. Mark, *Monatsh. Chem.*, **65**, 93(1934).

A15. W. Kuhn, *Kolloid. Z.*, **68**, 2(1934).

A16. H. Mark, private communication, May 2, 1983.

A17. P. J. Flory, *Principles of Polymer Chemistry*, Cornell University Press, Ithaca, NY, 1953.

A18. L. R. G. Treloar, *The Physics of Rubber Elasticity*, 3rd ed., Oxford, Clarendon Press, 1975.

A19. R. F. Boyer, *J. Macromol. Sci. Phys.*, **B12**, 253(1976).

A20. P. J. Flory, *Statistical Mechanics of Chain Molecules*, Wiley, New York, 1969.

A21. A. J. Staverman, *J. Polym. Sci.*, Symposium No. 51, 45 (1975).

A22. E. Guth and H. F. Mark, *J. Polym. Sci. B Polym. Phys.*, **29**, 627(1991).

APPENDIX 5.2   CALCULATIONS USING THE DIFFUSION COEFFICIENT

The diffusion of polystyrene of various molecular weights in high-molecular-weight polystyrene ($M_w = 2 \times 10^7$ g/mol) was measured by Mills et al. (B1) using forward recoil spectrometry. At 170°C, the diffusion coefficient was found to depend on the weight-average molecular weight as

$$D = 8 \times 10^{-3} M_w^{-2} \text{ cm}^2/\text{sec} \tag{A5.2.1}$$

If the concentration gradient of the diffusing species is given in concentration per unit distance, the unit

$$\frac{\text{mol}/\text{cm}^3}{\text{cm}} \times \frac{\text{cm}^2}{\text{sec}}$$

yields the flux in mol/(cm² · sec)—that is, the number of moles of polystyrene crossing a square centimeter of area per second.

Consider a polymer having a weight-average molecular weight of 1×10^6 g/mol. With a density of 1.05 g/cm³, a bulk concentration of about 1×10^{-6} mol/cm³ is obtained. Consider a diffusion over a 100-Å distance, or 1×10^{-6} cm from the bulk concentration to a zero concentration. In 1 sec, $8 \times 10^{-9} \cong 1 \times 10^{-8}$ mol will diffuse through a 1-cm² area. Since the original concentration was 1×10^{-6} mol/cm³, about 1% of the total volume diffuses in this time. In fact, this is most of the polymer that is lying on the surface of the hypothetical 1 cm³ under consideration.

Bulk polymeric materials being pressed together under modeling conditions require diffusion of the order of 100 Å to produce a significant number of entanglements, thereby fusing the interfacial boundary. In the commercial molding of polystyrene at 170°C, times of the order of 2 min might be employed. The largest portion of this time is actually required for heat transfer to be complete and for a uniform temperature to be achieved. Since the weight-average molecular weight of many polystyrenes is about 1×10^5 g/mol, the original boundary will be obliterated in a few seconds under these conditions. Thus the values obtained via polymer physics research confirm the values used in practice.

REFERENCE

B1. P. J. Mills, P. F. Green, C. J. Palmstrom, J. W. Mayer, and E. J. Kramer, *Appl. Phys. Lett.*, **45**(9), 957(1984).

6

THE CRYSTALLINE STATE

6.1 GENERAL CONSIDERATIONS

In the previous chapter, the structure of amorphous polymers was examined. In this chapter, the study of crystalline polymers is undertaken. The crystalline state is defined as one that diffracts x-rays and exhibits the first-order transition known as melting. In this manner, the crystalline state of polymers is like the crystalline state of smaller molecules.

However, polymers crystallized in the bulk state are never totally crystalline, a consequence of their long-chain nature and subsequent entanglements. The melting (fusion) temperature of the polymer, T_f, is always higher than the glass transition temperature, T_g. Thus the polymer may be either hard and rigid or flexible. An example of the latter is ordinary polyethylene, which has a T_g of about $-80°C$ and a melting temperature of about $+139°C$. At room temperature it forms a leathery product as a result.

The development of crystallinity in polymers depends on the regularity of structure in the polymer (see Chapter 2). Thus isotactic and syndiotactic polymers usually crystallize, whereas atactic polymers, with a few exceptions (where the side groups are small or highly polar), do not. Regular structures also appear in the polyamides (nylons), polyesters, and so on, and these polymers make excellent fibers.

Nonregularity of structure first decreases the melting temperature and finally prevents crystallinity. Mers of incorrect tacticity (see above) tend to destroy crystallinity, as does copolymerization. Thus statistical copolymers are generally amorphous. Blends of isotactic and atactic polymers show reduced crystallinity, with only the isotactic portion crystallizing. Under some circumstances, block copolymers containing a crystallizable block will crystallize; again, only the crystallizable block crystallizes.

Table 6.1 **Properties of selected crystalline polymers (1)**

Polymer	$T_f, °C$	$\Delta H_f, \dfrac{kJ}{mol}$
Polyethylene	139	7.87[a]
Poly(ethylene oxide)	66	8.29
it-Polystyrene	240	8.37
Poly(vinyl chloride)	212	3.28
Poly(ethylene terephthalate)[b]	265	24.1
Poly(hexamethylene adipamide)[c]	265	46.5
Cellulose tributyrate	207	12.6
cis-Polyisoprene[d]	28	4.40
Polytetrafluoroethylene[e]	330	5.74
it-polypropylene	171	8.79

[a] Per $-CH_2-CH_2-$. Note that values for $-CH_2-$ alone are sometimes reported.
[b] Dacron. [d] Natural rubber.
[c] Nylon. [e] Teflon.

Factors that control the melting temperature include polarity and hydrogen bonding as well as packing capability. Table 6.1 lists some important crystalline polymers and their melting temperatures (1).

6.1.1 Historical Aspects

Historically, the study of crystallinity of polymers was important in the proof of the macromolecular hypothesis, developed originally by Staudinger. If polymers were composed of long chains, how could they fit into unit cells that were known to be only several angstroms in size? The answer was that the unit cell contained only a few mers (2–7), which were repeated in adjacent unit cells.† If the adjacent unit cell is considered to repeat in the axial direction, its relationship to the whole chain is more easily visualized.

One of the first structures to be determined was of the natural polysaccharide cellulose. Its unit cell is illustrated in Figure 6.1 (3). In this case, the repeat unit is cellobiose, composed of two glucoside rings.

Of course, crystalline polymers constitute many of the plastics and fibers of commerce. Polyethylene is used in films to cover dry-cleaned clothes, and as water and solvent containers (such as wash bottles). Polypropylene makes a highly extensible rope, finding particularly important applications in the marine industry. Polyamides (nylons) and polyesters are used as both plastics and fibers. Their use in clothing is world famous. Cellulose, mentioned above, is used in clothing in both its native state (cotton) and its regenerated state (rayon). The film is called cellophane.

† If the whole polymer molecule was in a unit cell of a few Ångstroms on a side, the material would have a density of 50 times lead!

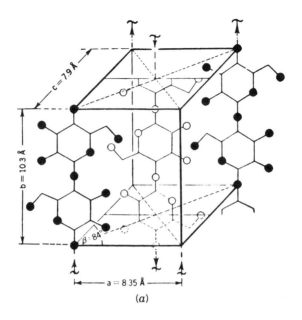

(a)

Parameter	Cellulose I[a]	Cellulose II[a]	Cellulose III[b]	Cellulose IV[a] Cellulose x
a-axis (Å)	8.35	8.02	7.74	8.12
b-axis (Å)	10.30	10.30	10.30	10.30
c-axis Å)	7.90	9.03	9.96	7.99
β (degrees)	83.3	62.8	58	90
Density (g/cm^3)	1.625	1.62	1.61	1.61

[a] Ø. Ellefsen, J. G. Ønnes, and N. Norman, *Acta Chem. Scand.*, **13**, 853(1959).
[b] C. Legrand, *J. Polym. Sci.*, **7**, 333(1951).

(b)

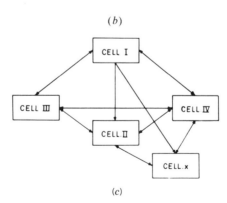

(c)

Figure 6.1 The crystalline structure of cellulose. (a) The unit cell of native cellulose, or cellulose I, as determined by x-ray analysis (2, 27, 28). (b) Unit cell dimensions of the four forms of cellulose. (c) Known pathways to alter the crystalline structure of cellulose. The lesser known form of cellulose x is also included (27).

6.1.2 Melting Phenomena

The melting of polymers may be observed by any of several experiments. For linear or branched polymers, the sample becomes liquid and flows. However, there are several possible complications to this experiment, which may make interpretation difficult. First of all, simple liquid behavior may not be immediately apparent because of the polymer's high viscosity. If the polymer is cross-linked, it may not flow at all. It must also be noted that amorphous polymers soften at their glass transition temperature, T_g, which is emphatically not a melting temperature, but which may resemble one, especially to the novice (see Chapter 8). If the sample does not contain colorants, it is usually hazy in the crystalline state because of the difference in refractive index between the amorphous and crystalline portions. On melting, the sample becomes clear, or more transparent.

The disappearance of crystallinity may also be observed in a microscope, for example, between crossed Nicols. The sharp x-ray pattern characteristic of crystalline materials gives way to amorphous halos at the melting temperature, providing one of the best experiments.

Another important way of observing the melting point is to observe the changes in specific volume with temperature. Since melting constitutes a

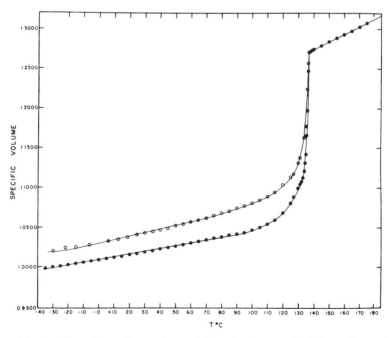

Figure 6.2 Specific volume – temperature relations for linear polyethylene. Open circles, specimen cooled relatively rapidly from the melt to room temperature before fusion experiments; solid circles, specimen crystallized at 130°C for 40 days, then cooled to room temperature prior to fusion (8).

Figure 6.3 The heat of fusion of poly(ethylene oxide) determined by DSC (9). The heat of fusion of the crystalline portion can be determined by melting point depression experiments, permitting an estimate of the percent crystallinity.

first-order phase change, a discontinuity in the volume is expected. Ideally, the melting temperature should give a discontinuity in the volume, with a concomitant sharp melting point. In fact, because of the very small size of the crystallites in bulk crystallized polymers (or alternatively, their imperfections), most polymers melt over a range of several degrees (see Figure 6.2) (8). The melting temperature is usually taken as the temperature at which the last trace of crystallinity disappears. This is the temperature at which the largest and/or most perfect crystals are melting.

Alternatively, the melting temperature can be determined thermally. Today, the differential scanning calorimeter (DSC) is popular, since it gives the heat of fusion as well as the melting temperature. Such an experiment is illustrated in Figure 6.3 for poly(ethylene oxide) (9). The heat of fusion, ΔH_f, is given by the area under the peak.

Further general studies of polymer fusion are presented in Sections 6.8 and 6.9, after the introduction of crystallographic concepts and the kinetics and thermodynamics of crystallization.

6.2   METHODS OF DETERMINING CRYSTAL STRUCTURE

6.2.1   A Review of Crystal Structure

Before beginning the study of the structure of crystalline polymers, the subject of crystallography and molecular order in crystalline substances is reviewed. Long before x-ray analysis was available, scientists had already deduced a great deal about the atomic order within crystals.

The science of geometric crystallography was concerned with the outward spatial arrangement of crystal planes and the geometric shape of crystals. Workers of that day arrived at three fundamental laws: (a) the law of constancy of interfacial angles, (b) the law of rationality of indices, and (c) the law of symmetry (10).

Briefly, the law of constancy of interfacial angles states that for a given substance, corresponding faces or planes that form the external surface of a crystal always intersect at a definite angle. This angle remains constant independent of the sizes of the individual faces.

The law of rationality of indices states that for any crystal a set of three coordinate axes can be chosen such that all the faces of the crystal will either intercept these axes at definite distances from the origin or be parallel to some of the axes. In 1784 Hauy showed that it was possible to choose among the three coordinate axes unit distances (a, b, c) not necessarily the same length. Furthermore, Hauy showed that it was possible to choose three coefficients for these three axes—m, n, and p—which are either integral whole numbers, infinity, or fractions of whole numbers, such that the ratio of the three intercepts of any plane in the crystal is given by (ma: nb: pc). The numbers m, n, and p are known as the Weiss indices of the plane in question. The Weiss indices have been replaced by the Miller indices, which are obtained by taking the reciprocals of the Weiss coefficients and multiplying through by the smallest number that will express the reciprocals as integers. For example, if a plane in the Weiss notation is given by a: ∞b: $\frac{1}{4}c$, the Miller indices become a: $0b$: $4c$, or more simply (104), which is the modern way of expressing the indices in the Miller system of crystal face notation.

The third law of crystallography states that all crystals of the same compound possess the same elements of symmetry. There are three types of symmetry—a plane of symmetry, a line of symmetry, and a center of symmetry (10). A plane of symmetry passes through the center of the crystal and divides it into two equal portions, each of which is the mirror image of the other. If it is possible to draw an imaginary line through the center of the crystal and then revolve the crystal about this line in such a way as to cause the crystal to appear unchanged two, three, four, or six times in 360° of revolution, then the crystal is said to possess a line of symmetry. Similarly, a crystal possesses a center of symmetry if every face has an identical atom at an equal distance on the opposite side of this center. On the basis of the total number of plane, line, and center symmetries, it is possible to classify the crystal types into six crystal systems, which may in turn be grouped into 32 classes and finally into 230 crystal forms.

The scientists of the pre-x-ray period postulated that any macroscopic crystal was built up by repetition of a fundamental structural unit composed of atoms, molecules, or ions, called the unit crystal lattice or space group. This unit crystal lattice has the same geometric shape as the macroscopic crystal. This line of reasoning led to the 14 basic arrangements of atoms in space, called space lattices. Among these are the familiar simple cubic, hexagonal, and triclinic lattices.

There are four basic methods in wide use for the study of polymer crystallinity: x-ray diffraction, electron diffraction, infrared absorption, and Raman spectra. The first two methods constitute the fundamental basis for crystal cell size and form, and the latter two methods provide a wealth of supporting data such as bond distances and intermolecular attractive forces. These several methods are now briefly described.

6.2.2 X-Ray Methods

In 1895, x-rays were discovered by Roentgen. The new x-rays were first applied to crystalline substances in 1912 and 1913, following the suggestion by Von Laue that crystalline substances ought to act as a three-dimensional diffraction grating for x-rays.

By considering crystals as reflection gratings for x-rays, Bragg (11) derived his now famous equation for the distance d between successive identical planes of atoms in the crystal:

$$d = \frac{n\lambda}{2 \sin \theta} \tag{6.1}$$

where λ is the x-ray wavelength, θ is the angle between the x-ray beam and these atomic planes, and n represents any whole number. It turns out that both the x-ray wavelength and the distance between crystal planes, d, are of the order of 1 Å. Such an analysis from a single crystal produces a series of spots.

However, not every crystalline substance can be obtained in the form of macroscopic crystals. This led to the Debye–Scherrer (12) method of analysis for powdered crystalline solids or polycrystalline specimens. The crystals are oriented at random so the spots become cones of diffracted beams that can be recorded either as circles on a flat photographic plate or as arcs on a strip of film encircling the specimen (see Figure 6.4) (13). The latter method permits the study of back reflections as well as forward reflections.

Basically, the intensity of the diffraction spot or line depends on the scattering power of the individual atoms, which in turn depends on the number of electrons in the atom. Other quantities of importance include the arrangement of the atoms with regard to the crystal planes, the angle of reflection, the number of crystallographically equivalent sets of planes contributing, and the amplitude of the thermal vibrations of the atoms. Both the intensities of the spots or arcs and their positions are required to calculate the crystal lattice, plus lots of imagination and hard work. The subject of x-ray analysis of crystalline materials has been widely reviewed (10, 13).

Figure 6.4   The Debye – Scherrer method for taking powder photographs. The angle RSX is 2θ, where θ is the angle of incidence on a set of crystal planes (13).

6.2.3   Electron Diffraction of Single Crystals

Electron microscopy provides a wealth of information about the very small, including a view of the actual crystal size and shape. In another mode of use, the electrons can be made to diffract, using their wavelike properties. In this regard they are made to behave like the neutrons considered earlier.

In the case of x-ray studies, the polymer samples are usually uniaxially oriented and yield fiber diagrams that correspond to single crystal rotation photographs. Electron diffraction studies utilize single crystals.

Since the polymer chains in single crystals are most often oriented perpendicular to their large flat surface, diffraction patterns perpendicular to the 001 plane are common. Tilting of the sample yields diffraction from other planes. The interpretation of the spots obtained utilizes Bragg's law in a manner identical to that of x-rays.

6.2.4   Infrared Absorption

Tadokoro (14) summarized some of the specialized information that infrared absorption spectra yield about crystallinity:

1. Infrared spectra of semicrystalline polymers include "crystallization-sensitive bands." The intensities of these bands vary with the degree of crystallinity and have been used as a measure of the crystallinity.

2. By measuring the polarized infrared spectra of oriented semicrystalline polymers, information about both the molecular and crystal structure can be obtained. Both uniaxially and biaxially oriented samples can be studied.

3. The regular arrangement of polymer molecules in a crystalline region can be treated theoretically, utilizing the symmetry properties of the chain or crystal. With the advent of modern computers, the normal modes of vibrations of crystalline polymers may be calculated and compared with experiment.

4. Deuteration of specific groups yields information about the extent of the contribution of a given group to specific spectral bands. This aids in the assignment of the bands as well as the identification of bands owing to the crystalline and amorphous regions.

6.2.5 Raman Spectra

Although Raman spectra have been known since 1928, studies on high polymers and other materials became popular only since the development of efficient laser sources. According to Tadokoro (14), some of the advantages of Raman spectra are the following:

1. Since the selection rules for Raman and infrared spectra are different, Raman spectra yield information complementary to the infrared spectra. For example, the $S — S$ linkages in vulcanized rubber and the $C = C$ bonds yield strong Raman spectra but are very weak or unobservable in infrared spectra.

2. Since the Raman spectrum is a scattering phenomenon, whereas the infrared methods depend on transmission, small bulk, powdered, or turbid samples can be employed.

3. On analysis, the Raman spectra provide information equivalent to very low-frequency measurements, even lower than 10 cm^{-1}. Such low-frequency studies provide information on lattice vibrations.

4. Polarization measurements can be made on oriented samples.

Of course, much of the above is widely practiced by spectroscopists on small molecules as well as big ones. Again, it must be emphasized that polymer chains are ordinary molecules that have been grown long in one direction.

6.3 THE UNIT CELL OF CRYSTALLINE POLYMERS

When polymers are crystallized in the bulk state, the individual crystallites are microscopic or even submicroscopic in size. They are an integral part of the solids and cannot be isolated. Hence studies on crystalline polymers in the bulk were limited to powder diagrams of the Debye–Scherer type, or fiber diagrams of oriented materials.

It was only in 1957 that Keller (15) and others discovered a method of preparing single crystals from very dilute solutions by slow precipitation. These too were microscopic in size (see Section 6.4). However, x-ray studies could now be carried out on single crystals, with concomitant increases in detail obtainable.

Of course a major difference between polymers and low-molecular-weight compounds relates to the very existence of the macromolecule's long chains.

These long chains traverse many unit cells. Their initial entangled nature impedes their motion, however, and leaves regions that are amorphous. Even the crystalline portions may be less than perfectly ordered.

This section describes the structure of the unit cell in polymers, principally as determined by x-ray analysis. The following sections describe the structure and morphology of single crystals, bulk crystallized crystallites, and spherulites and develop the kinetics and thermodynamics of crystallization.

6.3.1 Polyethylene

One of the most important polymers to be studied is polyethylene. Because of its simple structure, it has served as a model polymer in many laboratories. Also, its great commercial importance as a crystalline plastic has made the results immediately usable. It has been investigated both in the bulk and in the single-crystal state.

The unit cell structure of polyethylene was first investigated by Bunn (16). A number of experiments were reviewed by Natta and Corradini (17). The

Figure 6.5 A study of polyethylene single-crystal structure. (*a*) A single crystal of polyethylene, precipitated from xylene, as seen by electron microscopy. (*b*) Electron diffraction of the same crystal, with identical orientation. (*c*) Perspective view of the unit cell of polyethylene, after Bunn. (*d*) View along chain axis. This latter corresponds to the crystal and diffraction orientation in (*a*) and (*b*) (18). Courtesy of A. Keller and Sally Argon.

unit cell is orthorhombic, with cell dimensions of $a = 7.40$, $b = 4.93$, and $c = 2.534$ Å. The unit cell contains two mers (see Figure 6.5) (18). Not unexpectedly, the unit cell dimensions are substantially the same as those found for the normal paraffins of molecular weights in the range 300–600 g/mol. The chains are in the extended zigzag form; that is, the carbon–carbon bonds are trans rather than gauche. The zigzag form may also be viewed as a twofold screw axis.

The single-crystal electron diffraction pattern shown in Figure 6.5 was obtained by viewing the crystal along the c axis. Also shown is the single-crystal structure of polyethylene, which is typically diamond-shaped (see below). The unit cell is viewed from the c-axis direction, perpendicular to the diamonds.

6.3.2 Other Hydrocarbon Polymers

Because of the need for regularity along the chain, only vinyl polymers that are either isotactic or syndiotactic will crystallize. Thus isotactic polypropy-

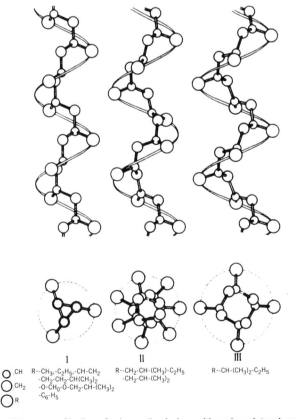

○	CH	R $\cdot$ -CH$_3$, -C$_2$H$_5$, -CH-CH$_2$	R $\cdot$ -CH$_2$-CH-(CH$_3$)-C$_2$H$_5$	R $\cdot$ -CH-(CH$_3$)$_2$-C$_2$H$_5$
○	CH$_2$	-CH$_2$-CH$_2$-CH(CH$_3$)$_2$	-CH$_2$-CH-(CH$_3$)$_2$	
		-O-CH$_0$-O-CH$_2$-CH-(CH$_3$)$_2$		
○	R	-C$_6$-H$_5$		

Figure 6.6 Possible types of helices for isotactic chains, with various lateral groups (19).

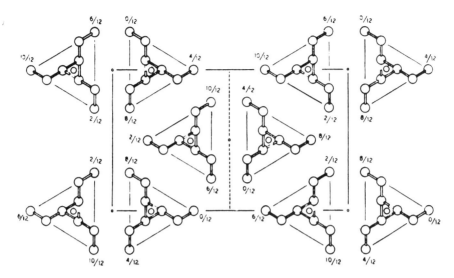

Figure 6.7 Enantiomorphous mode of packing polybutene-1 chains in a crystal (17).

lene crystallizes well and is a good fiber former, whereas atactic polypropylene is essentially amorphous.

The idea of a screw axis along the individual extended chains needs developing. Such chains may be viewed as having a n/p-fold helix, where n is the number of mer units and p is the number of pitches within the identity period. Of course, n/p will be a rational number. Some of the possible types of helices for isotactic polymers are illustrated in Figure 6.6 (19). Group I of Figure 6.6 has a helix that makes one complete turn for every three mer units, so $n = 3$ and $p = 1$. Group II shows seven mer units in two turns, so $n = 7$ and $p = 2$. Group III shows four mer units per turn.

Of course, both left- and right-handed helices are possible. The isotactic hydrocarbon polymers in question occur as enantiomorphic pairs that face each other, a closer packing being realized through the operation of a glide plane with translation parallel to the fiber axis. The enantiomorphic crystal structure of polybutene-1 is illustrated in Figure 6.7 (17). Note that better packing is achieved through the chains' having the opposite sense of helical twist.

6.3.3 Polar Polymers and Hydrogen Bonding

The hydrocarbon polymers illustrated above are nonpolar, being bonded together only by van der Waals-type attractive forces. When the polymers possess polar groups or hydrogen bonding capability, the most energetically favored crystal structures will tend to capitalize on these features. Figure 6.8 (20) illustrates the molecular organization within crystallites of a polyamide,

Figure 6.8 The hydrogen-bonded structure of nylon 66. The unit cell face is shown dotted (20).

known as nylon 66. The chains in the crystallites are found to occur as fully extended, planar zigzag structures.

X-ray analysis reveals that poly(ethylene terephthalate) (Dacron®) belongs to the triclinic system (21). The cell dimensions are $a = 4.56$, $b = 5.94$, $c = 10.75$ Å, with the angles being $\alpha = 98.5°$, $\beta = 118°$, $\gamma = 112°$. Both the polyamides and the aromatic polyesters are high melting polymers because of hydrogen bonding in the former case and chain stiffness in the latter case (see Table 6.1). As is well known, both of these polymers make excellent fibers.

The polyethers are a less polar group of polymers. Poly(ethylene oxide) will be taken as an example (22). Four (7/2) helical molecules pass through a unit cell with parameters $a = 8.05$, $b = 13.04$, $c = 19.48$ Å, and $\beta = 125.4°$ with the space group $P2_1/a - C_{2h}$. [Space groups are discussed by Tadokoro (23, 24). These symbols represent the particular one of the 230 possible space groups to which poly(ethylene oxide) belongs (see Section 6.2).]

Table 6.2 (23) summarizes the crystallographic data of some important polymers. Several polymers have more than one crystallographic form. Polytetrafluoroethylene, for example, undergoes a first-order crystal–crystal transition at 19°C.

6.3.4 Polymorphic Forms of Cellulose

A few words have already been said about the crystalline structure of cellulose (see Figure 6.1). The monoclinic unit cell structure illustrated for cellulose I was postulated many years ago by Meyer et al. (3–6) and has been confirmed many times. The b axis is the fiber direction, and the cell belongs to the space group P2, containing four glucose residues.

It is important to note that the structure shown in Figure 6.1a illustrates the chains running in alternating directions, up and down. While the x-ray results are not especially clear on this point, the assumption of all one way or of alternating opposite directions is very important in the development of crystalline models (see below).

As shown further in Figure 6.1b, c, four different polymorphic forms of crystalline cellulose exist. Cellulose I is native cellulose, the kind found in wood and cotton. Cellulose II is made either by soaking cellulose I in strong alkali solutions (e.g., making mercerized cotton) or by dissolving it in the viscose process, which makes the labile but soluble cellulose xanthate. The regenerated cellulose II products are known as rayon for the fiber form and cellophane for the film form. Cellulose III can be made by treating cellulose with ethylamine. Cellulose IV may be obtained by treatment with glycerol or alkali at high temperatures (25, 26).

Going from the polymorphic forms of cellulose back to cellulose I is difficult but can be accomplished by partial hydrolysis. The subject of the polymorphic forms of cellulose has recently been reviewed (27, 28).

6.3.5 Principles of Crystal Structure Determination

Through the use of x-ray analysis, electron diffraction, and the supporting experiments of infrared absorption and Raman spectroscopy, much information has been collected on crystalline polymers. Today, the data are analyzed through the use of computer techniques. Somewhere along the line, however, the investigator is required to use intuition to propose models of crystal structure. The proposed models are then compared to experiment. The models are then gradually refined to produce the structures given above. It must be emphasized, that the experiments do not yield the crystal structure; only researchers' imagination and hard work yield that.

However, it is possible to simplify the task. Natta and Corradini (19) postulated several principles for the determination of crystal structures,

Table 6.2 Selected crystallographic data (23)

Polymer	Crystal System, Space Group, Lattice Constants, and Number of Chains per Unit Cell[a]	Molecular Conformation	Crystal Density (g/cm^3)
Polyethylene $(-CH_2-CH_2-)_n$	Stable form, orthorhombic, $Pnam\text{-}D_{2h}^{16}$, $a = 7.417$ Å, $b = 4.945$ Å, $c = 2.547$ Å, $N = 2$	Planar zigzag (2/1)	1.00
	Metastable form, monoclinic, $C2/m - C_{2h}^3$, $a = 8.09$ Å, b (f.a.) $= 2.53$ Å, $c = 4.79$ Å, $\beta = 107.9°$, $N = 2$	Planar zigzag (2/1)	0.998
	High-pressure form, orthohexagonal (assumed), $a = 8.42$ Å, $b = 4.56$ Å, c (f.a.) has not been determined		
Polytetrafluoroethylene $(-CF_2-CF_2-)_n$	Below 19° C, pseudohexagonal (triclinic), $a' = b' = 5.59$ Å, $c = 16.88$ Å, $\gamma' = 119.3°$, $N = 1$	Helix (13/6) 163.5°	2.35
	Above 19°C, trigonal, $a = 5.66$ Å, $c = 19.50$ Å, $N = 1$	Helix (15/7) 165.8°	2.30
	High-pressure form I[12], orthorhombic, $Pnam\text{-}D_{2h}^{16}$, $a = 8.73$ Å, $b = 5.69$ Å, $c = 2.62$ Å, $N = 2$	Planar zigzag (2/1)	2.55
	High-pressure form II,[193] monoclinic, $B2/m\text{-}C_{2h}^3$, $a = 9.50$ Å, $b = 5.05$ Å, $c = 2.62$ Å, $\gamma = 105.5°$, $N = 2$	Planar zigzag (2/1)	2.74
it-Polypropylene $(-CH_2-\underset{\underset{CH_3}{\vert}}{CH}-)_n$	α-Form, monoclinic, $C2/c\text{-}C_{2h}^6$ of $Cc\text{-}C_s^4$, $a = 6.65$ Å, $b = 20.96$ Å, $c = 6.50$ Å, $\beta = 99°\ 20'$, $N = 4$	Helix (3/1) $(TG)_3$	0.936
	β-Form, hexagonal, $a = 19.08$ Å, $c = 6.49$ Å, $N = 9$	Helix (3/1) $(TG)_3$	0.922
	γ-Form, trigonal, $P3_121\text{-}D_3^4$ or $P3_221\text{-}D_3^6$, $a = 6.38$ Å, $c = 6.33$ Å, $N = 1$	Helix (3/1) $(TG)_3$	0.939

Polymer	Crystal structure	Conformation	Density

it-Polystyrene
 	
$(-CH_2-CH-)_n$
         $\vert$
         C_6H_5
	Trigonal, $R3c$-C_{3v}^6 or $R\bar{3}c$-D_{3d}^6, $a = 21.90$ Å,
	$c = 6.65$ Å, $N = 6$
		Helix (3/1)
		$(TG)_3$
			1.13

cis-1,4-Polyisoprene $(-CH_2-C(CH_3)=CH-CH_2-)_n$	Monoclinic, $P2_1/a$-C_{2h}^5, $a = 12.46$ Å, $b = 8.89$ Å, $c = 8.10$ Å, $\beta = 92°$, $N = 4$	cis-$ST\bar{S}$-cis-$\bar{S}TS$, (2/0)	1.02
Poly(vinyl chloride) $(-CH_2-CHCl-)_n$	Orthorhombic, $Pcam$-D_{2h}^{11}, $a = 10.6$ Å, $b = 5.4$ Å, $c = 5.1$ Å, $N = 2$	Planar zigzag	1.42
Polytetrahydrofuran $(-(CH_2)_4-O-)_n$	Monoclinic, $C2/c$-C_{2h}^6, $a = 5.59$ Å, $b = 8.90$ Å, $c = 12.07$ Å, $\beta = 134.2°$, $N = 2$	Planar zigzag (2/1)	1.11
Nylon 6 $(-(CH_2)_5-CONH-)_n$	α-Form, monoclinic, $P2_1$-C_2^2, $a = 9.56$ Å, b (f.a.) = 17.2 Å, $c = 8.01$ Å, $\beta = 67.5°$, $N = 4$	Planar zigzag (2/1)	1.23
	γ-Form, monoclinic, $P2_1/a$-C_{2h}^5, $a = 9.33$ Å, b (f.a.) = 16.88 Å, $c = 4.78$ Å, $\beta = 121°$, $N = 2$	Helix (2/1) $(T_4S T\bar{S})^2$	1.17
Nylon 66 $(-NH-(CH_2)_6NHCO-(CH_2)_4-CO-)_n$	α-Form, triclinic, $P\bar{1}$-C_i^1, $a = 4.9$ Å, $b = 5.4$ Å, $c = 17.2$ Å, $\alpha = 48.5°$, $\beta = 77°$, $\gamma = 63.5°$, $N = 1$	Planar zigzag (1/0)	1.24
	β-Form, triclinic, $P\bar{1}$-C_i^1, $a = 4.9$ Å, $b = 8.0$ Å, $c = 17.2$ Å, $\alpha = 90°$, $\beta = 77°$, $\gamma = 67°$, $N = 2$	Planar zigzag (1/0)	1.248
Nylon 610[†] $(-NH-(CH_2)_6-NHCO-(CH_2)_8-CO-)_n$	α-Form, triclinic, $P\bar{1}$-C_i^1, $a = 4.95$ Å, $b = 5.4$ Å, $c = 22.4$ Å, $\alpha = 49°$, $\beta = 76.5°$, $\gamma = 63.5°$, $N = 1$	Planar zigzag (1/0)	1.157
	β-Form, triclinic, $P\bar{1}$-C_i^1, $a = 4.9$ Å, $b = 8.0$ Å, $c = 22.4$ Å, $\alpha = 90°$, $\beta = 77°$, $\gamma = 67°$, $N = 2$	Planar zigzag (1/0)	1.196
Poly(ethylene terephthalate) $(-O-(CH_2)_2-O-CO-\!\!-\!\!\bigcirc\!\!-\!\!-CO-)_n$	Triclinic, $P\bar{1}$-C_i^1, $a = 4.56$ Å, $b = 5.94$Å, $c = 10.75$ Å, $\alpha = 98.5°$, $\beta = 112°$, $N = 1$	Nearly planar	1.455

[†]In this case, Nylon 6,10 may be clearer.

which introduce considerable order into the procedure. These are:

1. *The Equivalence Postulate*. It is possible to assume that all mer units in a crystal occupy geometrically equivalent positions with respect to the chain axis.

2. *The Minimum Energy Postulate*. The conformation of the chain in a crystal may be assumed to approach the conformation of minimum potential energy for an isolated chain oriented along an axis.

3. *The Packing Postulate*. As many elements of symmetry of isolated chain as possible are maintained in the lattice, so equivalent atoms of different mer units along an axis tend to assume equivalent positions with respect to atoms of neighboring chains.

The equivalence postulate is seen in the structures given in Figure 6.6. Here, the chain mers repeat their structure in the next unit cell.

Energy calculations made for both single molecules and their unit cells serve three purposes: (a) they clarify the factors governing the crystal and molecular structure already tentatively arrived at experimentally; (b) they suggest the most stable molecular conformation and its crystal packing starting from the individual mer chemical structure; and (c) they provide a collection of reliable potential functions and parameters for both intra- and intermolecular interactions based on well-defined crystal structures (23). An example of intermolecular interactions is hydrogen bonding in the polyamide structures described in Figure 6.8.

The packing postulate is seen at work in Figure 6.7, where enantiomorphic structures pack closer together in space than if the chains had the same sense of helical twist.

Lastly, one should not neglect the very simple but all important density. The crystalline cell is usually about 10% more dense than the bulk amorphous polymer. Significant deviations from this density must mean an incorrect model.

Because of the importance of polymer crystallinity generally, and the unit cell in particular, the subject has been reviewed many times (13, 17, 23, 29–38).

6.4 STRUCTURE OF CRYSTALLINE POLYMERS

6.4.1 The Fringed Micelle Model

Very early studies on bulk materials showed that some polymers were partly crystalline. X-ray line broadening indicated that the crystals were either very imperfect or very small (39). Assuming the latter, in 1928 Hengstenberg and Mark (40) estimated that the crystallites of ramie, a form of native cellulose, were about 55 Å wide and over 600 Å long by this method. It had already

Figure 6.9 The fringed micelle model. Each chain meanders from crystallite to crystallite, binding the whole mass together.

been established that the polymer chain passed through many unit cells. Because of the known high molecular weight, the polymer chain was calculated to be even longer than the crystallites. Hence it was reasoned that they passed in and out of many crystallites (30, 41–43). These findings led to the fringed micelle model.

According to the fringed micelle model, the crystallites are about 100 Å long (Figure 6.9). The disordered regions separating the crystallites are amorphous. The chains wander from the amorphous region through a crystallite, and back out into the amorphous region. The chains are long enough to pass through several crystallites, binding them together.

The fringe micelle model was used with great success to explain a wide range of behavior in semicrystalline plastics, and also in fibers. The amorphous regions, if glassy, yielded a stiff plastic. However, if they were above T_g, then they were rubbery and were held together by the hard crystallites. This model explains the leathery behavior of ordinary polyethylene plastics, for example. The greater tensile strength of polyethylene over that of low-molecular-weight hydrocarbon waxes was attributed to amorphous chains wandering from crystallite to crystallite, holding them together by primary bonds. The flexible nature of fibers was explained similarly; however, the chains were oriented along the fiber axis (see Section 6.3). The exact stiffness of the plastic or fiber was related to the degree of crystallinity, or fraction of the polymer that was crystallized.

Figure 6.10 Schematic view of a polyethylene single crystal exhibiting adjacent reentry. The orthorhombic subcell with dimensions a and b, typical of many n-paraffins, is illustrated below (38).

6.4.2 Polymer Single Crystals

Ideas about polymer crystallinity underwent an abrupt change in 1957 when Keller (15) succeeded in preparing single crystals of polyethylene. These were made by precipitation from extremely dilute solutions of hot xylene. These crystals tended to be diamond-shaped and of the order of 100–200 Å thick (see Figure 6.5) (17). Amazingly, electron diffraction analysis showed that the polymer chain axes in the crystal body were essentially perpendicular to the large, flat faces of the crystal. Since the chains were known to have contour lengths of about 2000 Å and the thickness of the single crystals was in the vicinity of 110–140 Å, Keller concluded that the polymer molecules in the crystals had to be folded upon themselves. These observations were immediately confirmed by Fischer (44) and Till (45).[†]

[†]The early literature reveals significant premonitions of this discovery. K. H. Storks, *J. Am. Chem. Soc.*, **60**, 1753(1938) suggested that the macromolecules in crystalline gutta percha were folded back and forth upon themselves in such a way that adjacent sections remained parallel. R. Jaccodine, *Nature (London)*, **176**, 305(1955) showed the spiral growth of polyethylene single crystals by a dislocation mechanism.

6.4.2.1 *The Folded Chain Model*

This led to the folded chain model, illustrated in Figure 6.10 (38). Ideally, the molecules fold back and forth with hairpin turns. While adjacent reentry has been generally confirmed by small-angle neutron scattering and infrared studies for single crystals, the present understanding of bulk crystallized polymers indicates a much more complex situation (see below).

Figure 6.10 uses polyethylene as the model material. The orthorhombic cell structure and the *a* and *b* axes are illustrated. The *c* axis runs parallel to the chains. The dimension ℓ is the thickness of the crystal. The predominant fold plane in polyethylene solution-grown crystals is along the (110) plane. Chain folding is also supported by NMR studies (see Section 6.7) (46–48).

For many polymers, the single crystals are not simple flat structures. The crystals often occur in the form of hollow pyramids, which collapse on drying. If the polymer solution is slightly more concentrated, or if the crystallization rate is increased, the polymers will crystallize in the form of various twins, spirals, and dendritic structures, which are multilayered (see Figure 6.11) (32). These latter form a preliminary basis for understanding polymer crystallization from bulk systems.

Simple homopolymers are not the only polymeric materials capable of forming single crystals. Block copolymers of poly(ethylene oxide) crystallize in the presence of considerable weight fractions of amorphous polystyrene (see Figure 6.12) (49) In this case, square-shaped crystals with some spirals are seen. The crystals reject the amorphous portion (polystyrene), which appears on the surfaces of the crystals.

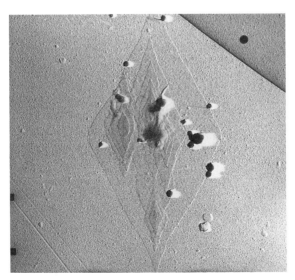

Figure 6.11   Single crystal of nylon 6 precipitated from a glycerol solution. The lamellae are about 60 Å thick. Black marks indicate 1 μm (32).

Figure 6.12 Single crystals of poly(ethylene oxide) – *block* – polystyrene diblock copolymers. (*a*) Optical micrograph. (*b*) Electron micrograph. M_n (PS) = 7.3 × 10³ g / mol; M_n (PEO) = 10.9 × 10³; weight fraction polystyrene is 0.34 (49).

Amorphous material also appears on the surfaces of homopolymer single crystals. As will be developed below, causes of this amorphous material range from chain-end cilia to irregular folding.

6.4.2.2 The Switchboard Model

In the switchboard model, the chains do not have a reentry into the lamellae by regular folding, but rather reenter more or less randomly (50). The model more or less resembles an old-time telephone switchboard. Of course, both the perfectly folded chain and switchboard models represent limiting cases. Real systems may combine elements of both. For bulk systems, this aspect is discussed in Section 6.7.

6.5 CRYSTALLIZATION FROM THE MELT

6.5.1 Spherulitic Morphology

In the previous sections it was observed that when polymers are crystallized from dilute solutions, they form lamellar-shaped single crystals. These crystals exhibit a folded chain habit and are of the order of 100–200 Å thick. From somewhat more concentrated solutions, various multilayered dendritic structures are observed.

Figure 6.13 Spherulites of low-density polyethylene, observed through crossed polarizers. Note characteristic Maltese cross pattern (32).

When polymer samples are crystallized from the bulk, the most obvious of the observed structures are the spherulites (31). As the name implies, spherulites are sphere-shaped crystalline structures that form in the bulk (see Figure 6.13) (32). One of the more important problems to be addressed concerns the form of the lamellae within the spherulite.

Spherulites are remarkably easy to grow and observe in the laboratory (33). Simple cooling of a thin section between crossed polarizers is sufficient, although controlled experiments are obviously more demanding. It is observed that each spherulite exhibits an extinction cross, sometimes called a Maltese cross. This extinction is centered at the origin of the spherulite, and the arms of the cross are oriented parallel to the vibration directions of the microscope polarizer and analyzer.

Usually the spherulites are really spherical in shape only during the initial stages of crystallization. During the latter stages of crystallization, the spherulites impinge on their neighbors. When the spherulites are nucleated simultaneously, the boundaries between them are straight. However, when the spherulites have been nucleated at different times, so that they are different in size when impinging on one another, their boundaries form

Figure 6.14 Surface replica of polyoxymethylene fractured at liquid nitrogen temperatures. Lamellae at lower left are oriented at an angle to the fracture surface. Lamellae elsewhere are nearly parallel to the fracture surface, being stacked up like cards or dishes in the bulk state. These structures closely resemble stacks of single crystals and led to ideas about chain folding in bulk materials (32).

hyperbolas. Finally, the spherulites form structures that pervade the entire mass of the material. The kinetics of spherulite crystallization are considered in Section 6.7.

Electron microscopy examination of the spherulitic structure shows that the spherulites are composed of individual lamellar crystalline plates (see Figure 6.14) (31). The lamellar structures sometimes resemble staircases, being composed of nearly parallel (but slightly diverging) lamellae of equal thickness.

X-ray microdiffraction (34) and electron diffraction (35) examination of the spherulites indicates that the c axis of the crystals is normal to the radial (growth) direction of the spherulites. Thus the c axis is perpendicular to the lamellae flat surfaces, showing the resemblance to single-crystal structures.

For some polymers, such as polyethylene (34), it was shown that their lamellae have a screwlike twist along their unit cell b axis, on the spherulite

Figure 6.15 Different types of light-scattering patterns are obtained from spherulitic poly-ethylene using (a) V_v and (b) H_v polarization (51).

radius. The distance corresponding to one-half of the pitch of the lamellar screw is just in accordance with the extinction ring interval visible on some photographs.

The growth and structure of spherulites may also be studied by small-angle light scattering (36, 37). The sample is placed between polarizers, a monochromatic or laser light beam is passed through, and the resultant scattered beam is photographed. Two types of scattering patterns are obtained, depending on polarization conditions (51). When the polarization of the incident beam and that of the analyzer are both vertical, it is called a V_v type of pattern. When the incident radiation is vertical in polarization but the analyzer is horizontal (polarizers crossed), an H_v pattern is obtained.

The two types of scattering patterns are illustrated in Figure 6.15 (51). These patterns arise from the spherulitic structure of the polymer, which is optically anisotropic, with the radial and tangential refractive indices being different.

The scattering pattern can be used to calculate the size of the spherulites (37) (see Figure 6.16). The maximum that occurs in the radial direction, U, is related to R, the radius of the spherulite by

$$U_{max} = \left(\frac{4\pi R}{\lambda} \right) \sin\left(\frac{\theta_{max}}{2} \right) = 4.1 \qquad (6.2)$$

where θ_{max} is the angle at which the intensity maximum occurs and λ is the wavelength. As the spherulites get larger, the maximum in intensity occurs at smaller angles.

Conversely, Stein (51) points out that in very rapidly crystallized polymers, spherulites are often not observed. The smaller amount of scattering observed results entirely from local structure. These structures are highly disordered.

Mandelkern recently drew a morphological map for polyethylene (52). He showed that the supermolecular structures become less ordered as the

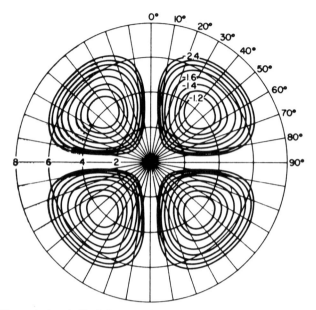

Figure 6.16 The calculated H_v light-scattering pattern for an idealized spherulite (37). Compare with the actual result, Figure 6.15*b*.

molecular weight is increased or the temperature of crystallization is decreased.

A model of the spherulite structure is illustrated in Figure 6.17 (38). The chain direction in the bulk crystallized lamellae is perpendicular to the broad plane of the structure, just like the dilute solution crystallized material.

The spherulite lamellae also contain low-angle branch points, where new lamellar structures are initiated. The new lamellae tend to keep the spacing between the crystallites constant.

While the lamellar structures in the spherulites are the analogue of the single crystals, the folding of the chains is much more irregular, as will be developed further in Section 6.6.2.3. In between the lamellar structures lies amorphous material. This portion is rich in components such as atactic polymers, low-molecular-weight material, or impurities of various kinds.

The individual lamellae in the spherulites are bonded together by tie molecules, which lie partly in one crystallite and partly in another. Sometimes these ties molecules are actually in the form of what are called intercrystalline links (53–56), which are long, threadlike crystalline structures with the *c* axis along their long dimension. These intercrystalline links are thought to be important in the development of the great toughness characteristic of semicrystalline polymers. They serve to tie the entire structure together by crystalline regions and/or primary chain bonds.

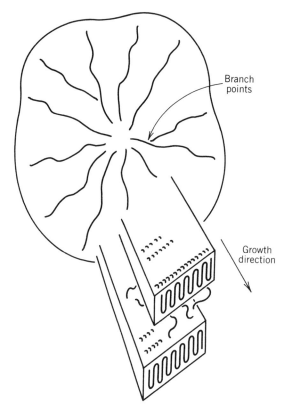

Figure 6.17 Model of spherulitic structure. Note growth directions and lamellar branch points, to fill space uniformly with crystalline material. After J. D. Hoffman et al. (38).

6.5.2 Mechanism of Spherulite Formation

On cooling from the melt, the first structure that forms is the single crystal. These rapidly degenerate into sheaflike structures during the early stages of the growth of polymer spherulites (see Figure 6.18) (31). These sheaflike structures have been variously called axialites (57) or hedrites (58). These transitional, multilayered structures represent an intermediate stage in the formation of spherulites (59). It is evident from Figure 6.18 that as growth proceeds, the lamellae develop on either side of a central reference plane indicated by *AB*. The lamellae fan out progressively and grow away from the plane as the structure begins to mature.

The sheaflike structures illustrated in Figure 6.18 are modeled in Figure 6.19 (31). As in Figure 6.18, both edge-on and flat-on views are illustrated. Figure 6.18*a* is modeled by Figure 6.19, row *a*, column III, and Figure 6.18*b* is modeled by row *b*, column III. Gradually, the lamellae in the hedrites

Figure 6.18 Electron micrographs of replicas of hedrites formed in the same melt-crystallized thin film of poly(4-methylpentene-1). (*a*) An edge-on view of a hedrite. Note the distinctly lamellar character and the "sheaflike" arrangement of the lamellae. (*b*) A flat-on view of a hedrite. Note the degenerate overall square outline of the object, whose lamellar texture is evident (31).

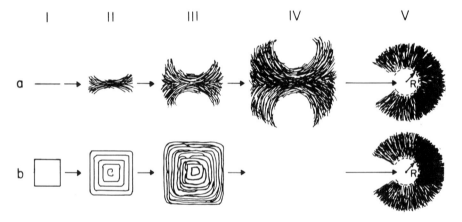

Figure 6.19 Schematic development of a spherulite from a chain-folded precursor crystal. Rows (*a*) and (*b*) represent, respectively, edge-on and flat-on views of the evolution of the spherulite (31).

diverge or fan outward in a splaying motion. Repeated splaying, perhaps aided by lamellae that are intrinsically curved, eventually leads to the spherical shape characteristic of the spherulite.

6.5.3 Spherulites in Polymer Blends and Block Copolymers

There are two cases to be considered. Either the two polymers composing the blend may be miscible and form one phase in the melt, or they are immiscible and form two phases. Martuscelli (60) pointed out that if the glass transition of the noncrystallizing component is lower than that of the crystallizing component (i.e., its melt viscosity will be lower, other things being equal), then the spherulites will actually grow faster, although the system is diluted. This is in general agreement with Section 6.6.2.2, which shows why crystallizable polymers containing low-molecular-weight fractions (which are not incorporated in the spherulite) also crystallize faster. Martuscelli also pointed out that the inverse was also true, especially if the noncrystallizing polymer was glassy at the temperature of crystallization.

The crystallization behavior is quite different if the two polymers are immiscible in the melt. Figure 6.20 (61) shows droplets of polyisobutylene dispersed in isotactic polypropylene. On spherulite formation, the droplets, which are noncrystallizing, become ordered within the growing arms of the crystallizing component.

PiB$_{MM}$ (80/20)   Tc=133°C

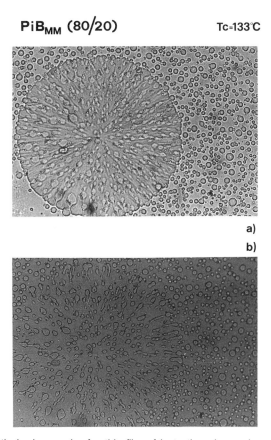

a)

b)

Figure 6.20 Optical micrograph of a thin film of isotactic polypropylene / polyisobutylene blend. (*a*) *it*-PP spherulite ($T_c = 133°C$), surrounded by melt blend at the early stage of crystallization. (*b*) The same region of film after melting of spherulite. Note multiphase morphology common to many blends (60).

Figure 6.21 Model of crystallizable triblock copolymer thermoplastic elastomer. The center block, amorphous, is rubbery, whereas the end blocks are crystalline (64).

Block copolymers also form spherulites (62–64). The morphology develops on a finer scale, however, because the domains are constrained to be of the order of the size of the individual blocks. In the case of a triblock copolymer, for example, the chain may be modeled as wandering from one lamella to another through an amorphous phase consisting of the center block (see Figure 6.21) (64).

Figure 6.22 (63) illustrates the spherulite morphology for poly(ethylene oxide)–*block*–polystyrene. Two points should be made. First, the glass transition temperature of the polystyrene component is higher than the temperature of crystallization (62); this particular sample was made by casting from chloroform. Second, the amount of polystyrene is small, only 19.6%. When the polystyrene component is increased, it disturbs the ordering process (see Figure 6.23) (63). This figure shows spheres rich in poly(ethylene oxide) lamellae but containing some polystyrene segments (dark spots) embedded in the poly(ethylene oxide) spherulites, as well as forming the more continuous phase. The size of the fine structure is of the order of a few hundred angstroms, because the two blocks must remain attached, even though they are in different phases.

Figure 6.22 An optical micrograph of a poly(ethylene oxide) – *block* – polystyrene copolymer containing 19.6% polystyrene (63). Crystals cast from chloroform and observed through crossed polarizers. White markers are 250 μm apart.

In the case where the polymer forms a triblock copolymer of the structure $A-B-A$, or a multiblock copolymer $(A-B)_m$, where A is crystalline at use temperature and B is rubbery (above T_g), then a thermoplastic elastomer is formed (see Figure 6.21). The material exhibits some degree of rubber elasticity at use temperature, the crystallites serving as cross-links. Above the melting temperature of the crystalline phase, the material is capable of flowing: that is, it is thermoplastic. It must be pointed out that a

Figure 6.23 A transmission electron micrograph of poly(ethylene oxide) – *block* – polystyrene, containing 70% polystyrene. The polystyrene phase is stained with OsO_4 (63).

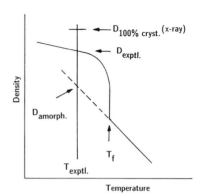

Figure 6.24 The experimental determination of the extent of polymer crystallinity using the density method.

very important kind of thermoplastic elastomer is when the *A* polymer is glassy rather than crystalline (see Section 9.16).

6.5.4 Percent Crystallinity in Polymers

As suggested above, most polymers are semicrystalline; that is, a certain fraction of the material is amorphous, while the remainder is crystalline. The reason why polymers fail to attain 100% crystallinity is kinetic, resulting from the inability of the polymer chains to completely disentangle and line up properly in a finite period of cooling or annealing.

There are several methods for determining the percent crystallinity in such polymers. The first involves the determination of the heat of fusion of the whole sample by calorimetric methods such as DSC; see Figure 6.3. The heat of fusion per mole of crystalline material can be estimated independently by melting point depression experiments; see Section 6.8.

A second method involves the determination of the density of the crystalline portion via x-ray analysis of the crystal structure, and determining the theoretical density of a 100% crystalline material. The density of the amorphous material can be determined from an extrapolation of the density from the melt to the temperature of interest; see Figure 6.24. Then the percent crystallinity is given by

$$\% \text{ Crystallinity} = \left[\frac{D_{\text{exptl}} - D_{\text{amorph}}}{D_{100\% \text{ cryst}} - D_{\text{amorph}}} \right] \times 100 \qquad (6.3)$$

where D_{exptl} represents the experimental density, and D_{amorph} and $D_{100\% \text{ cryst}}$ are the densities of the amorphous and crystalline portions, respectively.

A third method stems from the fact that the intensity of x-ray diffraction depends on the number of electrons involved and is thus proportional to the density. Besides Bragg diffraction lines for the crystalline portion, there is an amorphous halo caused by the amorphous portion of the polymer. This last occurs at a slightly smaller angle than the corresponding crystalline peak, because the atomic spacings are larger. The amorphous halo is broader than the corresponding crystalline peak, because of the molecular disorder.

Naturally, these methods will not yield the same answer for a given sample, but surprisingly good agreement is obtained. For many semicrystalline polymers, the crystallinity is in the range of 40–75%. Polymers such as polytetrafluoroethylene achieve 90% crystallinity, while poly(vinyl chloride) is often down around 15% crystallinity. The latter polymer is largely atactic, but short syndiotactic segments contribute greatly to its crystallinity. Of course, annealing usually increases crystallinity, as does orienting the polymer in fiber or film formation.

6.6 KINETICS OF CRYSTALLIZATION

During crystallization from the bulk, polymers form lamellae, which in turn are organized into spherulites or their predecessor structures, hedrites. This section is concerned with the rates of crystallization under various conditions of temperature, molecular weight, structure, and so on and the theories that provide not only an insight into the molecular mechanisms but considerable predictive power.

6.6.1 Experimental Observations of Crystallization Kinetics

It has already been pointed out that the volume changes on melting, usually increasing (see Figure 6.2). This phenomenon may be used to study the kinetics of crystallization. Figure 6.25 (65) illustrates the isothermal crystallization of poly(ethylene oxide) as determined dilatometrically. From Table 6.1 the melting temperature of poly(ethylene oxide) is 66° C, where the rate of crystallization is zero. The rate of crystallization increases as the temperature is decreased. This follows from the fact that the driving force increases as the sample is supercooled.

Crystallization rates may also be observed microscopically, by measuring the growth of the spherulites as a function of time. This may be done by optical microscopy, as has been done by Keith and Padden (66, 67), or by transmission electron microscopy of thin sections (68). The isothermal radial growth of the spherulites is usually observed to be linear (see Figure 6.26) (67). This implies that the concentration of impurity at the growing tips of the lamellae remains constant through the growth process. The more impurity, the slower the growth rate (Table 6.3) (67). However, linearity of growth rate is maintained. In such a steady state, the radial diffusion of rejected impurities is outstripped by the more rapidly growing lamellae, so that impurities diffuse aside and are trapped in interlamellar channels. In the case illustrated in Figure 6.26, the main "impurity" is low-molecular-weight atactic polypropylene.

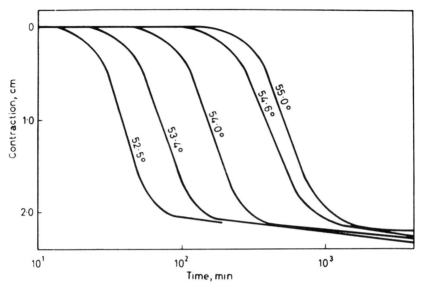

Figure 6.25   Dilatometric crystallization isotherms for poly(ethylene oxide), $M = 20,000$ g / mol. The Avrami exponent n falls from 4.0 to 2.0 as crystallization proceeds (65).

Figure 6.26   Spherulite radius as a function of time, grown isothermally (at 125°C) in a blend of 20% isotactic and 80% atactic ($M = 2600$) polypropylene. Note the linear behavior (67).

Table 6.3   Blends of unextracted isotactic polypropylene with atactic polypropylene (67)

Temperature (°C)	Radial Growth Rates (μm/min) for Various Compositions				
	100% Isotactic	90% Isotactic	80% Isotactic	60% Isotactic	40% Isotactic
120	29.4	29.4	26.4	22.8	21.2
125	13.0	12.0	11.0	8.90	8.57
131	3.88	3.60	3.03	2.37	2.40
135	1.63	1.57	1.35	1.18	1.12
Melting point (°C)	171	169	167	165	162

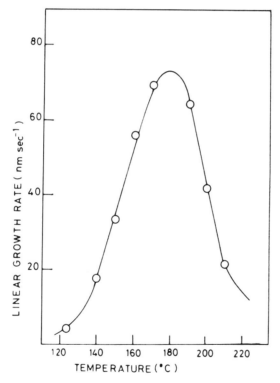

Figure 6.27 Plot of linear growth rate versus crystallization temperature for poly(ethylene terephthalate) (68). For this polymer, $T_m = 280°C$, and $T_g = 67°C$, at which points the rates of crystallization are theoretically zero.

When the radial growth rate is plotted as a function of crystallization temperature, a maximum is observed (see Figure 6.27) (68). As mentioned earlier, the increase in rate of crystallization as the temperature is lowered is controlled by the increase in the driving force. As the temperature is lowered still further, molecular motion becomes sluggish as the glass transition is approached, and the crystallization rate decreases again. Below T_g, molecular motion is so sluggish that the rate of crystallization effectively becomes zero.

6.6.2 Theories of Crystallization Kinetics

Section 6.4.2 described Keller's early preparations of single crystals from dilute solutions. Since the crystals were only about 100 Å thick and the chains were oriented perpendicular to the flat faces, Keller postulated that the chains had to be folded back and forth.

Similar structures, called lamellae, exist in the bulk state. While their folding is now thought to be much less regular, their proposed molecular organization remains similar. In the bulk state, however, these crystals are organized into the larger structures known as spherulites. Section 6.6.1 showed that the rate of radial growth of the spherulites was linear in time and that the rate of growth goes through a maximum as the temperature of crystallization is lowered. These several experimental findings form the basis for three theories of polymer crystallization kinetics.

The first of these theories is based on the work of Avrami (69–71), which adapts formulations intended for metallurgy to the needs of polymer science. The second theory was developed by Keith and Padden (66, 67), providing a qualitative understanding of the rates of spherulitic growth. Most recently, Hoffman and co-workers (38, 73–76) developed the kinetic nucleation theory of chain folding, which provides an understanding of how lamellar structures form from the melt. This theory continues to be developed even as this material is being written. Together, these theories provide insight into the kinetics not only of crystallization but also of the several molecular mechanisms taking part.

6.6.2.1 The Avrami Equation

The original derivations by Avrami (69–71) have been simplified by Evans (77) and put into polymer context by Meares (78) and Hay (79). In the following, it is helpful to imagine raindrops falling in a puddle. These produce expanding circles of waves that intersect and cover the whole surface. The drops may fall sporadically or all at once. In either case, they must strike the puddle surface at random points. The expanding circles of waves, of course, are the growth fronts of the spherulites, and the points of impact are the crystallite nuclei.

The probability p_x that a point P is crossed by x fronts of growing spherulites is given by an equation originally derived by Poisson (80):

$$p_x = \frac{e^{-E}E^x}{x!} \tag{6.4}$$

where E represents the average number of fronts of all such points in the system. The probability that P will not have been crossed by any of the fronts, and is still amorphous, is given by

$$p_0 = e^{-E} \tag{6.5}$$

since E^0 and 0! are both unity. Of course, p_0 is equal to $1 - X_t$, where X_t is the volume fraction of crystalline material, known widely as the degree of crystallinity. Equation (6.5) may be written

$$1 - X_t = e^{-E} \tag{6.6}$$

which for low degrees of crystallinity yield the useful approximation

$$X_t \cong E \tag{6.7}$$

For the bulk crystallization of polymers, X_t (in the exponent) may be considered the volume of crystallization material, V_t:

$$1 - X_t = e^{-V_t}. \tag{6.8}$$

The problem now resides on the evaluation of V_t. There are two cases to be considered: (a) the nuclei are predetermined—that is, they all develop at once on cooling the polymer to the temperature of crystallization—and (b) there is sporadic nucleation of the spheres.

For case (a), L spherical nuclei, randomly placed, are considered to be growing at a constant rate, g. The volume increase in crystallinity in the time period t to $t + dt$ is

$$dV_t = 4\pi r^2 L\, dr \tag{6.9}$$

where r represents the radius of the spheres at time t; that is,

$$r = gt \tag{6.10}$$

and

$$V_t = \int_0^1 4\pi g^2 t^2 Lg\, dt \tag{6.11}$$

Upon integration,

$$V_t = \tfrac{4}{3}\pi g^3 L t^3 \tag{6.12}$$

For sporadic nucleation the above argument is followed, but the number of spherical nuclei is allowed to increase linearly with time at a rate 1. Then spheres nucleated at time t_i will produce a volume increase of

$$dV_t = 4\pi g^2 (t - t_i)^2 l t g \, dt \tag{6.13}$$

Upon integration,

$$V_t = \tfrac{2}{3}\pi g^3 l t^4 \tag{6.14}$$

The quantities on the right of equations (6.12) and (6.14) can be substituted into equation (6.8) to produce the familiar form of the Avrami equation:

$$1 - X_t = e^{-Zt^n} \tag{6.15}$$

which is often written in the logarithmic form:

$$\ln(1 - X_t) = -Zt^n \tag{6.16}$$

The quantity Z is replaced by K in some books (78).

The above derivation suggests that the quantity n in equations (6.15) or (6.16) should be either 3 or 4. [If rates of crystallization are diffusion controlled, which occurs in the presence of high concentrations of noncrystallizable impurities (66, 67) $r = gt^{1/2}$, leading to half-order values of n.]

Both Z and n are diagnostic of the crystallization mechanism. The equation has been derived for spheres, discs, and rods, representing three-, two-, and one-dimensional forms of growth. The constants are summarized in Table 6.4 (79). It must be emphasized that the approximation given in equation (6.7) limits the equations to low degrees of crystallinity. In practice,

Table 6.4 The Avrami parameters for crystallization of polymers (79)

	Crystallization Mechanism	Avrami Constants		Restrictions
		Z	n	
Spheres	Sporadic	$2/3\pi g^3 l$	$4 \cdot 0$	3 dimensions
	Predetermined	$4/3\pi g^3 L$	$3 \cdot 0$	3 dimensions
Discs[a]	Sporadic	$\pi/3g^2 l d$	$3 \cdot 0$	2 dimensions
	Predetermined	$\pi g^2 L d$	$2 \cdot 0$	2 dimensions
Rods[b]	Sporadic	$\pi/4g l d^2$	$2 \cdot 0$	1 dimension
	Predetermined	$\tfrac{1}{2}\pi g L d^2$	$1 \cdot 0$	1 dimension

[a] Constant thickness d.
[b] Constant radius d.

Table 6.5   Range of the Avrami constant for typical polymers (79)

Polymer	Range of n	Reference
Polyethylene	2.6–4.0	(a)
Poly(ethylene oxide)	2.0–4.0	(b, c)
Polypropylene	2.8–4.1	(d)
Poly(decamethylene terephthalate)	2.7–4.0	(e)
it-Polystyrene	2.0–4.0	(f, g)

References: (a) W. Banks, M. Gordon, and A. Sharples, *Polymer*, **4**, 61, 289(1963). (b) J. N. Hay, M. Sabin, and R. L. T. Stevens, *Polymer*, **10**, 187(1969). (c) W. Banks and A. Sharples, *Makromol. Chem.*, **59**, 283(1963). (d) P. Parrini and G. Corrieri, *Makromol. Chem.*, **62**, 83(1963). (e) A. Sharples and F. L. Swinton, *Polymer*, **4**, 119(1963). (f) I. H. Hillier, *J. Polym. Sci.*, **A-2**(4), 1(1966). (g) J. N. Hay, *J. Polym. Sci.*, **A-3**, 433(1965).

the quantity n frequently decreases as the crystallization proceeds. Values for typical polymers are summarized in Table 6.5 (79).

If the system is considered as two-phased, then the volume of the amorphous phase is V_a and the volume of the crystalline phase is V_c. The total volume, V, is given by

$$V = X_t V_c + (1 - X_t)V_a \tag{6.17}$$

Then

$$1 - X_t = \frac{V - V_c}{V_a - V_c} \tag{6.18}$$

or, for dilatometric experiments,

$$1 - X_t = \frac{h_0 - h_t}{h_0 - h_\infty} \tag{6.19}$$

where h_0, h_t, and h_∞ represent dilatometric heights at time zero, time t, and the final dilatometric reading. Substitution of equation (6.19) into (6.16) yields a method of determining the constants Z and n experimentally (e.g., see Figure 6.25).

6.6.2.2   Keith–Padden Kinetics of Spherulitic Crystallization

Although the Avrami equation provides useful data on the overall kinetics of crystallization, it provides little insight as to the molecular organization of the crystalline regions, structure of the spherulites, and so on.

Section 6.5 described how the spherulites are composed of lamellar structures that grow outward radially. The individual chains are folded back and forth tangentially to the growing spherical surface of the spherulite (see

CHAIN AXIS IN
LAMELLAR
FIBERS

GROWTH
DIRECTION

DISORDERED
INTERFIBRILLAR
MELT

Figure 6.28   Schematic representation of the distribution of residual melt and disordered material among lamellae in a spherulite (67).

Figure 6.28) (67). Normally, the rate of growth in the radial direction is constant until the spherulites meet (see Figure 6.26). As the spherulites grow, the individual lamellae branch. Impurities, atactic components, and so on become trapped in the interlamellar regions.

The first theory to address the kinetics of spherulitic growth in crystallizing polymers directly was developed by Keith and Padden (66, 67, 81). According to Keith and Padden (66), a parameter of major significance is the quantity

$$\delta = \frac{D}{G} \tag{6.20}$$

where D is the diffusion coefficient for impurity in the melt and G represents the radial growth rate of a spherulite. The quantity δ, whose dimension is that of length, determines the lateral dimensions of the lamellae, and that noncrystallographic branching should be observed when δ becomes small enough to be commensurate with the dimensions of the disordered regions on their surfaces. Thus δ is a measure of the internal structure of the spherulite, or its coarseness.

By logarithmic differentiation of equation (6.20),

$$\frac{1}{\delta}\left(\frac{d\delta}{dT}\right) = \frac{1}{D}\left(\frac{dD}{dT}\right) - \frac{1}{G}\frac{dG}{dT} \tag{6.21}$$

The derivative dD/dT always has a positive value. However, dG/dT may be positive or negative (see Figure 6.27). The coarseness of the spherulites depends on which of the two terms on the right of equation (6.21) is the larger. If the quantity on the right-hand side of the equation is positive, an increase in coarseness is expected as the temperature is increased.

The radial growth rate, G, may be described by the equation

$$G = G_0 e^{\Delta E/RT} e^{-\Delta F^*/RT} \tag{6.22}$$

where ΔF^* is the free energy of formation of a surface nucleus of critical size, and ΔE is the free energy of activation for a chain crossing the barrier to the crystal. Equation (6.22) allows the temperature dependence of spherulite growth rates to be understood in terms of two competing processes. Opposing one another are the rate of molecular transport in the melt, which increases with increasing temperature, and the rate of nucleation, which decreases with increasing temperature (see Figure 6.27). According to Keith and Padden, diffusion is the controlling factor at low temperatures, whereas at higher temperatures the rate of nucleation dominates. Between these two extremes, the growth rate passes through a maximum where the two factors are approximately equal in magnitude.

6.6.2.3   *Hoffman's Nucleation Theory*

The major shortcoming of the Keith–Padden theory resides in its qualitative nature. Although great insight into the morphology of spherulites was attained, little detail was given concerning growth mechanisms, particularly the thermodynamics and kinetics of the phenomenon.

More recently, Hoffman and co-workers attacked the kinetics of polymer crystallization anew (38, 72–76). Hoffman began with the assumption that chain folding and lamellar formation are kinetically controlled, the resulting crystals being metastable. The thermodynamically stable form is the extended chain crystal, obtainable by crystallizing under pressure (82).

The basic model is illustrated in Figure 6.29 (38), where ℓ is the thin dimension of the crystal, x the large dimension, σ_e the fold surface interfacial free energy, and σ the lateral surface interfacial free energy, a single chain making up the entire crystal. The free energy of formation of a single chain-folded crystal may be set down in the manner of Gibbs as

$$\Delta\phi_{\text{crystal}} = 4x\ell\sigma + 2x^2\sigma_e - x^2\ell(\Delta f) \tag{6.23}$$

where the quantity Δf represents the bulk free energy of fusion, which can be approximated from the entropy of fusion, ΔS_f, by assuming that the heat

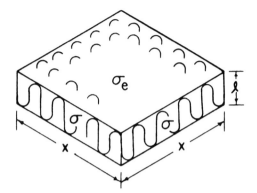

Figure 6.29 Thin chain-folded crystal showing σ and σ_e (schematic) (38).

of fusion, Δh_f, is independent of the temperature:

$$\Delta f = \Delta h_f - T\,\Delta S_f = \Delta h_f - \frac{T\,\Delta h_f}{T_f^0} = \frac{\Delta h_f(\Delta T)}{T_f^0} \qquad (6.24)$$

At the melting temperature of the crystal, the free energy of formation is zero. For $x \gg \ell$,

$$T_f = T_f^0\left[1 - \frac{2\sigma_e}{\Delta h_f \ell}\right] \qquad (6.25)$$

Equation (6.25) yields the melting point depression in terms of fundamental parameters. The quantity σ_e may be interpreted in terms of the fold structure. For the actual crystallization, chains are added from the melt or solution to the surface of the crystal defined by the area $x\ell$.

6.6.2.4 An Example Calculation

Find the fold surface interfacial free energy of a new polymer with the following characteristics: the actual melting temperature is 350°C, while after extended annealing, it melts at 360°C. X-ray analysis shows its lamellae are 150 Å thick. The heat of fusion is 12.0 kJ/mol. (See Table 6.1.)
    Equation (6.25) can be rewritten

$$\sigma_e = \frac{\Delta h_f \ell}{2}\left(1 - \frac{T_f}{T_f^0}\right)$$

Substituting the given values,

$$\sigma_e = \frac{12.0 \text{ kJ/mol} \times 1.50 \times 10^{-6} \text{ cm}}{2}\left(1 - \frac{623 \text{ K}}{633 \text{ K}}\right)$$

$$= 1.42 \times 10^{-7} \text{ kJ} \cdot \text{cm/mol}$$

This polymer has a mer molecular weight of 100 g/mol and a density of 1.0 g/cm^3. Then,

$$100 \text{ g/mol} \times 1 \text{ cm}^3/\text{g} = 100 \text{ cm}^3/\text{mol}$$

and

$$\sigma_e = \frac{1.42 \times 10^{-7} \text{ kJ} \cdot \text{cm/mol}}{100 \text{ cm}^3/\text{mol}} = 1.42 \times 10^{-9} \text{ kJ/cm}^2$$

6.6.2.5   *Three Regimes of Crystallization Kinetics*

Hoffman defined three regimes of crystallization kinetics from the melt, which differ according to the rate that the chains are deposited on the crystal surface.

Regime I. One surface nucleus causes the completion of the entire substrate of length L (see Figure 6.30) (73). Many molecules may be required to complete L. The term "surface nucleus" refers to a segment of a chain sitting down on a preexisting crystalline lamellar structure, as opposed to the nucleus, which initiates the lamellae from the melt in the first place.

These nuclei are deposited sporadically in time on the substrate at a rate i per unit length in a manner that is highly dependent on the temperature. Substrate completion at a rate g begins at the energetically favorable niche that occurs on either side of the surface nucleus, chain folding assumed.

As illustrated in Figure 6.30, the overall growth rate is given by G, and g is the substrate completion rate. The quantities a_0 and b_0 refer to the molecular width and layer thickness, respectively. The quantity l_g^* refers to the initial fold thickness of the lamellae. The portion of chain occupying the length l_g^* is called a stem.

Again noting Figure 6.30, the "reeling in" or reptation rate r (see Section 5.4) is given by

$$r = \left(\frac{\ell_g^*}{a_0}\right)g \tag{6.26}$$

A significant question, still being debated in the literature, is whether or not reptation type diffusion is sufficiently rapid to supply chain portions as required (83). An alternate theory, proposed by Yoon and Flory (83) suggests that disengagement of the macromolecule from its entanglements with other

Figure 6.30 Surface nucleation and substrate completion with reptation in regime I, where one surface nucleus deposited at rate i causes completion of substrate of length L, giving overall growth rate $G_1 = b_0 iL$. Multiple surface nuclei occur in regime II (not shown) and lead to $G_{\parallel} = b_0(2ig)^{1/2}$, where g is the substrate completion rate. The substrate completion rate, g, is associated with a "reeling in" rate $r = (\ell_g^*/a_0)g$ for the case of adjacent reentry (73).

chains in the melt is necessary for regular folding. In this theory, the 100–200 skeletal bonds corresponding to one traversal of the crystal lamella readily undergo the conformational rearrangements required for their deposition in the growth layer. On the other hand, Hoffman (73) concludes that the reptation rate characteristic of the melt is fast enough to allow a significant degree of adjacent reentry "regular" folding during crystallization.

According to Hoffman, the overall growth rate is given by

$$G_1 = b_0 iL = b_0 a_0 n_s i \tag{6.27}$$

where G_1 is the growth rate in regime I, n_s represents the number of stems of width a_0 that make up this length, and i is the rate of deposition of surface nuclei.

The free energy of crystallizing ν stems and $\nu_f = \nu - 1$ folds can be generalized from equation (6.23), using Figure 6.30:

$$\Delta\phi_\nu = 2b_0 \ell \sigma + 2\nu_f a_0 b_0 \sigma_e - \nu a_0 b_0 \ell \, \Delta f \tag{6.28}$$

Figure 6.31 Free energy of formation of a chain-folded surface nucleus as a function of the number of stems. The diagram shows the relationship of the rate constants A_0, B_1, A, and B, to the free energy of the processes involved in the formation and growth of the surface nucleus as it spreads in the g direction (38).

which for ν large becomes

$$\Delta\phi_\nu = 2b\ell_g^*\sigma + \nu a_0 b_0\left(2\sigma_e - \ell_g^*\,\Delta f\right) \qquad (6.29)$$

The total flux of polymer from the melt to the crystal may be calculated as follows. The quantity S represents the net flow of polymer from sites in the liquid ($\nu = 0$) to the first stem of polymer on the crystal, $\nu = 1$. The quantity N_0 represents the number of chains in the liquid, N_1 the number of molecules having one stem on the crystal, and so on. The corresponding rate constants are A_0, B_1, and A and B (see Figure 6.31) (38). These quantities are equated to exponential functions derived from the model in Figure 6.30. Finally, the growth rate in regime I can be expressed (84) as

$$G_{\mathrm{I}} = \left(\frac{C_{\mathrm{I}}}{n}\right)\exp\left(\frac{Q_D^*}{RT}\right)\exp\left(-\frac{K_{g(\mathrm{I})}}{T(\Delta T)}\right) \qquad (6.30)$$

where C_{I} is a preexponential factor for regime I, n is the number of $-CH_2-$ units in a polymer chain, or mers, Q_D^* is the activation energy for steady-state reptation, $K_{g(\mathrm{I})}$ is the nucleation constant for regime I, and ΔT

Figure 6.32 Model for regime II growth showing multiple nucleation. The quantity S_k represents the mean separation between the primary nuclei, and S_n denotes the mean distance between the associated niches. The primary nucleation rate is i, and the substrate completion rate is g. The overall observable growth rate is G_{II}. Reptation tube contains molecule being reeled at rate r onto substrate (72).

represents the melting temperature minus the actual temperature of crystallization, or the degree of undercooling. Numerical examples are given below.

Regime II. In this regime, multiple surface nuclei occur on the same crystallizing surface, because the rate of nucleation is larger than the rate crystallization of each molecule. This, in turn, results from the larger undercooling necessary to reach regime II. As in regime I, each molecule is assumed to fold back and forth to give adjacent reentry (see Figure 6.32) (72). An important parameter in this regime is the niche separation distance, S_n.

The rate of growth of the lamellae is

$$G_{\text{II}} = b_0(2ig)^{1/2} \tag{6.31}$$

The number of nucleation sites per centimeter in regime II is given by

$$N_k = \left(\frac{i}{2g}\right)^{1/2} \tag{6.32}$$

Then, the mean separation distance between the sites is

$$S_k = \frac{1}{N_k} = \left(\frac{2g}{i}\right)^{1/2} \tag{6.33}$$

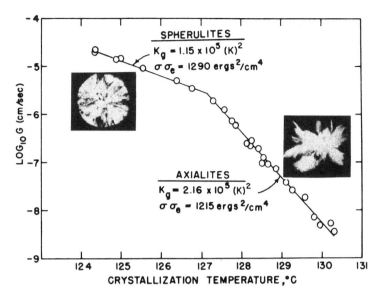

Figure 6.33 Change in growth kinetics from regime I to regime II in melt-crystallized polyethylene. Growth rate in melt-crystallized polyethylene is shown as a function of the crystallization temperature for fraction with molecular weight of 30,600 g/mol. The pronounced change in slope is accompanied by a change in morphology, as shown by optical micrographs of the type from which the growth rate data were obtained (38).

Regimes I and II differ morphologically as well as kinetically. In regime I, polyethylene usually forms axialites. In regime II, normal spherulites are formed. The rates of growth show different slopes, as illustrated in Figure 6.33 (38). For very high or very low molecular weights, Hoffman et al. (85) found that the rates of growth of polyethylenes gave curves, rather than a sharp break at the dividing point between the two regimes.

Regime III. This regime becomes important when the niche separation characteristic of the substrate in regime II approaches the width of a stem. In this regime, the crystallization rate is very rapid. The growth rate is given by

$$G_{III} = b_0 iL = b_0 i n_s a_0 \qquad (6.34)$$

where n_s is the mean number of stems laid down in the niche adjacent to the newly nucleated stem.

In regime III, the chains do not undergo adjacent reentry into the lamellae but rather have only a few folds before entering the amorphous phase. Then they are free to reenter the same lamella via a type of switchboard model, or go on to the next lamella.

Based on equation (6.30) and its equivalents for regimes II and III, exact equations for the growth rate of polyethylene as a function of molecular weight have been worked out. These are summarized in Table 6.6. Thus,

Table 6.6 Crystallization rates for polyethylene (72)

Regime I. Between $T = 145$ and $129°C$

$$G_1 = \frac{5.00 \times 10^{13}}{n_z} e^{-7000/RT} e^{-1.800 \times 10^5 / T(\Delta T)}$$

Regime II. Between $T = 129$ and $122°C$

$$G_{II} = \frac{3.34 \times 10^7}{n_z} e^{-7000/RT} e^{-0.900 \times 10^5 / T(\Delta T)}$$

Regime III. Below $T = 122°C$

$$G_{III} = \frac{6.20 \times 10^{11}}{n_z} e^{-7000/RT} e^{-1.800 \times 10^5 / T(\Delta T)}$$

given the z-average number of $-CH_2-$ groups and the temperature of crystallization, the exact rates of growth of the lamellae in the spherulites (or axialites) may be determined.

As the temperature is lowered through regimes I, II, and III, substrate completion rates per chain decrease. However, more chains are crystallizing simultaneously. At temperatures approaching T_g, de Gennes reptation is severely limited, as illustrated in Figure 6.27.

6.7 THE REENTRY PROBLEM IN LAMELLAE

In the above, the lamellae were assumed to be formed through regular adjacent reentry, although it was recognized that this was an oversimplification. The concept of the switchboard and folded-chain models were briefly developed in Section 6.4. The question of the molecular organization within polymer single crystals as well as the bulk state has dogged polymer science since the discovery of lamellar-shaped single crystals in 1957 (15). Again, x-ray and other studies show that the chains are perpendicular to the lamellar surface (see Section 6.4.2).

Since the chain length far exceeds the thickness of the crystal, the chains must either reenter the crystal or go elsewhere. However, the relative merits of the switchboard versus folded-chain models remained substantially unresolved for several years for lack of appropriate instrumentation.

6.7.1 Infrared Spectroscopy

Beginning in 1968, Tasumi and Krimm (86) undertook a series of experiments using a mixed crystal infrared spectroscopy technique. Mixed single crystals of protonated and deuterated polymer were made by precipitation from dilute solution. The characteristic crystal field splitting in the infrared spectrum was measured and analyzed to determine the relative locations of the chain stems of one molecule, usually the deuterated portion, in the crystal lattice. The main experiments involved blending protonated and deuterated polyethylenes (87–89).

The main findings were that folding takes place with adjacent reentry along (110) planes for dilute solution-grown crystals. In addition, it was also concluded that there is a high probability for a molecule to fold back along itself on the next adjacent (110) plane.

Melt-crystallized polyethylene was shown to be organized differently, with a much lower (if any) extent of adjacent reentry (89). However, significant undercooling was required to prevent segregation of the deuterated species from the ordinary, hydrogen-bearing species.[†] Since the experimental rate of cooling was estimated to be 1°C per minute down to room temperature, it may be that some crystallization occurred in all three regimes (see Section 6.6.2.3).

6.7.2 Carbon-13 NMR

Additional evidence for chain folding in solution-grown crystals comes from carbon-13 NMR studies of partially epoxidized 1,4-trans-polybutadiene crystals (46, 47). This polymer was crystallized from dilute heptane solution and oxidized with *m*-chloroperbenzoic acid. This reaction is thought to epoxidize the amorphous portions present in the folds, while leaving the crystalline stem portions intact.

The result was a type of block copolymer with alternating epoxy and double-bonded segments. NMR analyses showed that for the two samples studied the chain-folded portion was about 2.4 and 3 mers thick, whereas the stems were 15.2 and 40.8 mers thick, respectively. Since the number of mer units to complete the tightest fold in this polymer has been calculated to be about three (48), the NMR study strongly favors a tight adjacent reentry fold model for single crystals.

Broad line proton NMR (90) and Raman analyses (91) of polyethylenes indicated three major regions for crystalline polymers: the crystalline region, the interfacial or interzonal region, and the amorphous or liquidlike region. For molecular weight of 250,000 g/mol, the three regions were 75, 10, and 15% of the total, respectively. The presence of the interfacial regions reduces the requirements for chain folding (92).

6.7.3 Small-Angle Neutron Scattering

6.7.3.1 Single-Crystal Studies

With the advent of small-angle neutron scattering (SANS), the several possible modes of chain reentry could be put to the test anew (76, 93–105).

[†]In fact, one early problem that continues to plague many studies is that deuterated polyethylene tends to phase-separate from ordinary, protonated polymer, even though they are chemically identical. The cause has been related to slightly different crystallization rates owing to polydeuteroethylene melting 6°C lower than ordinary polyethylene.

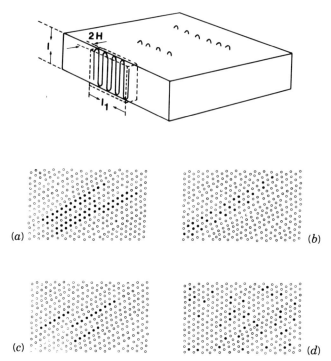

Figure 6.34 Models of stem reentry for chain sequences in a lamellar-shaped crystal. (a) Regular reentry with superfolding; (b) partial nonadjacency (stem dilution) as required by closer matching of the experimental data in accord with Yoon and Flory; (c) adjacent stem positions without superfolding; (d) the switchboard model. All reentry is along the (110) plane; superfolding is along adjacent (110) planes. View is from the (001) plane, indicated by dots (106).

Sadler and Keller (103–105) prepared blends of deuterated and normal (protonated) polyethylene and crystallized them from dilute solution. The radius of gyration, R_g, was determined as a function of molecular weight.

The several models possible are illustrated in Figure 6.34 (106). For adjacent reentry (Figure 6.34c), R_g should vary as $M^{1.0}$ for high enough molecular weights, since a type of rod would be generated. For the switchboard model (Figure 6.34d), R_g is expected to vary close to $M^{0.5}$, since the chains would be expected to be nearly Gaussian in conformation. The several possible relationships between R_g and M are set out in Table 6.7 (107).

For solution-grown crystals, Sadler and Keller (104) found that R_g depended on M only to the 0.1 power. Such a situation could arise only if the stems folded up on themselves beyond a certain number of entries (see Figure 6.34), called superfolding. However, the 0.1 power dependence appears to hold only for intermediate molecular weight ranges. For low enough M, there should not be superfolding. For high enough molecular weight, a square plate with a 0.5 power dependence would be generated.

Table 6.7   Relationships among geometric shape, R_g, and M (107)

Geometric Shape	R_g Equals	Molecular Weight Dependence
Sphere	$D/\sqrt{20/3}$	$M^{1/3}$
Rod	$L/\sqrt{12}$	M^1
Random coil	$r/\sqrt{6}$	$M^{0.5}$
Rectangular plate	$(b^2 + l^2)^{1/2}/\sqrt{12}$	M^{variable}
Square plate	$A^{1/2}/\sqrt{6}$	$M^{0.5}$

Symbols: D, diameter; L, length of rod; r, end-to-end distance; A, area; b, width of plate; l, length of plate.

Recent quantitative calculations of the absolute scattering intensities expected from various crystallite models for single crystals by Keller (105) and Yoon and Flory (108–110) (Figure 6.35) on polyethylene suggested that the model for adjacent reentry does not correlate with experiment. Rather, Yoon and Flory put forward a model requiring a stem dilution by a factor of 2–3. The calculated scattering functions are shown in Figure 6.35; this leads to the model in Figure 6.34b. This last suggests a type of skip mechanism, with two or three chains participating.

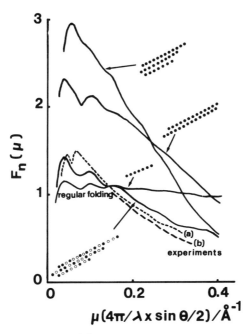

Figure 6.35   Scattering functions $F_n(\mu)$ calculated for various arrangements of $n_s = 40$ PED stems, each stem containing 80 bonds, in a PEH single-crystal matrix (105).

6.7.3.2 Melt-Crystallized Polymers

Upon crystallization from the melt, an entirely different result emerges. Experiments by Sadler and Keller (103–106) showed that nearly random stem reentry was most likely; that is, some type of switchboard model was correct. Quantitative calculations by Yoon and Flory (108–110) and by Dettenmaier et al. (111, 112) on melt-crystallized polyethylene (113) and isotactic polypropylene (114) also showed that adjacent reentry should occur only infrequently on cooling from the melt.

Three regions of space were defined by Yoon (109): a crystalline lamellar region about 100 Å thick, an interfacial region about 5–15 Å thick, and an amorphous region about 50 Å thick. A nearby reentry model constrains the chain within the interfacial layer during the irregular folding process. The calculated reentry dimensions for solution and melt-crystallized polyethylene are summarized in Table 6.8 (109). A major problem, of course, was that the samples had to be severely undercooled to prevent segregation of the two species during crystallization. Undoubtedly, large portions of the crystallization took place in regime III.

Crist et al. (115) deuterated or protonated a slightly branched polybutadiene to produce a type of polyethylene. Blends of these two materials were used in SANS studies. The scattering curves indicated identical dimensions (R_g values) for both the melt and melt-crystallized materials; these were the same as expected for Flory θ-solvent values for polyethylene. Using wide-angle neutron scattering, Wignall et al. (96) concluded that the number of stems that could be regularly folded had an upper limit of about four.

Of course, other crystallizable polymers have been studied (97–102). These include polypropylene (97–100), poly(ethylene oxide) (101), and isotactic polystyrene (102). Wignall et al. (96) have summarized the values of $R_g/M_w^{1/2}$ in both the melt and crystalline states (see Table 6.9). Most interestingly, the dimensions in the crystalline state and in the melt state are virtually identical for all these polymers. None of these data show a decrease in R_g on crystallization. That the R_g values in the melt and in the crystallized material are the same all but rules out regular folding. In fact, the data for poly(ethylene oxide) (101) shows a slight increase, if anything. By way of summarizing the above studies on melt-crystallized polymers, it seems that

Table 6.8   Polyethylene lamellar reentry dimensions (109)

Case	Reentry Statistics	
	Probability of Reentering Same Crystal	Average Displacement on Reentry
Solution-crystallized	1.0	10–15 Å
Melt-crystallized	0.7	25–30 Å

Table 6.9 Comparison of molecular dimensions in molten and crystallized polymers (96)

Polymer	Method of Crystallization	$R_g/M_w^{1/2}\mathrm{\AA}/(\mathrm{g/mol})^{1/2}$ Melt	Crystallized
Polyethylene	Rapidly quenched from melt	0.46	0.46
Polypropylene	Rapidly quenched	0.35	0.34
	Isothermally crystallized at 139°C	0.35	0.38
	Rapidly quenched from melt and subsequently annealed at 137°C	0.35	0.36
Poly(ethylene oxide)	Slowly cooled	0.42	0.52
Isotactic polystyrene	Crystallized at 140°C (5 hr)	0.26–0.28[a]	0.24–0.27
	Crystallized at 140°C (5 hr) then at 180°C (50 min)		0.26
	Crystallized at 200° (1 hr)	0.22[a]	0.24–0.29

[a]Dimensions in the melt were not available. The values quoted are for atactic polystyrene annealed in the same way as the crystalline material.

adjacent reentry occurs much less than in solution-crystallized polymers. Some experiments suggest very little adjacent reentry.

Hoffman (72) and Frank (116) point out, however, that some folding is required. Alternatively, the crystals must have the chains at an oblique angle to the crystal surface. If neither condition is met, a serious density anomaly at the crystalline–amorphous interface is predicted, the density being too high. These workers point out that the difficulty can be mitigated by interspersing some tight folds between (or among) longer loops in the amorphous phase (see Figure 6.36) (116). Frank (116) derived a general equation that combines the probability of back folding, p, and the obliquity angle, θ, with the crystalline stem length, l, and the contour length of the chain, L, to yield the minimum conditions to prevent an anomalous density in the amorphous region:

$$\left(1 - p - \frac{2l}{L}\right)\cos\theta \leq \frac{3}{10} \qquad (6.35)$$

The above findings led to two different models. In 1980, Dettenmaier et al. (111) proposed their solidification model, whereby it was assumed that crystallization occurred by a straightening out of short coil sequences without a long-range diffusion process. Thus these sequences of chains crystallized where they stood, following a modified type of switchboard model (50). This was the first model to illustrate how R_g values could remain virtually unchanged during crystallization.

On the other hand, Hoffman (72) showed that the density of the amorphous phase is better accounted for by having at least about 2/3 adjacent

(a)

(b)

Figure 6.36 Alternative resolutions of the density paradox $(1 - p - 2\ell/L)\cos\theta < 3/10$ (116): (a) increased chain folding beyond critical value and (b) oblique angle crystalline stems to reduce amorphous chain density at interface.

reentries, which he calls the variable cluster model. An illustration of how a chain can crystallize with a few folds in one lamella, then move on through an amorphous region to another lamella, where it folds a few more times and so on, is illustrated in Figure 6.37 (107). Thus a regime III crystallization according to the variable cluster model will substantially retain its melt value of R_g.

The general conclusion from these studies is that the fringed micelle model fails, because it predicts that the density of the amorphous polymer at both ends of the crystal will be higher than that of the crystal itself. While the chains could be laid over at a sharp angle (Figure 6.36b), the most viable alternative is to introduce a significant amount of chain folding.

It is of interest to compare the results of this modern research with Hosemann's paracrystalline model, first published in 1962. As illustrated in Figure 6.38 (117), this model emphasizes lattice imperfections and disorder, as might be expected from regime III crystallization. This model also serves as a bridge between the concepts of crystalline and amorphous polymers (see Figure 5.3). More recent research by Hosemann has continued to examine the partially ordered state (118, 119).

Figure 6.37 The variable cluster model, showing how a chain can crystallize from the melt with some folding and some amorphous portions and retain, substantially, its original dimensions and its melt radius of gyration (107).

A.	amorphous „phase"
C.F.	clustered fibrils (hot stretched)
C.G.	crystal growth in bulk material
E.	end of a chain
F.P.	four-point-diagram
L.B.	long backfolding (Flory)
M.F.	migrating fold
P.	paracrystalline layerlattice
S.	straight chains
S.B.	short backfolding (Keller)
S.C.	single crystals
S.F.	single fibrils (cold stretched)
S.H.	shearing region
S.T.	Statton model
V.	„voids"

MODEL OF LINEAR POLYETHYLENE

Figure 6.38 The paracrystalline model of Hosemann (117). Amorphous structures are illustrated in terms of defects. A radius of gyration approaching amorphous materials might be expected.

By way of summary, for dilute solution-grown crystals a modified regular reentry model fits best, with the same molecule forming a new stem either after immediate reentry or after skipping over one or two nearest-neighbor sites. For melt-formed crystals, the concept of folded chains is considerably modified. Since active research in this area is now in progress, perhaps a more definitive set of conclusions will be forthcoming.

It must be remembered that the formation of lamellae, whether with adjacently folded chains or with a switchboardlike structure, is kinetically controlled by the degree of undercooling and finite rates of molecular motion. The thermodynamically most stable crystal form is thought to have extended chains.

6.7.4 Extended Chain Crystals

In the above, it was pointed out that polymer chains fold during crystallization because of kinetic circumstances. Wunderlich (120, 121) pointed out that in thermodynamic equilibrium the crystalline state has an extended chain macroconformation. This extended chain conformation can be brought about by annealing for long periods of time, particularly under pressure. During this time, the fold period of the lamellae gradually increases.

For example, extended chain single crystals of polyethylene were grown at 227°C and 4800 atm for 20 hr, followed by cooling at 1.6°C/hr to room temperature (122). Thicknesses in the molecular chain direction up to 30,000 Å were found. Since the weight-average molecular weight of the polymer was 80,000 g/mol, the crystals may have contained more than one extended chain. The sample had a density of 0.995 g/cm³ at 25°C, corresponding to 97.5% crystallinity. Some aspects of this work have been reviewed (123).

6.8 THERMODYNAMICS OF FUSION

In the previous sections it was shown that the formation of lamellae with folded chains was essentially a kinetically controlled phenomenon. This section treats the free energy of polymer crystallization and melting point depression.

Melting is a first-order transition, ordinarily accompanied by discontinuities of such functions as the volume and the enthalpy. Ideal and real melting in polymers is illustrated in Figure 6.39. Ideally, polymers should exhibit the behavior shown in Figure 6.39a, where the volume increases a finite amount exactly at the melting (fusion) temperature, T_f. (The subscript M, for melting, is also in wide use. In this text, M represents mixing.) Note that the coefficient of thermal expansion also increases above T_f. Owing to the range of crystallite sizes and degrees of perfection in the real case, a range of melting temperatures is usually encountered, as shown experimentally in Figure 6.2.

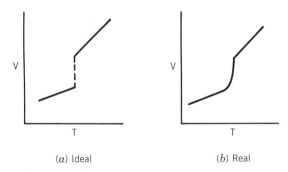

(a) Ideal   (b) Real

Figure 6.39 Dilatometric behavior of polymer melting.

The free energy of fusion, ΔG_f, is given by the usual equation,

$$\Delta G_f = \Delta H_f - T\Delta S_f \tag{6.36}$$

where ΔH_f and ΔS_f represent the molar enthalpy and entropy of fusion. At the melting temperature, ΔG_f equals zero, and

$$T_f = \frac{\Delta H_f}{\Delta S_f} \tag{6.37}$$

Thus a smaller entropy or a larger enthalpy term raises T_f. Since highly polar molecules such as polyamides tend to have large values of ΔH_f (see Table 6.1), they have high melting temperatures.

6.8.1 Theory of Melting Point Depression

The melting point depression in crystalline substances from the pure state T_f^0 is given by the general equation (124)

$$\frac{1}{T_f} - \frac{1}{T_f^0} = -\frac{R}{\Delta H_f} \ln a \tag{6.38}$$

where a represents the activity of the crystal in the presence of the impurity.

The thermodynamics of melting in polymers was developed by Flory and his co-workers (125–127). To a first approximation, the melting point depression depends on the mole fraction of impurity, X_B, the mole fraction of crystallizable polymer being X_A. Substituting X_A for a in equation (6.38),

$$\frac{1}{T_f} - \frac{1}{T_f^0} = -\frac{R}{\Delta H_f} \ln X_A \tag{6.39}$$

For small values of X_B,

$$-\ln X_A = -\ln(1 - X_B) \cong X_B \tag{6.40}$$

In the following, ΔH_f is the heat of fusion per mole of crystalline mers. There are three important cases in which the melting temperature may be depressed. If X_B represents the mole fraction of noncrystallizable comonomer incorporated in the chain,

$$\frac{1}{T_f} - \frac{1}{T_f^0} = \frac{R}{\Delta H_f} X_B \tag{6.41}$$

The mer unit at the end of the chain must always have a different chemical structure from those of the mers along the chain. Thus end mers constitute a special type of impurity, and the melting point depends on the molecular weight. If M_0 is the molecular weight of the end mer (and assuming that both ends are identical), the mole fraction of the chain ends is given approximately by $2M_0/M_n$. Thus

$$\frac{1}{T_f} - \frac{1}{T_f^0} = \frac{R}{\Delta H_f} \frac{2M_0}{M_n} \tag{6.42}$$

which predicts that the highest possible melting temperature will occur at infinite molecular weight.

If a solvent or plasticizer is added, the case is slightly more complicated. Here, the molar volume of the solvent, V_1, and the molar volume of the polymer repeat unit, V_u, cannot be assumed to be equal. Also, the interaction between the polymer and the solvent needs to be taken into account. The result may be written (127)

$$\frac{1}{T_f} - \frac{1}{T_f^0} = \frac{R}{\Delta H_f} \frac{V_u}{V_1} (v_1 - \chi_1 v_1^2) \tag{6.43}$$

where v_1 represents the volume fraction of diluent, and χ_1 is the Flory solvent interaction parameter. The quantity χ_1 and the second virial coefficient are closely related (see Section 3.3.2). If the molar volumes of the solvent molecules and the polymer mers are approximately equal, and v_1 is small (< 0.1), equation (6.43) degenerates into the form given by equation (6.41).

The quantity χ_1 has been interpreted in several ways (125, 126). Principally, it is a function of the energy of mixing per unit volume. For calculations involving plasticizers, the form using the solubility parameters δ_1 and δ_2 is particularly easy to use (128):

$$\chi_1 = \frac{(\delta_1 - \delta_2)^2 V_1}{RT} \tag{6.44}$$

Corresponding relations for the depression of the glass transition temperature, T_g, by plasticizer are given in Section 8.8.1.

Corresponding equations for the dependence of the melting point on pressure were derived by Karasz and Jones (129). For a pressure P_f,

$$P_f - P_f^0 = \frac{RT_f}{\Delta V_f} \frac{V_u}{V_1} (v_1 - \chi_1 v_1^2) \tag{6.45}$$

where ΔV_f is the volume change on fusion.

6.8.2 A Worked Example

Suppose we swell poly(ethylene oxide) with 10% of benzene. What will be the new melting temperature, if it melted at 66°C dry?

Equation (6.43) can be solved, with the aid of Tables 3.1, 3.2, 3.4, and 6.1. First, the molar volumes of the polymer and the solvent must be computed: for benzene, with six carbons and six hydrogens, $M = 78$ g/mol. Its density is 0.878 g/cm^3. Division yields 88 cm^3/mol. For poly(ethylene oxide), the mer molecular weight is 44 g/mol, and its density is 1.20 g/cm^3, yielding a molar volume of 36.6 cm^3/mol. Equation (6.43), with appropriate numerical values, is

$$\frac{1}{T_f} = -\frac{R}{\Delta H_f}\left(\frac{V_u}{V_1}\right)(v_1 - \chi_1 v_1^2) + \frac{1}{T_f^0}$$

$$\frac{1}{T_f} = +\frac{0.083 \text{ kJ/mol} \cdot \text{K}}{8.29 \text{ kJ/mol}}\left(\frac{36.6 \text{ cm}^3/\text{mol}}{88 \text{ cm}^3/\text{mol}}\right)(0.1 - 0.19 \times 0.1^2) + \frac{1}{339 \text{ K}}$$

$$T_f = 297.8 \text{ K or } 24.8°\text{C}$$

If room temperature is about 25°C, then the crystallinity will be destroyed by the benzene plasticizer. If v_1 and/or χ_1 are small, the term $\chi_1 v_1^2$ will be small compared to v_1 and can be ignored.

6.8.3 Experimental Thermodynamic Parameters

First T_f^0 may be determined either directly on the pure polymer or by a plot of $1/T_f$ versus v_1. The latter is very useful in the case where the polymer decomposes below its melting temperature.

Once T_f^0 is determined, equations (6.43) and (6.44) permit the calculation of both the Flory interaction parameter and the heat of fusion of the polymer from the slope and intercept of a plot of $(1/T_f - 1/T_f^0)/v_1$ against v_1/T_f (see Figure 6.40) (126). The heat of fusion determined in this way measures only the crystalline portion. If heat of fusion data are compared with corresponding data obtained by DSC (see Figure 6.3), which measures the heat of fusion for the whole polymer, the percent crystallinity may be obtained.

6.8.4 Entropy of Melting

In classical thermodynamics, the change in Gibbs free energy is zero at the melting point,

$$\Delta G_f = \Delta H_f - T\Delta S_f = 0 \tag{6.46}$$

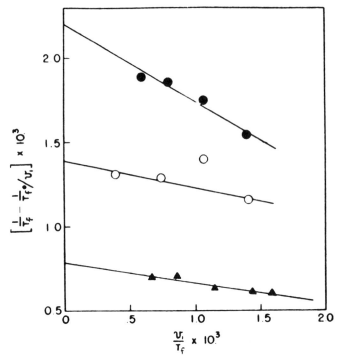

Figure 6.40 Determination of heats of fusion and χ, using a plot of $(1/T_f - 1/T_f^0)/v_1$ against v_1/T_f for mixtures of cellulose tributyrate with various diluents: hydroquinone nonomethyl ether (solid circles), dimethyl phthalate (open circles), ethyl laurate (triangles) (126). The use of T_f in the denominator of the x-axis is arbitrary, making the lines straighter. It should be divided out before numerical calculations are completed.

where $T = T_f$. For polyethylene, per $-CH_2-$ group, with $\Delta H_f = 3.94$ kJ/mol, see Table 6.1. Then,

$$\Delta S_f = \Delta H_f/T_f = 9.61 \text{ J/mol} \cdot \text{K} \qquad (6.47)$$

at a T_f of 410 K.

Statistical thermodynamics asks how many conformational changes are involved in the melting process. If the polyethylene is crystallized in the all trans conformation, and two gauche plus the trans are possible in the melt (see Section 2.1.2), then the polymer goes from one possible conformation to three on melting (130, 131). From equation (3.13),

$$\Delta S = R \ln \Omega = R \ln 3 = 9.13 \text{ J/mol} \cdot \text{K} \qquad (6.48)$$

The agreement between the classical and statistical results, equations (6.47) and (6.48), is seen to be excellent, noting the approximations involved.

A more general statistical thermodynamic theory can be obtained with the quasi-lattice models; see Figure 3.3. If a coordination number z of the lattice is assumed, then there are $z - 1$ choices of where to put the next bond in the chain. (This is a little smaller with excluded volume considered.) Then, the entropy varies with dimensionality of the quasi-lattice according to

$$3 - D \qquad \Delta S_f = R \ln 5 = 13.4 \text{ J/mol} \cdot \text{K} \qquad (6.49)$$

$$2 - D \qquad \Delta S_f = R \ln 3 = 9.13 \text{ J/mol} \cdot \text{K} \qquad (6.50)$$

$$1 - D \qquad \Delta S_f = R \ln 1 = 0 \qquad (6.51)$$

If more than one group needs to be considered, the values are multiplicative, rather than additive. Taking two —CH_2— groups, for example, yields nine conformations, rather than six. The general equation can be written

$$\Delta S = R \ln(A^a B^b C^c \cdots) \qquad (6.52)$$

where $A, B, C, \ldots$ are the number of ways the various moieties can be arranged in space, and $a, b, c, \ldots$ are the number of appearances of the moiety in each mer.

Using the ΔH_f values from Figure 6.40 and other studies, Mandelkern and Flory (126) determined the entropy of fusion [see equation (6.37)]. Values for several polymers are illustrated in Table 6.10 (126). While the entropies per mer varied widely, as might be expected from the enormous differences in the sizes and structures of the units, values divided by the number of chain bonds about which free rotation is permitted gave more nearly uniform values. According to Flory (132), the configurational entropy of fusion per segment should be $R \ln(Z' - 1)$, where Z' is the coordination number of the lattice. Values of E.U./No. bonds permitting rotation in Table 6.10 are in rough agreement with Flory's calculation. Based on equa-

Table 6.10   Entropies of fusion for various polymers (126)

Polymer	Repeating Unit	Entropy of fusion, J/mol · K	
		E.u./mol of Repeating Unit	E.u./No. Bonds Permitting Rotation
Polyethylene	—CH_2—	8.37	8.37
Cellulose tributyrate	—$C_{18}H_{28}O_{18}$—	25.9	13.0
Poly(decamethylene sebacate)		145	6.3
—O—$(CH_2)_{10}$—O—CO—$(CH_2)_8$CO—			
Poly(N,N'-sebacoylpiperazine)		57.3	5.0

$$-N \begin{array}{c} \diagup CH_2CH_2 \diagdown \\ \diagdown CH_2CH_2 \diagup \end{array} N-CO(CH_2)_8CO-$$

tion (6.37), it is easy to understand why large heats of fusion produce high melting polymers (see Table 6.1).

In experiments such as the above, heating and cooling are usually done very slowly. Therefore regime I structures may predominate. The melting temperature is higher under these conditions than when cooling or heating is rapid, in which case regime II and III kinetics apply.

6.8.5 Crystalline – Amorphous Polymer Blends

The above discussion relates primarily to addition of low-molecular-weight species to a high-molecular-weight polymer. When the added species is also polymeric, quite different analytical relationships hold, because of the reduced entropy of mixing.

According to the Flory–Huggins approximation (see Section 3.3), the chemical potential μ_2 per mole of crystallizable polymer units in the mixture relative to the pure liquid chemical potential, μ_2^0, can be written

$$\mu_2 - \mu_2^0 = \frac{RTV_2}{V_1}\left[\frac{\ln v_2}{x_2} + \left(\frac{1}{x_2} - \frac{1}{x_1}\right)(1 - v_2) + \chi_1(1 - v_2)^2\right] \quad (6.53)$$

where the subscript 1 is identified with the amorphous polymer and 2 with the crystalline polymer, v represents the volume fraction, V is the molar volume of the repeating units, and x is essentially the degree of polymerization (133).

The difference in the chemical potential between a crystalline polymer unit μ_2^c and the counterpart unit in the pure liquid state can be written (134)

$$\mu_2^c - \mu_2^0 = -\left(\Delta H_{f_2} - T\Delta S_{f_2}\right)$$

$$= -\Delta H_{f_2}\left(1 - \frac{T}{T_f^0}\right) \quad (6.54)$$

If one assumes that at the melting point, T_f, of the mixture the chemical potentials of the crystalline component in the crystalline and liquid phases are identical, one derives from equations (6.53) and (6.54)

$$\frac{1}{T_f} - \frac{1}{T_f^0} = -\frac{RV_2}{\Delta H_{f_2} V_1}\left[\frac{\ln v_2}{x_2} + \left(\frac{1}{x_2} - \frac{1}{x_1}\right)(1 - v_2) + \chi_1(1 - v_2)^2\right]$$

$$(6.55)$$

In the present case of two polymers, both x_1 and x_2 are very large compared to unity. Therefore equation (6.55) reduces to

$$\frac{1}{T_f} - \frac{1}{T_f^0} = -\frac{RV_2}{\Delta H_{f_2} V_1} \chi_1 (1 - v_2)^2 \tag{6.56}$$

If $V_1 \cong V_2$, which is often the case, a further simplification is possible.

$$\frac{1}{T_f} - \frac{1}{T_f^0} = -\frac{R\chi_1}{\Delta H_{f_2}} (1 - v_2)^2 \tag{6.57}$$

Equations (6.56) and (6.57) describe the melting point depression due to mixing of a crystalline polymer with an amorphous polymer. The important point is the role played by χ_1; in these equations the melting point depression behavior is specified by the heat of mixing rather than the entropy of mixing. Equations (6.56) and (6.57) state that melting point depressions can be realized only if χ_1 is negative. As described in Chapter 4, positive values of χ_1 for mixtures of two high polymers results in phase separation. In the case of a crystalline–amorphous polymer blend, phase separation would not be expected to result in large melting point depressions.

The above mathematics can be checked in an interesting way. If $x_1 = 1$ and x_2 is large, which is the case for a low-molecular-weight solvent, equation (6.55) can be written

$$\frac{1}{T_f} - \frac{1}{T_f^0} = \frac{RV_2}{\Delta H_{f_2} V_1} \left[(1 - v_2) - \chi_1 (1 - v_2)^2 \right] \tag{6.58}$$

which has the same form as equation (6.43), as it must, noting that $1 - v_2 = v_1$.

Returning to equation (6.56), χ_1 can be written in the form

$$\chi_1 = \frac{BV_1}{RT} \tag{6.59}$$

If B is positive, it may be equated to the square of the differences between the solubility parameters [equation (6.44)]. No such interpretation is possible if B is negative. However, χ_1 can be negative, indicating an exothermic heat of mixing.

Substitution of equation (6.59) into equation (6.56) yields the following form:

$$\left(\frac{1}{T_f} - \frac{1}{T_f^0} \right) = -\frac{BV_2 v_1^2}{\Delta H_{f_2} T_f} \tag{6.60}$$

which, as before, suggests a plot of $(1/T_f - 1/T_f^0)/v_1$ versus v_1/T_f.

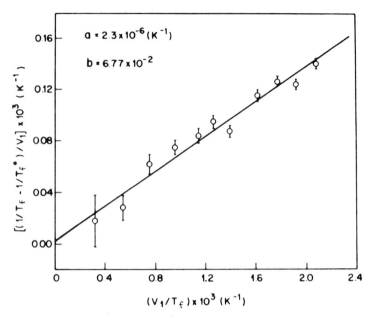

Figure 6.41   Determination of the thermodynamics of polymer blending by melting point depression for the system poly(vinylidene fluoride) – *blend* – poly(methyl methacrylate) (134).

Poly(vinylidene fluoride) crystallizes, and it is totally miscible with poly(methyl methacrylate). The melting point depression of the former by the latter is shown in Figure 6.41 (134), plotted according to the above relationships. The heat of fusion for poly(vinylidene fluoride) is estimated to be 1.6 kcal/mol, and χ_1 is calculated to be -0.295. The negative value for χ_1 indicates that the polymer pair can form a thermodynamically stable solution at temperatures above the melting point; indeed, this pair is miscible in all proportions.

6.8.6   The Hoffman – Weeks Equilibrium Melting Temperature

There are several definitions of the equilibrium melting temperature currently in use. According to equation (6.42), the highest melting temperature (and presumably the equilibrium melting temperature) is reached at infinite molecular weight. Another definition assumes infinitely thick crystalline lamellae (135), see Section 6.7.4.

According to Hoffman and Weeks (136), the equilibrium melting temperature of a polymer, T_f^*, is defined as the melting point of an assembly of crystals, each of which is so large that surface effects are negligible and that each such large crystal is in equilibrium with the normal polymer liquid. Furthermore, the crystals at the melting temperature must have the equilibrium degree of perfection consistent with the minimum free energy at T_f^*.

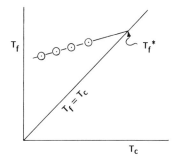

Figure 6.42   Idealized Hoffman – Weeks plot, showing the extrapolation to T_f^*.

While this definition holds for most pure compounds, polymers as ordinarily crystallized tend to melt below T_f^* because the crystals are small and all too imperfect. Thus the temperature of crystallization has an important influence on the experimentally observed melting point. Hoffman and Weeks (136) found the following relation to hold:

$$T_f^* - T_f = \phi'(T_f^* - T_c) \qquad (6.61)$$

where ϕ' represents a stability parameter that depends on crystal size and perfection. The quantity ϕ' may assume all values between 0 and 1, where $\phi' = 0$ implies that $T_f = T_f^*$, whereas $\phi' = 1$ implies $T_f = T_c$. Therefore crystals are most stable at $\phi' = 0$ and inherently unstable at $\phi' = 1$. Values of ϕ' near $\frac{1}{2}$ are common.

The experiment generally involves rapidly cooling the polymer from the melt to some lower temperature, T_c, where it is then crystallized isothermally. For higher crystallization temperatures, the polymer forms more perfect crystals; that is, it is better annealed. Hence its melting temperature increases.

To determine T_f^*, a plot of T_c versus T_f is prepared (136) (Figure 6.42). A line is drawn where $T_c = T_f$. The experimental data are extrapolated to the intersection with the line. The temperature of intersection is T_f^*.

6.9   EFFECT OF CHEMICAL STRUCTURE ON THE MELTING TEMPERATURE

The actual values of the enthalpy and entropy of fusion are, of course, controlled by the chemical structure of the polymer. The most important inter- and intramolecular structural characteristics include structural regularity, bond flexibility, close packing ability, and interchain attraction (137, 138). In general, high melting points are associated with highly regular structures, rigid molecules, close packing capability, strong interchain attraction, or several of these factors combined.

The effect of structural irregularities can be illustrated by a study of polyesters (138) having the general structure

$$\left(\!\!\begin{array}{c}O\\||\\C\end{array}\!\!-\!\!\left\langle\bigcirc\right\rangle\!\!-\!\!\begin{array}{c}O\\||\\C\end{array}\!\!-\!\!O\!-\!R\!-\!O\!\!\right)_{\!n} \qquad (6.62)$$

which is the general structure of Dacron®. The melting temperature depends on the regularity of the group R. For aliphatic groups, the size and regularity of R are both important:

R	T_f, °C
$-CH_2-CH_2-$	265
$(CH_2)_3$	220

$-CH_2-CH-$
     $\|$
    CH_3
	Noncrystalline

The irregularity of the atactic propylene unit destroys crystallinity entirely.

The effect of bond flexibility may also be examined utilizing the polyester structure (138). In this case, substitutions in the rigid aromatic group are made:

$$\left(\!O\!-\!\begin{array}{c}O\\||\\C\end{array}\!\!-\!R'\!-\!\begin{array}{c}O\\||\\C\end{array}\!\!-\!O\!-\!CH_2\!-\!CH_2\!\right)_{\!n} \qquad (6.63)$$

R'	T_f, °C
$-\!\left\langle\bigcirc\right\rangle\!-$	265
$-\!\left\langle\bigcirc\right\rangle\!\!-\!\!\left\langle\bigcirc\right\rangle\!-$	355
$(CH_2)_4$	50

The flexible aliphatic group, although having dimensions similar to those of the phenyl group, has a much lower melting temperature. It should be pointed out that the aliphatic polyesters, first synthesized by Carothers (139), failed as clothing fibers because they melted during washing or ironing. Aromatic polyesters (140) as well as aliphatic nylons achieved the necessary high melting temperatures.

Interchain forces can be illustrated by the following substitutions (138) of increasingly polar groups:

$$\left(\!\!\!\begin{array}{c} O \\ \| \\ O-C \end{array}\!\!\!-\!\!\!\bigcirc\!\!\!-R''-\!\!\!\bigcirc\!\!\!-\!\!\!\begin{array}{c} O \\ \| \\ C \end{array}\!\!\!-O-CH_2-CH_2\!\!\!\right)_{\!n} \quad (6.64)$$

R''	$T_f, °C$
$+CH_2\!\!\rightarrow_{\!4}$	170
$-O-CH_2-CH_2-O-$	240
$-NH-CH_2-CH_2-NH-$	273

Similar effects are caused by bulky substituents and by odd or even numbers of carbon atoms in hydrocarbon segments of the chain. Generally, bulky groups lower T_f, because they separate the chains. The odd or even number of carbon atoms affects the regularity of packing. Of course, the frequency of occurrence of polar groups is very important. As the length of the aliphatic group R in equation (6.62) is increased, the melting point gradually approaches that of polyethylene, or about 139°C.

In each of the cases above, of course, the crystal structure is governed by the principles laid down in Section 6.3.5 and elsewhere. Generally, those structures that are most tightly bonded, fit the most closely together, and are held in place the most rigidly will have the highest melting temperatures.

6.10 FIBER FORMATION AND STRUCTURE

The synthetic fibers of today, the nylons, polyesters, rayons, and so on, are manufactured by a process called spinning. Spinning involves extrusion through fine holes known as spinnerets. Immediately after the spinning process, the polymer is oriented by stretching or drawing. This both increases polymer chain orientation and degree of crystallinity. As a result, the modulus and tensile strength of the fiber are increased.

Fiber manufacture is subdivided into three basic methods, melt spinning, dry spinning, and wet spinning; see Table 6.11 (141). Melt spinning is the simplest but requires that the polymer be stable above its melting temperature. Nylon 66 is a typical example. Basically, the polymer is melted and forced through spinnerets, which may have from 50 to 500 holes. The fiber diameter immediately after the hole and before attenuation begins is larger than the hole diameter. This is called die swell, which is due to a relaxation of the viscoelastic stress-induced orientation in the hole, see Section 5.4 and Figure 10.20.

Table 6.11 Spinning processes (141)

| | | Solution Spinning | |
| | | Wet Spinning | |
Melt Spinning	Dry Spinning	Coagulation	Regeneration
Nylon	Cellulose acetate		Viscose rayon
Polyester	Cellulose triacetate		Cupro
Polyethylene	Acrylic	Acrylic	
	Modacrylic	Modacrylic	
Polypropylene	Aramid	Aramid	
PVDC	Elastane	Elastane	
	PVC	PVC	
	Vinylal		

During the cooling process, the fiber is subjected to a draw-down force, which introduces the orientation. Additional orientation may be introduced later by stretching the fiber to a higher draw ratio.

In dry spinning, the polymer is dissolved in a solvent. A typical example is polyacrylonitrile dissolved in dimethylformamide to 30% concentration. The polymer solution is extruded through the spinnerets, after which the solvent is rapidly evaporated (Figure 6.43) (141). After the solvent is evaporated, the fiber is drawn as before.

In wet spinning, the polymer solution is spun into a coagulant bath. An example is a 7% aqueous solution of sodium cellulose xanthate (viscose), which is spun into a dilute sulfuric acid bath, also containing sodium sulfate and zinc sulfate (141a). The zinc ions form temporary ionic cross-links between the xanthate groups, holding the chains together while the sulfuric acid, in turn, removes the xanthate groups, thus precipitating the polymer. After orientation and so on, the final product is known as rayon.

6.10.1 X-Ray Fiber Diagrams

For the purpose of x-ray analyses, the samples should be as highly oriented and crystalline as possible. Since these are also the conditions required for strong, high-modulus fibers, basic characterization and engineering requirements are almost identical.

Figure 6.44 (23, 142, 143) illustrates a typical x-ray fiber diagram, for polyallene, $+CH_2-C(=CH_2)+_n$. The actual fiber orientation is vertical. The most intense diffractions are on the equitorial plane; note the 110 and 200 reflections. Note the rather intense amorphous halo, appearing inside the 011 reflection.

Because of the imperfect orientation of the polymer in the fiber, arcs are seen, rather than spots. The variation in the intensity over the arcs can be used, however, to calculate the average orientation.

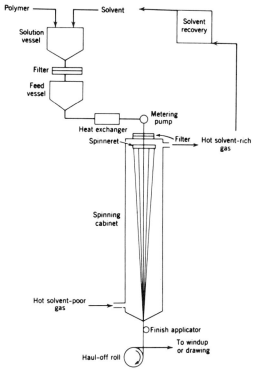

Figure 6.43 In typical dry-spinning operations, hot gas is used to evaporate the solvent in the spinning cabinet. The fibers are simultaneously oriented (141).

A further complication in the interpretation of the fiber diagram arises because it actually is a "full rotation photograph." In these ways, fiber diagrams differ from those of single crystals. Vibrational analyses via infrared and Raman spectroscopy studies play an important role in the selection of the molecular model, as described above.

6.10.2 Natural Fibers

Natural fibers were used long before the discovery of the synthetics in the 20th century. Natural fibers are usually composed of either cellulose or protein, as shown in Table 6.12. Animal hair fibers belong to a class of proteins known as keratin, which serve as the protective outer layer of the higher vertebrates. The silks are partly crystalline protein fibers. The crystalline portions of these macromolecules are arranged in antiparallel pleated sheets, a form of the folded-chain lamellae (144).

The morphology of the natural fibers is often quite complex; see Figure 6.45 (145). The cellulose making up these trachieds is a polysaccharide; see Table 1.4. The crystalline portion of the cellulose making up the trachieds is

Figure 6.44 X-Ray fiber diagram of polyallene (*a*), and its indexing (*b*) (23).

highly oriented, following the various patterns indicated in Figure 6.45. The winding angles of the cellulose form the basis for a natural composite of great strength and resilience. A similar morphology exists in cotton cellulose.

The fibrous proteins (keratin) are likewise highly organized; see Figure 6.46 (146). Proteins are actually polyamide derivatives, a copolymeric form of nylon 2, where the mers are amino acids. For example, the structure of the

Table 6.12 Chemical nature of natural fibers

Cellulose	Protein
Cotton	Wool
Tracheid (Wood)	Hair
Flax	Silk
Hemp	Spider Webs
Coir	
Ramie	
Jute	

Figure 6.45 The cell walls of a tracheid or wood fiber have several layers, each with a different orientation of microfibrils (145). ML, middle lamella, composed of lignin; P, primary wall; S_1, S_2, S_3, layers of the secondary wall; W, warty layer. The lumen in the interior of the warty layer is used to transport water.

amino acid phenylalanine in a protein may be written

$$-N-\overset{\overset{\displaystyle H}{|}}{\underset{\underset{\displaystyle CH_2-\bigcirc}{|}}{C}}-\overset{\overset{\displaystyle H}{|}}{C}-\overset{\overset{\displaystyle O}{\parallel}}{C}- \tag{6.65}$$

There are some 23 amino acids in nature. These are organized into an α-helix in the fibrous proteins, which in turn are combined to form protofibrils as shown. In addition to being crystalline, the fibrous proteins are cross-linked though disulfide bonds contained in the cystine amino acid mer,

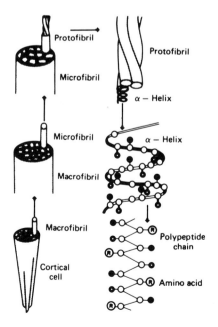

Figure 6.46 A wool-fiber cortical cell is a complex structure, being composed ultimately of proteins (146).

which is especially high in keratin. Animal tendons, composed of collagen, another fibrous protein, have also been shown to have a surprisingly complex hierarchical structure (147).

6.11 THE HIERARCHICAL STRUCTURE OF POLYMERIC MATERIALS

While all polymers are composed of long-chain molecules, it is the organization of such materials at higher and higher levels which progressively determines their properties and ultimately determines their applications. Table 6.13 summarizes the hierarchical structure of the several classes of polymers according to size range. The largest synthetic crystalline polymer structures are the spherulites, while the largest structures of the natural polymers are the walls of the whole (living) cell. Not shown are the amorphous natural polymers, the most important being starch, another polysaccharide.

The hierarchical structure of polymers has been reviewed by Baer and co-workers (148, 149). They emphasize that the occurrence of crystallinity in synthetic polymers requires a sufficiently stereoregular chemical structure, so that the chain molecules can pack closely in parallel orientation. Structure at the nanometer level is generally determined by the pronounced tendency of these chains to crystallize by folding back and forth within crystals of thin lamellar habit. Chain folding itself is a kinetic phenomenon. Given sufficient annealing time, the lamellar thickness increases until the chains are all straight at thermodynamic equilibrium.

When crystallization occurs in flowing solutions or melts, fine fibrous crystals may be produced that consist of highly extended chains oriented axially. This will be recognized as the basis for the supermolecular structure in fibers. Even after prolonged crystallization from the melt, there always remains an appreciable fraction of a disordered phase known as amorphous

Table 6.13 Hierarchical structure of polymers

Size Range	Synthetic		Natural	
	Amorphous	Crystalline	Cellulosic	Protein
Ångstroms	Mer	Mer	Glucose	Amino Acid
μm	Chain coil Physical entanglement Crosslink	Stem	Stem	α-Helix
Microns	Multicomponent morphologies	Lamellae Spherulite	Lamellae Microfibril	Protofibril Microfibril
100 Microns	—	—	Fiber (cell)	Microfibril Cell

polymer. If the polymer can crystallize, the appearance of amorphous material is also a kinetic phenomenon, which at equilibrium should disappear. However, the physical entanglements and already existing crystalline structure may slow further crystallization down to substantially zero.

While amorphous materials are significantly less organized than their crystalline counterparts, entanglements, branches, and cross-links significantly control their properties. When two polymers are blended, they generally phase-separate to create a supermolecular morphology, which may be of the order of tens of micrometers, while graft and block copolymers exhibit morphologies of the order of hundreds of angstroms.

6.12 HOW DO YOU KNOW IT'S A POLYMER?

Suppose a polymer scientist watches a demonstration. The contents of two bottles are mixed, and a white precipitate appears. "Here, I have created a new polymer for you," the demonstrator announces. The polymer scientist examines the two bottles, which are unlabeled. The demonstrator mysteriously leaves the building. What experiments should be performed to determine if the white precipitate is polymeric? (A separate, but equally important, question asks: What is its structure?)

In 1920, such questions were far from trivial; note Appendix 5.1. Even today, questions of this nature need to be addressed. For example, the structure of ordinary (window) glass is considered polymeric by some, and not by others. It surely exhibits a glass transition (see Chapter 8) and has a high viscosity, but also apparently has a time variable structure.

There are several types of experiment that should be performed in order to ascertain that the material is (or is not) polymeric. A key experiment is the determination of the molecular weight. If the molecular weight is above about 25,000 g/mol, most scientists will consider that evidence in favor of a polymeric structure. However, it may be colloidal, as a sulfur or silver sol, and not polymeric.

A next series of experiments involves ordinary chemistry, as outlined in Section 2.2. Elemental analysis, characterization of degradation products, end-group analysis, and determination of the probable mer structure are all important in solving the puzzle.

Perhaps the material can be reacted or degraded to form a soluble compound, which can then be characterized. An example is cellulose, a natural polymer that was part of the earliest discussions. Insoluble itself, it can be acetylated to form soluble cellulose acetate for molecular weight determination, by osmometry at the time. Then it was subjected to degradation to glucose and cellobiose, which were determined to be the monomer and dimer, respectively.

Morawetz, in his history of polymer science (150), takes the reader back to before the Macromolecular hypothesis. Cis-polyisoprene, natural rubber, was investigated around the turn of the 20th century. Morawetz writes of the destructive distillation of rubber to isoprene, and of the ozonolysis of rubber to levulinic aldehyde. These studies led to the dimer ring structure of dimethylcyclooctadiene:

$$
\begin{array}{c}
\text{CH}_3 \qquad\qquad\qquad \text{CH}_3 \\
| \qquad\qquad\qquad\qquad | \\
\ldots\text{C}\!-\!\text{CH}_2\!-\!\text{CH}_2\!-\!\text{CH}\ldots\text{C}\!-\!\text{CH}_2\!-\!\text{CH}_2\!-\!\text{C}\ldots \\
\|\qquad\qquad\qquad\qquad\|\qquad\| \\
\ldots\text{CH}\!-\!\text{CH}_2\!-\!\text{CH}_2\!-\!\text{C}\ \ldots\text{CH}\!-\!\text{CH}_2\!-\!\text{CH}_2\!-\!\text{C}\ldots \\
|\qquad\qquad\qquad\qquad\qquad| \\
\text{CH}_3 \qquad\qquad\qquad \text{CH}_3
\end{array}
\qquad (6.66)
$$

The strength of the affinity of the carbon–carbon double bonds was supposed to hold the material together with partial valences. Only much later was it determined that the structure was actually a long chain of isoprene mers, connected in the 1,4-positions, with all the double bonds in the cis-configuration.

Such structures imply the confusion then reigning between colloids and polymers. Both were large structures. What was the difference between a sulfur colloidal dispersion and a rubber solution? The answer was found after many years of research.

Another class of experiments that are important include FT-IR, Raman spectroscopy, and NMR, which provide physical evidence as to where each atom is located and what moieties are present; see Section 2.2. In combination with a known high molecular weight and chemical analyses, large parts of the puzzle can be put together.

If the material is crystalline, x-ray and electron diffraction experiments can help decide the relative location of large sections of the chain. Are the presumed "mers" lined up in such a fashion that they can be reasonably bonded together? What is the relation between one cell and its neighbors?

Returning to our mysterious demonstration, the material could have been nylon 66, synthesized via interfacial polymerization. Then, elemental analysis would show the percent nitrogen, carbon, oxygen, and hydrogen, and each of the methods described above would be brought to play their roles. If, however, it was concentrated hydrochloric acid and concentrated sodium hydroxide that were mixed, the result would eventually be shown to be sodium chloride, ordinary salt! Philosophically, of course, conclusions are the product of reasoning power. It must be remembered that experiments provide evidence, and nothing else.

REFERENCES

1. J. Brandrup and E. H. Immergut, Eds., *Polymer Handbook*, 2nd ed., Wiley-Interscience, New York, 1975.
2. K. H. Meyer and H. Mark, *Ber. Deutsch. Chem. Ges.*, **61**, 593(1928).

3. H. Mark and K. H. Meyer, *Z. Phys. Chem.* **B2**, 115(1929).

4. K. H. Meyer and H. Mark, *Z. Phys. Chem.*, **B2**, 115(1929).

5. K. H. Meyer and L. Misch, *Ber. Deutsch. Chem. Ges., B.*, **70B**, 266(1937).

6. K. H. Meyer and L. Misch, *Helv. Chim. Acta*, **20**, 232(1937).

7. H. Mark, presented at the American Chemical Society meeting, Seattle, Washington, March 1983.

8. L. Mandelkern, *Rubber Chem. Tech.*, **32**, 1392(1959).

9. R. B. Jones and L. H. Sperling, unpublished.

10. S. H. Maron and C. F. Prutton, *Principles of Physical Chemistry*, 4th ed., MacMillan, London, 1965, Chap. 2.

11. W. L. Bragg, *Proc. Camb. Philos. Soc.*, **17**, 43(1913).

12. P. Debye and P. Scherrer, *Phys. Z.*, **17**, 277(1916).

13. C. W. Bunn, *Chemical Crystallography*, Oxford University Press, London, 1945, p. 109.

14. H. Tadokoro, *Structure of Crystalline Polymers*, Wiley-Interscience, New York, 1979, Ch. 5.

15. A. Keller, *Philos. Mag.*, **2**, 1171(1957).

16. C. W. Bunn, *Trans. Faraday Soc.*, **35**, 482(1939).

17. G. Natta and P. Corradini, *Rubber Chem. Tech.*, **33**, 703(1960).

18. Courtesy of Dr. A. Keller and Sally Argon.

19. G. Natta and P. Corradini, *J. Polym. Sci.*, **39**, 29(1959).

20. D. R. Holmes, C. W. Bunn, and D. J. Smith, *J. Polym. Sci.*, **17**, 159(1955).

21. R. de P. Daubeny, C. W. Bunn, and C. J. Brown, *Proc. R. Soc. (Lond.)*, **A226**, 531(1954).

22. Y. Takahashi and H. Tadokoro, *Macromolecules*, **6**, 672(1973).

23. H. Tadokoro, *Structure of Crystalline Polymers*, Wiley-Interscience, New York, 1979.

24. H. Tadokoro, Y. Chatani, T. Yoshihara, S. Tahara, and S. Murahashi, *Makromol. Chem.*, **73**, 109(1964).

25. B. C. Rånby, *Acta Chem. Scand.*, **6**, 101, 116(1952).

26. L. Loeb and L. Segal, *J. Polym. Sci.*, **14**, 121(1954).

27. O. Ellefsen and B. A. Tonnesen, in *Cellulose and Cellulose Derivatives*, Part IV, N. M. Bikales and L. Segal, Eds., Wiley-Interscience, New York, 1971, p. 151.

28. J. A. Hawsman and W. A. Sisson, *Cellulose and Cellulose Derivatives*, Part I, E. Ott, H. M. Spurlin, and M. W. Grafflin, Eds., Wiley-Interscience, New York, 1954, p. 231.

29. J. W. S. Hearle, *Polymers and Their Properties*, Vol. I, *Fundamentals of Structure and Mechanics*, Ellis Horwood Ltd., Chichester, England, 1982.

30. J. W. S. Hearle and R. H. Peters, *Fiber Structure*, Textile Institute, Manchester, England, 1963.

31. F. Khoury and E. Passaglia, in *Treatise on Solid State Chemistry*, Vol. 3, *Crystalline and Noncrystalline Solids*, N. B. Hannay, Ed., Plenum, New York, 1976, Ch. 6.

32. P. H. Geil, *Polymer Single Crystals*, Interscience, New York, 1963.

33. F. W. Billmeyer, Jr., P. H. Geil, and K. R. Van der Weg, *J. Chem. Ed.*, **37**, 460(1960).

34. Y. Fujiwara, *J. Appl. Polym. Sci.*, **4**, 10(1960).

35. E. J. Roche, R. S. Stein, and E. L. Thomas, *J. Polym. Sci. Polym. Phys. Ed.*, **18**, 1145(1980).

36. R. S. Stein and A. Misra, *J. Polym Sci. Polym. Phys. Ed.*, **18** 327(1980).

37. R. S. Stein, in *Rheology, Theory and Applications*, Vol. 5, F. R. Eirich, Ed., Academic New York 1969, Chap. 6.

38. J. D. Hoffman, G. T. Davis, and J. I. Lauritzen, Jr., in *Treatise on Solid State Chemistry*, Vol. 3, *Crystalline and Noncrystalline Solids*, N. B. Hannay, Ed., Plenum, New York, 1976, Chap. 7.

39. W. A. Sisson, in *Cellulose and Cellulose Derivatives*, Interscience, New York, 1943, pp. 203–285.

40. J. Hengstenberg and J. Mark, *Z. Kristallogr.*, **69**, 271(1928).

41. W. O. Statton, *J. Polym. Sci.*, **20C**, 117(1967).

42. K. Herrmann and O. Gerngross, *Kautschuk*, **8**, 181(1932).

43. K. Herrmann, O. Gerngross, and W. Abitz, *Z. Phys. Chem.*, **10**, 371(1930).

44. E. W. Fischer, *Z. Naturforsch.*, **12a**, 753(1957).

45. P. H. Till, Jr., *J. Polym. Sci.*, **24**, 301(1957).

46. F. A. Bovey, *Org. Coat. Appl. Polym. Sci Prepr.*, **48**(1), 76(1983).

47. F. C. Schilling, F. A. Bovey, S. Tseng, and A. E. Woodward, *Macromolecules*, **16**, 808(1983).

48. T. Oyama, K. Shiokawa, and Y. Murata, *Polym. J.*, **6**, 549(1974).

49. A. J. Kovacs, J. A. Manson, and D. Levy, *Kolloid Z.*, **214**, 1(1966).

50. P. J. Flory, *J. Am. Chem. Soc.*, **84**, 2857(1962).

51. R. S. Stein, *J. Chem. Ed.*, **50** 748(1973).

52. L. Mandelkern, in *Physical Properties of Polymers*, J. E. Mark, A. Eisenberg, W. W. Graessley, L. Mandelkern, and J. L. Koenig, Eds., American Chemical Society, Washington, DC, 1984.

53. H. D. Keith, F. J. Padden, Jr., and R. G. Vadimsky, *J. Polym. Sci.*, *A-2*, **4**, 267(1966).

54. H. D. Keith, F. J. Padden, and R. G. Vadimsky, *J. Appl. Phys.*, **42**, 4585(1971).

55. Y. Hase and P. H. Geil, *Polym. J.* (*Jpn.*), **2**, 560, 581(1971).

56. F. Rybnikar and P. H. Geil, *J. Macromol. Sci. Phys.*, **B7**, 1(1973).

57. D. C. Bassett, A. Keller, and S. Mitsuhashi, *J. Polym. Sci.*, **A1**, 763(1963).

58. P. H. Geil, in *Growth and Perfection of Crystals*, R. H. Doremus, B. W. Roberts, and D. Turnbull, Eds., Wiley, New York, 1958, pp. 579–585.

59. H. D. Keith, *J. Polym. Sci.*, **A2**, 4339(1964).

60. E. Martuscelli, Multicomponent Polymer Blends Symposium, Capri, Italy, May 1983.

61. E. Martuscelli, *Polym. Eng. Sci.*, **24**, 563(1984).

62. A. J. Kovacs, *Chim. Ind. Genie Chim.*, **97**, 315(1967).

63. R. G. Crystal, P. F. Erhardt, and J. J. O'Malley, in *Block Copolymers*, S. L. Aggarwal, Ed., Plenum, New York, 1970.

64. K. E. Hardenstine, C. J. Murphy, R. B. Jones, L. H. Sperling, and G. E. Manser, *J. Appl. Polym. Sci.*, **30**, 2051(1985).

65. J. N. Hay and M. Sabin, *Polymer*, **10**(3), 203(1969).

66. H. D. Keith and F. J. Padden, Jr., *J. Appl. Phys.*, **35**, 1270(1964).

67. H. D. Keith and F. J. Padden, Jr., *J. Appl. Phys.*, **35**, 1286(1964).

68. L. H. Palys and P. J. Phillips, *J. Polym. Sci. Polym. Phys. Ed.*, **18**, 829(1980).

69. M. Avrami, *J. Chem. Phys.*, **7**, 1103(1939).

70. M. Avrami, *J. Chem. Phys.*, **8**, 212(1940).

71. M. Avrami, *J. Chem. Phys.*, **9**, 177(1941).

72. J. D. Hoffman, *Polymer*, **24**, 3(1983).

73. J. D. Hoffman, *Polymer*, **23**, 656(1982).

74. E. A. DiMarzio, C. M. Guttman, and J. D. Hoffman, *Faraday Discuss. Chem. Soc.*, **68**, 210(1979).

75. C. M. Guttman, J. D. Hoffman, and E. A. DiMarzio, *Faraday Discuss. Chem. Soc.*, **68**, 297(1979).

76. J. D. Hoffman, C. M. Guttman, and E. A. DiMarzio, *Faraday Discuss. Chem. Soc.*, **68**, 177(1979).

77. U. R. Evans, *Trans. Faraday Soc.*, **41**, 365(1945).

78. P. Meares, *Polymers: Structure and Bulk Properties*, Van Nostrand, New York, 1965, Chap. 5.

79. J. N. Hay, *Br. Polym. J.*, **3**, 74(1971).

80. S. D. Poisson, *Recherches sur la Probabilite des Judgements en Matiere Criminelle et en Matiere Civile*, Bachelier, Paris, 1837, p. 206.

81. H. D. Keith and F. J. Padden, Jr., *J. Appl. Phys.*, **34**, 2409(1963).

82. B. Wunderlich and L. Melillo, *Makromol. Chem.*, **118**, 250(1968).

83. D. O. Yoon and P. J. Flory, *Faraday Discuss. Chem. Soc.*, **68**, 288(1979).

84. J. D. Hoffman, lecture, London, June 1983.

85. J. D. Hoffman, L. J. Frolen, G. S. Ross, and J. I. Lauritzen, Jr., *J. Res. NBS*, **79A**(6), 671(1975).

86. M. Tasumi and S. Krimm, *J. Polym. Sci.*, *A-2*, **6**, 995(1968).

87. M. I. Bank and S. Krimm, *J. Polym. Sci.*, *A-2*, **7**, 1785(1969).

88. S. Krimm and T. C. Cheam, *Faraday Discuss. Chem. Soc.*, **68**, 244(1979).

89. X. Jing and S. Krimm, *Polym. Lett.*, **21**, 123(1983).

90. R. Kitamarn, F. Horii, and S. H. Hyon, *J. Polym. Sci. Polym. Phys. Ed.*, **15**, 821(1977).

91. M. Glotin and L. Mandelkern, *Colloid Polym. Sci.*, **260**, 182(1982).

92. L. Mandelkern, in *Physical Properties of Polymers*, J. E. Mark, A. Eisenberg, W. W. Graessley, L. Mandelkern, and J. L. Koenig, Eds. American Chemical Society, Washington, DC, 1984.

93. M. Stamm, E. W. Fischer, M. Dettenmaier, and P. Convert, *Faraday Discuss. Chem. Soc.*, **68**, 263(1979).

94. D. G. H. Ballard, A. N. Burgess, T. L. Crawley, G. W. Longman, and J. Schelten, *Faraday Discuss. Chem. Soc.*, **68**, 279(1979).

95. M. Stamm, *J. Polym. Sci. Polym. Phys. Ed.*, **20**, 235(1982).

96. G. D. Wignall, L. Mandelkern, C. Edwards, and M. Glotin, *J. Polym. Sci. Polym. Phys. Ed.*, **20**, 245(1982).

97. D. M. Sadler and R. Harris, *J. Polym. Sci. Polym. Phys. Ed.*, **20**, 561(1982).

98. J. Schelten, G. D. Wignall, D. G. H. Ballard, and G. W. Longman, *Polymer*, **18**, 1111(1977).

99. J. Schelten, A. Zinken, and D. G. H. Ballard, *Colloid Polym. Sci.*, **259**, 260(1981).

100. D. G. H. Ballard, P. Cheshire, G. W. Longman, and J. Schelten, *Polymer*, **19**, 379(1978).

101. E. W. Fischer, M. Stamm, M. Dettenmaier, and P. Herschenraeder, *Polym. Prepr. Am. Chem. Soc. Div. Polym. Chem.*, **20**(1), 219(1979).

102. J. M. Guenet, *Polymer*, **22**, 313(1981).

103. D. M. Sadler and A. Keller, *Macromolecules*, **10**, 1128(1977).

104. D. M. Sadler and A. Keller, *Science*, **203**, 263(1979).

105. A. Keller, *Faraday Discuss. Chem. Soc.*, **68**, 145(1979).

106. D. M. Sadler and A. Keller, *Polymer*, **17**, 37(1976).

107. L. H. Sperling, *Polym. Eng. Sci.*, **24**, 1(1984).

108. D. Y. Yoon and P. J. Flory, *Polymer*, **18**, 509(1977).

109. D. Y. Yoon, *J. Appl. Cryst.*, **11**, 531(1978).

110. D. Y. Yoon and P. J. Flory, *Faraday Discuss. Chem. Soc.*, **68**, 289(1979).

111. M. Dettenmaier, E. W. Fischer, and M. Stamm, *Colloid Polym. Sci.*, **258**, 343(1979).

112. M. Stamm, E. W. Fischer, and M. Dettenmaier, *Faraday Discuss. Chem. Soc.*, **68**, 263(1979).

113. J. Schelten, D. G. H. Ballard, G. D. Wignall, G. W. Longman, and W. Schmatz, *Polymer*, **17**, 751(1976).

114. D. G. H. Ballard, P. Cheshire, G. W. Longman, and J. Schelten, *Polymer*, **19**, 379(1978).

115. B. Crist, W. W. Graessley, and G. D. Wignall, *Polymer*, **23**, 1561(1982).

116. F. C. Frank, *Faraday Discuss. Chem. Soc.*, **68**, 7(1979).

117. R. Hosemann, *Polymer*, **3**, 349(1962).

118. F. J. Balta-Calleja and R. Hosemann, *J. Appl. Crystallogr.*, **13**, 521(1980).

119. R. Hosemann, *Colloid Polym. Sci.*, **260**, 864(1982).

120. B. Wunderlich, in *Macromolecular Physics*, Vol. I, Academic Press, Orlando, 1973.

121. B. Wunderlich and L. Melillo, *Makromol. Chem.*, **118**, 250(1968).

122. E. Hellmuth 2nd, and B. Wunderlich, *J. Appl. Phys.*, **36**, 3039(1965).

123. B. Wunderlich, *J. Polym. Sci. Symp.*, **43**, 29(1973).

124. W. J. Moore, *Physical Chemistry*, 4th ed., Prentice-Hall, Englewood Cliffs, NJ, 1972, p. 134.

125. P. J. Flory, *J. Chem. Phys.*, **17**, 223(1949).

126. L. Mandelkern and P. J. Flory, *J. Am. Chem. Soc.*, **73**, 3206(1951).

127. L. Mandelkern, R. R. Garrett, and P. J. Flory, *J. Am. Chem. Soc.*, **74**, 3949(1952).

128. G. M. Bristow and W. F. Watson, *Trans. Faraday Soc.*, **54**, 1731(1958).

129. F. E. Karasz and L. D. Jones, *J. Phys. Chem.*, **71**, 2234(1967).

130. A. V. Tobolsky, *Properties and Structure of Polymers*, Wiley, New York, 1960.

131. M. Warner, in *Side-Chain Liquid Crystals*, C. B. McArdle, Ed., Blackie, Glasgow, 1989.

132. P. J. Flory, *J. Chem. Phys.*, **10**, 51(1942).

133. R. L. Scott, *J. Chem. Phys.*, **17**, 279(1949).

134. T. Nishi and T. T. Wang, *Macromolecules*, **8**, 909(1975).

135. B. Wunderlich, *Macromolecular Physics*, Vol. 1, Academic Press, Orlando, 1973.

136. J. D. Hoffman and J. J. Weeks, *J. Res. Natl. Bur. Stand.*, **66A**, 13(1962).

137. R. E. Wilfong, *J. Polym. Sci.*, **54**, 385(1961).

138. R. W. Lenz, *Organic Chemistry of Synthetic High Polymers*, Interscience, New York, 1967, pp. 91–95.

139. W. H. Carothers, *J. Am. Chem. Soc.*, **51**, 2548, 2560(1929).

140. J. R. Whinfield, *Nature*, **158**, 930(1946).

141. J. E. McIntire and M. J. Denton, in *Encyclopedia of Polymer Science and Engineering*, Vol. 6, J. I. Kroschwitz, Ed., Wiley, New York, 1986.

141a. J. W. Schappel and G. C. Bockno, in "Cellulose and Cellulose Derivatives," N. M. Bikales and L. Segal, Eds., Vol. 5, Part 5, High Polymer Series, Wiley-Interscience, New York, 1971.

142. H. Tadokoro, Y. Takahasi, S. Otsuka, K. Mori, and F. Imaizumi, *J. Polym. Sci.*, *3B*, **3B**, 697(1965).

143. H. Tadokoro, M. Kobayaski, K. Mori, Y. Takahashi, and S. Taniyama, *J. Polym. Sci. C*, **22**, 1031(1969).

144. L. H. Sperling and C. E. Carraher, in *Encyclopedia of Polymer Science and Engineering*, 2nd ed., Vol. 12, J. I. Kroschwitz, Ed., Wiley, New York, 1988.

145. G. Tsoumis, *Wood as a Raw Material*, Pergamon, New York, 1968.

146. W. S. Boston, in *Encylopedia of Textiles*, *Fibers*, *and Nonwoven Fabrics*, M. Grayson, Ed., Wiley-Interscience, New York, 1984.

147. K. Kastelic, A. Galeski, and E. Baer, *Conn. Tiss. Res.*, **6**, 11(1978).

148. E. Baer, A. Hiltner, and H. D. Keith, *Science*, **235**, 1015(1987).

149. E. Baer, *Sci. Am.* **254**(10), 179(1986).

150. H. Morawetz, *Polymers: The Origins and Growth of a Science*, Wiley, New York, 1985.

GENERAL READING

C. W. Bunn, *Chemical Crystallography*, Oxford University Press, London, 1945.

J. M. G. Cowie, *Polymers: Chemistry and Physics of Modern Materials*, Intext Educational Materials, New York, 1973.

R. D. Deanin, *Polymer Structure, Properties, and Applications*, Cahners Books, Boston, 1972.

M. Doi and S. F. Edwards, *The Theory of Polymer Dynamics*, Oxford, New York, 1986.

P. J. Flory, *Principles of Polymer Chemistry*, Cornell University Press, Ithaca, NY, 1953.

P. H. Geil, *Polymer Single Crystals*, Interscience, New York, 1963.

J. W. S. Hearle, *Polymers and Their Properties*, Ellis Horwood, Chichester, England, 1982.

L. Mandelkern, *Crystallization of Polymers*, McGraw-Hill, New York, 1964.

J. E. Mark, A. Eisenberg, W. W. Graessley, L. Mandelkern, and J. L. Koenig, *Physical Properties of Polymers*, American Chemical Society, Washington, DC, 1984.

P. Meares, *Polymers: Structure and Bulk Properties*, Van Nostrand, London, 1965.

Organization of macromolecules in the condensed phase, *Faraday Discuss. Chem. Soc.*, **68**, (1979).

F. Rodriguez, *Principles of Polymer Systems*, 2nd ed., McGraw-Hill, New York, 1982.

H. Staudinger, *Die Hochmolekularen Organische Verbindung*, Springer-Verlag, Berlin, 1932.

H. Tadokoro, *Structure of Crystalline Polymers*, Wiley-Interscience, New York, 1979.

B. Wunderlich, *Macromolecular Physics*, Vol. I, *Crystal Structure, Morphology, Defects*, Academic Press, Orlando, 1973; Vol. II, *Crystal Nucleation, Growth, Annealing*, Academic Press, Orlando, 1976.

STUDY PROBLEMS

1. Based on the unit cell structure for cellulose I, calculate its theoretical crystal density. (See Figure 6.1.)

2. If polyethylene of z-average molecular weight 30,000 g/mol is cooled from the melt at 1°C/min and 100°C/min, estimate the fractions of polymer crystallized in regimes I, II, and III. [*Hint:* Assume an instantaneous nucleation density of $10^4/cm^3$.]

3. A difficultly crystallizable high-molecular-weight polymer was finally crystallized in regime I. Compare and contrast the properties of the crystallized polymer and the amorphous polymer at the same temperature and pressure. Specifically, how do the densities, radii of gyration, and morphology via optical microscopy differ?

4. Calculate ΔH_f and χ_1 for the three diluents of cellulose tributyrate illustrated in Figure 6.40. Are these good or poor solvents for cellulose tributyrate? Given that the solubility parameter of dimethyl phthalate is 10.7 $(cal/cm^3)^{1/2}$, calculate the solubility parameter for the polymer.

5. Polymers are supposed to consist of long chains, yet the unit cell, by x-ray studies, is about the same size as those of ordinary molecules, containing only relatively few atoms. How can this be?

6. Compare and contrast the Avrami, Keith–Padden, and Hoffman theories of crystallization.

7. If single crystals of polyethylene were carefully laid in a mat, heated briefly above T_f (interdiffusion time < reptation time), and recooled to form a monolithic solid, how would the product compare mechanically to normal, extruded polyethylene?

8. Given the unit cell structure of polyethylene (Figure 6.5), compute the theoretical density of the 100% crystalline product. [*Hint:* see Table 6.2.]

9. Equation (6.43) shows corrections to the melting point depression due to mismatch of molar volumes of solvent and mer, and the polymer–solvent interaction parameter. Should the corresponding equations for copolymers and finite molecular weight be corrected similarly? See equations (6.41) and (6.42). If so, derive suitable relations.

10. Poly(decamethylene adipate), density = 0.99 g/cm^3, was mixed with various quantities of dimethylformamide $(CH_3)_2NCHO$, $d = 0.9445$ g/cm^3 and the melting temperatures observed:[†]

v_1	$T_f, °C$
0.078	72.5
0.202	66.5
0.422	61.5
0.603	57.5

(a) What is the melting point of the pure polymer? (b) What is the heat of fusion of poly(decamethylene adipate), and the value of the χ_1?

11. What spherulite radius can be calculated from Figure 6.16?

12. Devise an NMR experiment to study chain folding in (a) cellulose triacetate, (b) isotactic polystyrene, and (c) transpolyisoprene (Gutta percha). [*Hint:* What chemical modifications, if any, are required?]

13. Compare infrared, NMR, and SANS results on chain folding in single crystals. Can you devise a new experiment to investigate the problem?

14. Read an original paper published in the last 12 months on crystalline polymer behavior or theory, and write a brief report on it in your own words. Cite the authors and exact reference. Does it support the present text? Add new ideas or data? Contradict present theories or ideas?

15. Single crystals of polyethylene are grown from different molecular weight materials from $M = 2000$ to 5×10^7 g/mol. The crystals are all 150 Å thick, with adjacent reentry and superfolding after each 20 stems. How

[†]L. Mandelkern, R. R. Garrett, and P. J. Flory, *J. Am. Chem. Soc.*, **74**, 3949(1952).

does R_g depend on M in this region? Plot the results. What dependence of R_g on M is predicted as M goes to infinity?

16. The lattice constants of orthorhombic polyethylene have been determined as a function of temperature[†]

	LATTICE CONSTANTS, Å		
T, K	a	b	c
4	7.121	4.851	2.548
77	7.155	4.899	2.5473
293	7.399	4.946	2.543
303	7.414	4.942	2.5473

(a) What is the theoretical volume coefficient of expansion of 100% crystalline polyethylene? (The volume coefficient of expansion is given by

$$\alpha = \frac{1}{V}\left(\frac{\partial V}{\partial T}\right)_P$$

where the change in volume V is measured as a function of temperature T at constant pressure P.) (b) Why are the c-axis lattice constants substantially independent of the temperature?

17. You were handed bottles of dimethylcyclooctadiene and cis-polyisoprene, but they became mixed up. What experiments would you perform to identify them?

18. Toothbrushes are available with soft, medium, and hard bristles, generally made of nylon. (a) Propose at least two distinctly different ways by which the bristles could be controlled to provide soft, medium, or hard performance. (b) If you had bristles from different toothbrushes, what tests or experiments would you perform to determine the ways that were actually used to control the hardness? (c) Why nylon?

19. You are handed a glass fiber reinforced sample. "Do not damage this very valuable material," your boss asks, "but we need to know what is in it!" What nondestructive experiments would you perform *in situ* to determine the chemical structure, crystallinity (if any), orientation, and so on in the material?

[†]H. Tadokoro, *Structure of Crystalline Polymers*, Wiley-Interscience, New York, 1979, p. 375.

7

POLYMERS IN THE LIQUID CRYSTALLINE STATE

7.1 DEFINITION OF A LIQUID CRYSTAL

Liquid crystals, LCs, are substances that exhibit long-range order in one or two dimensions, but not all three. Both small molecules and polymers may exist in the liquid crystalline state, but generally special chemical structures are required. Thus, while amorphous substances are more or less entirely disordered (see Chapter 5) and crystalline materials are ordered in all three dimensions (see Chapter 6), the LCs lie in-between in properties. Stated differently, LCs may differ from true crystals in the absence of long-range molecular order in directions transverse to the domain axes, or in the imperfection of molecular alignment along these axes. LC polymers are a relatively new discovery dating from about 1950 (1), with the field growing extremely rapidly. Engineering advances include new classes of high-modulus fibers, high-temperature plastics, and a host of new electronic and data storage materials.

The formation of liquid crystals is a direct consequence of molecular asymmetry. It arises because two molecules cannot occupy the same space at the same time and is largely entropically derived.

A simple example of a "liquid crystal" may be considered using a sink filled with water. Onto the water surface, toothpicks are slowly added without overlap. At first, the toothpicks are random in order. However, as the surface becomes more concentrated, the toothpicks begin to align into small groups more or less side by side. Now there are two phases present: the dilute, disordered phase, and the LC phase. The reason why the toothpicks order themselves is to prevent one toothpick from having to go through (or over) another. Much of what follows can use this ordering behavior as a model.

7.2 ROD-SHAPED CHEMICAL STRUCTURES

In the preceding chapters the concept of the random coil was developed, where axial correlation of mers is only of very short distance; see Figure 5.1. Now a number of polymers have been synthesized which are substantially rod-shaped. An example is poly(p-phenylene terephthalamide), PPD-T, with the chemical structure

$$
\left[NH-\!\!\bigcirc\!\!-NH-\!\!\overset{O}{\underset{O}{\underset{\|}{C}}}-\!\!\bigcirc\!\!-\overset{O}{\underset{\|}{C}} \right]_n
\tag{7.1}
$$

Because the amide groups are located in the trans positions on the phenylene rings, there is minimal bond rotation out of the chain axis.

Another example is a copolymer of p-hydroxybenzoic acid, p,p'-biphenol, and terephthalic acid, creating the polyester structure

$$
\left[O-\!\!\bigcirc\!\!-\overset{O}{\underset{\|}{C}}-O-\!\!\bigcirc\!\!-\!\!\bigcirc\!\!-O-\overset{O}{\underset{\|}{C}}-\!\!\bigcirc\!\!-\overset{O}{\underset{\|}{C}} \right]_n
\tag{7.2}
$$

Again, the ester groups, together with the para-bonded phenylene rings, form strictly rodlike structures.

7.3 LIQUID CRYSTALLINE MESOPHASES

Liquid crystalline structures can be organized into several classes, much the same as crystalline materials are organized into body-centered, triclinic, and so on lattices. The major LC mesophase topologies are shown in Figure 7.1 (2). The nematic LCs are organized in one dimension only, with their chains lying parallel to each other at equilibrium. (The toothpicks described earlier also formed a nematic mesophase.) The smectic LCs are ordered in two dimensions. Polymer scientists now recognize a large number of smectic mesophases. Figure 7.1b illustrates the smectic A and C structures, two of the more important mesophases. The cholesteric mesophase, Figure 7.1a, is like a two-dimensional twisted nematic mesophase. The discotic mesophase resembles stacks of dishes or coins.

The common feature of the structures shown in Figure 7.1 is asymmetry of shape, manifested either as rods of axial ratio greater than about 3, or by thin platelets of biaxial order. Of course, actual materials are made up of molecules. Some LC chemical structures of low molecular weight are shown in Figure 7.2 (2a). The reader will imagine how the various chemical struc-

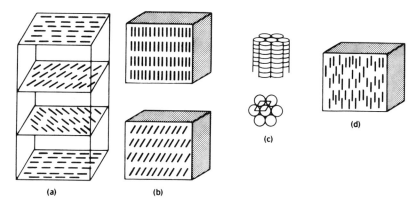

Figure 7.1 Liquid crystal structures: (*a*) cholesteric (*b*) smectic A (top) and smectic C (bottom), (*c*) discotic, and (*d*) nematic (2).

Figure 7.2 Asymmetric organic molecules in the form of rods or plates may form liquid crystal structures (2a).

tures fit into the indicated mesophases of Figure 7.1. Thus the old concept of all matter being composed only of crystalline liquid, and gaseous phases is obsolete. Besides amorphous solids [such as polystyrene or poly(methyl methacrylate)], there are the liquid crystalline structures.

A LC-forming polymer may exhibit multiple mesophases at different temperatures or pressures. As the temperature is raised, the polymer then goes through multiple first-order transitions from a more ordered to a less ordered state. Scientists speak of a "clearing temperature," where the last (or only) LC phase gives way to the isotropic melt or solution.

A phase diagram for a mesogenic side chain polymer and a related monomer is illustrated in Figure 7.3 (3, 4). Note that the monomer by itself crystallizes. The nematic mixture clears in the temperature range of 360–370 K.

The various mesophasic transitions are first-order transitions, because volume and heat capacity go through discontinuities when plotted against temperature or pressure; see Section 6.8. (The glass transition is ideally a

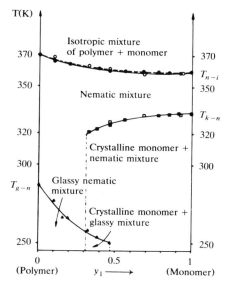

Figure 7.3 Phase diagrams illustrate the expected behavior of a material as a function of temperature and composition. Here, a mixture of the polymer

and the monomer

are shown. The value y_1 is the weight fraction. The full symbols represent DSC measurements; open symbols represent data obtained by polarizing microscope (3, 4).

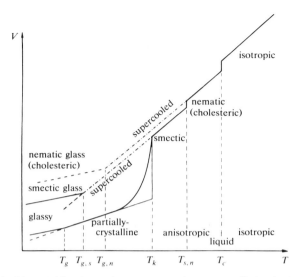

Figure 7.4 Liquid crystal-forming polymers may undergo many first-order transitions. Here, as the temperature is raised, the polymer first melts to a smectic structure, then to a nematic structure, and then to an isotropic melt (3).

second-order transition; see Sections 8.2 and 8.6.3.) The various transitions that LC polymers undergo as the temperature increases are illustrated schematically in Figure 7.4 (3). Here, a semicrystalline polymer melts first to a smectic mesophase, then to a nematic mesophase, and finally to the isotropic liquid state as the temperature increases. Both the volume and the enthalpic jumps through the several transitions are additive; that is, the sum of the changes exactly equals that of going directly from the crystalline state to the isotropic melt.

For example, the kinetics of forming many LC mesophases is fairly slow. Then, by rapid heating, the polymer can be made to skip one or more mesophases, going directly to the melt state. In that case, the volume and enthalpy changes will exactly equal the sum of the changes of going through each of the mesophases, if the heating rate is slower.

Again, it must be emphasized that not all polymers go through LC mesophases. Those polymers that form random coils melt directly to the isotropic liquid state. Some portion of the chain or side chain must be rod- or disk-shaped to form a LC mesophase.

7.4 LIQUID CRYSTAL CLASSIFICATION

Section 7.3 classified LCs according to their mesophase structure (2). Another important classification method divides LC polymers into lyotropic, thermotropic, and mesogenic side group compositions. The lyotropic LCs form ordered states in concentrated solutions, similar again to the toothpick analogy previously described. Frequently, rod-shaped polymer chains are dissolved to concentrations of about 30% for processing purposes, particularly for fiber formation. The main reason for using solutions rather than melts is that the polymers in question decompose below their very high melting temperatures. Thermotropic LCs exist as ordered melts, being able to remain in the LC state without decomposition. The mesogenic side-chain

Table 7.1 Important polyamides yielding liquid crystalline mesophases (5)

Table 7.2 Lyotropic solutions of polyheterocyclic compounds (2)

Compound	Structure	Lyotropic Solution	Reference
Poly(1,4-phenylene-2,6-benzobisimidazole)		methanesulfonic acid	(a)
Poly(1,4-phenylene-2,6-benzobisoxazole) (PBO)		methanesulfonic acid chlorosulfonic acid 100% sulfuric acid	(b, c) (d)
Poly(1,4-phenylene-2,6-benzobisthiazole) (PBT)		5–10% in polyphosphoric acid methanesulfonic acid	(e) (c, e, f)
Poly[2,6-(1,4-phenylene)-4-phenylquinoline]		1.0–1.5% in *m*-cresol-di-*m*-cresyl phosphate	(g)
Poly[1,1'-(4,4'-biphenylene)-6, 6'-bis(4-phenylquinoline)]		> 9% in *m*-cresol-di-*m*-cresyl phosphate	(g)
Poly[2,5(6)-benzimidazole] (AB–PBI)		methanesulfonic acid	(h)

References: (a) T. E. Helminiak, F. E. Arnold, and C. L. Benner, *Polym. Prepr. Am. Chem. Soc. Div. Polym. Chem.*, **16**(2), 659(1975). (b) G. C. Berry, P. M. Cotts, and S.-G. Chu, *Br. Polym. J.*, **13**, 47(1981); C. P. Wong, H. Ohnuma, and G. C. Berry, *J. Polym. Sci. Polym. Symp.*, **65**, 173(1978). (c) J. F. Wolfe, B. H. Loo, and F. E. Arnold, *Macromolecules*, **14**, 915(1981). (d) E. W. Choe and S. N. Kim, *Macromolecules*, **14**, 920(1981). (e) J.F. Wolfe, B. H. Loo, and E. R. Sevilla, *Polym. Prepr. Am. Chem. Soc. Div. Polym. Chem.*, **22**(1), 60(1981). (f) S.-G. Chu, S. Venkatraman, G. C. Berry, and Y. Einaga, *Macromolecules*, **14**, 939(1981). (g) P. D. Sybert, W. H. Beever, and J. K. Stille, *Macromolecules*, **14**, 493(1981). (h) A. Wereta, Jr. and M. T. Gehatia, *Polym. Eng. Sci.*, **18**, 204(1978).

compositions usually have random coil backbone chains, with rod-shaped side groups. It is these latter that form the LC structures in this case. The mesogenic side-chain polymers are actually a subclass of the thermotropic LCs, since they are usually processed in the melt state.

7.4.1 Lyotropic Liquid Crystalline Chemical Structures

Most often, the lyotropic LC polymers form nematic mesophases. Most of the polymers in this class are aromatic polyamides with aromatic ring structures, as shown in Table 7.1 (5). Several of the polymers in Table 7.1 (see page 284) form very high-modulus fibers; see Chapter 11.

Heterocyclic polymers yield materials with outstanding high-temperature performance; see Table 7.2 (see page 284) (2). Many of these polymers have ladder or semiladder chemical structures. If one bond in a ladder structure is broken by heat or oxidation, the chain may retain its original molecular

Table 7.3   Melting point versus structure for selected thermotropic LC polyesters (9)

Bulky Side Group	T (Melting) °C
	> 600
	≈ 340

Bent Monomer	T (Melting) °C
	≈ 400
	≈ 350

Flexible Group	T (Melting) °C
	> 400
	≈ 210

weight. If a single carbon atom chain is oxidized, usually the chain is degraded, with concomitant loss of properties.

Some natural polymers also form lyotropic systems. These include cellulose derivatives, cellulose being a stiff molecule, but not entirely rod shaped. Another polymer is the polypeptide poly(γ-benzyl L-glutamate) (6, 7),

$$\left(\begin{array}{c} H \quad\quad O \\ | \quad\quad || \\ N-CH-C \\ | \\ (CH_2)_2 \\ | \\ O = C-O-CH_2- \end{array} \right)_n \bigcirc \tag{7.3}$$

which forms birefringent, anisotropic phases in organic solutions. The shape of these polypeptides is in the form of α-helices, with regular intramolecular hydrogen bonding and association of the solvent with the substituents to stabilize the structure in the form of a rod.

7.4.2 Thermotropic Liquid Crystalline Chemical Structures

Most of the thermotropic LC polymers are aromatic copolyesters, some being terpolymers (8). Simple homopolymer aromatic polyesters generally have too high a melting temperature to form thermotropic mesophases without decomposition, but copolymerization reduces their melting temperatures; see Section 6.8.1. In order to reduce the melting temperature of these materials,

Side chain LC polymers   Main chain LC polymers

▭ Mesogenic group
〰〰 Flexible spacer

Figure 7.5 One way of classifying thermotropic LC polymers is to examine whether the mesogenic unit is in the main chain or in the side chain (10).

Table 7.4 Examples of smectic polymers by lengthening the segments A and B

Number	Polymer	A	Phase Transitions (K)	$\Delta \bar{H}_{LC-i}$ (J/g)
1	$\cdots -CH_2-\underset{\displaystyle (CH_2)_2}{\overset{\displaystyle CH_3}{\underset{\displaystyle |}{\overset{\displaystyle |}{C}}}}-\cdots$	$\text{---}COO\text{---}\bigcirc\text{---}OCH_3$	g 369, n 394, i	2.3
		$-OC_6H_{13}$	g 410, s 451, i	11.5
3	$\cdots-\underset{\displaystyle (CH_2)_3-O-}{\overset{\displaystyle CH_3}{\underset{\displaystyle |}{\overset{\displaystyle |}{Si}}}}-O-\cdots$	$\text{---}OCH_3$	g 288, n 334, i	2.2
4		$-OC_6H_{13}$	g 288, s 385, i	11.6

Table 7.4 (*Continued*)

Number	Polymer B	Phase Transitions (K)	$\Delta \bar{H}_{LC-i}$ (J/g)
1	$\cdots -CH_2-\overset{\displaystyle CH_3}{\underset{\displaystyle \vert}{C}}-\cdots$ $COO-(CH_2)_2-O$—⬡—COO—⬡—$CH=N$—⬡—CN	g 361, n 580, i	—
2	$-(CH_2)_6$	g 324, s 607, i	—
3	$\cdots -CH_2-\overset{\displaystyle CH_3}{\underset{\displaystyle \vert}{C}}-\cdots$ $COO-(CH_2)_2-O$—⬡—⬡—OCH_3	g 393, n, 425, i	2.8
4	$-(CH_2)_6$	g 119, s 136, i	7.1

g, glassy; n, nematic; s, smectic; i, isotropic.

289

several techniques are available; see Table 7.3 (9). These include (9)

1. Copolymerization of several mesogenic monomers, which produces random copolymers with depressed melting temperatures.
2. Use of monomers with bulky side groups, which prevents close packing in the LC mesophases, sometimes referred to as "frustrated chain packing."
3. Use of bent comonomers to interrupt the order in the system.
4. Incorporation of flexible spacers, to decrease rigidity. This permits the development of bends or elbows in the chain.

Flexible spacers placed in the main chain achieve both improved solubility and the lowering of transition temperatures (10, 11). The flexible spacer may be hydrocarbon based, as in Table 7.3 (see page 286), or a short series of dimethyl siloxane groups. The use of spacers is illustrated schematically for both main-chain and side-chain mesogenic units in Figure 7.5 (see page 286) (10).

It must be emphasized that neither the lyotropic nor the thermotropic polymers are ordinarily used in the liquid crystalline state. The LC state is highly convenient for processing, yielding highly ordered structures with low viscosities. However, both types are highly crystalline in actual usage, going from the LC to the crystalline state by either removing the solvent or cooling the system. However, the side-chain mesogenic materials usually remain in the LC state for their intended uses.

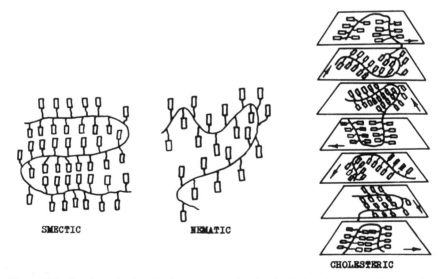

SMECTIC   NEMATIC   CHOLESTERIC

Figure 7.6 A schematic showing how mesogenic LC side-chain polymers can be organized into different mesophases (12).

7.4.3 Side-Chain Liquid Crystalline Chemical Structures

Polymers here have rod- or disk-shaped side groups placed on ordinary random coil polymers, frequently acrylics or siloxanes; see Table 7.4 (see pages 288 and 289) (3). The first attempts to form side-chain LC structures involved attachment of short rod-shaped moieties directly to the main chain (Figure 7.6) (12). These efforts were disappointing. Two reasons were identified (3):

1. The glass transition of the polymers is always at much higher temperatures than for the corresponding polymers without mesogenic side chains.
2. During the polymerization process, the LC packing of the monomers was destroyed by steric requirements.

It was reasoned by Plate and Shibaev (13) that if the mesogenic moieties were to be linked to the backbone through alkyl spacer groups, they should develop a LC structure similar to that attained before polymerization. The effects of backbone type, spacer group, and length of the "free" substituent at the end of the mesogenic moiety all must be considered. For example, as the length of the spacer group is increased, the low enthalpies of phase transformation of the nematic polymers become larger and the better ordered smectic phases form in their place. The structure of these LC phases

Figure 7.7 Oriented smectic LC polymers yield very distinctive x-ray patterns. Here, the smectic A mesophase (S_A) is shown for polymers having the structures (13)

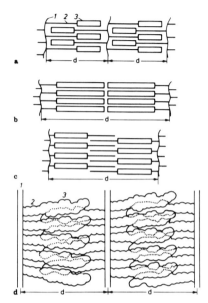

Figure 7.8 LC side chains can be packed in several arrangements, even considering only the smectic A mesophase: (*a*) single-layer packing, (*b*) two-layer packing, (*c*) packing with overlapping alkyl "tails," and (*d*) packing with partial overlapping of the mesogenic side chains. 1, Main chain; 2, spacer; 3, mesogenic group; *d*, repeat distance, as revealed by x-rays (13).

was determined by x-ray analysis, as shown in Figure 7.7 (see page 291) (13). Similar considerations apply to the "free" substituent.

The packing arrangement of these materials is illustrated in Figure 7.8 (13). The side chains can be arranged in either single-layer or double-layer packing arrangements, Figures 7.8*a* and 7.8*b*, respectively. Packing with partial overlap is also possible; see Figures 7.8*c* and 7.8*d*. The selection of the probable packing mode was based on the *d* spacing shown in Figure 7.8, which corresponds to the thickness of the smectic layer.

In contrast to the backbone types of LC, the side-chain types exhibit neither high modulus nor high strength. However, they are proving highly interesting for their structural and optical properties, especially as transformed by electric and magnetic fields; see Section 12.3.

7.5 THERMODYNAMICS AND PHASE DIAGRAMS

7.5.1 Historical Aspects

Liquid crystals were first noted for organic molecules about 100 years ago, when Reinitzer (14) observed a peculiar melting behavior of a number of cholesterol esters. He found that the crystals of the substances melted sharply to form an opaque melt instead of the usual clear melt. The conclusion Reinitzer drew from this observation was that some type of order still existed in the molten state. Furthermore, Reinitzer observed that the opacity vanished at a higher temperature, called the clearing temperature

(15). Shortly thereafter, Lehmann (16) reported that ammonium oleate and p-azoxyphenetole exhibited turbid states between the truly crystalline and the truly isotropic fluid state. To describe the strange behavior of the new state, Lehmann introduced the term "Flussige Kristalle," or liquid crystals. As mentioned previously, the first polymeric liquid crystal discovered was poly(γ-benzyl L-glutamate) (1), in 1950.

In 1956, Paul Flory contributed a pair of papers based on statistical thermodynamics, which proved to have enormous predictive power (17, 18). The model was a rod of x isodiametric segments, each of a size that occupied one lattice cell. The quantity x also constitutes the axial ratio of the molecule. The molecules were assumed to lie at some arbitrary angle from the preferred axis. The positioning probabilities were formulated as the product of the number of lattice sites available to the initial segment of the chain times the probabilities that the sites required for each successive segment of the chain would be unoccupied, and hence accessible (19). This was entirely an entropic calculation; heats of mixing were ignored. Most importantly, a transition from complete disorder to partial order was predicted to occur abruptly and discontinuously beyond a critical volume concentration, v_2^*, of polymer,

$$v_2^* = \frac{8}{x}\left(1 - \frac{2}{x}\right) \tag{7.4}$$

which is approximately equal to $8/x$ for high values of x. Equation (7.4) yields the threshold volume fraction for appearance of a stable anisotropic phase.

The first *synthetic* LC polymers, poly(p-benzamide) and poly(p-phenylene terephthalamide), followed in 1966 and 1968, respectively (20, 21); the latter is known as Kevlar™. Thus theoretical papers based only on the appearance of scant data on protein materials led the way to major advances in the synthetic areas.[†]

7.5.2 Importance of the χ_1 Parameter

The statistical thermodynamic theory described above pertains to rods devoid of interactions other than the short-range repulsions, which preclude intrusion of one rod on the space occupied by another. Thus theoretical deductions stem solely from the geometrical aspects of the molecules. Flory then introduced the polymer–solvent interaction parameter, χ_1, for lyotropic systems. The resulting phase diagram is illustrated in Figure 7.9 (19). The narrow biphasic gap is little affected by the interactions for negative values of

[†] It must be pointed out that Flory made a large number of theoretical predictions, which, together with his outstanding research in the field of polymers, led to his winning the Nobel prize in chemistry in 1974.

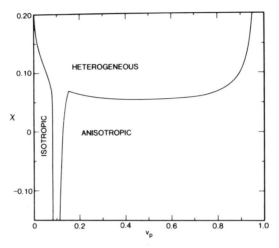

Figure 7.9 Phase diagram for rods of axial ratio $x = 100$. The positions of the coexisting phases are determined by the value of the parameter χ. The minimum of the shallow concave portion of the diagram is a critical point marking the emergence of two anisotropic phases, in addition to the isotropic phase appearing on the left of the figure (19).

χ_1, because the rods are attracted to the solvent. If, however, χ_1 is positive, a critical point emerges at $\chi_1 = 0.055$. For values of χ_1 immediately above this critical limit, the shallow concave curve delineates the loci of coexisting anisotropic phases, these being in addition to the isotropic and nematic phases of lower concentration within the narrow biphasic gap on the left. At $\chi_1 = 0.070$, a triple point is reached at the cusp, where three phases coexist, dilute isotropic, dilute anisotropic, and concentrated anisotropic. The heterogeneous regions contain both isotropic and anisotropic phase domains.

7.6   MESOPHASE IDENTIFICATION IN THERMOTROPIC POLYMERS

A major question in the liquid crystal field relates to the identification of the various mesophases. Several methods may be used (22):

1. Optical pattern or texture observations with a polarizing microscope. Isotropic liquids have no texture, while both regular crystals and liquid crystals do. However, the texture of each type of phase is different.

2. Nematic and smectic phases can be distinguished by DSC on the basis of the magnitude of the enthalpy change accompanying the transition to the isotropic phase.

3. Miscibility with known liquid crystals, to form isomorphous mesophases; see Figure 7.3.

Figure 7.10 Optical microscopy reveals a Schlieren texture for a copolyester formed by the transesterification of poly(ethylene-1,2-diphenoxyethane-p,p'-dicarboxylate) with p-acetoxy-benzoic acid. Recognizable singularities $S = \pm \frac{1}{2}$ are marked with arrows. Crossed polarizers, 260°C, ×200 (23).

4. Possibilities of inducing significant molecular orientation by either supporting surface treatments, or external electrical or magnetic fields.

5. Small-angle x-ray studies can be used to establish molecular long-range order; see Figure 7.7.

6. Small-angle light scattering, particularly between crossed nicols, yields different patterns for different mesophases.

The most widely used method is optical microscopy, especially between crossed polarizers. It is used to identify the appearance of mesophases and transitions between the various mesophases and isotropic materials.

Many liquid crystalline polymers exhibit Schlieren textures, which display dark brushes. These correspond to extinction positions of the mesophase. At certain points, two or more dark brushes meet at points called disclinations; see Figure 7.10 (23). The disclination strength is calculated from the number of dark brushes meeting at one point:

$$|S| = (\text{number of brushes meeting})/4 \qquad (7.5)$$

The sign of S is positive when the brushes turn in the same direction as the rotated polarizers, and negative when they turn in the opposite direction. Mesophase identification by this procedure requires

$$S = \pm\tfrac{1}{2}; \pm 1 \qquad \text{Nematic}$$

$$S = \pm 1 \qquad\qquad \text{Smectic}$$

Thus a nematic phase is indicated by a mixture of two and four point disclinations, while a smectic phase exhibits only four point disclinations.

Disclinations are like dislocations in crystalline solids, where domains of differing orientations meet. The disclinations cause distortion of the director field of the polymer chains, giving rise to an excess free energy of the liquid crystal material (24).

7.7 FIBER FORMATION

7.7.1 Viscosity of Lyotropic Solutions

Rod-shaped polymer chains make unusually strong fibers, primarily because it is easy to orient the chains in the fiber direction, without folding; see Chapter 11. Important examples are the aromatic polyamides, which are

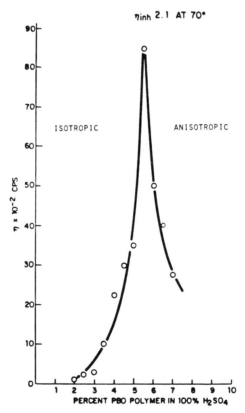

Figure 7.11   The viscosity of rod-shaped polymers usually goes through a maximum as the chains organize from the isotropic state to a mesostate. Data for poly(p-phenylene benzo-bisoxazole) is concentrated sulfuric acid (26).

soluble in high-strength sulfuric acid or in mixed solvents containing N,N-dimethylacetamide. As the concentration of the polymer is increased, the viscosity first increases, then shows an abrupt decrease; see Figure 7.11 (25, 26). This has been interpreted (26) as the formation of the nematic mesophase, v_2^* having been exceeded; see equation (7.1). Of course, the mesophase is optically anisotropic, providing a quantitative measure of the orientation. Interestingly, the viscosity of lyotropic solutions in shear flow is *lower* than that of random coil solutions of the same molecular weight and concentration. The reason is the lack of molecular entanglements in the nematic mesophase. Thus, from a strict engineering point of view, it cost less energy to pump the solutions around, and the resulting fibers are stronger.

7.7.2 Molecular Orientation

The general orientation of the nematic morphology is increased by the shearing action of flow accompanying fiber spinning. Thus, at the instant the concentrated polymer solution exits the spinnerette, ideally all the molecules are highly oriented in the direction of flow. Referring to the toothpick model introduced earlier, all the toothpicks are now aligned in the flow direction, rather than having domains of differing orientation, as when the system is at rest.

It must be emphasized that after the solvent is removed, the polymer crystallizes. Thus the aromatic polyamide fibers, while existing in liquid crystal mesophases during preparation, are highly crystalline when fully formed. Again, because of the lack of chain folds and other imperfections, the fibers have higher moduli and higher strength.

7.8 COMPARISON OF MAJOR POLYMER TYPES

7.8.1 Molecular Conformation

The dilute solution, conformation, and bulk behaviors of rigid rods are compared with random coils in Table 7.5 (27, 28). Also included are the extended chains, both those chains that are stiffer than usual and those rod-shaped chains that contain flexible spacers. For example, v_2^* is higher for the extended chain than the rigid rod, as might be expected by the decreased order, while this feature is missing entirely from the random coil solutions. The Mark–Houwink value of a [see equation (3.81)] reaches a theoretical maximum value of 2.0 for rods, while stiff chains such as cellulose trinitrate have $a = 1.0$. Random coils, sometimes called flexible chains in this context, have values in the range of 0.5–0.8, depending on how thermodynamically "good" the solvent is. Spheres, which follow the Einstein viscosity relationship, have $a = 0$, the viscosity of the solution ideally depending only on the concentration of the solution and not on the size of the spheres.

Table 7.5   Comparison of major polymer types (28)

Property	Rigid Rod	Extended Chain	Random Coils
LC v_2^*			
critical concentration, %	4–5	14–15	None
Dilute solution			
conformation	Rod	Worm	Coil
T_g	No[†]	No[†]	Yes
T_m	No[*]	No[*]	Yes
Persistence			
length, Å	> 500	90–130	10
Mark–Houwink			
a value	1.8	1.0	0.5–0.8
Catenation			
angle, degrees	180	150–162	109–120
Max [η],			
dl/g	48–60	15–25	0.1–3

[*] May decompose before melting.
[†] May be difficult to observe if highly crystalline.

Figure 7.12 Rigid-chain and semiflexible-chain ordered assemblies as the concentration of the polymer is increased (29).

The relative orientation of rigid-rod polymers is schematically compared with that of the semiflexible polymers in Figure 7.12 (29). Here v_2^* is depicted as being between 5 and 10% concentration, with nearly total alignment occurring above 15% concentration.

Typically, rigid-rod systems such as the aromatic polyamides or the polyheterocyclic polymers do not exhibit a glass transition, because their amorphous fraction is negligibly small, and they decompose before their melting temperature. By comparison, many of the crystalline random coil polymers, such as the aliphatic nylons, exhibit normal melting phenomena, as well as glass transitions. Thus the properties of the new rigid-rod polymers are quite different from those of random coil polymers.

7.8.2 Properties in Bulk

These last three chapters delineate the three bulk forms of polymers: amorphous, crystalline, and liquid crystalline. How are they best distinguished in the laboratory? Key experiments involve x-ray diffraction, optical microscopy, and methods of observing transitions. Typical results are as follows:

Experiment	Amorphous	Crystalline	Liquid Crystalline
X-ray	Amorphous halo	3-D order	1- or 2-D order
Polarizing microscope	no texture	spherulites	Schlieren texture
DSC, dilatometric, etc.	only T_g	T_g and T_f	T_g and two or more first-order transitions

7.9 BASIC REQUIREMENTS FOR LIQUID CRYSTAL FORMATION

A liquid crystal must satisfy three basic requirements, regardless of molecular size or shape:

1. There must be a first-order transition between the true crystalline state at the lower temperature bound leading to the liquid crystalline state, and another first-order transition leading to the isotropic liquid state (or another liquid crystal state) at the upper temperature bound of the liquid crystalline state.

2. A liquid crystal must exhibit one- or two-dimensional order only; true crystals have three-dimensional order, and the isotropic liquid is completely disordered.

3. A liquid crystalline material must display some degree of fluidity, although for polymers the viscosity may be high.

Usually, the conditions for liquid crystal formation are best met when the molecules have at least some portion of their structure in the form of rods or disks. However, this is not an absolute requirement, and a surprisingly large number of polymers are now being found to exhibit liquid crystalline phases. While truly rigid-rod polymers ordinarily do not have a glass transition, if some degree of flexibility is built into the polymer chain, there may be one.

The backbone types of polymeric liquid crystals may be lyotropic, requiring a solvent for entrance into the liquid crystalline phase, or thermotropic, requiring heat for entrance into the liquid crystalline phase. However, when dry and at service temperature, both types ordinarily exhibit three-dimensional crystalline order. The side-chain types of liquid crystalline polymers, however, are being researched extensively for uses while actually in the liquid crystalline state.

REFERENCES

1. A. Elliott and E. J. Ambrose, *Discuss. Faraday Soc.*, **9**, 246(1950).
2. S. L. Kwolek, P. W. Morgan, and J. R. Schaefgen, in *Encyclopedia of Polymer Science and Engineering*, 2nd ed., Vol. 9, J. I. Kroschwitz, Ed., Wiley, New York, 1987.
2a. J. L. White, *J. Appl. Polym. Sci. Polym. Symp.*, **41**, 3(1985).
3. H. Finkelmann and G. Rehage, *Adv. Polym. Sci.*, **60/61**, 99(1984).
4. H. Finkelmann, J. J. Kock, and G. Rehage, *Mol. Cryst. Liq. Cryst.*, **89**, 23(1982).
5. J. R. Schaefgen, T. I. Bair, J. W. Ballou, S. L. Kwolek, P. W. Morgan, M. Panar, and J. Zimmerman, in *Ultra-High Modulus Polymers*, A. Ciferri and I. M. Ward, Eds., Applied Science, London, 1979.
6. W. G. Miller, C. C. Wu, G. L. Antee, J. H. Rai, and K. G. Goebel, *Pure Appl. Chem.*, **38**, 37(1974).
7. R. Sakamoto, *Colloid Polym. Sci.*, **262**, 788(1984).
8. A. J. East, L. F. Charbonneau, and G. W. Calundann, U.S. Pat. No. 4,330,457 (1982).
9. G. Huynh-Ba and E. F. Cluff, in *Polymeric Liquid Crystals*, A. Blumstein, Ed., Plenum, New York, 1985.
10. C. K. Ober, J. I. Jin, and R. W. Lenz, in *Liquid Crystal Polymers I*, N. A. Plate, Ed., Advances in Polymer Science Vol. 59, Springer-Verlag, Berlin, 1984.
11. W. J. Jackson and F. H. Kuhfuss, *J. Polym. Sci. Polym. Chem. Ed.*, **14**, 2043(1976).

12. N. A. Plate, R. V. Talroze, and V. P. Shibaev, in *Polymer Yearbook 3*, R. A. Pethrick, Ed., Harwood, Chur, United Kingdom, 1986.

13. N. A. Plate and V. P. Shibaev, *Comb-Shaped Polymers and Liquid Crystals*, Plenum Press, New York, 1987, translated from the Russian by S. L. Schnur.

14. F. Reinitzer, *Monatsh. Chem.*, **9**, 421(1888).

15. J. H. Wendorff, in *Liquid Crtystalline Order in Polymers*, A. Blumstein, Ed., Academic Press, Orlando, 1978.

16. O. Lehmann, *Z. Kristallogr. Kristallgeom. Kristallphys. Kristallchem.*, **18**, 464(1890).

17. P. J. Flory, *Proc. R. Soc. London Ser. A.*, **234**, 60(1956).

18. P. J. Flory, *Proc. R. Soc. London Ser. A.*, **234**, 73(1956).

19. P. J. Flory, in *Liquid Crystal Polymers I*, M. Gordon, Ed., Advances in Polymer Science, Vol. 59, Springer-Verlag, Berlin, 1984.

20. S. L. Kwolek, B.P. 1,198,081 (priority June 13, 1966).

21. S. L. Kuolek, B.P. 1,283,064 (priority June 12, 1968).

22. C. Noel, in *Polymeric Liquid Crystals*, A. Blumstein, Ed., Plenum, New York, 1985.

23. C. Noel, F. Laupretre, C. Friedrich, B. Fayolle, and L. Bosio, *Polymer*, **25**, 808(1984).

24. T. Hashimoto, A. Nakai, T. Shiwaku, H. Hasegawa, S. Rojstaczer, and R. S. Stein, *Macromolecules*, **22**, 422(1989).

25. D. B. DuPre, "Liquid Crystals," in *Encyclopedia of Chemical Technology*, 3rd ed., Vol. 14, M. Grayson, Ed., Wiley, New York, 1981, p. 417.

26. E. W. Choe and S. N. Kim, *Macromolecules*, **14**, 920(1981).

27. J. F. Wolfe, B. H. Loo, and F. E. Arnold, *Macromolecules*, **14**, 915(1981).

28. J. F. Wolfe, presented at Pacifichem, Honolulu, December 1989.

29. D. R. Ulrich, *Polymer*, **28**, 533(1987).

GENERAL READING

A. Blumstein, Ed., *Liquid Crystalline Order in Polymers*, Academic Press, Orlando, 1978.

A. Blumstein, Ed., *Polymeric Liquid Crystals*, Plenum Press, New York, 1985.

A. Ciferri, W. R. Krigbaum, and R. B. Meyer, Eds., *Polymer Liquid Crystals*, Academic Press, Orlando, 1982.

P. G. de Gennes, *The Physics of Liquid Crystals*, Oxford Press, New York, 1975.

S. L. Kwolek, P. W. Morgan, and J. R. Schaefgen, in *Encyclopedia of Polymer Science and Engineering*, Vol. 9, J. I. Kroschwitz, Ed., Wiley, New York, 1987.

C. B. McArdle, Ed., *Side Chain Liquid Crystal Polymers*, Chapman and Hall, New York, 1989.

N. A. Plate, Ed., *Liquid Crystal Polymers I*, Advances in Polymer Science Vol. 59, Springer-Verlag, Berlin, 1984.

N. A. Plate, Ed., *Liquid Crystal Polymers II/III*, Advances in Polymer Science Vol. 60/61, Springer-Verlag, Berlin, 1984.

N. A. Plate and V. P. Shibaev, *Comb-Shaped Polymers and Liquid Crystals*, Plenum Press, New York, 1987.

N. A. Plate, R. V. Talroze, and V. P. Shibaev, in *Polymer Yearbook 3*, R. A. Pethrick, Ed., Harwood, Chur, United Kingdom, 1986.

V. V. Shilov, *Structure of Polymeric Liquid Crystals*, Naukova Dumka, Kiev, 1990.

D. R. Ulrich, *Polymer*, **28**, 533(1987).

G. Vertogen and W. H. de Jeu, *Thermotropic Liquid Crystals, Fundamentals*, Springer-Verlag, Berlin, 1988.

R. A. Weiss and C. K. Ober, Eds., *Liquid-Crystalline Polymers*, ACS Books, Washington, DC, 1990.

J. L. White and S. Onagi, Eds., *U.S.–Japan Seminar on Polymer Liquid Crystals*, Applied Polymer Symposium No. 41, Wiley, New York, 1985.

STUDY PROBLEMS

1. As the temperature is raised, some polymers melt from a regular three-dimensional crystal to a smectic phase, then to a nematic phase, and then finally to an isotropic melt. How would the appropriate DSC curve look for such a polymer?

2. Based on Figure 7.9, discuss whether lyotropic polymer systems exhibit two or more phases simultaneously and, if so, under what conditions?

3. There are three major types of polymers: semicrystalline, liquid crystalline, and amorphous. (a) Give an example of each, naming the polymer. (b) Given an unknown polymer, what one experiment would you perform to identify which type you had? Give an illustration of the data that you might obtain. (c) What major behavioral differences would you expect among the three types?

4. (a) Compare and contrast the dilute solution behavior of rigid rods and random coils. (b) Why is the viscosity higher for rigid rods in the case of dilute solutions, but lower in the case of concentrated solutions?

5. Read any scientific or engineering paper written in the last three years on the topic of liquid crystalline polymers, and briefly report on its findings. How does it update this textbook?

6. In an actual experiment, 5.9 cm × 1.5 mm × 1 mm Forster^{TM} flat toothpicks were added to a water surface one at a time with no overlapping, with mild agitation. A distinct nematic phase formed at about a total surface concentration of 3%. Thereafter, the remaining disordered phase maintained a 0.8% surface concentration up to a total surface concentration of 14%. How would these toothpicks behave if they were first cut in half? In fourths? (Try the experiment in a sink!)

8

GLASS–RUBBER TRANSITION BEHAVIOR

The state of a polymer depends on the temperature and on the time allotted to the experiment. While this is equally true for semicrystalline or amorphous polymers, although in different ways, the discussion in this chapter centers on amorphous materials.

At low enough temperatures, all amorphous polymers are stiff and glassy. This is the glassy state, sometimes called the vitreous state, especially for inorganic materials. On warming, the polymers soften in a characteristic temperature range known as the glass–rubber transition region. Here, the polymers behave in a leathery manner. The importance of the glass transition in polymer science was stated by Eisenberg:[†] "The glass transition is perhaps the most important single parameter which one needs to know before one can decide on the application of . . . non-crystalline polymers . . . ."

For amorphous polymers, the glass transition temperature, T_g, constitutes their most important mechanical property. In fact, upon synthesis of a new polymer, the glass transition temperature is among the first properties measured. This chapter describes the behavior of amorphous polymers in the glass transition range, emphasizing the onset of molecular motions associated with the transition. Before beginning the main topic, two introductory sections are presented. The first defines a number of mechanical terms that will be needed, and the second describes the mechanical spectrum encountered as a polymer's temperature is raised.

[†]In J. E. Mark, A. Eisenberg, W. W. Graessley, L. Mandelkern, and J. L. Koenig, *Physical Properties of Polymers*, American Chemical Society, Washington, DC, 1984.

303

8.1 SIMPLE MECHANICAL RELATIONSHIPS

Terms such as "glassy," "rubbery," and "viscous" imply a knowledge of simple material mechanical relationships. Although such information is usually obtained by the student in elementary courses in physics or mechanics, the basic relationships are reviewed here because they are used throughout the text. More detailed treatments are available (1–4).

8.1.1 Modulus

8.1.1.1 Young's Modulus

Hook's law assumes perfect elasticity in a material body. Young's modulus, E, may be written

$$\sigma = E\varepsilon \tag{8.1}$$

where σ and ε represent the tensile stress and strain, respectively. Young's modulus is a fundamental measure of the stiffness of the material. The higher its value, the more resistant the material is to being stretched.

The tensile stress is defined in terms of force per unit area. If the sample's initial length is L_0 and its final length is L, then the strain is $\varepsilon = (L - L_0)/L_0$ [†] (see Figure 8.1).

The forces and subsequent work terms have some good simple examples. Consider a postage stamp, which weighs about 1 g and is about 1 cm × 1 cm in size. It requires about 1 dyne of force to lift it from the horizontal position to the vertical position, as in turning it over. One dyne of force through 1 cm gives 1 erg, the amount of work done. Modulus is usually reported in dynes/cm^2, in terms of force per unit area. Frequently, the pascal unit of modulus is used, 10 dynes/cm^2 = 1 pascal.

8.1.1.2 Shear Modulus

Instead of elongating (or compressing!) a sample, it may be subjected to various shearing or twisting motions (see Figure 8.1). The ratio of the shear stress, f, to the shear strain, s, defines the shear modulus, G:

$$G = \frac{f}{s} \tag{8.2}$$

These and other mechanical terms are summarized in Table 8.1.

[†] In more graphic language, the amount of stress applied is measured by the amount of grunting the investigator does, and the strain is measured by the sample's groaning.

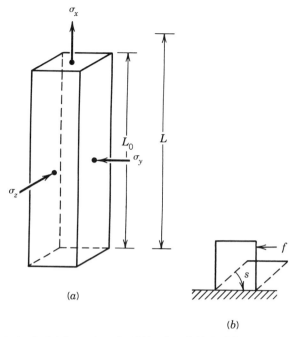

Figure 8.1 Mechanical deformation of solid bodies. (*a*) Triaxial stresses on a material body undergoing elongation. (*b*) Simple shear deformation.

Table 8.1 Some mechanical terms

Term	Definition
σ	Stress
ε	Strain
E	Young's modulus
G	Shear modulus
B	Bulk modulus
ν	Poisson's ratio
η	Coefficient of viscosity
J	Tensile compliance
s	Shear strain
f	Shear stress
R	Gas constant, 8.31×10^7 (dynes $\cdot$ cm)/(mol $\cdot$ K)
t	Time

8.1.2 Newton's Law

The equation for a perfect liquid exhibiting a viscosity, η, may be written

$$f = \eta \left(\frac{ds}{dt} \right) \tag{8.3}$$

where f and s represent the shear stress and strain, respectively, and t is the time. For simple liquids such as water or toluene, equation (8.3) reasonably describes their viscosity, especially at low shear rates. For larger values of η, flow is slower at constant shear stress. While neither equation (8.1) nor equation (8.3) accurately describes polymer behavior, they represent two important limiting cases.

The basic definition of viscosity should be considered in terms of equation (8.3). Consider two 1-cm^2 planes 1 cm apart imbedded in a liquid. If it takes 1 dyne of force to move one of the planes 1 cm/sec relative to the other in a shearing motion, the liquid has a viscosity of 1 poise. Viscosity is also expressed in pascal-seconds, with 1 Pa · sec = 10 poise.

8.1.3 Poisson's Ratio

When a material body is elongated (or undergoes other modes of deformation), in general the volume changes, usually increasing when elongational strains are applied. Poisson's ratio, ν, is defined as

$$-\nu\varepsilon_x = \varepsilon_y = \varepsilon_z \tag{8.4}$$

where the strain ε_x is applied in the x direction and the strains ε_y and ε_z are responses in the y and z directions, respectively (see Figure 8.1). Table 8.2 summarizes the behavior of ν under several circumstances.

For analytical purposes, Poisson's ratio is defined on the differential scale. If V represents the volume,

$$V = xyz \tag{8.5}$$

Table 8.2   Values of Poisson's ratio

Value	Interpretation
0.5	No volume change during stretch
0.0	No lateral contraction
0.49–0.499	Typical values for elastomers
0.20–0.40	Typical values for plastics

Then

$$\frac{d \ln V}{d \ln x} = \frac{d \ln x}{d \ln x} + \frac{d \ln y}{d \ln x} + \frac{d \ln z}{d \ln x} \tag{8.6}$$

and

$$-\frac{d \ln y}{d \ln x} = -\frac{d \ln z}{d \ln x} = \nu \tag{8.7}$$

Since $d \ln x/d \ln x = 1$, for no volume change $\nu = 0.5$ (see Table 8.2).

On extension, plastics exhibit considerable volume increases, as illustrated by the values of ν in Table 8.2. The physical separation of atoms provides a major mechanism for energy storage and short-range elasticity.

8.1.4 The Bulk Modulus and Compressibility

The bulk modulus, B, is defined as

$$B = -V \left(\frac{\partial P}{\partial V} \right)_T \tag{8.8}$$

where P is the hydrostatic pressure. Normally, a body shrinks in volume on being exposed to increasing external pressures, so that the term $(\partial P/\partial V)_T$ is negative.

The inverse of the bulk modulus is the compressibility, β,

$$\beta = \frac{1}{B} \tag{8.9}$$

which is strictly true only for a solid or liquid in which there is no time-dependent response. Bulk compression usually does not involve long-range conformational changes, but rather a forcing together of the chain atoms. Of course, materials ordinarily exist under a hydrostatic pressure of 1 atm at sea level.

8.1.5 Relationships Between *E, G, B*, and ν

A three-way equation may be written relating the four basic mechanical properties:

$$E = 3B(1 - 2\nu) = 2(1 + \nu)G \tag{8.10}$$

Any two of these properties may be varied independently, and, conversely, a knowledge of any two defines the other two. As an especially important relationship, when $\nu \cong 0.5$,

$$E \cong 3G \qquad (8.11)$$

which defines the relationship between E and G to a good approximation for elastomers.

Equation (8.10) can also be used to evaluate Poisson's ratio for elastomers. Rearranging the two left-hand portions, we have

$$1 - 2\nu = \frac{E}{3B} = \frac{\beta E}{3} \qquad (8.12)$$

Because the quantity $1 - 2\nu$ is close to zero for elastomers (but cannot be exactly so), exact evaluation of ν depends on the evaluation of the right-hand side of equation (8.12). Values in the literature for elastomers vary from 0.49 to 0.49996 (4a).

Thus, in contrast to plastics, separation of the atoms plays only a small role in the internal storage of energy. Instead, conformational changes in the chains come to the fore, the major subject of Chapter 9.

8.1.6 Compliance Versus Modulus

If the modulus is a measure of the stiffness or hardness of an object, its compliance is a measure of softness. In regions far from transitions, the elongational compliance, J, is defined as

$$J \simeq 1/E \qquad (8.13)$$

For regions in or near transitions, the relationship is more complex. Ferry has reviewed this topic (2).

8.1.7 Numerical Values for *E*

Before proceeding with the description of the temperature behavior of polymers, it is of interest to establish some numerical values for Young's modulus (see Table 8.3). Polystyrene represents a typical glassy polymer at room temperature. It is about 40 times as soft as elemental copper, however.

Table 8.3   Numerical values of Young's modulus

Material	E (dyne/cm^2)	E (Pa)
Copper	1.2×10^{12}	1.2×10^{11}
Polystyrene	3×10^{10}	3×10^{9}
Soft rubber	2×10^{7}	2×10^{6}

Soft rubber, exemplified by such materials as rubber bands, is nearly 1000 times softer still. Perhaps the most important observation from Table 8.3 is that the modulus varies over wide ranges, leading to the wide use of logarithmic plots to describe the variation of modulus with temperature or time.

8.1.8 Storage and Loss Moduli

The quantities E and G refer to quasistatic measurements. When cyclical or repetitive motions of stress and strain are involved, it is more convenient to talk about dynamic mechanical moduli. The complex Young's modulus has the formal definition

$$E^* = E' + iE'' \tag{8.14}$$

where E' is the storage modulus and E'' is the loss modulus. Note that $E = |E^*|$. The quantity i represents the square root of minus one. The storage modulus is a measure of the energy stored elastically during deformation, and the loss modulus is a measure of the energy converted to heat. Similar definitions hold for G^*, J^*, and other mechanical quantities.

8.2 FIVE REGIONS OF VISCOELASTIC BEHAVIOR

Viscoelastic materials simultaneously exhibit a combination of elastic and viscous behavior. While all substances are viscoelastic to some degree, this behavior is especially prominent in polymers. Generally, viscoelasticity refers to both the time and temperature dependence of mechanical behavior.

The states of matter of low-molecular-weight compounds are well known: crystalline, liquid, and gaseous. The first-order transitions that separate these states are equally well known: melting and boiling. Another well-known first-order transition is the crystalline–crystalline transition, in which a compound changes from one crystalline form to another.

By contrast, no high-molecular-weight polymer vaporizes to a gaseous state; all decompose before the boiling point. In addition, no high-molecular-weight polymer attains a totally crystalline structure, except in the single-crystal state (see Section 6.4.2).

In fact, many important polymers do not crystallize at all but form glasses at low temperatures. At higher temperatures, they form viscous liquids. The transition that separates the glassy state from the viscous state is known as the glass–rubber transition. According to theories to be developed later, this transition attains the properties of a second-order transition at very slow rates of heating or cooling.

Before entering into a detailed discussion of the glass transition, the five regions of viscoelastic behavior are briefly discussed to provide a broader

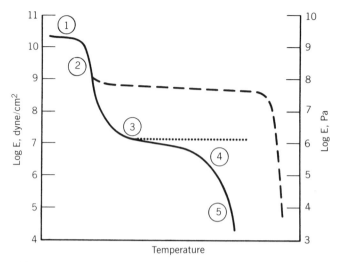

Figure 8.2 Five regions of viscoelastic behavior for a linear, amorphous polymer. Also illustrated are effects of crystallinity (dashed line) and cross-linking (dotted line).

picture of the temperature dependence of polymer properties. In the following, quasistatic measurements of the modulus at constant time, perhaps 10 or 100 sec, and the temperature being raised 1°C/min will be assumed.

8.2.1 The Glassy Region

The five regions of viscoelastic behavior for linear amorphous polymers (3, 3a, 5–7) are shown in Figure 8.2. In region 1, the polymer is glassy and frequently brittle. Typical examples at room temperature include polystyrene (plastic) drinking cups and poly(methyl methacrylate) (Plexiglas®) sheets).

Young's modulus for glassy polymers just below the glass transition temperature is surprisingly constant over a wide range of polymers, having the value of approximately 3×10^{10} dynes/cm^2 (3×10^9 Pa). In the glassy state, molecular motions are largely restricted to vibrations and short-range rotational motions.

Tobolsky (3) provides a simple approximate explanation for the constant modulus values in the glassy state. Starting with the Lennard-Jones potential (8), which describes the energy of interaction between a pair of isolated molecules, he calculated the formula for the molar lattice energy between polymer segments. Finally, the value of the bulk modulus, B, was calculated in terms of the cohesive energy density, CED, which represents the energy theoretically required to move a detached segment into the vapor phase. This, in turn, is related to the square of the solubility parameter:

$$B = 8.04(\text{CED}) = 8.04\delta^2 \qquad (8.15)$$

The factor 8.04 arises from Lennard-Jones considerations (3a).

Using polystyrene as an example, its CED is 83 cal/cm^3 (3.47×10^9 ergs/cm^3). Its Poisson ratio is approximately 0.30. From equation (8.10), $E \simeq 1.2B$, and hence

$$E \simeq 9.6(\text{CED}) \tag{8.16}$$

For polystyrene, a value of $E = 3.3 \times 10^{10}$ dynes/cm^2 is calculated, which is surprisingly close to the value obtained via modulus measurements, 3×10^{10} dynes/cm^2. It should be noted that many hydrocarbon and not-too-polar polymers have CED values within a factor of 2 of the value for polystyrene.

8.2.2 The Glass Transition Region

Region 2 in Figure 8.2 is the glass transition region. Typically, the modulus drops a factor of about 1000 in a 20–30°C range. The behavior of polymers in this region is best described as leathery, although a few degrees of temperature change will obviously affect the stiffness of the leather.

For quasistatic measurements such as illustrated in Figure 8.2, the glass transition temperature, T_g, is often taken at the maximum rate of turndown of the modulus at the elbow, where $E \cong 10^{10}$ dyn/cm^2. Often the glass transition temperature is defined as the temperature where the thermal expansion coefficient (Section 8.3) undergoes a discontinuity. Enthalpic and dynamic definitions are given in Section 8.2.9.) Other, more precise definitions are given in Section 8.5.

Qualitatively, the glass transition region can be interpreted as the onset of long-range, coordinated molecular motion. While only 1–4 chain atoms are involved in motions below the glass transition temperature, some 10–50 chain atoms attain sufficient thermal energy to move in a coordinated manner in the glass transition region (7a, 9–12) (see Table 8.4) (7a, 13). The number of chain atoms (10–50) involved in the coordinated motions was deduced by observing the dependence of T_g on the molecular weight between cross-links, M_c. When T_g became relatively independent of M_c in a plot of T_g versus M_c, the number of chain atoms was counted. It should be emphasized that these results are tenuous at best.

The glass transition temperature itself varies widely with structure and other parameters, as will be discussed later. A few glass transition temperatures are shown in Table 8.4. Interestingly, the idealized map of polymer

Table 8.4 Glass transition parameters (7a, 13)

Polymer	T_g, °C	Number of Chain Atoms Involved
Poly(dimethyl siloxane)	−127	40
Poly(ethylene glycol)	−41	30
Polystyrene	+100	40–100
Polyisoprene	−73	30–40

behavior shown in Figure 8.2 can be made to fit any of these polymers merely by moving the curve to the right or left, so that the glass transition temperature appears in the right place.

8.2.3 The Rubbery Plateau Region

Region 3 in Figure 8.2 is the rubbery plateau region. After the sharp drop that the modulus takes in the glass transition region, it becomes almost constant again in the rubbery plateau region, with typical values of 2×10^7 dynes/cm^2 (2×10^6 Pa). In the rubbery plateau region, polymers exhibit long-range rubber elasticity, which means that the elastomer can be stretched, perhaps several hundred percent, and snap back to substantially its original length on being released.

Two cases in region 3 need to be distinguished:

1. The polymer is linear. In that case, the modulus will drop off slowly, as indicated in Figure 8.2. The width of the plateau is governed primarily by the molecular weight of the polymer; the higher the molecular weight, the longer the plateau (see Figure 8.3) (14).

An interesting example of such a material is unvulcanized natural rubber. When Columbus came to America (15), he found the American Indians playing ball with natural rubber. This product, a linear polymer of very high molecular weight, retains its shape for short durations of time. However, on standing overnight, it creeps, first forming a flat spot on the bottom, and eventually flattening out like a pancake. (See Section 9.2 and Chapter 10.)

2. The polymer is cross-linked. In this case, the dotted line in Figure 8.2 is followed, and improved rubber elasticity is observed, with the creep portion suppressed. The dotted line follows the equation $E = 3nRT$, where n is the number of active chain segments in the network and RT represents the gas

Figure 8.3 Effect of molecular weight on length of plateau (3a).

constant times the temperature; see equation (9.36). An example of a cross-linked polymer above its glass transition temperature obeying this relationship is the ordinary rubber band. Cross-linked elastomers and rubber elasticity relationships are the primary subjects of Chapter 9.

The rapid, coordinated molecular motion in this region is governed by the principles of reptation and diffusion laid down in Section 5.4. Thus, when the elastomer is stretched, the chains deform with a series of rapid motions of the de Gennes type. The model must be altered slightly for cross-linked systems, for then the chain ends are bound at the cross-links. The motion is thought to become a more complex affair involving the several chain segments that are bound together.

So far, the discussion has been limited to amorphous polymers. If a polymer is semicrystalline, the dashed line in Figure 8.2 is followed. The height of the plateau is governed by the degree of crystallinity. This is so because of two reasons: first, the crystalline regions tend to behave as a filler phase, and second, because the crystalline regions also behave as a type of physical cross-link, tying the chains together.

The crystalline plateau extends until the melting point of the polymer. The melting temperature, T_f, is always higher than T_g, T_g being from one-half to two-thirds of T_f on the absolute temperature scale (see Section 8.9.3 for further details).

8.2.4   The Rubbery Flow Region

As the temperature is raised past the rubbery plateau region for linear amorphous polymers, the rubbery flow region is reached—region 4. In this region, the polymer is marked by both rubber elasticity and flow properties, depending on the time scale of the experiment. For short time scale experiments, the physical entanglements are not able to relax, and the material still behaves rubbery. For longer times, the increased molecular motion imparted by the increased temperature permits assemblies of chains to move in a coordinated manner (depending on the molecular weight), and hence to flow (see Figure 8.3) (3a). An example of a material in the rubbery flow region is Silly Putty®, which can be bounced like a ball (short-time experiment) or pulled out like taffy (a much slower experiment).

It must be emphasized that region 4 does not occur for cross-linked polymers. In that case, region 3 remains in effect up to the decomposition temperature of the polymer (Figure 8.2).

8.2.5   The Liquid Flow Region

At still higher temperatures, the liquid flow region is reached—region 5. The polymer flows readily, often behaving like molasses. In this region, as an idealized limit, equation (8.3) is obeyed. The increased energy allotted to the chains permits them to reptate out through entanglements rapidly and flow as individual molecules.

For semicrystalline polymers, the modulus depends on the degree of crystallinity. The amorphous portions go through the glass transition, but the crystalline portion remains hard. Thus a composite modulus is found. The melting temperature is always above the glass transition temperature (see below). At the melting temperature, the modulus drops rapidly to that of the corresponding amorphous material, now in the liquid flow region. It must be mentioned that modulus and viscosity are related through the molecular relaxation time, also discussed below.

8.2.6   Effect of Plasticizers

Polymers are frequently plasticized to "soften" them. These plasticizers are usually small, relatively non-volatile molecules which dissolve in the polymer, separating the chains from each other and hence making reptation easier. In the context of Figure 8.2, the glass transition temperature is lowered, and the rubbery plateau modulus is lowered. If the polymer is semi-crystalline, the plasticizer reduces the melting temperature and/or reduces the extent of crystallinity.

An example is poly(vinyl chloride), which has a T_g of $+80$ C. Properly plasticized, it has a T_g of about $+20$ C or lower, forming the familiar "vinyl." A typical plasticizer is dioctyl phthalate, with a solubility parameter of 8.7 $(cal/cm^3)^{1/2}$, fairly close to that of poly(vinyl chloride), 9.6 $(cal/cm^3)^{1/2}$, see Table 3.2. A significant parameter in this case is thought to be hydrogen bonding between the hydrogen on the same carbon as the chlorine and the ester group on the dioctyl phthalate. Other factors influencing the glass transition temperature are considered in Sections 8.6.3.2, 8.7, and 8.8.

8.2.7   Definitions of the Terms "Transition," "Relaxation," and "Dispersion"

The term "transition" refers to a change of state induced by changing the temperature or pressure.

The term "relaxation" refers to the time required to respond to a change in temperature or pressure. It also implies some measure of the molecular motion, especially near a transition condition. Frequently, an external stress is present, permitting the relaxation to be measured. For example, one could state that $1/e$ (0.367) of the polymer chains respond to an applied stress in 10 sec at the glass transition temperature, providing a simple molecular definition.

The term "dispersion" refers to the emission or absorption of energy—that is, a loss peak—at a transition. In practice, the literature sometimes uses these terms somewhat interchangeably.

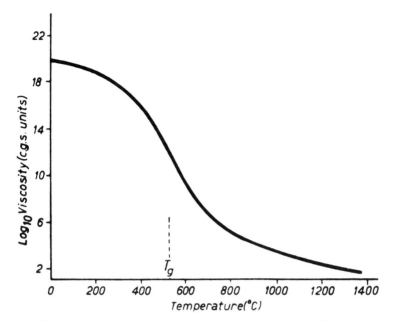

Figure 8.4 Viscosity – temperature relation of a soda – lime – silica glass (15b). Soda – lime – silica glass is one of the commonly used glasses for windows and other items.

8.2.8   Melt Viscosity Relationships near T_g

The above discussion emphasizes changes in the modulus with temperature. Equally large changes also take place in the viscosity of the polymer. In fact, the term "glass transition" refers to the temperature in which ordinary glass softens and flows (15a). (Glass is an inorganic polymer, held together with both covalent $-Si-O-Si-$ bonds and ionic bonds.) In this case, viscoelastic region 3 is virtually absent.

The viscosity–temperature relationship of glass is shown in Figure 8.4 (15b). A criterion sometimes used for T_g for both inorganic and organic polymers is the temperature at which the melt viscosity reaches a value of 1×10^{13} poises (15c) on cooling.

8.2.9   Dynamic Mechanical Behavior Through the Five Regions

The change in the modulus with temperature has already been introduced (see Section 8.2.2). More detail about the transitions is available through dynamic mechanical measurements, sometimes called dynamic mechanical spectroscopy (DMS) (see Section 8.1.8).

While the temperature dependence of G' is similar to that of G, the quantity G'' behaves quite differently (see Figure 8.5) (7). The loss quantities behave somewhat like the absorption spectra in infrared spectroscopy, where

Measurement
  Methods

Diletometry

DSC

Differential
Scanning
Calorimetry

DMA
DMS

Dynamic
  mechanical
(Analytical Spectroscopy)

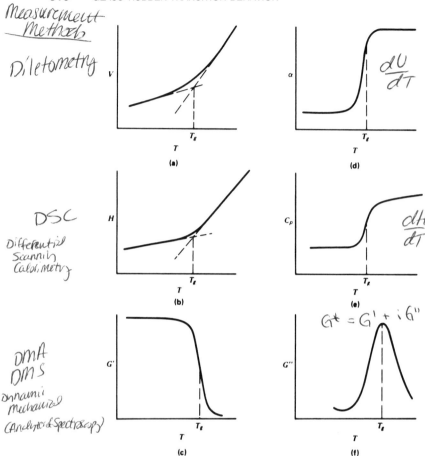

$G^* = G' + iG''$

Figure 8.5 Idealized variations in volume, V, enthalpy, H, and storage shear modulus, G' as a function of temperature. Also shown are α, the volume coefficient of expansion, and C_p, the heat capacity, which are, respectively, the first derivatives of V and H with respect to temperature, and the loss shear modulus, G'' (7).

the energy of the electromagnetic radiation is just sufficient to cause a portion of a molecule to go to a higher energy state. (Infrared spectrometry is usually carried out by varying the frequency of the radiation at constant temperature.) This exact analogue is frequently carried out for polymers also (see below). Of course, in DMS, the energy is imparted by mechanical waves. The subject has recently been reviewed (15a).

Measurements by DMS refer to any one of several methods where the sample undergoes repeated small-amplitude strains in a cyclic manner. Molecules perturbed in this way store a portion of the imparted energy elastically and dissipate a portion in the form of heat (1–4). The quantity E', Young's storage modulus, is a measure of the energy stored elastically, whereas E'', Young's loss modulus, is a measure of the energy lost as heat (see Figure 8.6) (16).

$$E^* = E' + iE''$$

Figure 8.6 Simplified definition of E' and E'' (16). When a viscoelastic ball is dropped onto a perfectly elastic floor, it bounces back to a height E', a measure of the energy stored elastically during the collision between the ball and the floor. The quantity E'' represents the energy lost as heat during the collision.

Another equation in wide use is

$$\frac{E''}{E'} = \tan \delta \qquad (8.17)$$

where $\tan \delta$ is called the loss tangent, δ being the angle between the in-phase and out-of-phase components in the cyclic motion. Tan δ also goes through a series of maxima. The maxima in E'' and $\tan \delta$ are sometimes used as the definition of T_g. For the glass transition, the portion of the molecule excited may be from 10 to 50 atoms or more (see Table 8.4).

The dynamic mechanical behavior of an ideal polymer is illustrated in Figure 8.7 (see page 318). The storage modulus generally follows the behavior of Young's modulus as shown in Figure 8.2. In detail, it is subject to equation (8.14), so that the storage modulus is slightly smaller, depending on the value of E''.

The quantities E'' and $\tan \delta$ display decided maxima at T_g, the $\tan \delta$ maximum appearing several degrees centigrade higher than the E'' peak. Also shown in Figure 8.7 is the β peak, generally involving a smaller number of atoms. The area under the peaks, especially when plotted with a linear y axis, is related to the chemical structure of the polymer (16a). The width of the transition and shifts in the peak temperatures of E'' or $\tan \delta$ are sensitive guides to the exact state of the material, molecular mixing in blends, and so on. The smaller transitions, such as the β peak, usually appear at other temperatures, such as the Schatzki crankshaft transition (see Section 8.4.1), or at higher temperatures (see Section 8.4.2).

Also included in the x axis of Figure 8.7 are the log time and the minus log frequency axes. As discussed in Sections 8.5.3 and 8.6.1.2, the minus log

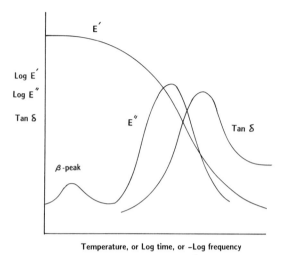

Temperature, or Log time, or –Log frequency

Figure 8.7   Schematic of dynamic mechanical behavior.

frequency dependence of the mechanical behavior takes the same form as temperature. As the imposed frequency on the sample is raised, it goes through the glass transition in much the same way as when the temperature is lowered. The theory of the glass transition will be discussed in Section 8.6.

For small stresses and strains, if both the elastic deformation at equilibrium and the rate of viscous flow are simple functions of the stress, the polymer is said to exhibit linear viscoelasticity. While most of the material in this chapter refers to linear viscoelasticity, much of the material in Chapter 9 concerns nonlinear behavior.

A principal value of the loss quantities stems from their sensitivity to many other types of transitions besides the glass transition temperature. The maxima in E'', G'', tan δ, and so on provide a convenient and reproducible measure of each transition's behavior. From an engineering point of view, the intensity of the loss quantities can be utilized in mechanical damping problems such as vibration control.

In the following section, several types of instrumentation used to measure the glass transition and other transitions are discussed.

8.3   METHODS OF MEASURING TRANSITIONS IN POLYMERS

The glass transition and other transitions in polymers can be observed experimentally by measuring any one of several basic thermodynamic, physical, mechanical, or electrical properties as a function of temperature. It will be remembered that, in first-order transitions such as melting and boiling, there is a discontinuity in the volume–temperature plot. For second-order transitions such as the glass transition, a change in slope occurs, as illustrated in Figure 8.5 (7).

The volumetric coefficient of expansion, α, is defined as

$$\alpha = \frac{1}{V}\left(\frac{\partial V}{\partial T}\right)_p \tag{8.18}$$

where V is the volume of the material, and α has the units K^{-1}. While this quantity increases as the temperature increases beyond T_g, it usually changes over a range of 10–30°C. The elbow shown in Figure 8.5a is not sharp. Similar changes occur in the enthalpy, H, and the heat capacity at constant pressure (C_p) (Figures 8.5b and 8.5e, respectively).

8.3.1 Dilatometry Studies

There are two ways of characterizing polymers via dilatometry. The most obvious is volume–temperature measurements (17), where the polymer is confined by a liquid and the change in volume is recorded as the temperature is raised (see Figure 8.8). The usual confining liquid is mercury, since it does not swell organic polymers and has no transition of its own through most of the temperature range of interest. In an apparatus such as is shown in Figure 8.8, outgassing is required.

Figure 8.8 A mercury-based dilatometer (17). Bulb A contains the polymer (about 1 g), capillary B is for recording volume changes (Hg + polymer), G is a capillary for calibration, sealed at point F. After packing bulb A, the inlet is constricted at E, C contains weighed mercury to fill all dead space, and D is a second constriction.

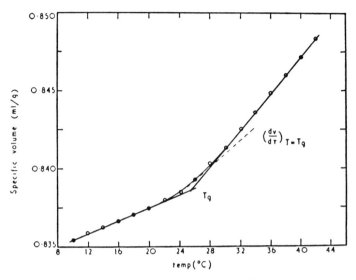

Figure 8.9 Dilatometric studies on branched poly(vinyl acetate) (17).

The results may be plotted as specific volume versus temperature (see Figure 8.9) (17). Since the elbow in volume–temperature studies is not sharp (all measurements of T_g show a dispersion of some 20–30°C), the two straight lines below and above the transition are extrapolated until they meet; that point is usually taken as T_g. Dilatometric and other methods of measuring T_g are summarized in Table 8.5.

The straight lines above and below T_g in Figure 8.9, of course, yield the volumetric coefficient of expansion, α. The linear coefficient of expansion, β, can also be employed (18).

Dilatometric data agree well with modulus–temperature studies, especially if the heating rates and/or length of times between measurements are controlled. (Raising the temperature 1°C/min roughly corresponds to a 10-sec mechanical measurement.) Besides being a direct measure of T_g, dilatometric studies provide free volume information, of use in theoretical studies of the glass transition phenomenon (see Section 8.6.1). The subject of glass transitions, and dilatometric studies in particular, has recently been reviewed (6, 7, 16, 26, 27).

8.3.2 Thermal Methods

Two closely related methods dominate the field—the older method, differential thermal analysis (DTA), and the newer method, differential scanning calorimetry (DSC). Both methods yield peaks relating to endothermic and exothermic transitions and show changes in heat capacity. The DSC method also yields quantitative information relating to the enthalpic changes in the polymer (19–21) (see Table 8.5).

Table 8.5   Methods of measuring the glass transition

Method	Representative Instrumentation	Reference
1. Dilatometry		
(a) Volume–temperature	Polymer confined by mercury (homemade)	(17)
(b) Linear expansivity	TMS + computer	(18)
2. Thermal		
(a) DSC	Perkin–Elmer DSC I	(19)
	DuPont 990 DSC	(20)
	Mettler DSC TA3300	
(b) DTA	DuPont 900	(21)
(c) Calorimetric, C_p	Perkin–Elmer DSC II	(18)
3. Mechanical		
(a) Static	Gehman	(22)
	Clash–Berg	
(b) Dynamic	Rheovibron DDV II (Toyo Instrument Co.)	(23)
	Viscoanalyzer (Metrovib)	
	Torsional Braid Analysis (Plastics Analysis Consultants)	(24)
	Quest (Anatrol)	
	Torsional pendulum (homemade)	(25)
	Dynamic Mechanical Thermal Analyzer (TA Instruments)	(25a)
	Dynastat Mark II (Imass Corp.)	(22a)
4. Dielectric and magnetic		
(a) Dielectric loss	General Radio Capacitance Bridge	(26)
	DuPont Dielectric Analyzer, DEA 2970	
(b) Broad-line NMR	JOEL JNH 3H60 Spectrometer	(36)
5. Melt viscosity	Rheometrics Mechanical Spectrometer	
	Weissenberg Rheogoniometer	

In making DTA measurements, the temperature of the sample is compared with a reference material. This material should not have a transition in the temperature range of interest and frequently is powdered alumina. Both the sample and the reference material are heated at a uniform rate from a block on which they rest (see Figure 8.10) (1). Usually, a heating rate of 10–20°C/min is employed. Owing to differences in heat capacity, the sample and the reference material maintain slightly different temperatures, characterized by ΔT. The quantity ΔT changes, because of changes in the specific heat, for T_g, or the presence of an endotherm or exotherm.

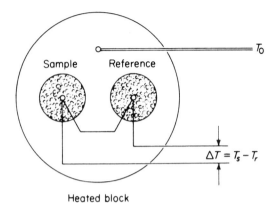

Heated block

Figure 8.10 Schematic of differential thermal analysis (DTA) apparatus (1). Heated block is programmed so that T_0 increases linearly with time. The difference between the sample temperature (T_s) and the reference temperature (T_r), ΔT, is recorded as a function of T_0.

The DSC method uses a servo system to supply energy at a varying rate to the sample and the reference, so that the temperatures of the two stay equal. The DSC output plots energy supplied against average temperature. By this improved method, the areas under the peaks can be directly related to the enthalpic changes quantitatively.

As illustrated by Figure 8.11 (27a), T_g can be taken as the temperature at which one-half of the change in heat capacity, ΔC_p, has occurred. There is a

Figure 8.11 Example of a differential scanning calorimetry trace of poly(ethylene terephthalate – co – p-oxbenzoate), quenched, reheated, cooled at 0.5 K/min through the glass transition, and reheated for measurement at 10 K/min (27a). T_g is taken at the temperature at which half the increase in heat capacity has occurred. The width of the transition is indicated by ΔT.

hysteresis peak associated with the transition which appears frequently, but apparently not all the time. Blanchard et al. (27b) also found such a peak for polystyrene and concluded that it was associated with the transition itself, perhaps destroying some residual order in an otherwise amorphous polymer. At this time, it is not known whether the peak is caused by a true effect associated with T_g or by a thermal lag associated with overshooting of equilibrium conditions (see Section 8.6.2). In some cases, the hysteresis peak might be related to physical aging, see Section 10.2.6.

A related technique, thermogravimetric analysis (TGA), must be introduced at this point if only to prevent confusion. In using TGA, the weight of the sample is recorded continuously as the temperature is raised. Volatilization, dehydration, oxidation, and other chemical reactions can easily be recorded, but the simple transitions are missed, as no weight changes occur.

8.3.3 Mechanical Methods

Since the very notion of the glass–rubber transition stems from a softening behavior, the mechanical methods provide the most direct determination of the transition temperature. Two fundamental types of measurement prevail —the static or quasistatic methods, and the dynamic methods.

Results of the static type of measurement have already been shown in Figure 8.2. For amorphous polymers and many types of semicrystalline polymers in which the crystallinity is not too high, stress relaxation, Gehman, and/or Clash–Berg instrumentation provide rapid and inexpensive scans of the temperature behavior of new polymers before going on to more complex methods.

Several instruments are employed to measure the dynamic mechanical spectroscopy (DMS) behavior (See Table 8.5). The Rheovibron (27c) requires a sample that is self-supporting and that yields absolute values of the storage modulus and tan δ. The value of E'' is calculated by equation (8.17). Typical data are shown in Figure 8.12 (28). Although the instrument operates at several fixed frequencies, 110 Hz is most often employed. The sample size is

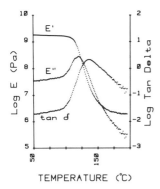

Figure 8.12 Dynamic mechanical spectroscopy on polystyrene cross-linked with 2% divinyl benzene. Data taken with a Rheovibron at 110 Hz. Note that the tan δ peak occurs at a slightly higher temperature than the E'' peak (28). Experiment run by J. J. Fay, Lehigh University.

about that of a paper match stick. This method provides excellent results with thermoplastics (23) and preformed polymer networks (28).

An increasingly popular method for studying the mechanical spectra of all types of polymers, especially those that are not self-supporting, is torsional braid analysis (TBA). In this case, the monomer, prepolymer, polymer solution, or melt is dipped onto a glass braid, which supports the sample. The braid is set into a torsional motion. The sinusoidal decay of the twisting action is recorded as a function of time as the temperature is changed (24, 29–32). Because the braid acts as a support medium, the absolute magnitudes of the transitions are not obtained; only their temperatures and relative intensities are recorded. The TBA method appears to have largely replaced the torsional pendulum (25).

Figure 8.13 shows typical TBA data for cellulose triacetate. Also shown by way of comparison are DTA and TGA results. In Figure 8.13, p represents the period of oscillation—that is, the inverse of the natural frequency of the sample, braid, and attachments. The quantity $1/p^2$ is a measure of the stiffness of the system, proportional to the modulus. Since the modulus of the glass braid stays nearly constant through the range of measurement, all changes are representative of the polymer. The quantity n represents the number of oscillations required to reduce the angular amplitude, A, by a fixed ratio. In this case, $A_i/A_{i+n} = 20$ (31). By way of comparison, earlier measurements via volume–temperature had placed T_g at 172°C (33, 34) and T_f at 307°C (35), in good agreement with the TBA results in Figure 8.13.

The DTA results in Figure 8.13 show T_f at 290°C and an exothermic then endothermic decomposition at 356°C and 400°C, respectively, corresponding to the weight loss shown by the TGA study. Cellulose triacetate is known to have three second-order transitions (34); it is not clear whether the lower temperature transitions associated with the DTA plot represent these or other motions.

8.3.4 Dielectric and Magnetic Methods

As stated previously, part of the work performed on a sample will be converted irreversibly into random thermal motion by excitation of the appropriate molecular segments. In Section 8.3.3, the loss maxima so produced through mechanical means were used to characterize the glass transition. The two important electromagnetic methods for the characterization of transitions in polymers are dielectric loss (26) and broad-line nuclear magnetic resonance (NMR) (36–38).

The dielectric loss constant, ε'', or its associated tan δ can be measured by placing the sample between parallel plate capacitors and alternating the electric field. Polar groups on the polymer chain respond to the alternating field. When the average frequency of molecular motion equals the electric field frequency, absorption maxima will occur.

Figure 8.13 Comparison of torsional braid analysis, differential thermal analysis, and thermogravimetric analysis data for cellulose triacetate. The bottom figure shows the twisting of the sample in the absence of oscillations as a result of expansion or contraction of the sample at T_g and T_f (32).

If the dielectric measurements are carried out at the same frequency as the DMS measurements, the transitions will occur at the same temperatures (see Figure 8.14) (38). The glass transition for the polytrifluorochloroethylene at 52°C shown is due to static measurements. The values at 100°C shown are close to those reported for dynamic measurements (13a).

Broad-line NMR measurements depend on the fact that hydrogen nuclei, being simply protons, possess a magnetic moment and therefore precess about an imposed alternating magnetic field, especially at radio frequencies.

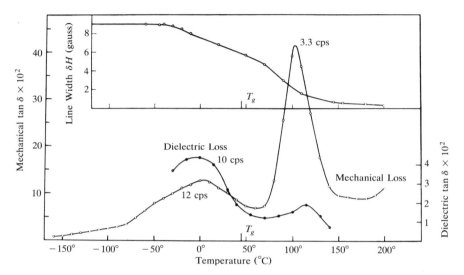

Figure 8.14   Mechanical and dielectric loss tangent tan δ and NMR absorption line width δH (maximum slope, in gauss) of polytrifluorochloroethylene (Kel-F) (38).

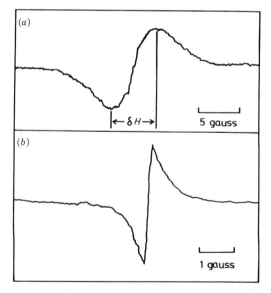

Figure 8.15   Broad-line NMR spectra of a cured epoxy resin. (a) Broad line at 291 K; (b) motionally narrowed line at 449 K (T_g + 39 K) (36).

Stronger interactions exist between the magnetic dipoles of different hydrogen nuclei in polymers below the glass transition temperature, resulting in a broad signal. As the chain segments become more mobile with increasing temperature through T_g, the distribution of proton orientations around a given nucleus becomes increasingly random, and the signal sharpens. The behavior of NMR spectra below and above T_g is illustrated in Figure 8.15 (36). The narrowing of the line width through the glass transition for polytrifluorochloroethylene is shown in the inset of Figure 8.14. A number of other methods of observing T_g are discussed by Boyer (7).

8.3.5 A Comparison of the Methods

All the methods of measuring T_g depend on either a basic property or some derived property. The principal ones have already been discussed.

Basic Property	Derived Property
Volume	Refractive index
Modulus	Penetrometry
Dielectric loss	Resistivity

From a practical point of view, since the derived property frequently represents the quantity of interest, it is measured rather than the basic property.

The methods most commonly used at the present time (7) include direct-recording DSC units, the Rheovibron, and the torsional braid. The special value of the mechanical units lies in the fact that loss and storage moduli are frequently of prime engineering value. Thus the instrument supplies basic scientific information about the transitions while giving information about the damping and stiffness characteristics.

On the other hand, DSC supplies thermodynamic information about T_g. Of particular interest is the change in the heat capacity, which reflects fundamental changes in molecular motion. Thus values of C_p are of broad theoretical significance, as described in Section 8.6.

8.4 OTHER TRANSITIONS AND RELAXATIONS

As the temperature of a polymer is lowered continuously, the sample may exhibit several second-order transitions. By custom, the glass transition is designated the α transition, and successively lower temperature transitions are called the $\beta, \gamma, \ldots$ transitions. One important second-order transition appears above T_g, designated the T_{ll} (liquid–liquid) transition. Of course, if the polymer is semicrystalline, it will also melt at a temperature above T_g.

8.4.1 The Schatzki Crankshaft Mechanism

8.4.1.1 Main-Chain Motions

There appear to be two major mechanisms for transitions in the glassy state (41, 7, 39). For main-chain motions in hydrocarbon-based polymers such as polyethylene, the Schatzki crankshaft mechanism (40), Figure 8.16 (41), is thought to play an important role. Schatzki showed that eight $-CH_2-$ units could be lined up so that the 1–2 bonds and the 7–8 bonds form a colinear axis. Then, given sufficient free volume, the intervening four $-CH_2-$ units rotate more or less independently in the manner of an old-time automobile crankshaft. It is thought that at least four $-CH_2-$ units in succession are required for this motion. The transition of polyethylene occurring near $-120°C$ is thought to involve the Schatzki mechanism.

It is interesting to consider the basic motions possible for small hydrocarbon molecules by way of comparison. At very low temperatures, the CH_3- groups in ethane can only vibrate relative to the other. At about 90 K ethane undergoes a second-order transition as detected by NMR absorption (41a), and the two CH_3- units begin to rotate freely, relative to one another. For propane and larger molecules, the number of motions becomes more complex (41b), as now three-dimensional rotations come into play. One might imagine that octane itself might have the motion illustrated in Figure 8.16 as one of its basic energy absorbing modes.

8.4.1.2 Side-Chain Motions

The above considers main-chain motions. Many polymers have considerable side-chain "foliage," and these groups can, of course, have their own motions.

A major difference between main-chain and side-chain motions is the toughness imparted to the polymer. Low-temperature main-chain motions act to absorb energy much better than the equivalent side-chain motions, in the face of impact blows. When the main-chain motions absorb energy under these conditions, they tend to prevent main-chain rupture. (The temperature of the transition actually appears at or below ambient temperature, noting the equivalent "frequency" of the growing crack. The frequency dependence is discussed in Section 8.5.) Toughness and fracture in polymers are discussed in Chapter 11.

Figure 8.16 Schatzki's crankshaft motion (40) requires at least four $-CH_2-$ groups in succession. As illustrated, for eight $-CH_2-$ groups, bonds 1 and 7 are colinear and intervening $-CH_2-$ units can rotate in the manner of a crankshaft (41).

8.4.2 The T_{ll} Transition

As illustrated in Figure 8.17 (42), the T_{ll} transition occurs above the glass transition and is thought to represent the onset of the ability of the entire polymer molecule to move as a unit (7, 43, 44). Above T_{ll}, physical entanglements play a much smaller role, as the molecule becomes able to translate as a whole unit.

Although there is much evidence supporting the existence of a T_{ll} (43–45), it is surrounded by much controversy (46–49). Reasons include the strong dependence of T_{ll} on molecular weight and an analysis of the equivalent behavior of spring and dashpot models (see Section 10.1). The critics contend that T_{ll} is an instrumental artifact produced by the composite nature of the specimen in torsional braid analysis (TBA), since TBA instrumentation is the principal method of studying this phenomenon (see Figure 8.17).

Many polymers show evidence of several transitions besides T_g. Table 8.6 summarizes the data for polystyrene, including the proposed molecular mechanisms for the several transitions.

Figure 8.17 Thermomechanical spectra (relative rigidity and logarithmic decrement versus temperature (K) of anionic polystyrene, $M_n = 20,200$ (42).

Table 8.6 Multiple transitions in amorphous polystyrene (26)

Temperature	Transitions	Suggested Mechanisms
433 K (160°C)	T_{ll}	Liquid$_1$ → liquid$_2$
373 K (100°C)	T_g	Long-range chain motions
325 ± 4 K (~ 50°C)	β	Torsional vibrations of phenyl groups
130 K	γ	Motions due to four carbon backbone moieties
38–48 K	δ	Oscillation or wagging of phenyl groups

8.5 TIME AND FREQUENCY EFFECTS ON RELAXATION PROCESSES

So far, the discussion has implicitly assumed that the time (for static) or frequency (for dynamic) measurements of T_g were constant. In fact, the observed glass transition temperature depends very much on the time allotted to the experiment, becoming lower as the experiment is carried out slower.

For static or quasistatic experiments, the effect of time can be judged in two ways: (a) by speeding up the heating or cooling rate, as in dilatometric experiments, or (b) by allowing more time for the actual observation. For example, in measuring the shear modulus by Gehman instrumentation, the sample may be stressed for 100 sec rather than 10 sec before recording the angle of twist.

In the case of dynamic experiments, especially where the sample is exposed to a sinusoidal motion, the frequency of the experiment can be varied over wide ranges. For dynamic mechanical spectroscopy, the frequency range can be broadened further by changing instrumentation. For example, DMS measurements in the 20,000-Hz range can best be carried out by employing sound waves. In dielectric studies, the frequency of the alternating electric field can be varied.

The inverse of changing the frequency of the experiment, making measurements as a function of time at constant temperature, is called stress relaxation or creep and is discussed in Section 8.5.2 and Chapter 10. In the following paragraphs, a few examples of time and frequency effects are given.

8.5.1 Time Dependence in Dilatometric Studies

It was pointed out in Section 8.3.1 that the elbow in volume–temperature studies constitutes a fundamental measure of the glass transition temperature, since the coefficient of expansion increases at T_g. The heating or cooling rate is critical in determining exactly where the transition will be observed, however, as illustrated in Figure 8.18 (50). Here, decreasing the

Figure 8.18 Dilatometric studies of poly(vinyl acetate) as a function of cooling rate. The glass transition temperature T_g' is lower than T_g, because a lower cooling rate (1°C per 100 hr versus 0.02 hr) was employed (50). Also shown is the cubic coefficient of expansion, α, measured at the 0.02-hr cooling rate.

cooling rate of poly(vinyl acetate) by a factor of 50 reduces the glass transition temperature by about 8°C. Similarly, T_g varies with heating rate in DSC studies.

8.5.2 Time Dependence in Mechanical Relaxation Studies

If a polymer sample is held at constant strain and measurements of stress are recorded as a function of time, stress relaxation of the type shown in Figure 8.19 (50a) will be observed. The shape of the curve shown in Figure 8.19 bears comparison with those in Figure 8.2. In the present case, log time has replaced temperature in the x axis, but the phenomenon is otherwise similar. As time is increased, more molecular motions occur, and the sample softens.

It must be emphasized that the sample softens only after the time allowed for relaxation. For example, on a given curve of the type shown in Figure 8.19, the modulus might be 1×10^7 dynes/cm^2 after 10 years, showing a rubbery behavior. Someone coming up and pressing his thumb into the material after 10 years will report the material as being much harder; however, it must be remembered that pressing one's thumb into a material is

Figure 8.19 Master curve for polyisobutylene (50a). The shift factor X, selected here so that $\log X_{298} = 0$, is equivalent to the experimentally determined WLF A_t (see Section 8.6.1.2).

a short-time experiment, of the order of a few seconds. (This assumes physical relaxation phenomena only, not true chemical degradation.)

The slope corresponding to the glass transition has been quantitatively treated by Aklonis and co-workers (50b–d) for relaxation phenomena. Aklonis defined a steepness index (SI) as the maximum of the negative slope of a stress relaxation curve in the glass transition region. They found that while polyisobutylene has a SI of about 0.5, poly(methyl methacrylate) has a value of about 1.0, and polystyrene is close to 1.5. Aklonis treated the data theoretically, using the Rouse–Bueche–Zimm bead and spring model (50e–g), based in turn on the Debye damped torsional oscillator model (50h, 50i). Aklonis concluded that values of SI equal to 0.5 represented a predominance of intramolecular forces and that a SI of 1.5 represented a predominance of intermolecular forces. A SI of 1.0 was an intermediate case. Stress relaxation is treated in greater detail in Chapter 10.

8.5.3 Frequency Effects in Dynamic Experiments

The loss peaks such as those illustrated in Figure 8.7 and Figure 8.12 can be determined as a function of frequency. The peak frequency can then be plotted against $1/T$ to obtain apparent activation energies.

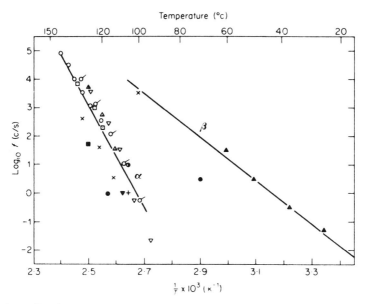

Figure 8.20   The frequency dependence of the α and β transitions of polystyrene (39a). Composite of both dynamic mechanical and dielectric studies of several researchers.

Figure 8.20 (39a) shows the $\alpha^\dagger$ transition, T_g, of polystyrene increasing steadily in temperature as the frequency of measurement is increased. Both DMS and dielectric measurements are included. Since T_g is usually reported at 10 sec (or 1×10^{-1} Hz), a glass transition temperature of 100°C may be deduced from Figure 8.20, which is, in fact, the usually reported T_g.

The straight line for the α relaxation process, as drawn, corresponds to an apparent activation energy of 84 kcal/mol (51). The β relaxation possesses a corresponding apparent energy of activation of 35 kcal/mol. Section 8.6.2.3 discusses methods of calculating these values. As a first approximation for many polymers, $T_\beta \cong 0.75 T_g$ (51a).

The WLF equation (Section 8.6.1.2) says that T_g will change 6–7°C per decade of frequency. Figure 8.19 yields about 5°C change in T_g for a factor of 10 increase in the time scale, and Figure 8.20 yields about 7.5°C per decade. Obviously, this depends on the apparent energy of activation of the individual polymer, but many of the common carbon-backbone polymers have similar energies of activation. The effect of frequency on mechanical behavior is discussed further in Section 10.3.

---

†The peaks in the loss spectrum are sometimes labeled $\alpha, \beta, \gamma, \ldots$, with α being the highest temperature peak.

While each of these second-order transitions has a frequency dependence, the corresponding first-order melting transition for semicrystalline polymers does not.

8.6   THEORIES OF THE GLASS TRANSITION

The basic experimental behavior of polymers near their glass transition temperatures was explored in the preceding phenomenological description. In Section 8.5, T_g was shown to decrease steadily as the time allotted to the experiment was increased. One may raise the not so hypothetical question, is there an end to the decrease in T_g as the experiment is slowed? How can the transition be explained on a molecular level? These are the questions to which the theories of the glass transition are addressed.

The following paragraphs describe three main groups of theories of the glass transition (2, 26, 52): the free-volume theory, the kinetic theory, and the thermodynamic theory. Although these three theories may at first appear to be as different as the proverbial three blind men's description of an elephant, they really examine three aspects of the same phenomenon and can be successfully unified, if only in a qualitative way.

8.6.1   The Free-Volume Theory

As first developed by Eyring (53) and others, molecular motion in the bulk state depends on the presence of holes, or places where there are vacancies or voids (see Figure 8.21). When a molecule moves into a hole, the hole, of course, exchanges places with the molecule, as illustrated by the motion indicated in Figure 8.21. (This model is also exemplified in the children's game involving a square with 15 movable numbers and one empty place; the object of the game is to rearrange the numbers in an orderly fashion.) With real materials, Figure 8.21 must be imagined in three dimensions.

Although Figure 8.21 suggests small molecules, a similar model can be constructed for the motion of polymer chains, the main difference being that

Figure 8.21   A quasicrystalline lattice exhibiting vacancies, or holes. Circles represent molecules; arrow indicates molecular motion.

more than one "hole" may be required to be in the same locality, as cooperative motions are required (see reptation theory, Section 5.4). Thus for a polymeric segment to move from its present position to an adjacent site, a critical void volume must first exist before the segment can jump.

The important point is that molecular motion cannot take place without the presence of holes. These holes, collectively, are called free volume. One of the most important considerations of the theory discussed below involves the quantitative development of the exact free-volume fraction in a polymeric system.

8.6.1.1 *T_g as an Iso–Free-Volume State*

In 1950, Fox and Flory (54) studied the glass transition and free volume of polystyrene as a function of molecular weight and relaxation time. For infinite molecular weight, they found that the specific free volume, v_f, could be expressed above T_g as

$$v_f = K + (\alpha_R - \alpha_G)T \qquad (8.19)$$

where K was related to the free volume at 0 K, and α_R and α_G represented the cubic (volume) expansion coefficients in the rubbery and glassy states, respectively (see Section 8.3.1). Fox and Flory found that below T_g the same specific volume–temperature relationships held for all the polystyrenes, independent of molecular weight. From this study, they concluded that (a) below T_g the local conformational arrangement of the polymer segments was independent of both molecular weight and temperature, and (b) the glass transition temperature was an iso–free-volume state.

Simha and Boyer (55) then postulated that the free volume at $T = T_g$ should be defined as

$$v - (v_{0,R} + \alpha_G T) = v_f \qquad (8.20)$$

Figure 8.22 illustrates these quantities.

Substitution of the quantity

$$v = v_{0,R} + \alpha_R T \qquad (8.21)$$

leads to the relation

$$(\alpha_R - \alpha_G)T_g = K_1 \qquad (8.22)$$

In the above, v is the specific volume, and $v_{0,G}$ and $v_{0,R}$ are the volumes extrapolated to 0 K using α_G and α_R as the coefficients of expansion, respectively. Based on the data in Table 8.7 (see page 337) (55), Simha and

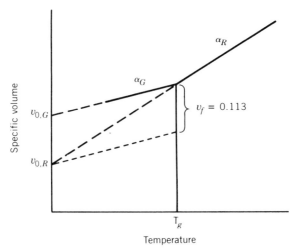

Figure 8.22 A schematic diagram illustrating free volume as calculated by Simha and Boyer.

Boyer concluded that

$$(\alpha_R - \alpha_G)T_g = 0.113 \tag{8.23}$$

Equation (8.23) leads directly to the finding that the free volume at the glass transition temperature is indeed a constant, 11.3%. This is the largest of the theoretical values derived, but the first. (Many simple organic compounds have a 10% volume increase on melting, it should be pointed out.) Other early estimates placed the free volume at about 2% (52, 54).

The use of α_G in equation (8.20) results from the conclusion that expansion in the glassy state occurs at nearly constant free volume; hence $\alpha_G T$ is proportional to the occupied volume. The use of $\alpha_R T_g$ in Table 8.7 arises from the less exact, but simpler relationship

$$\alpha_R T_g = K_2 = 0.164 \tag{8.24}$$

The quantities K_1 and K_2 provide a criterion for the glass temperature, especially for new polymers or when the value is in doubt. This latter arises in systems with multiple transitions, for example, and in semicrystalline polymers, where T_g may be lessened or obscured.

The relation expressed in equation (8.23), though approximate, has been a subject of more recent research. Simha and co-workers (56a, 56b) found equation (8.23) still acceptable, while Sharma et al. (56c) found $\alpha_R - \alpha_G$ roughly constant at 3.2×10^{-4} deg^{-1} (see below).

Table 8.7 Test of the glass temperature as an iso–free-volume state[a] (55)

Polymer	T_g, K	$\alpha_R \times 10^4$ K^{-1}	$\alpha_R T_g$ (K_2)	$(\alpha_R - \alpha_G)$ $\times 10^4$ K^{-1}	$(\alpha_R - \alpha_G)T_g$ (K_1)	Reference
Polyethylene	143	13.5	0.192	6.8–8.0	0.097–0.113	(b)
Poly(dimethyl siloxane)	150	12	0.180	9.3	0.140	(c)
		8.12	0.122	5.4	0.081	(d)
Polytetrafluoroethylene	160	8.3	0.133	7.0	0.112	(e)
Polybutadiene	188	7.8	0.147	5.8	0.109	(c)
Polyisobutylene	199.4	6.18	0.123	4.70	0.094	(f)
Hevea rubber	201	6.16	0.124	4.1	0.082	(c)
Polyurethane	213	8.02	0.171	6.04	0.129	(g)
Poly(vinylidene chloride)	256	5.7	0.146	4.5	0.115	(h)
Poly(methyl acrylate)	282	5.6	0.158	2.9	0.082	(c)
Poly(vinyl acetate)	302	5.98	0.18	3.9	0.118	(c)
Poly(4-methyl pentene-1)	302	7.61	0.23	3.78	0.114	(i)
Poly(vinyl chloride)	355	5.2	0.185	3.1	0.110	(c)
Polystyrene	373	5.5	0.205	3.0	0.112	(c)
Poly(methyl methacrylate)	378	5.0	0.189	3.05	0.115	(c)
	378	4.60	0.174	2.45	0.093	(j)

[a] α_R is the coefficient of cubical expansion above T_g, α_G the same below T_g, both calculated for 100% amorphous polymer.
[b] Estimating from Quinn and Mandelkern, *J. Am. Chem. Soc.*, **80**, 3178(1958), $\alpha_G = 1.2 \times 10^{-4}$. If sample of Marlex 50 is 80% crystalline, these values are multiplied by 5. A variety of sources suggest $\alpha_{melt} = 8.8$–10×10^{-4}, $\alpha_{crystal} = 2 \times 10^{-4}$.
[c] L. A. Wood, *J. Polym. Sci.*, **28**, 319(1958).
[d] K. E. Polmateer and M. J. Hunter, *J. Appl. Polym. Sci.*, **1**, 3(1959).
[e] Private communication from H. Starkweather, E. I. duPont de Nemours and Company, Wilmington, Delaware, to Simha and Boyer.
[f] Private communication from F. Reding, Union Carbide Chemicals Company, South Charleston, West Virginia, to Simha and Boyer.
[g] Private communication from A. J. Havlek, Jet Propulsion Laboratory, California Institute of Technology, Pasadena, California, to Simha and Boyer. This polyurethane was made from polypropylene glycol and toluene diisocyanate.
[h] R. F. Boyer and R. S. Spencer, *J. Appl. Phys.*, **15**, 398(1944), Table II.
[i] C. W. Bunn, *J. Polym. Sci.*, **16**, 339(1955).
[j] S. Rogers and L. Mandelkern, *J. Phys. Chem.*, **61**, 985(1957).

8.6.1.2 The WLF Equation

Section 8.2.7 illustrates how polymers soften and flow at temperatures near and above T_g. Flow, a form of molecular motion, requires a critical amount of free volume. This section considers the analytical relationships between polymer melt viscosity and free volume, particularly the WLF (Williams–Landel–Ferry) equation. The WLF equation is derived here because the free volume at T_g arises as a fundamental constant. The application of the WLF equation to viscosity and other polymer problems is considered in Chapter 10.

Early work of Doolittle (56) on the viscosity, η, of nonassociated pure liquids such as n-alkanes led to an equation of the form

$$\ln \eta = B(v_0/v_f) + \ln A \tag{8.25}$$

where A and B are constants and v_0 is the occupied volume, and as before v_f is the specific free volume. The Doolittle equation can be derived by considering the molecular transport of a liquid consisting of hard spheres (57–61).

An important consequence of the Doolittle equation is that it provides a theoretical basis for the WLF equation (62). One derivation of the WLF equation begins with a consideration of the need of free volume to permit rotation of chain segments, and the hindrance to such rotation caused by neighboring molecules.

The quantity P is defined as the probability of the barriers to rotation per unit time, cooperative motion, or reptation being surmounted (63). An Arrhenius-type relationship is assumed, where ΔE_{act} is the free energy of activation of the process:

$$P = \exp(-\Delta E_{act}/kT) \tag{8.26}$$

Of course, P increases with temperature.

Next, the time ("time scale") of the experiment is considered. Long times, t, allow for greater probability of the required motion, and P also increases. The theory assumes that tP must reach a certain value for the onset of the motion, and for the associated transition to be recorded:

$$\ln tP = \text{constant} = -\Delta E_{act}/kT + \ln t \tag{8.27}$$

hence

$$\ln t = \text{constant} + \Delta E_{act}/kT \tag{8.28}$$

Equation (8.28) equates the logarithm of time with an inverse function of the temperature. Taking the differential,

$$\Delta \ln t = -\frac{\Delta E_{act}}{kT^2} \Delta T \tag{8.29}$$

the relationships becomes clearer: an increase in the logarithm of time is equivalent to a decrease in the absolute temperature. This must be understood in the context of the time–temperature relationship for the onset of a particular cooperative motion.

The quantity ΔE_{act} is associated with free volume and qualitatively would be expected to decrease as the fractional free volume increases. It is assumed that

$$\frac{\Delta E_{act}}{kT} = \frac{B'}{f} \qquad (8.30)$$

where B' is a constant and f is the fractional free volume. Noting the similarity of form between equations (8.30) and (8.25), B' is taken as equal to B. Then, instead of the Arrhenius relation,

$$P = \exp\left(-\frac{B}{f}\right) \qquad (8.31)$$

The quantity tP still remains constant for the particular set of properties to be observed (not necessarily T_g):

$$\ln tP = \text{constant} = -\frac{B}{f} + \ln t \qquad (8.32)$$

Taking the differential,

$$\Delta \ln t = B\Delta\left(\frac{1}{f}\right) \qquad (8.33)$$

which states that a change in the fractional free volume is equivalent to a change in the logarithm of the time scale of the event to be observed.

In Section 8.6.1.1, it was concluded (54) that the expansion in the glassy state occurs at constant free volume. (Actually, free volume must increase slowly with temperature, even in the glassy state.) As illustrated in Figures 8.9 and 8.22, the coefficient of expansion increases at T_g, allowing for a steady increase in free volume above T_g. Setting α_f equal to the expansion coefficient of the free volume, and f_0 as the fractional free volume at T_g or other point of interest, the dependence of the fractional free volume on temperature may be written

$$f = f_0 + \alpha_f(T - T_0) \qquad (8.34)$$

where T_0 is a generalized transition temperature. Equation (8.33) may be differentiated,

$$\Delta \ln t = B\left(\frac{1}{f} - \frac{1}{f_0}\right) \qquad (8.35)$$

Substituting equation (8.34) into (8.35),

$$\Delta \ln t = B\left[\frac{1}{f_0 + \alpha_f(T - T_0)} - \frac{1}{f_0}\right] \tag{8.36}$$

Cross-multiplying:

$$\Delta \ln t = B\left\{\frac{f_0 - [f_0 + \alpha_f(T - T_0)]}{f_0[f_0 + \alpha_f(T - T_0)]}\right\} \tag{8.37}$$

$$\Delta \ln t = -\frac{B\alpha_f(T - T_0)/f_0}{f_0 + \alpha_f(T - T_0)} \tag{8.38}$$

Dividing by α_f:

$$\Delta \ln t = -\frac{(B/f_0)(T - T_0)}{f_0/\alpha_f + (T - T_0)} \tag{8.39}$$

Consider the meaning of $\Delta \ln t$:

$$\Delta \ln t = \ln t - \ln t_0 = \ln(t/t_0) = \ln A_T \tag{8.40}$$

where A_T is called the reduced variables shift factor (1–4). The quantity A_T will be shown to relate not only to the time for a transition with another time but also to many other time-dependent quantities at the transition temperature and another temperature. The most important of these quantities is the melt viscosity, described below and in Section 10.4.

The theoretical form of the WLF equation can now be written:

$$\ln A_T = -\frac{(B/f_0)(T - T_0)}{f_0/\alpha_f + (T - T_0)} \tag{8.41}$$

Or in log base 10 form,

$$\log A_T = -\frac{B}{2.303 f_0}\left[\frac{(T - T_0)}{f_0/\alpha_f + (T - T_0)}\right] \tag{8.42}$$

Equations (8.41) and (8.42) show that a shift in the log time scale will produce the same change in molecular motion as will the indicated nonlinear change in temperature.

The derivation leading to equations (8.41) and (8.42) suggests a generalized time dependence. Before proceeding with an interpretation of the constants in these equations, it is useful to consider the derivation originally presented by Williams, Landel, and Ferry (62).

Beginning with the Doolittle equation, equation (8.25), they note that for small v_f,

$$\frac{v_f}{v_0} \simeq \frac{v_f}{v_0 + v_f} = f \tag{8.43}$$

where $v_0 + v_f$ is the specific volume, and equation (8.43) provides a quantitative definition for f. Equation (8.25) may now be written in terms of the melt viscosity,

$$\ln \eta = \ln A + \frac{B}{f} \tag{8.44}$$

Subtracting conditions at T_0 (or T_g):

$$\ln \eta - \ln \eta_0 = \ln A - \ln A + \frac{B}{f} - \frac{B}{f_0} \tag{8.45}$$

$$\ln\left(\frac{\eta}{\eta_0}\right) = B\left(\frac{1}{f} - \frac{1}{f_0}\right) \tag{8.46}$$

Furthermore, the viscosity is a time (shear rate)-dependent quantity,

$$\ln(\eta/\eta_0) = \ln A_T = \ln(t/t_0) \tag{8.47}$$

Noting equation (8.40), this leads directly back to equation (8.35). Thus equations (8.41) and (8.42) follow directly from the original Doolittle equation, although in a somewhat more limited form.

Now, the constants in equation (8.42) may be evaluated. Experimentally, for many linear amorphous polymers above T_g, independent of chemical structure,

$$\log\left(\frac{\eta}{\eta_g}\right) = -\frac{17.44(T - T_g)}{51.6 + T - T_g} \tag{8.48}$$

where T_0 has been set as T_g. (For T_0 equal to an arbitrary temperature, T_s, about 50°C above T_g, the constants in the WLF equation read

$$\log\left(\frac{\eta}{\eta_s}\right) = -\frac{8.86(T - T_s)}{101.6 + T - T_s} \tag{8.49}$$

in an alternately phrased mode of expression.) Comparing equation (8.48) with (8.42), we have

$$\frac{B}{2.303f_0} = 17.44 \tag{8.50}$$

$$\frac{f_0}{\alpha_f} = 51.6 \tag{8.51}$$

Here, three unknowns and two equations are shown, which can be solved by assigning the constant B a value of unity (62), consistent with the viscosity data of Doolittle. Then $f_0 = 0.025$, and $\alpha_f = 4.8 \times 10^{-4}$ deg^{-1}.

The value of α_f may be verified in a rough way through equation (8.23). Here, if the free volume is constant in the α_G region, then $\alpha_R - \alpha_G \simeq \alpha_f$. The value of $\alpha_f = 4.8 \times 10^{-4}$ deg^{-1} leads to a temperature of $-38°$C, a temperature at least in the range of the T_g's observed for many polymers. Sharma et al. (56c) found $\alpha_f = 3.2 \times 10^{-4}$ deg^{-1}.

The finding of $f_0 = 0.025$ is more significant. It assigns the value of the free volume at the T_g of any polymer at 2.5%. This approximate value has stood the test of time. Wrasidlo (52) suggested a value of 2.35%, based on more recent thermodynamic data, in relatively good agreement with the WLF value.

For numerical results, it must be emphasized that the WLF equation is good for the range T_g to $T_g + 100$. In equations (8.48) and (8.49), T must be larger than T_g or T_0. Its power lies in its generality: no particular chemical structure is assumed other than a linear amorphous polymer above T_g. For a generation of polymer scientists and rheologists, the WLF equation has provided a mainstay both in utility and theory.

8.6.1.3   An Example of WLF Calculations

The WLF equation, equation (8.48), can be used to calculate melt viscosity changes with temperature. Suppose a polymer has a glass transition temperature of 0°C. At 40°C, it has a melt viscosity of 2.5×10^5 poises (P) (2.5×10^6 Pa · sec). What will its viscosity be at 50°C?

First calculate η_g:

$$\log\left(\frac{\eta}{\eta_g}\right) = -\frac{17.44(T - T_g)}{51.6 + (T - T_g)}$$

$$\log \eta_g = \log 2.5 \times 10^5 + \frac{17.44(313 - 273)}{51.6 + (313 - 273)}$$

$$\log \eta_g = 13.013$$

Polymers often have melt viscosities near 10^{13} P at their glass transition temperature, 1.03×10^{13} P in this case.

Now, calculate the new viscosity:

$$\log \eta = 13.013 - \frac{17.44(323 - 273)}{51.6 + (323 - 273)}$$

$$\eta = 2.69 \times 10^4 \text{ P} \quad \text{or} \quad 2.69 \times 10^5 \text{ Pa} \cdot \text{sec}.$$

Thus a 10°C increase in temperature has decreased the melt viscosity by approximately a factor of 10 in this range. The WLF equation can be used to calculate shift factors (Section 8.6.1.2), failure envelopes (Section 11.2.5.2), and much more.

8.6.2 The Kinetic Theory of the Glass Transition

The free-volume theory of the glass transition, as developed in Section 8.6.1, is concerned with the introduction of free volume as a requirement for coordinated molecular motion. The WLF equation also serves to introduce some kinetic aspects. For example, if the time frame of an experiment is decreased by a factor of 10 near T_g, equations (8.47) and (8.48) indicate that the glass transition temperature should be raised by about 3°C:

$$\lim_{T \to T_g} \left(\frac{\log A_T}{T - T_g} \right) = -0.338 \tag{8.52}$$

$$T - T_g = \frac{-1.0}{-0.338} = +3.0 \tag{8.53}$$

For larger changes in the time or frequency frame, values of 6–7°C are obtained from equation (8.48), in agreement with experiment. For example, if $A_T = 1 \times 10^{-10}$, an average value of 6.9°C per decade change in T_g is obtained. The kinetic theory of the glass transition, to be developed in this section, considers the molecular and macroscopic response within a varying time frame.

8.6.2.1 *Energy States and Heat Capacity*

Much more information about molecular motion can be obtained by considering the kinetics of vitrification, or freezing in. A general approach to the freezing-in process assumes two possible states on a molecular scale with a difference in molar energy, ε_h (see Figure 8.23). An interesting point is the prediction of the characteristic maximum in the heat capacity (64), already illustrated in Figure 8.11. The reality of the maximum in the heat capacity, however, is still controversial (65, 66).

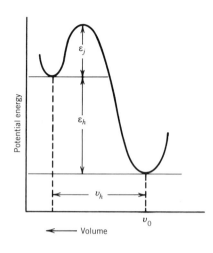

Figure 8.23 Kinetics of bulk flow by altering hole content. v_h = molar volume of a hole; ε_h = molar excess energy over "no-hole" situation; ε_j = activation energy for the disappearance of a hole. (See Section 8.6.2.2.)

8.6.2.2 Relaxation Time and Hole Content

The theory of holes in matter was applied by Eyring (67) to the theory of the glass transition. In this theory, molecular-size holes assume some semblance of what is normally attributed to matter: v_h is the "molar volume" of a hole, and ε_h represents the molar excess of energy over the "no-hole" situation. The activation energy for the disappearance of a hole is represented by ε_j, and the partition function of the holes by Q^h. The partition function of the activated state is represented by Q^{Ξ} (see Figure 8.23).

One may consider the process of adding or subtracting holes to a system, which resembles the mixing or demixing of solution components. When a hole is removed, it creates an elastic strain, which propagates to the surface with the velocity of elastic waves. One may say that the hole evaporated. Of course, one by one, molecules are moving inward to fill the hole space. Eventually, the excess energy is taken up by the decreased liquid lattice volume. The material is said to be in the glassy state if the number of holes and their spatial positions become frozen in; that is, the molecules are unable to move from their locations into a hole.

The equilibrium number of holes, N_h^*, has been expressed (68–70) as

$$N_h^* = N_0(v_0/v_h)e^{-\varepsilon_h/RT} \tag{8.54}$$

where N_0 and v_0 are the number of moles of repeating units in question and the molar volume per repeating unit. The quantity N_0 is set equal to unity in all further calculations, so that all quantities are calculated for 1 mol of repeating units (64).

The relaxation time for the disappearance of holes, τ_h, may be expressed as (68, 69)

$$\tau_h = \frac{h}{kT}\frac{Q^h}{Q^{\equiv}}e^{\varepsilon_j/RT} \tag{8.55}$$

where k and h are the Boltzmann constant and the Planck constant, respectively.

Since the change in the number of holes on going through the glass transition yields the change in the heat capacity, ΔC_{pg}, it is possible to calculate ε_h from experimental determinations (70) of ΔC_p. The quantities v_0/v_h can also be determined experimentally using extrapolated or measured values of the density and cohesive energy (70).

After some further analysis (52, 64, 68, 71), numerical values may be deduced for the individual quantities. A short list of these is collected in Table 8.8. Perhaps especially interesting are the quantities v_0/v_h and n_0/n_h, which shows that the number of holes per mole of monomer is larger than the volume of holes per mole of volume occupied by molecules, v_0, indicating that the holes are smaller than the mers. Also, ε_j is two orders of magnitude larger than ε_h (see Figure 8.23).

In his hole theory of liquids and glasses, Nose (72) made two assumptions of a somewhat different nature regarding the properties of holes: (a) the change in cell volume with pressure and temperature is essentially independent of the hole fraction, and (b) the free energy is expressed by two parts: (i) internal degrees of freedom, which are independent of volume, and (ii) external degrees of freedom, which depend on the hole fraction, the mean distance between nearest-neighbor segments, and the temperature. Using this theory, Nose examined the change in heat capacity, ΔC_p, and derived the general relationship

$$\frac{TV(\Delta\alpha)^2}{\Delta C_p^{\text{inter}}\,\Delta\beta} = 1 \tag{8.56}$$

where T is the temperature, V is the volume, $\Delta\alpha$ and $\Delta\beta$ are the differences in the thermal expansion coefficient and compressibility between the liquid (rubbery) and glassy states, respectively, and $\Delta C_p^{\text{inter}}$ is the part of ΔC_p attributable to intersegmental interactions.

8.6.2.3 The Apparent Activation Energy of the Process

The apparent activation energy of molecular relaxation near T_g was deduced by Kovacs (50):

$$E_a = 2.303R\left[\frac{d\log A_T}{d(1/T)}\right] = -RT^2\left[\frac{d\log A_T}{dT}\right] \tag{8.57}$$

Table 8.8 Thermodynamic properties of the glass transition (52)

Polymer	T_g, K	$\alpha_R \times 10^4$, deg^{-1}	$\alpha_G \times 10^4$, deg^{-1}	v_g, cm^3/g	ΔC_p, cal/(mol·deg)	v_0/v_h	$\phi_1^* \times 10^2$	n_0/n_h	ε_h, cal/mol	ε_j, kcal/mole of holes
Polyethylene	140	5.31	2.01	1.035	2.37	6.85	0.762	20.1	1093	—
Polyisoprene	201	6.16	2.06	1.038	7.35	14.70	1.59	4.18	1254	—
Polystyrene	368	5.50	1.80	0.973	6.35	8.82	2.91	3.76	1867	157.6
Poly(vinyl chloride)	352	4.67	1.40	0.723	4.44	6.77	1.88	7.67	2101	—
Poly(vinyl acetate)	302	5.98	2.07	0.839	8.61	13.29	1.95	3.76	1783	—
Poly(methyl methacrylate)	378	4.60	2.15	0.870	7.0	10.31	2.55	3.70	2016	—
Poly(ethylene terephthalate)	337	4.50	1.80	0.771	15.51	21.69	1.87	2.41	1999	—
Poly(2,6-dimethyl phenylene oxide)	480	5.13	2.04	0.970	6.90	9.38	3.72	2.75	2212	—
Poly(butylene oxide)	185	—	—	1.00	13.5	—	—	—	—	—
Poly(dimethyl siloxane)	150	8.5	4.5	0.905	5.4	12.00	1.24	6.61	1009	—
Cellulose tricaproate	223	5.5	3.7	—	—	—	—	—	—	—

Key: T_g, glass transition temperature; α_R, specific volume expansion coefficient in the liquid (rubbery) state; α_G, specific volume expansion coefficient in the glass state; v_g, specific volume of polymer at T_g; ΔC_p, change in heat capacity at T_g; n_0/n_h, ratio of monomer units to holes according to (71); ϕ_1^*, specific free volume according to (71) (above); v_0/v_h, occupied volume divided by hole volume, according to (68); ε_h, molar hole energy according to (71); ε_j, activation energy of holes.

Equation (8.57) leads to activation energies of about 210 kcal/mol for polystyrene, evaluated at $T = T_g$.

8.6.2.4 Determination of the Segment Size in Motion

An interesting explanation of the size of the segment that moves independently above T_g is due to Erying (72a). Reference is made to his model of the lattice structure of a liquid, containing some unoccupied sites or holes. For a low-molecular-weight liquid, the Arrhenius equation

$$\eta = A e^{\overline{E}/RT} \qquad (8.58)$$

is followed, where A is a constant and $\overline{E}$ represents the activation energy for viscous flow. The quantity $\overline{E}$ is related to the latent heat of vaporization of the liquid, since motion of the molecule out of its surroundings is part of both processes. As the molecular weights of homologous series are increased, $\overline{E}$ first rises proportionally to the heat of vaporization, then levels off. Eventually, a value of $\overline{E}$ is reached that is independent of the chain length. From this information, it may be concluded that the unit of flow is much smaller than the long polymer chain. Variously, it is estimated to be in the range of 5–50 carbon atoms (see Section 8.2).

8.6.3 Thermodynamic Theory of T_g

All noncrystalline polymers display what appears to be a second-order transition in the Ehrenfest sense (73): the temperature and pressure derivatives of both volume and entropy are discontinuous when plotted against T or P, although the volume and the entropy themselves remain continuous (see Figure 8.5).

In Section 8.6.2, it was argued that the transition is primarily a kinetic phenomenon because (a) the temperature of the transition can be changed by changing the time scale of experiment, slower measurements resulting in lower T_g's, and (b) the measured relaxation times near the transition approach the time scale of the experiment.

One can ask what equilibrium properties these glass-forming materials have, even if it is necessary to postulate infinite time scale experiments. A thermodynamically based answer was provided by Gibbs and DiMarzio (74–78), based on a lattice model.

8.6.3.1 The Gibbs and DiMarzio Theory

Gibbs and DiMarzio (74–78) argued that even though the observed glass transitions are indeed a kinetic phenomenon, the underlying true transitions can nevertheless possess equilibrium properties, even if they are difficult to realize. At infinitely long times, they predict a true second-order transition, when the material finally reaches equilibrium. In infinitely slow experiments,

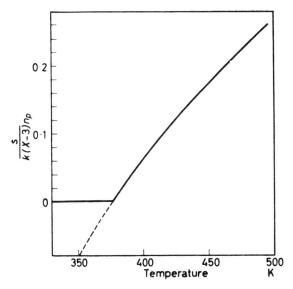

Figure 8.24   Schematic diagram of the conformational entropy of a polymer as a function of temperature according to the Gibbs – DiMarzio theory (76).

a glassy phase will eventually emerge whose entropy is negligibly higher than that of the crystal. The temperature dependence of the entropy at the approach of T_g is shown in Figure 8.24. This true T_g, designated T_2, as will be shown later, lies some 50°C below the T_g observed at ordinary times.

The central problem of the Gibbs–DiMarzio theory is to find the configurational partition function, Q, from which the expression for the configurational entropy can be calculated. In a manner similar to the kinetic theory described in Section 8.6.2, hindered rotation in the polymer chain is assumed to arise from two energy states: ε_1 is associated with one possible orientation, and ε_2 is associated with all the remaining orientations; $\varepsilon_1 - \varepsilon_2 = \Delta E$, a coordination number, Z, of 4 being assumed. The intermolecular energy is given by the hole energy, α.

The application of the partition function assumes an equilibrium state:

$$Q = \sum_{f, n_x, n_0} W(f_1 n_x \ldots, f_i n_x \ldots, n_0)$$
$$\times \exp\left[- \frac{E(f_1 n_x \ldots, f_i n_x \ldots, n_0)}{kT} \right] \qquad (8.59)$$

where $f_i n_x$ is the number of molecules packed in conformation i, and W is the total number of ways that the n_x (x degree of polymerization) molecules can be packed into $xn_x + n_0$ sites on the quasilattice, with n_0 being the number of holes. An expression for W was derived earlier by Flory (79) for

n_x polymer chains and n_0 solvent molecules, which was used by Gibbs and DiMarzio in their calculations (see Section 3.3).

Once Q is formulated [see equation (8.59)], statistical thermodynamics provides the entropy:

$$S = kT\left(\frac{\partial \ln Q}{\partial T}\right)_{V,n} + k \ln Q \qquad (8.60)$$

from which all necessary calculations can be made (75). This theory has been applied to the variation of the glass temperature with the molecular weight (75) (see Section 8.7), random copolymer composition (80, 81) (see Section 8.8), plasticization (82), extension (78), and cross-linking (78). This last is briefly explored in Section 8.6.3.2.

8.6.3.2 Effect of Cross-link Density on T_g

The criterion of the second-order transition temperature is that the temperature-dependent conformational entropy, S_c, becomes zero. If S_0 is the conformational entropy for the un-cross-linked system, and ΔS_R is the change in conformational entropy due to adding cross-links (78),

$$S_c = S_0 + \Delta S_R = 0 \qquad (8.61)$$

Since cross-linking decreases the conformational entropy, qualitatively it may be concluded that the transition temperature is raised. The final relation may be written

$$\frac{T(\chi') - T(0)}{T(0)} = \frac{KM\chi'/\gamma}{1 - KM\chi'/\gamma} \qquad (8.62)$$

where χ' is the number of cross-links per gram, M is the mer molecular weight, and γ is the number of flexible bonds per mer, backbone, and side chain. The quantity K is found by experiment and, interestingly enough, appears independent of the polymer (see Table 8.9).

Table 8.9 Constants for cross-link effect on T_g (78, 83)

Polymer	γ	M/γ	$K \times 10^{23}$	$Z \times 10^4$
Natural rubber	3	22.7	1.30	3.2
Polystyrene	2	52	1.20	4.6
Poly(methyl methacrylate)	4	25	1.38	1.8

Table 8.10   Glass transition theory box scores

Theory	Advantages	Disadvantages
Free-volume theory	1. Time and temperature of visco-elastic events related to T_g 2. Coefficients of expansion above and below T_g related	1. Actual molecular motions poorly defined
Kinetic theory	1. Shifts in T_g with time frame quantitatively determined 2. Heat capacities determined	1. No T_g predicted at infinite time scales
Thermodynamic theory	1. Variation of T_g with molecular weight, diluent, and cross-link density predicted 2. Predicts true second-order transition temperature	1. Infinite time scale required for measurements 2. True second-order transition temperature poorly defined

An alternate relation dates back to Ueberreiter and Kanig (11)

$$\Delta T_{g,c} = ZD \tag{8.63}$$

where the change in the glass temperature with increasing cross-linking, $\Delta T_{g,c}$, is equal to the cross-link density, D, times a constant, Z. Recently, Glans and Turner (83) compared equations (8.62) and (8.63), using cross-linked polystyrene. The glass transition elevation was observed via DSC analysis (an endothermal peak was reported at T_g; see Section 8.6.2). Plots of straight lines were obtained for $\Delta T_{g,c}$ versus cross-link density, verifying equation (8.63). Some values for Z, with D in units of moles per gram and ΔT_c in K, are also shown in Table 8.9.

8.6.3.3   A Summary of the Glass Transition Theories

In the above, three apparently disparate theories of the glass transition were presented. The basic thrust of each is summarized conveniently here and in Table 8.10.

1. The free-volume theory introduces free volume in the form of segment-size voids as a requirement for the onset of coordinated molecular motion. This theory provides relationships between coefficients of expansion below and above T_g and yields equations relating viscoelastic motion to the variables of time and temperature.

2. The kinetic theory defines T_g as the temperature at which the relaxation time for the segmental motions in the main polymer chain is of the

same order of magnitude as the time scale of the experiment. The kinetic theory is concerned with the rate of approach to equilibrium of the system, taking the respective motions of the holes and molecules into account. The kinetic theory provides quantitative information about the heat capacities below and above the glass transition temperature and explains the 6–7°C shift in the glass transition per decade of time scale of the experiment.

3. The thermodynamic theory introduces the notion of equilibrium and the requirements for a true second-order transition, albeit at infinitely long time scales. The theory postulates the existence of a true second-order transition, which the glass transition approaches as a limit when measurements are carried out more and more slowly. It successfully predicts the variation of T_g with molecular weight and cross-link density (see Section 8.7), diluent content, and other variables.

A summary of the free-volume numbers of the various theories can be made:

Theory	Free Volume Fraction
WLF	0.025
Hirai and Eyring	0.08
Miller[†]	0.12
Simha-Boyer	0.113

The analytical development of these theories illustrates the power of statistical thermodynamics in providing solutions to important polymer problems. However, much remains to be done. It has been said that less than 5% of all fundamental knowledge has been wrested from nature. This is certainly true in the study of polymer glass transitions. Insofar as research in this area remains highly active, it is highly probable that new insight will provide an integrated theory in the near future. Attempts to do so up until now are summarized in the following section.

8.6.3.4 A Unifying Treatment

Adam and Gibbs (82) attempted to unify the theories relating the rate effect of the observed glass transition and the equilibrium behavior of the hypothetical second-order transition. They proposed the concept of a "cooperatively rearranging region," defined as the smallest region capable of conformational change without a concomitant change outside the region. At T_2, this region becomes equal to the size of the sample, since only one conformation is available.

Adam and Gibbs rederived the WLF equation, putting it in terms of the potential energy hindering the cooperative rearrangement per mer, the molar

[†]A. A. Miller, *J. Chem. Phys.*, **49**, 1393(1968); *J. Polym. Sci.*, **A-2**(6), 249, 1161(1968).

conformational entropy, and the change in the heat capacity at T_g. By choosing the temperature T in the WLF equation to be T_s [see equation (8.49)] and suitable rearrangements of the WLF formulation to isolate T_2, they find that

$$\frac{T_g}{T_2} = 1.30 \pm 8.4\% \tag{8.64}$$

for a wide range of glass-forming systems, both polymeric and low molecular weight.

For low-temperature elastomers such as the polybutadiene family, $T_g \cong$ 200 K. According to equation (8.64), $T_2 \cong 154$ K, or about 50 K below T_g. According to the WLF equation, equation (8.48), the viscosity becomes infinite at $T - T_g = -51.6°C$, which is about the same number. Although this simplified approach yields less quantitative agreement at higher temperatures, the ideas still are interesting.

8.7 EFFECT OF MOLECULAR WEIGHT ON T_g

8.7.1 Linear Polymers

Studies of the increase in T_g with increasing polymer molecular weight date back to the works of Ueberreiter in the 1930s (84). The theoretical analysis of Fox and Flory (54) (see Section 8.6.1.1) indicated that the general relationship between T_g at a molecular weight M was related to the glass temperature at infinite molecular weight, $T_{g,\infty}$, by

$$T_g = T_{g,\infty} - \frac{K}{(\alpha_R - \alpha_G)M} \tag{8.65}$$

with K being a constant depending on the polymer. Equation (8.65) follows from the decrease in free volume with increasing molecular weight, caused in turn by the increasing number of connected mers in the system, and decreased number of end groups.

The ubiquitous polystyrene seems to have been investigated more than any other polymer (27b, 54, 84). DSC data, first extrapolated to low heating rate, are shown in Figure 8.25 (27b). (These data also show an endothermic peak at T_g; see earlier discussions.) The equation for slow heating rates may be expressed

$$T_g = 106°C - \frac{2.1 \times 10^5}{M_n} \tag{8.66}$$

Figure 8.25 The glass transition temperature of polystyrene as a function of $1 / M_n$ (27b).

For heating rates normally encountered (54)

$$T_g = 100°C - \frac{1.8 \times 10^5}{M_n} \tag{8.67}$$

The molecular weight in equation (8.67) is for fractionated polystrene. For slow experiments, these equations suggest a 6°C increase in T_g at infinite molecular weight.

More recently, Cowie (84a) studied the variation of T_g with molecular weight from oligomers up. He noted that with a semilog plot of T_g versus log DP (degree of polymerization), T_g appeared substantially invariant with log DP above a DP of some several hundred (Figure 8.26). The critical DP between regions I and II depends only on the T_g of the polymer and is

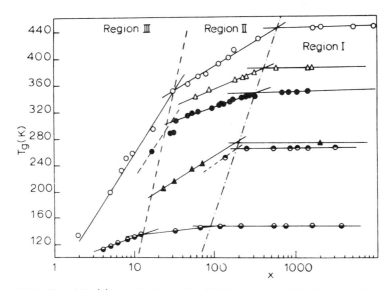

Figure 8.26 Plot of T_g (K) as a function of log DP, the number of backbone chain atoms / bonds, for poly(α-methylstyrene) (○); poly(methylmethacrylate) (△); poly(vinyl chloride) (●); isotactic polypropylene (▲); atactic polypropylene (◐); and poly(dimethylsiloxane) (◒) (84a).

independent of the structure, following the equation, in K:

$$T_g(\infty) = 372.6 \log DP_c - 595 \tag{8.68}$$

The line between regions II and III, dividing the oligomers from the polymers, is suggestive of the number of backbone atoms involved in the cooperative motions at T_g (see Table 8.4).

8.7.2 Effect of T_g on Polymerization

According to equation (8.65), the glass transition depends on the molecular weight. What happens during an isothermal polymerization? When the polymerization begins, the monomers are always in the liquid state. Sometimes, however, the system may go through T_g and the polymer may vitrify as the reaction proceeds. Since molecular motion is much reduced when the system is below T_g, the reaction substantially stops.

Two conditions can be distinguished. First, during a chain polymerization, the monomer effectively acts like a plasticizer for the nascent polymer. An example relates to the emulsion polymerization of polystyrene, often carried out at about 80°C. The reaction will not proceed quite to 100% conversion, because the system vitrifies.

Second, during stepwise polymerization, the molecular weight is continually increasing. An especially interesting case involves gelation. Taking epoxy polymerization as an example, the resin[†] is simultaneously polymerizing and cross-linking (see Section 3.7.3).

Gillham (84b–g) pointed out the need to postcure the polymer above $T_{g\infty}$, the glass transition temperature of the fully cured[‡] system. He developed a time–temperature–transformation (TTT) reaction diagram that may be used to provide an intellectual framework for understanding and comparing the cure and glass transition properties of thermosetting systems. Figure 8.27 illustrates the TTT diagram. Besides $T_{g\infty}$, the diagram also displays $_{gel}T_g$, the temperature at which gelation and vitrification occur simultaneously, and T_{g0}, the glass transition temperature of the reactants. The particular S-shaped curve between T_{g0} and $T_{g\infty}$ results because the reaction rate is increased with increasing temperature. At a temperature intermediate between $_{gel}T_g$ and $T_{g\infty}$, the reacting mass first gels, forming a network. Then it vitrifies, and the reaction stops, incomplete. To the novice, the reaction products may appear complete. This last may result in material failure if the temperature is suddenly raised.

The TTT diagram explains why epoxy and similar reactions are carried out in steps, each at a higher temperature. The last step, the postcure, must be done above $T_{g\infty}$. Other points shown in Figure 8.27 include the devitrification

---

[†]Resin is an early term for polymer, often used with epoxies.
[‡]Cure is an early term for cross-linking, also frequently used with epoxies.

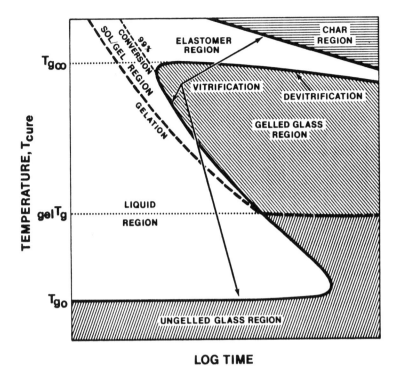

LOG TIME

Figure 8.27 The thermosetting process as illustrating by the time – temperature – transformation reaction diagram (84g).

region, caused by degradation, and the char region, at still higher temperatures.

8.8   EFFECT OF COPOLYMERIZATION ON T_g

The above discussion relates to simple homopolymers. Addition of a second component may take the form of copolymerization or polymer blending. Addition of low-molecular-weight compounds results in plasticization. Experimentally, two general cases may be distinguished: where one phase is retained and where two or more phases result.

8.8.1   One-Phase Systems

Based on the thermodynamic theory of the glass transition, Couchman derived relations to predict the T_g composition dependence of binary mixtures of miscible high polymers (85) and other systems (86–88). The treatment that follows is easily generalized to the case for random copolymers (85).

Consider two polymers (or two kinds of mers, or one mer and one plasticizer) having pure-component molar entropies denoted as S_1 and S_2, and their respective mole fractions (moles of mers for the polymers) as X_1 and X_2. The mixed system molar entropy may be written

$$S = X_1 S_1 + X_2 S_2 + \Delta S_m \tag{8.69}$$

where ΔS_m represents the excess entropy of mixing. For later convenience, S_1 and S_2 are referred to their respective pure-component glass transition temperatures of T_{g1} and T_{g2}, when their values are denoted as S_1^0 and S_2^0.

Heat capacities are of fundamental importance in glass transition theories, because the measure of the heat absorbed provides a direct measure of the increase in molecular motion. The use of classical thermodynamics leads to an easy introduction of the pure-component heat capacities at constant pressure, C_{p1} and C_{p2}:

$$S = X_1 \left\{ S_1^0 + \int_{T_{g1}}^{T} C_{p1}\, d \ln T \right\} + X_2 \left\{ S_2^0 + \int_{T_{g2}}^{T} C_{p2}\, d \ln T \right\} + \Delta S_m \tag{8.70}$$

The mixed-system glass transition temperature, T_g, is defined by the requirement that S for the glassy state be identical to that for the rubbery state, at T_g. This condition and the use of appropriate superscripts G and R lead to the equation

$$
X_1^G \left\{ S_1^{0,G} + \int_{T_{g1}}^{T_g} C_{p1}^G\, d \ln T \right\} + X_2^G \left\{ S_2^{0,G} + \int_{T_{g2}}^{T_g} C_{p2}^G\, d \ln T \right\} + \Delta S_m^G
$$
$$
= X_1^R \left\{ S_1^{0,R} + \int_{T_{g1}}^{T_g} C_{p1}^R\, d \ln T \right\} + X_2^R \left\{ S_2^{0,R} + \int_{T_{g2}}^{T_g} C_{p2}^R\, d \ln T \right\} + \Delta S_m^R
$$
$$\tag{8.71}$$

Since $S_i^{0,G} = S_i^{0,R}$ $(i = 1, 2)$ and $X_i^G = X_i^R = X_i$, equation (8.71) may be simplified:

$$
X_1 \left\{ \int_{T_{g1}}^{T_g} \left(C_{p1}^G - C_{p1}^R \right) d \ln T \right\}
$$
$$
+ X_2 \left\{ \int_{T_{g2}}^{T_g} \left(C_{p2}^G - C_{p2}^R \right) d \ln T \right\} + \Delta S_m^G - \Delta S_m^R = 0 \tag{8.72}
$$

In regular small-molecule mixtures, ΔS_m is proportional to $X \ln X + (1 - X)\ln(1 - X)$, where X denotes X_1 and X_2. Similar relations hold for polymer–solvent (plasticizer) and polymer–polymer combinations. Combined with the continuity relation, $\Delta S_m^G = \Delta S_m^R$. For random copolymers, these quantities are also equal. Then

$$X_1 \int_{T_{g1}}^{T_g} \Delta C_{p1}\, d\ln T + X_2 \int_{T_{g2}}^{T_g} \Delta C_{p2}\, d\ln T = 0 \tag{8.73}$$

where Δ denotes transition increments. Again, the increase in the heat capacity at T_g reflects the increase in the molecular motion and the increased temperature rate of these motions.

After integration, the general relationship emerges,

$$X_1 \Delta C_{p1} \ln\left(\frac{T_g}{T_{g1}}\right) + X_2 \Delta C_{p2} \ln\left(\frac{T_g}{T_{g2}}\right) = 0 \tag{8.74}$$

For later convenience the X_i are exchanged for mass (weight) fractions, M_i (recall that the ΔC_{pi} are then per unit mass), and equation (8.74) becomes

$$\ln T_g = \frac{M_1 \Delta C_{p1} \ln T_{g1} + M_2 \Delta C_{p2} \ln T_{g2}}{M_1 \Delta C_{p1} + M_2 \Delta C_{p2}} \tag{8.75}$$

or, equivalently,

$$\ln\left(\frac{T_g}{T_{g1}}\right) = \frac{M_2 \Delta C_{p2} \ln(T_{g2}/T_{g1})}{M_1 \Delta C_{p1} + M_2 \Delta C_{p2}} \tag{8.76}$$

Equation (8.76) is shown to fit T_g data of thermodynamically miscible blends (see Figure 8.28). Four particular nontrivial cases of the general mixing relation may be derived.

Making use of the expansions of the form $\ln(1+x) = x$, for small x, and noting that T_{g1}/T_{g2} usually is not greatly different from unity yield

$$T_g \simeq \frac{M_1 \Delta C_{p1} T_{g1} + M_2 \Delta C_{p2} T_{g2}}{M_1 \Delta C_{p1} + M_2 \Delta C_{p2}} \tag{8.77}$$

which has the same form as the Wood equation (89), originally derived for random copolymers.

If $\Delta C_{pi} T_{gi} = \text{constant}$(56a–c, 90), the familiar Fox equation (91) appears after suitable cross multiplying:

$$\frac{1}{T_g} = \frac{M_1}{T_{g1}} + \frac{M_2}{T_{g2}} \tag{8.78}$$

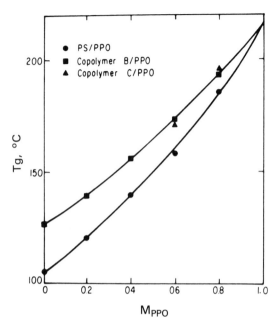

Figure 8.28   Glass-transition temperatures, T_g, of poly(2,6-dimethyl-1,4-phenylene oxide) – *blend* – polystyrene (PPO / PS) blends versus mass fraction of PPO, M_{PPO}. The full curve was calculated from equation (8.76) as circles. $\Delta C_{p1} = 0.0671$ cal $K^{-1} \cdot g^{-1}$, $\Delta C_{p2} = 0.0528$ cal $K^{-1} \cdot g^{-1}$; $T_{g1} = 378$ K, $T_{g2} = 489$ K. PPO was designated as component 2 (85, 92).

The Fox equation (91) was also originally derived for random copolymers. This equation predicts the typically convex relationship obtained when T is plotted against M_2 (see Figure 8.28). If $\Delta C_{p1} \simeq \Delta C_{p2}$, the equation of Pochan et al. (93) follows from equation (8.75):

$$\ln T_g = M_1 \ln T_{g1} + M_2 \ln T_{g2} \tag{8.79}$$

Finally, if both pure-component heat capacity increments have the same value and the log functions are expanded,

$$T_g = M_1 T_{g1} + M_2 T_{g2} \tag{8.80}$$

which predicts a linear relation for the T_g of the blend, random copolymer, or plasticized system. This equation usually predicts T_g too high. Equations (8.78) and (8.80) are widely used in the literature. Couchman's work (85–88) shows the relationship between them. Previously, they were used on a semiempirical basis.

These equations also apply to plasticizers, a low-molecular-weight compound dissolved in the polymer. In this case, the plasticizer behaves as a compound with a low T_g. The effect is to lower the glass transition temperature. A secondary effect is to lower the modulus, softening it through much of the temperature range of interest. An example is the plasticization of poly(vinyl chloride) by dioctyl phthalate to make compositions known as "vinyl."

8.8.2 Two-Phase Systems

Most polymer blends, as well as their related graft and block copolymers and interpenetrating polymer networks, are phase-separated (94) (see Section 4.3). In this case, each phase will exhibit its own T_g. Figure 8.29 (95, 96) illustrates two glass transitions appearing in a series of triblock copolymers of different overall compositions. The intensity of the transition, especially in the loss spectra (E''), is indicative of the mass fraction of that phase.

The storage modulus in the plateau between the two transitions depends both on the overall composition and on which phase is continuous. Electron microscopy shows that the polystyrene phase is continuous in the present case. As the elastomer component increases (small spheres, then cylinders, then alternating lamellae), the material gradually softens. When the rubbery

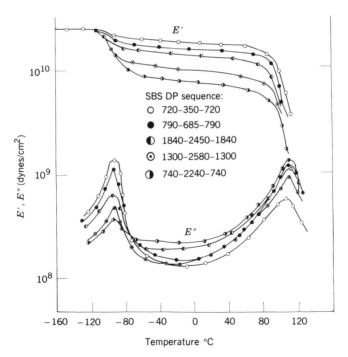

Figure 8.29 Dynamic mechanical behavior of polystyrene – *block* – polybutadiene – *block* – polystyrene, a function of the styrene – butadiene mole ratio (95, 96).

Table 8.11 Phase composition of epoxy / acrylic simultaneous interpenetrating networks (97)

Glycidyl Methacrylate[a] (%)	Dispersed Phase Weight Fraction		Matrix Phase Weight Fraction	
	PnBA[b]	Epoxy	PnBA	Epoxy
0	0.97	0.03	0.09	0.91
0.3	0.82	0.18	0.12	0.88
3.0	—	—	0.30	0.70

[a]Grafting mer, increases mixing.
[b]Poly(n-butyl acrylate).

phase becomes the only continuous-phase, the storage modulus will decrease to about 1×10^8 dynes/cm^2.

If appreciable mixing between the component polymers occurs, the inward shift in the T_g of the two phases can each be expressed by the equations of Section 8.8.1 (97). Using equation (8.78), the extent of mixing within each phase in a simultaneous interpenetrating network of an epoxy resin and poly(n-butyl acrylate) was calculated (see Table 8.11). The overall composition was 80/20 epoxy/acrylic, and glycidyl methacrylate is shown to enhance molecular mixing between the chains.

8.9 EFFECT OF CRYSTALLINITY ON T_g

The previous discussion centered on amorphous polymers, with atactic polystyrene being the most frequently studied polymer. Semicrystalline polymers such as polyethylene or polypropylene or of the nylon and polyester types also exhibit glass transitions, though only in the amorphous portions of these polymers. The T_g is often increased in temperature by the molecular-motion restricting crystallites. Sometimes T_g appears to be masked, especially for high crystalline polymers.

Boyer (7) points out that many semicrystalline polymers appear to possess two glass temperatures: (a) a lower one, $T_g(L)$, which refers to the completely amorphous state and which should be used in all correlations with chemical structure (this transition correlates with the molecular phenomena discussed in previous sections), and (b) an upper value, $T_g(U)$, which occurs in the semicrystalline material and varies with extent of crystallinity and morphology.

8.9.1 The Glass Transition of Polyethylene

Linear polyethylene, frequently referred to as polymethylene, offers a complete contrast with polystyrene in that it has no side groups and has a high

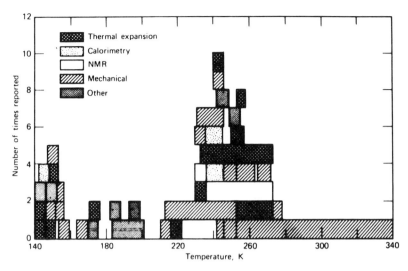

Figure 8.30 Histogram showing number of times a given value of T_g for linear polyethylene has been reported in the literature by various standard methods indicated (98).

degree of crystallinity, usually in excess of 80%. Because of the high degree of crystallinity, molecular motions associated with T_g are partly masked, leading to a confusion with other secondary transitions (see Figure 8.30) (98). Thus various investigators consider the T_g of polyethylene to be in three different regions: $-30°C$, $-80°C$, or $-128°C$.

Davis and Eby support the $-30°C$ value on the basis of volume–time measurements; Stehling and Mandelkern (99) favor the $-128°C$ value based on mechanical measurements. Illers (100) and Boyer (7) support the value of $-80°C$ based on extrapolations of completely amorphous ethylene–vinyl acetate copolymer data with copolymer–T_g relationships. Boyer (7) supports the position that $-80°$ is $T_g(L)$ and $-30°C$ represents $T_g(U)$. The transition at $-128°C$ is thought to be related to the Schatzki crankshaft motion (Section 8.4.1), although the situation apparently is more complicated (100).

8.9.2 The Nylon Family Glass Transition

Two subfamilies of aliphatic nylons exist:

$$\left[NH{\left(CH_2 \right)}_n NH - \overset{\overset{\displaystyle O}{\|}}{C}{\left(CH_2 \right)}_m \overset{\overset{\displaystyle O}{\|}}{C} \right]_{DP} \tag{8.81}$$

from diacids and dibases, and

$$
\left[NH\text{---}(CH_2)_{n-1}\overset{\overset{\displaystyle O}{\displaystyle \|}}{C} \right]_{DP}
\tag{8.82}
$$

originating from ω-amino acids. Both subfamilies are semicrystalline; of course, they form commercially important fibers.

The usually stated T_g range is $T_g \simeq +40°C$ for nylon 612 to $T_g \simeq 60°C$ for nylon 6 (7); however, T_g depends on the crystallinity of the particular sample. N-methylated nylons, with a lower hydrogen bonding, have lower T_g's (101). As n and m increase in equations (8.81) and (8.82), the structure becomes more polyethylene-like, and T_g gradually decreases. Interestingly, when $n > 4$, there is a characteristic mechanical loss peak at about $-130°C$, again suggestive of the Schatzki motion (Section 8.4.1).

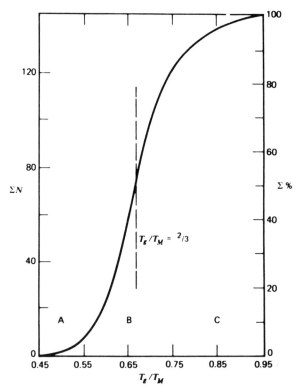

Figure 8.31 Range in T_g / T_f values found in the literature. Region A contains unsubstituted polymers. Region C includes poly(α-olefins) with long side chains. Region B contains the majority of vinyl, vinylidene, and condensation polymers. The left ordinate is cumulative number, N, and the right ordinate is cumulative percentage of all examples reported as having the indicated T_g / T_f values (7).

8.9.3 Relationships Between T_g and T_f

The older literature (102) suggested two relationships between T_g and T_f: $T_g/T_f \simeq \frac{1}{2}$ for symmetrical polymers, and $T_g/T_f \simeq \frac{2}{3}$ for nonsymmetrical polymers. Though definitions of symmetry differ, one method uses the appearance of atoms down the chain: if a central portion of the chain appears the same when viewed from both ends, it is symmetrical. However, even from the beginning, there were many exceptions to the above. The only rule obeyed in this regard is that T_g is always lower than T_f for homopolymers. This is because (a) the same kinds of molecular motion should occur at T_g and T_f, and (b) short-range order exists at T_g, but long-range order exists at T_f.

More recently, Boyer (7) has prepared a cumulative plot of T_g/T_f (see Figure 8.31). Region A (the old $T_g/T_f \simeq \frac{1}{2}$) contains most of the polymers which are free from side groups other than H and F (and hence symmetrical) and contain such polymers as polyethylene, poly(oxymethylene), and poly(vinylidene fluoride). Region B contains most of the common vinyl, vinylidene, and condensation polymers such as the nylons. About 55% of all measured polymers lie in the band $T_g/T_f = 0.667 \pm 0.05$ (7). Region C contains poly(α-olefins) with long alkyl side groups as well as other nontypical polymers such as poly(2,6-dimethylphenylene oxide), which has T_g/T_f approximately equal to 0.93. For an unknown polymer, then, the relationship $T_g/T_f = \frac{2}{3}$ is a good way of providing an estimate of one transition if the temperature of the other is known.

8.10 DEPENDENCE OF T_g ON CHEMICAL STRUCTURE

In Section 8.9, some effects of crystallinity and hydrogen bonding on T_g were considered. The effect of molecular weight was discussed in Section 8.7, and the effect of copolymerization was discussed in Section 8.8. This section discusses the effect of chemical structure in homopolymers.

Boyer (103) suggested a number of general factors that affect T_g (see Table 8.12. In general, factors that increase the energy required for the onset of molecular motion increase T_g; those that decrease the energy requirements lower T_g.

Table 8.12 Factors affecting T_g (103)

Increase T_g	Decrease T_g
Intermolecular forces	In-chain groups promoting
High CED	flexibility (double-bonds
Intrachain steric hindrance	and ether linkages)
Bulky, stiff side groups	Flexible side groups
	Symmetrical substitution

8.10.1 Effect of Aliphatic Side Groups on T_g

In monosubstituted vinyl polymers and at least some other classes of polymers, flexible pendant groups reduce the glass transition of the polymer by acting as "internal diluents," lowering the frictional interaction between chains. The total effect is to reduce the rotational energy requirements of the backbone.

The aliphatic esters of poly(acrylic acid) (104), poly(methacrylic acid) (105), and other polymers (106) (see Figure 8.32) (26) show a decline in T_g as the number of $-CH_2-$ units in the side group increases. At still longer aliphatic side groups, T_g increases as side-chain crystallization sets in, impeding chain motion. In this latter composition range, the materials feel waxy. In the ultimate case, of course, the polymer would behave like slightly diluted polyethylene. For cellulose triesters (34), the minimum in T_g is observed at the triheptanoate, probably because of the increased basic backbone stiffness.

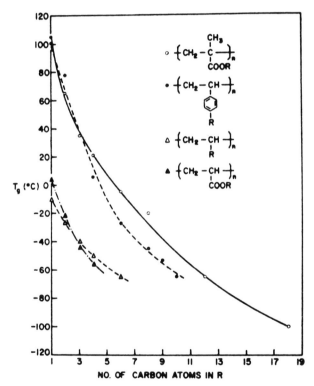

Figure 8.32 Effect of side-chain lengths on the glass transition temperatures of polymethacrylates [○ (S. S. Rogers and L. Mandelkern, *J. Phys. Chem.*, **61**, 985, 1957)]; poly-*p*-alkyl styrenes [● (W. G. Barb, *J. Polym. Sci.*, **37**, 515, 1959)]; poly-*α*-olefins [△ (M. L. Dannis, *J. Appl. Polym. Sci.*, **1**, 121, 1959; K. R. Dunham, J. Vandenbergh, J. W. H. Farber, and L. E. Contois., *J. Polym. Sci.*, **1A**, 751, 1963)]; and polyacrylates [▲ (J. A. Shetter, *Polym. Lett.*, **1**, 209, 1963)] (26).

Table 8.13   Effect of tacticity on the glass transition temperatures of polyacrylates and polymethacrylates (108)

	T_g (°C)				
	Polyacrylates		Polymethacrylates		
Side Chain	Isotactic	Dominantly Syndiotactic	Isotactic	Dominantly Syndiotactic	100% Syndiotactic
Methyl	10	8	43	105	160
Ethyl	−25	−24	8	65	120
n-Propyl	—	−44	—	35	—
Iso-Propyl	−11	−6	27	81	139
n-Butyl	—	−49	−24	20	88
Iso-Butyl	—	−24	8	53	120
Sec-Butyl	−23	−22	—	60	—
Cyclo-Hexyl	12	19	51	104	163

8.10.2   Effect of Tacticity on T_g

The discussion so far in this chapter has assumed atactic polymers, which with a few exceptions are amorphous. Other stereo isomers include isotactic and syndiotactic polymers (see Section 2.3).

The effect of tacticity on T_g may be significant, as illustrated in Table 8.13 (107, 108). Karasz and MacKnight (108) noted that the effect of tacticity on T_g is expected in view of the Gibbs–DiMarzio theory (Section 8.6.3.1). In disubstituted vinyl polymers, the energy difference between the two predominant rotational isomers is greater for the syndiotactic configuration than for the isotactic configuration. In monosubstituted vinyl polymers, where the other substituent is hydrogen, the energy difference between the rotational states of the two pairs of isomers is the same. Thus the acrylates in Table 8.13 have the same T_g for the two isomers, whereas the methacrylates show distinctly different T_g's, with the isotactic form always having a lower T_g than the syndiotactic form.

8.11   EFFECT OF PRESSURE ON T_g

The above discussion has assumed constant pressure at 1 atm. Since an increased pressure causes a decrease in the total volume [see equation (8.8)], an increase in T_g is expected based on the prediction of decreased free volume.

Tamman and Jellinghaus (109) showed that a plot of volume versus pressure at a temperature near the transition shows an elbow reminiscent of the volume–temperature plot (see Figure 8.9). If the temperature is raised at elevated pressures, T_g will in fact show a corresponding increase (see Figure 8.33) (110).

The results in Figure 8.33 can easily be interpreted in terms of the free-volume theory of T_g. In developing the WLF equation (Section 8.6.1.2), it was shown that the free-volume fraction at any temperature above T_g could be expressed $f = f_0 + \alpha_f(T - T_g)$. If the free-volume compressibility is κ_f, then

$$f_{t,p} = f_0 + \alpha_f \left[T - T_g(0) \right] - \kappa_f P \tag{8.83}$$

where $T_g(0)$ refers to the glass transition at zero pressure. Under particular glass transition temperature and pressure conditions, $f_{t,p} = f_0$ and equation (8.83) becomes

$$\alpha_f \left[T_g - T_g(0) \right] = \kappa_f P \tag{8.84}$$

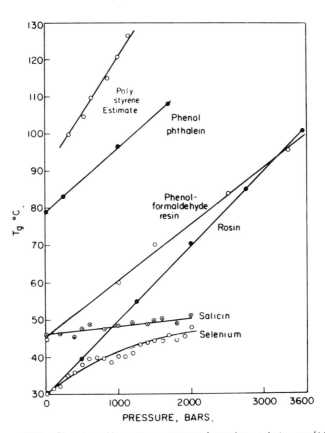

Figure 8.33 Glass transition versus pressure for various substances (110).

By differentiating with respect to pressure (111–114)

$$\left(\frac{\partial T_g}{\partial P}\right)_f = \frac{\Delta \kappa_f}{\Delta \alpha_f} \tag{8.85}$$

a relation strongly reminiscent of Ehrenfest's (73) relation for the change of a second-order transition temperature with pressure,

$$\frac{TV\Delta\alpha}{\Delta C_p} = \frac{\Delta\kappa}{\Delta\alpha} = \frac{dT_g}{dP} \tag{8.86}$$

where the Δ sign refers to changes from below to above T_g [see also equation (8.74)]. Several representative values of $\partial T_g/\partial P$ are shown in Table 8.14 (26). Since $\Delta\alpha \simeq \alpha_f \simeq 4.8 \times 10^{-4}$ deg^{-1}, $\kappa_f \simeq \Delta\kappa$ may be estimated. For polystyrene, Table 8.14 predicts a T_g rise of 31°C for a rise in pressure to 1000 atm, in agreement with Figure 8.33.

For polyurethanes, Quested et al. (115) found that $\Delta\kappa_f/\Delta\alpha_f$ was greater than dT_g/dP, except at pressures close to 1 bar. At high pressures, dT_g/dP reached a limiting value of 10.4°C/kbar. The effect of pressure has been studied for ultrasonic velocities (116) and fracture stress differences (117).

In the above, it was demonstrated that an increase in pressure can bring about vitrification. This result is important in engineering operations such as molding or extrusion, where operation too close to T_g (1 bar) can result in a stiffening of the material.

Table 8.14 **Pressure coefficients of the glass transition temperatures for selected materials (26)**

Material	$T_g(°C)$	dT_g/dP (K/atm)
Natural rubber	−72	0.024
Polyisobutylene	−70	0.024
Poly(vinyl acetate)	25	0.022
Rosin	30	0.019
Selenium	30	0.015–0.004[a]
Salicin	46	0.005
Phenolphthalein	78	0.019
Poly(vinyl chloride)	87	0.016
Polystyrene	100	0.031
Poly(methyl methacrylate)	105	0.020–0.023
Boron trioxide	260	0.020

[a]The variation is probably due to the different compressibilities of ring and chain material.

Thus we may refer to a glass transition pressure. In a broader sense, the glass transition is multidimensional. We could also refer to the glass transition molecular weight (Section 8.7), the glass transition concentration (for diluted or plasticized species), and so forth.

8.12 DAMPING AND DYNAMIC MECHANICAL BEHAVIOR

When the loss modulus or loss tangent is high, as in the glass transition region, the polymers are capable of damping out noise and vibrations, which, after all, are a particular form of dynamic mechanical motion. This section describes some of the aspects of behavior of a polymer under sinusoidal stresses at constant amplitude.

If an applied stress varies with time in a sinusoidal manner, the sinusoidal stress may be written

$$\sigma = \sigma_0 \sin \omega t \tag{8.87}$$

where ω is the angular frequency in radians, equal to $2\pi \times$ frequency. For Hookian solids, with no energy dissipated, the strain is given by

$$\varepsilon = \varepsilon_0 \sin \omega t \tag{8.88}$$

For real materials, the stress and strain are not in phase, the strain lagging behind the stress by the phase angle δ. The relationships among these parameters are illustrated in Figure 8.34. Of course, the phase angle defines an in-phase and out-of-phase component of the stress, σ' and σ'', as defined in Section 8.1.

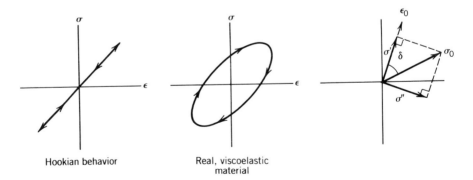

Hookian behavior   Real, viscoelastic
                         material

Figure 8.34 Simple dynamic relationships between stress and strain, illustrating the role of the phase angle.

Then the relationships between the in-phase and out-of-phase components and δ are given by

$$\sigma' = \sigma_0 \cos \delta \tag{8.89}$$

$$\sigma'' = \sigma_0 \sin \delta \tag{8.90}$$

The dynamic moduli may now be written

$$E' = \frac{\sigma'}{\varepsilon_0} = E^* \cos \delta \tag{8.91}$$

$$E'' = \frac{\sigma''}{\varepsilon_0} = E^* \sin \delta \tag{8.92}$$

$$E^* = \frac{\sigma_0}{\varepsilon_0} = \left(E'^2 + E''^2 \right)^{1/2} \tag{8.93}$$

In terms of complex notation,

$$E^* = E' + iE'' \tag{8.94}$$

(see Section 8.1) and

$$E = |E^*| = \left(E'^2 + E''^2 \right)^{1/2} \tag{8.95}$$

where, of course, $E''/E' = \tan \delta$.

Again, the logarithmic decrement, Δ, may be defined as the natural logarithm of the amplitude ratio between successive vibrations (see Section 8.3.4). This is a measure of damping. The quantity $\tan \delta$ is related to the logarithmic decrement,

$$\Delta \cong \pi \tan \delta$$

$$\Delta \cong \frac{\pi E''}{E'} \tag{8.96}$$

The heat energy per unit volume per cycle is $H = \pi E'' \varepsilon_0^2$, where ε_0 is the maximum amplitude, and Hooke's law is valid. Further relationships are given in Section 10.2.4, where damping is related to models. Hence both the loss tangent and the logarithmic decrement are proportional to the ratio of energy dissipated per cycle to the maximum potential energy stored during a cycle.

These loss terms are at a maximum near the glass transition or a secondary transition. This phenomenon is widely used in engineering for the construction of objects subject to making noise or vibrations. Properly protected by high damping polymers, car doors close more quietly, motors make less noise, and mechanical damage to bridges through vibrations is reduced.

Figure 8.35 Damping as a function of temperature for several polymers. Composition A is an IPN. Note that it damps nearly evenly over a broad temperature range. Compositions B – F are various homopolymers and copolymers that damp over narrow temperature ranges (118).

It must be mentioned that the glass transition can be broadened or shifted through various chemical and physical means. These include the use of plasticizers, fillers, or fibers, or through the formation of interpenetrating polymer networks. In this last case, the very small phases formed and molecular chains trapped by cross-links promote juxtaposition of the two molecular species. Then flexible molecules "rub up against" stiffer molecules, and a very broad glass transition may result, spanning the range of the two homopolymer transitions. Thus with *cross*–poly(ethyl acrylate)–*inter*–*cross*–poly(methyl methacrylate), which is composed of chemically isomeric polymers, glass transition behavior 100°C wide is obtained (28, 118, 119) (see Figure 8.35) (118). Conversely, objects such as rubber tires must be built of low damping elastomers, lest they overheat in service and blow out.

8.13   DEFINITIONS OF ELASTOMERS, PLASTICS, ADHESIVES, AND FIBERS

This chapter began with an overall approach to the behavior of polymers as a function of temperature. Cast in another way, at ambient temperatures, a polymer will be above, below, or at its glass transition temperature, with concomitant properties. Other ways of dividing polymers are according to the presence or absence of crystallinity (Chapter 6) or according to the presence or absence of cross-linking (Chapter 9).

In a certain simplified sense, definitions are given below that will help identify polymers found in the everyday world with their scientific properties.

1. An elastomer is a cross-linked, amorphous polymer above its T_g. An example is the common rubber band.

2. An adhesive is a linear or branched amorphous polymer above its T_g. It must be able to flow on a molecular scale to "grip" surfaces. (This definition is not to be confused with polymerizable adhesive materials, present in monomeric form. These are "tacky" or "sticky" only in the partly polymerized state. Frequently, they are cross-linked "thermoset," finally. Contact with the surface to be adhered must be made before gelation, in order to work.) An example is the postage stamp adhesive, composed of linear poly(vinyl alcohol), which is plasticized by water (or saliva) from below its T_g to above its T_g. On migration of the water away from the adhesive surface, it "sticks."

3. A plastic is usually below its T_g if it is amorphous. Crystalline plastics may be either above or below their T_g's.

4. Fibers are always composed of crystalline polymers. Apparel fiber polymers are usually close to T_g at ambient temperatures to allow flexibility in their amorphous portions.

5. House coatings and paints based on either oils or latexes are usually close to T_g at ambient temperatures. A chip of such material is usually flexible but not rubbery, being in the leathery region (see Section 8.2.2).

REFERENCES

1. F. Rodriguez, *Principles of Polymer Systems*, 2nd ed., McGraw-Hill, New York, 1982, Chaps. 8 and 15.

2. J. D. Ferry, *Viscoelastic Properties of Polymers*, 3rd ed., Wiley, New York, 1980, Chap. 1

3. A. V. Tobolsky, *Properties and Structure of Polymers*, Wiley, New York, 1960, pp. 71–78.

3a. G. V. Vinogradov, E. A. Ozyara, A. Y. Malkin, and V. A. Grechanovskii, *J. Polym. Sci.*, **A-2**(9), 1153(1971).

4. J. J. Aklonis, W. J. MacKnight, and M. C. Shen, *Introduction to Polymer Viscoelasticity*, Wiley, New York, 1972, (a) Chap. 4.

4a. H. Breuer, NATO Short Course on Polymer Blends, London, England, July, 1984.

5. A. V. Tobolsky and J. R. McLoughlin, *J. Polym. Sci.*, **8**, 543(1952).

6. J. J. Aklonis, *J. Chem. Ed.*, **58**(11), 892(1981).

7. R. F. Boyer, in *Encyclopedia of Polymer Science and Technology*, Suppl. Vol. 2, N. M. Bikales, Ed., Interscience, New York, 1977, p. 745; (a) pp. 822–823.

8. J. E. Lennard-Jones, *Proc. R. Soc. Lond.*, **A106**, 463(1924).

9. D. Katz and I. G. Zervi, *J. Polym. Sci.*, **46C**, 139(1974).

10. D. Katz and G. Salee, *J. Polym. Sci.*, **A-2**(6), 801(1968).

11. K. Ueberreiter and G. Kanig, *J. Chem. Phys.*, **18**, 399(1950).

12. G. M. Martin and L. Mandelkern, *J. Res. Natl. Bur. Stand.*, **62**, 141(1959).

13. J. Brandruys and E. H. Immergut, Eds., *Polymer Handbook*, 2nd ed., Wiley, New York, 1975; III-139; (a) III-150.

14. A. V. Tobolsky and H. Yu, unpublished, see ref. 3a.

15. H. J. Stern, *Rubber: Natural and Synthetic*, 2nd ed., Palmerton Publishing, New York, 1967, Chap. 1.

15a. A. F. Yee and M. T. Takemori, *J. Polym. Sci. Polym. Phys. Ed.*, **20**, 205(1982).

15b. G. O. Jones, *Glass*, Methuen, London, 1956.

15c. R. F. Boyer and R. S. Spencer, *J. Appl. Phys.*, **16**, 594(1945).

16. L. H. Sperling, *J. Polym. Sci. Polym. Symp.*, **60**, 175(1977).

16a. M. C. O. Chang, D. A. Thomas, and L. H. Sperling, *J. Polym. Sci. Part B Polym. Phys.*, **26**, 1627(1988).

17. P. Meares, *Trans. Faraday Soc.*, **53**, 31(1957).

18. F. C. Chen, C. L. Choy, S. P. Wong, and K. Young, *Polymer*, **21**, 1139(1980).

19. J. M. Widmaier and G. C. Meyer, *Macromolecules*, **14**, 450(1981); *Rubber Chem. Technol.*, **54**(5), 940(1981).

20. C. Kow, M. Morton, and L. J. Fetters, *Rubber Chem. Technol.*, **55**(1), 245(1982).

21. K. C. Frisch, D. Klempner, S. Migdal, H. L. Frisch, and H. Ghiradella, *Polym. Eng. Sci.*, **14**, 76(1974).

22. J. K. Yeo, L. H. Sperling, and D. A. Thomas, *Polym. Eng. Sci.*, **21**(11), 696(1981).

22a. C. M. Roland and C. A. Trask, in "Sound and Vibration Damping with Polymers," R. D. Corsaro and L. H. Sperling, Eds., ACS Books, Washington, D.C., 1990.

23. S. L. Kim, M. Skibo, J. A. Manson, and R. W. Hertzberg, *Polym. Eng. Sci.*, **17**(3) 194(1977).

24. J. K. Gillham, *Polym. Eng. Sci.*, **16**, 353(1976).

25. L. E. Nielsen and R. Buchdahl, *SPE J.*, **9**, 16(1953).

25a. R. P. Kusy and A. R. J. Greenberg, *J. Thermal Anal.*, **18**, 117(1980).

26. M. C. Shen and A. Eisenberg, *Prog. Solid State Chem.*, **3**, 407(1966); *Rubber Chem. Technol.*, **43**, 95(1970). *Rubber Chem. Technol.*, **43**, 156(1970).

27. A. Eisenberg and M. Shen, *Rubber Chem. Technol.*, **43**, 156(1970).

27a. W. Meesiri, J. Menczel, U. Guar, and B. Wunderlich, *J. Polym. Sci. Polym. Phys. Ed.*, **20**, 719(1982).

27b. L. P. Blanchard, J. Hesse, and S. L. Malhorta, *Can. J. Chem.*, **52**, 3170(1974).

27c. (a) M. Takayanagi, *Proc. Polym. Phys.* (*Jpn.*), 1962–1965. (b) Toya Baldwin Co., Ltd. Rheovibron Instruction Manual, 1969.

28. L. H. Sperling, in *Sound and Vibration Damping with Polymers*, R. D. Corsaro and L. H. Sperling, Eds., ACS Books Symp. Ser. 424, American Chemical Society, Washington, DC, 1990.

29. Bordon Award Symposium honoring J. K. Gillham, all the papers in *Polym. Eng. Sci.*, **19**(10) (1979).

30. J. K. Gillham, *Polym. Eng. Sci.*, **19**, 749(1979).

31. J. K. Gillham, *CRC Crit. Rev. Macromol. Sci.*, **1**, 83(1972).

32. J. K. Gillham, *AICHE J.*, **20**, 1066(1974).

33. J. Russell and R. G. Van Kerpel, *J. Polym. Sci.*, **25**, 77(1957).

34. A. F. Klarman, A. V. Galanti, and L. H. Sperling, *J. Polym. Sci.*, **A-2**(7), 1513(1969).

35. C. J. Malm, J. W. Mench, D. L. Kendall, and G. D. Hiatt, *Ind. Eng. Chem.*, **43**, 688(1951).

36. L. Banks and B. Ellis, *J. Polym. Sci. Polym. Phys. Ed.*, **20**, 1055(1982).

37. H. G. Elias, *Macromolecules: Structure and Properties*, Vol. 1, Plenum, New York, 1977, Chap. 10.

38. N. Saito, K. Okano, S. Iwayanagi, and T. Hideshima, in *Solid State Physics*, Vol. 14, F. Seitz and D. Turnbull, Eds., Academic Press, Orlando, 1963, p. 344.

39. N. G. McCrum, B. E. Read, and G. Williams, *Anelastic and Dielectric Effects in Polymeric Solids*, Wiley, New York, 1967.

39a. Figure 10.35.

40. (a) T. F. Schatzki, *J. Polym. Sci.*, **57**, 496(1962); (b) J. J. Aklonis and W. J. MacKnight, *Introduction to Polymer Viscoelasticity*, 2nd ed., Wiley-Interscience, New York, 1983, p. 81.

41. H. A. Flocke, *Kolloid Z.*, **180** 118(1962).

41a. H. S. Gutawsky, G. B. Kistiakowsky, G. E. Pake, and E.M. Purcell, *J. Chem. Phys.*, **17**(10), 972(1949); see also J. G. Powles and H. S. Gutowsky, *J. Chem. Phys.*, **21**, 1695(1953), and W. P. Slichter and E. R. Mandell, *J. Appl. Phys.*, **29**, 1438(1958).

41b. J. V. Koleske and J. A. Faucher, *Polym. Eng. Sci.*, **19**(10), 716(1979).

42. S. J. Stadnicki, J. K. Gillham, and R. F. Boyer, *J. Appl. Polym. Sci.*, **20**, 1245(1976).

43. J. K. Gillham, J. A. Benci, and R. F. Boyer, *Polym. Eng. Sci.*, **16**, 357(1976).

44. R. F. Boyer, *Polym. Eng. Sci.*, **19**(10), 732(1979).

45. S. Hedvat, *Polymer*, **22**, 774(1981).

46. G. D. Patterson, H. E. Bair, and A. Tonelli, *J. Polym. Sci. Polym. Symp.*, **54**, 249(1976).

47. L. E. Nielsen, *Polym. Eng. Sci.*, **17**, 713(1977).

48. R. M. Neumann, G. A. Senich, and W. J. MacKnight, *Polym. Sci. Eng.*, **18**, 624(1978).

49. J. Heijboer, *Polym. Eng. Sci.*, **19**(10), 664(1979).

50. A. J. Kovacs, *J. Polym. Sci.*, **30**, 131(1958).

50a. E. Catsiff and A. V. Tobolsky, *J. Polym. Sci.*, **19**, 111(1956).

50b. V. B. Rele and J. J. Aklonis, *J. Polym. Sci.*, **46C**, 127(1974).

50c. K. C. Lin and J. J. Aklonis, *Polym. Sci. Eng.*, **21**, 703(1981).

50d. J. J. Aklonis, *IUPAC Proceedings*, University of Massachusetts, Amherst, July 12–16, 1982, p. 834.

50e. P. E. Rouse, *J. Chem. Phys.*, **21**, 1272(1953).

50f. F. Bueche, *J. Chem. Phys.*, **22**, 603(1954).

50g. B. H. Zimm, *J. Chem. Phys.*, **24**, 269(1956).

50h. A. V. Tobolsky and D. B. DuPre, *Adv. Polym. Sci.*, **6**, 103(1969).

50i. A. V. Tobolsky and J. J. Aklonis, *J. Phys. Chem.*, **68**, 1970(1964).

51. R. F. Boyer, in *Encyclopedia of Polymer Science and Technology*, Vol. 13, N. M. Bikales, Ed., Interscience, New York, 1970, p. 277.

51a. R. Boyer, in *Encyclopedia of Polymer Science Technology* Suppl. Vol. II, p. 765, Wiley, New York, 1977.

52. W. Wrasidlo, *Thermal Analysis of Polymers, Advances in Polymer Science*, Vol. 13, Springer-Verlag, New York, 1974, p. 3.

53. H. Eyring, *J. Chem. Phys.*, **4**, 283(1936).

54. T. G. Fox and P. J. Flory, *J. Appl. Phys.*, **21**, 581(1950); T. G. Fox and P. J. Flory, *J. Polym. Sci.*, **14**, 315(1954).

55. R. Simha and R. F. Boyer, *J. Chem. Phys.*, **37**, 1003(1962).

56. A. K. Doolittle, *J. Appl. Phys.*, 1471(1951).

56a. R. Simha and C. E. Weil, *J. Macromol. Sci. Phys.*, **B4**, 215(1970).

56b. R. F. Boyer and R. Simha, *J. Polym. Sci.*, **B11**, 33(1973).

56c. S. C. Sharma, L. Mandelkern, and F. C. Stehling, *J. Polym. Sci.*, **B10**, 345(1972).

57. D. Turnbull and M. H. Cohen, *J. Chem. Phys.*, **31**, 1164(1959).

58. D. Turnbull and M. H. Cohen, *J. Chem. Phys.*, **34**, 120(1961).

59. F. Bueche, *J. Chem. Phys.*, **21**, 1850(1953).

60. F. Bueche, *J. Chem. Phys.*, **24**, 418(1956).

61. F. Bueche, *J. Chem. Phys.*, **30**, 748(1959).

62. M. L. Williams, R. F. Landel, and J. D. Ferry, *J. Am. Chem. Soc.*, **77**, 3701(1955).

63. E. H. Andrews, *Fracture in Polymers*, American Elsevier, New York, 1968, pp. 9–16.

64. B. Wunderlich, D. M. Bodily, and M. H. Kaplan, *J. Appl. Phys.*, **55**(1), 95(1964).

65. M. V. Vol'kenshtein and O. B. Ptitsyn, *Z. Tekh, Fiz.*, **26**, 2204(1956); *Soviet Phys. Tech. Phys.*, **1**, 2138(1957).

66. A. Q. Tool, *J. Am. Ceram. Soc.*, **29**, 240(1946).

67. H. Eyring, *Chem. Phys.*, **4**, 283(1936).

68. N. Hirai and H. Eyring, *J. Appl. Phys.*, **29**, 810(1958).

69. N. Hirai and H. Eyring, *J. Polym. Sci.*, **37**, 51(1959).

70. B. Wunderlich, *J. Phys. Chem.*, **64**, 1052(1960).

71. G. Kanig, *Kolloid Z. Z. Polym.*, **233**, 829(1969).

72. T. Nose, *Polym. J.*, **2**, 437(1971).

72a. S. Glasstone, K. J. Laidler, and H. Eyring, *The Theory of Rate Processes*, McGraw-Hill, New York, 1941.

73. P. Ehrenfest, *Leiden Comm. Suppl.*, 756(1933).

74. J. H. Gibbs, *J. Chem. Phys.*, **25**, 185(1956).

75. J. H. Gibbs and E. A. DiMarzio, *J. Chem. Phys.*, **28**, 373(1958).

76. J. H. Gibbs, in *Modern Aspects of the Vitreous State*, J. D. Mackenzie, Ed., Butterworth, London, 1960.

77. E. A. DiMarzio and J. H. Gibbs, *J. Polym. Sci.*, **A1**, 1417(1963).

78. E. A. DiMarzio, *J. Res. Natl. Bur. Stds.*, **68A**, 611(1964).

79. P. J. Flory, *Proc. R. Soc. (Lond.)*, **A234**, 60(1956).

80. E. A. DiMarzio and J. H. Gibbs, *J. Polym. Sci.*, **40**, 121(1959).

81. E. A. DiMarzio and J. H. Gibbs, *J. Polym. Sci.*, **1A**, 1417(1963).

82. G. Adam and J. H. Gibbs, *J. Chem. Phys.*, **43**, 139(1965).

83. J. H. Glans and D. T. Turner, *Polymer*, **22**, 1540(1981).

84. E. Jenckel and K. Ueberreiter, *Z. Phys. Chem.*, **A182**, 361(1938).

84a. J. M. G. Cowie, *Eur. Polym. J.*, **11**, 297(1975).

84b. J. K. Gillham, *Polym. Eng. Sci.*, **19**, 676(1979).

84c. J. K. Gillham, in *Development in Polymer Characterization–3*, J. V. Dawkins, Ed., Applied Science Publishers, London, 1982, Chap. 5, pp. 159–227.

84d. J. K. Gillham, in *The Role of the Polymer Matrix in the Processing and Structural Properties of Composite Materials*, J. C. Seferis and L. Nicolais, Eds., Plenum, New York, 1983, pp. 127–145.

84e. J. B. Enns and J. K. Gillham, in *Polymer Characterization: Spectroscopic, Chromatographic, and Physical Instrumental Methods*, C. D. Craver, Ed., Advances in Chemistry Series No. 203, American Chemical Society, Washington, DC, 1983, pp. 27–63.

84f. J. B. Enns and J. K. Gillham, *J. Appl. Polym. Sci.*, **28**, 2567(1983).

84g. J. K. Gillham, *Encyclopedia of Polym. Sci. Tech.*, **4**, 519(1986).

85. P. R. Couchman, *Macromolecules*, **11**, 1156(1978).

86. P. R. Couchman, *Polym. Eng. Sci.*, **21**, 377(1981).

87. P. R. Couchman, *J. Mater. Sci.*, **15**, 1680(1980).

88. P. R. Couchman and F. E. Karasz, *Macromolecules*, **11**, 117(1978).

89. J. M. Bardin and D. Patterson, *Polymer*, **10**, 247(1969); L. A. Wood, *J. Polym. Sci.*, **28**, 319(1958).

90. R. F. Boyer, *J. Macromol. Sci. Phys.*, **7**, 487(1973).

91. T. G. Fox, *Bull. Am. Phys. Soc.*, **1**, 123(1956).

92. J. R. Fried, F. E. Karasz, and W. J. MacKnight, *Macromolecules*, **11**, 150(1978).

93. J. M. Pochan, C. L. Beatty, and D. F. Hinman, *Macromolecules*, **11**, 1156(1977).

94. J. A. Manson and L. H. Sperling, *Polymer Blends and Composites*, Plenum, New York, 1976.

95. M. Matsuo, *Jpn. Plastics*, **2**, 6(1958).

96. M. Matsuo, T. Ueno, H. Horino, S. Chujya, and H. Asai, *Polymer*, **9**, 425(1968).

97. P. R. Scarito and L. H. Sperling, *Polym. Eng. Sci.*, **19**, 297(1979).

98. G. T. Davis and R. K. Eby, *J. Appl. Phys.*, **44**, 4274(1973).

99. F. C. Stehling and L. Mandelkern, *Macromolecules*, **3**, 242(1970).

100. K. H. Illers, *Kolloid Z. Z. Polym.*, **190**, 16(1963); **231**, 622(1969); **250**, 426(1972).

101. G. Champetier and J. P. Pied, *Makromol. Chem.*, **44**, 64(1961).

102. R. F. Boyer, *J. Appl. Phys.*, **25**, 825(1954).

103. R. F. Boyer, *Rubber Chem. Technol.*, **36**, 1303(1963).

104. J. A. Shetter, *Polym. Lett.*, **1**, 209(1963).

105. S. S. Rogers and L. Mandelkern, *J. Phys. Chem.*, **61**, 985(1957).

106. W. G. Barb, *J. Polym. Sci.*, **37**, 515(1957).

107. S. Bywater and P. M. Toporawski, *Polymer*, **13**, 94(1972).

108. F. E. Karasz and W. T. MacKnight, *Macromolecules*, **1**, 537(1968).

109. G. Tamman and W. Jellinghaus, *Ann. Phys.*, [5] **2**, 264(1929).

110. A. Eisenberg, *J. Phys. Chem.*, **67**, 1333(1963).

111. J. D. Ferry and R. A. Stratton, *Kolloid Z.*, **171**, 107(1960).

112. J. M. O'Reilly, *J. Polym. Sci.*, **57**, 429(1962).

113. J. E. McKinney, H. V. Belcher, and R. S. Marvin, *Trans. Soc. Rheol.*, **4**, 347(1960).

114. M. Goldstein, *J. Chem. Phys.*, **39**, 3369(1963).

115. D. L. Quested, K. P. Pae, B. A. Newman, and J. I. Scheinbaum, *J. Appl. Phys.*, **51**(10), 5100(1980).

116. D. L. Quested and K. D. Pae, *Ind. Eng. Prod. Res. Dev.*, **22**, 138(1983).

117. Y. Kaieda and K. D. Pae, *J. Mater. Sci.*, **17**, 369(1982).

118. L. H. Sperling, T. W. Chiu, R. G. Gramlich, and D. A. Thomas, *J. Paint Technol.*, **46**, 47(1974).

119. J. A. Grates, D. A. Thomas, E. C. Hickey, and L. H. Sperling, *J. Appl. Polym. Sci.*, **19**, 1731(1975).

GENERAL READING

J. J. Aklonis and W. J. MacKnight, *Introduction to Polymer Viscoelasticity*, 2nd ed., Wiley-Interscience, New York, 1983.

J. A. Brydson, "Plastics Materials," 5th Ed., Butterworths, London, 1989.

R. D. Corsaro and L. H. Sperling, *Sound and Vibration Damping with Polymers*, ACS Books Symp. Ser. No. 424, American Chemical Society, Washington, DC, 1990.

J. D. Ferry, *Viscoelastic Properties of Polymers*, 3rd ed., Wiley, New York, 1980.

L. E. Nielsen, *Mechanical Properties of Polymers and Composites*, Vols. I and II, Dekker, New York, 1974.

J. P. Silbia, Ed., *A Guide to Materials Characterization and Chemical Analysis*, VCH, New York, 1988.

A. V. Tobolsky, *Properties and Structure of Polymers*, Wiley, New York, 1960.

STUDY PROBLEMS

1. Name the five regions of viscoelastic behavior, and give an example of a commercial polymer commonly used in each region.

2. Name and give a one-sentence definition of the three theories of the glass transition.

3. Polystyrene homopolymer has a $T_g = 100°C$, and polybutadiene has a $T_g = -90°C$. Estimate the T_g of a 50/50 w/w statistical copolymer, poly(styrene–*stat*–butadiene).

4. A new linear amorphous polymer has a T_g of $+10°C$. At $+25°C$, it has a melt viscosity of 6×10^8 poises. Estimate its viscosity at $40°C$.

5. Define the following terms: free volume; loss modulus; tan δ; stress relaxation; plasticizer; Schatzki crankshaft motions; A_t; WLF equation; compressibility; Young's modulus.

6. As the newest employee of Polymeric Industries, Inc., you are attending your first staff meeting. One of the company's most respected chemists is speaking:
   "Yesterday, we completed the preliminary evaluation of the newly synthesized thermoplastic, poly(wantsa cracker). The polymer has a melt viscosity of 1×10^6 poises at 140°C. Our characterization laboratory reported a glass transition temperature of 110°C."
   "You know our extruder works best at 2×10^3 poises," broke in the mechanical engineer, "and you know poly(wantsa cracker) degrades at 160°C. Therefore, we won't be able to use poly(wantsa cracker)!"
   As you reach for your trusty calculator to estimate the melt viscosity of poly(wantsa cracker) at 160°C to make a reasoned decision of your own, you realize suddenly that all eyes are on you.
   (a) What is the melt viscosity of poly(wantsa cracker) at 160°C?
   (b) Can Polymeric Industries use the polymer? If not, what can they do to the polymer to increase usability?
   (c) What is the structure of poly(wantsa cracker), anyway?

7. Draw a log E–temperature plot for a linear, amorphous polymer.
   (a) Indicate the position and name the five regions of viscoelastic behavior.
   (b) How is the curve changed if the polymer is semicrystalline?
   (c) How is it changed if the polymer is cross-linked?
   (d) How is it changed if the experiment is run faster—that is, if measurements are made after 1 sec rather than 10 sec?
   In parts (b), (c), and (d), separate plots are required, each change properly labeled. E stands for Young's modulus.

8. During a coffee break, two chemists, three chemical engineers, and an executive vice president began discussing plastics. "Now everyone knows that such materials as plastics, rubber, fibers, paints, and adhesives have very little in common," began the executive vice president. "For example, nobody manufactures plastics and paints with the same equipment. ..."
   Even though you are the most junior member of the group, you interrupt; "That last may be so, but all those materials are closely related because ..."

Complete the statement in 100 words or less. If you think that the above materials are in fact *not* related, you have 100 words to prove the executive vice president correct.

9. Briefly discuss the salient points in the derivation of the WLF equation.

10. Prepare a "box score" table, laying out the more important advantages and disadvantages of the three theories of the glass transition.

11. A new polymer has been synthesized in your laboratory, and you are proudly discussing the first property studies when your boss walks in. "We need a polymer with a cubic coefficient of thermal expansion of less than 4×10^{-4} deg^{-1} at 50°C. Can we consider your new stuff?" Your technician hands you the sheet of paper with available data:

$$T_g = 100°C$$

$$\alpha_R = 5.5 \times 10^{-4} \text{ deg}^{-1} \text{ at } 150°C$$

Your boss adds, "By the way, the Board meets in 30 minutes. Any answer by then would surely be valuable." You begin to tear your hair out by its roots, wondering how you can solve this one so fast without going back into the lab, since there really isn't much time.

12. The T_g of poly(vinyl acetate) is listed as 29°C. If 5 mol% of divinyl benzene is copolymerized in the polymer during polymerization, what is the new glass transition temperature?

13. Noting the instruments mentioned in Section 8.3, what instrument would you most like to have in your laboratory if you were testing (a) each of the three theories of T_g, (b) the molecular weight dependence of T_g, (c) the effect of cross-linking on T_g. Defend your choice.

14. A new atactic polymer has a T_g of 0°C. Your boss asks, "If we made the isotactic form, what is its melting temperature likely to be?" Suddenly, you remember that back in college you took physical polymer science . . . .

15. Rephrase the definitions in Section 8.13, and use other examples.

16. Your assistant rushes in with a new polymer. "It softens at 50°C," she says.
"Is it a glass transition or a melting temperature?" you ask.
"How would I know?" she answers. "I never took physical polymer science!"
Describe two simple but foolproof experiments to distinguish between the two possibilities.

17. A piece of polystyrene is placed under 100 atm pressure at room temperature. What is the fractional volume decrease?

18. Given: an epoxy/acrylate blend, 60/40 overall composition. The loss moduli are determined on the blend and pure components, with the

following results:

Temperature, °C

Questions: (a) What is the composition of each phase? (b) What is the volume fraction of each of the two phases? (All quantities in mass; assume densities are equal.)

19. A rubber ball is dropped from a height of 1 yard and bounces back 18 in. Assuming a perfectly elastic floor, approximately how much did the ball heat up? The heat capacity, C_p, of SBR rubber is about 1.83 kJ kg^{-1} · K^{-1}.

20. Write a 100–125-word essay on the importance of free volume in polymer science. This essay is to be accompanied by at least one figure, construction, or equation illustrating your thought train.

21. A new polymer was found to soften at 50°C. Several experiments were performed to determine if the softening was a glass transition or a melting point.
    (a) In interpreting the results for each experiment, was it a glass transition? a melting transition? cannot be determined for sure? or was there some mistake in the experiment?
    (b) What is your reasoning for each decision?

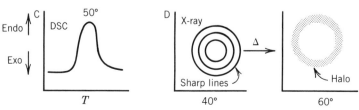

APPENDIX 8.1   MOLECULAR MOTION NEAR
THE GLASS TRANSITION[†]

The following experiment is intended for use as a classroom demonstration and takes about 30 min. Place the Superball® in the liquid nitrogen 30 min before class begins. Have a second such ball for comparison of properties.

Experiment

Time: Actual laboratory time about 1 h

Level: Physical Chemistry

Principles Illustrated:

1. The onset of molecular motion in polymers.
2. The influence of molecular motion on mechanical behavior.

Equipment and Supplies:

1 solid rubber ball (a small Superball® is excellent)

1 hollow rubber ball (optional)

1 Dewar flask of liquid nitrogen, large enough to hold above rubber balls

1 ladle or spoon with long handle, to remove frozen balls (alternately, tie balls with long string)

1 yardstick (or meter stick)

1 clock (a watch is fine)

1 hard surface, suitable for ball bouncing (most floors or desks are suitable)

First, place the yardstick vertical to the surface and drop Superball® from the top height; record percent recovery (bounce). Remove Superball® from liquid nitrogen, and record percent bounce immediately and each succeeding minute (or more often) for the first 15 min, then after 20 and 25 min and each 5 min thereafter until recovery equals that first obtained.

For the hollow rubber ball, first observe toughness at room temperature, then cool in liquid nitrogen, then throw hard against the floor or wall. Observe the glassy behavior of the pieces and their behavior as they warm up. Experimenter should wear safety glasses.

For extra credit, obtain three small dinner bells, coat two of them with any latex paint. (Most latex paints have T_g near room temperature.) After drying, ring the bells. Place one coated bell in a freezer, and compare its behavior cold to the room-temperature coated bell. How can you explain the difference observed?

[†]Reprinted in part from L. H. Sperling, *J. Chem. Ed.*, **59**, 942(1982).

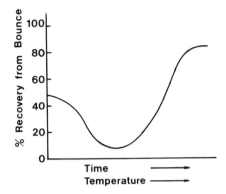

Figure A8.1.1   Ball bounce recovery.

Another related experiment involves dipping adhesive tape into liquid nitrogen. Outdoor gutter drain tapes are excellent because of their size. Compare the stickiness of the tape before and after freezing. (Note again the definition of an adhesive.)

On warming the frozen solid rubber ball, the percent recovery (bounce) versus time will go through a minimum at T_g, as shown in Figure A8.1.1.

Below T_g, the ball is glassy and bounces much like a marble. At T_g, the bounce is at a minimum owing to conversion of kinetic energy to heat. (The ball actually warms up slightly.) Above T_g, normal rubber elasticity and bounce characteristics are observed.

For an experiment emphasizing rolling friction near T_g, see G. B. Kauffman, S. W. Mason, and R. B. Seymour, *J. Chem. Ed.*, **67**, 199(1990).

9

CROSS-LINKED POLYMERS AND RUBBER ELASTICITY

9.1 CROSS-LINKS AND NETWORKS

During reaction, polymers may be cross-linked to several distinguishable levels. At the lowest level, branched polymers are formed. At this stage, the polymers remain soluble, sometimes known as the sol stage. As cross-links are added, clusters form, and cluster size increases. Eventually, the structure becomes infinite in size; that is, the composition gels. At this stage, a Maxwellian demon could, in principle, traverse the entire macroscopic system stepping on one covalent bond after another. Continued cross-linking produces compositions where, eventually, all the polymer chains are linked to other chains at multiple points, producing, in principle, one giant covalently bonded molecule. This is commonly called a polymer network.

9.1.1 The Sol – Gel Transition

The reaction stage referred to as the sol–gel transition (1–3) is called the gel point. At the gel point, the viscosity of the system becomes infinite, and the equilibrium modulus climbs from zero to finite values. In simple terms, the polymer goes from being a liquid to being a solid. There are three different routes for producing cross-linked polymers:

1. Step polymerization reactions, where little molecules such as epoxies (oxiranes) react with amines, or isocyanates react with polyols to form short, branched chains, eventually condensing it into epoxies or

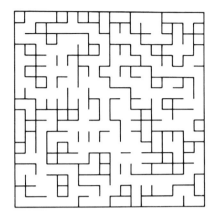

Figure 9.1 A square lattice example of percolation, at the gel point (6). Note structures that span the whole "sample."

polyurethanes, respectively. Schematically,

$$nA \sim A + mB \overset{B}{\underset{\wr}{\sim}} B \rightarrow$$

$$BA \sim AB \sim$$

$$\sim AB \overset{\wr}{\underset{}{\sim}} BA \sim AB \sim \qquad (9.1)$$

2. Chain polymerization, with multifunctional molecules present. An example is styrene polymerized with divinyl benzene.
3. Postpolymerization reactions, where a linear (or branched) polymer is cross-linked after synthesis is complete. An example is the vulcanization of rubber with sulfur, which will be considered further below.

The general features of structural evolution during gelation are described by percolation (or connectivity) theory, where one simply connects bonds (or fills sites) on a lattice of arbitrary dimension and coordination number (4–6). Figure 9.1 (6) illustrates a two-dimensional system at the gel point. It must be noted that gels at and just beyond the gel point usually coexist with sol clusters. These can also be seen in Figure 9.1. It is common to speak of the conversion factor, p, which is the fraction of bonds that have been formed between the mers of the system; see Section 3.7. For the two-dimensional schematic illustrated in Figure 9.1, $p = \frac{1}{2}$ yields the gel point.

9.1.2 Micronetworks

A special type of network exists, which involves only one or a few polymer chains. Thus microgels may form during specialized reaction conditions where only a few chains are interconnected.

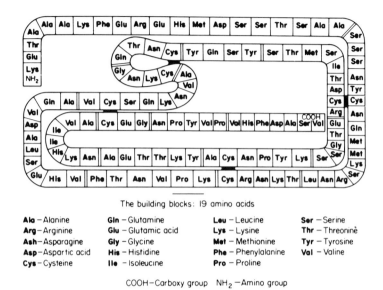

The building blocks: 19 amino acids

Ala – Alanine	**Gln** – Glutamine	**Leu** – Leucine	**Ser** – Serine
Arg – Arginine	**Glu** – Glutamic acid	**Lys** – Lysine	**Thr** – Threonine
Ash – Asparagine	**Gly** – Glycine	**Met** – Methionine	**Tyr** – Tyrosine
Asp – Aspartic acid	**His** – Histidine	**Phe** – Phenylalanine	**Val** – Valine
Cys – Cysteine	**Ile** – Isoleucine	**Pro** – Proline	

COOH – Carboxy group   NH$_2$ – Amino group

Figure 9.2 The structure of ribonuclease A, a micronetwork of one chain. Note cysteine – cysteine bonds (8).

Globular proteins constitute excellent examples of one-molecule micronetworks (7), where a single polymer chain is intramolecularly cross-linked; see Figure 9.2 (8). Here, disulfide bonds help keep the three-dimensional structure required for protein activity.

Such proteins may be denatured by any of several mechanisms, especially heat. Thus when globular proteins are cooked, the intramolecular cross-links become delocalized, forming intermolecular bonds instead. This is the major difference between raw egg white and hard-boiled egg white, for example.

Fully cross-linked on a macroscopic scale, polymer networks fall into different categories. It is convenient to call such polymers, which are used below their glass transition temperature, thermosets, which are usually plastics incapable of further flow. On the other hand, those networks that are above their glass transition temperature are rubbery unless very heavily cross-linked. While all these kinds of cross-linked polymers are important, the remainder of this chapter is devoted primarily to amorphous, continuous polymer networks above their glass transition temperature, the regime of rubber elasticity.

While the rubber elasticity theory to be described below presumes a randomly cross-linked polymer, it must be noted that each method of network formation described above has distinctive nonuniformities, which can lead to significant deviation of experiment from theory. For example, chain polymerization leads first to microgel formation (9, 10), where several chains bonded together remain dissolved in the monomer. On continued polymerization, the microgels grow in number and size, eventually forming a macroscopic gel. However, excluded volume effects, slight differences in reactivity between the monomer and cross-linker, and so on lead to systematic variations in cross-link densities at the 100–500-Å level.

9.2 HISTORICAL DEVELOPMENT OF RUBBER

A simple rubber band may be stretched several hundred percent; yet on being released, it snaps back substantially to its original dimensions. By contrast, a steel wire can be stretched reversibly for only about a 1% extension. Above that level, it undergoes an irreversible deformation and then breaks. This long-range, reversible elasticity constitutes the most striking property of rubbery materials. Rubber elasticity takes place in the third region of polymer viscoelasticity (see Section 8.2) and is especially concerned with cross-linked amorphous polymers in that region.

Columbus, on his second trip to America, found the American Indians playing a game with rubber balls (11, 12) made of natural rubber. These crude materials were un-cross-linked but of high molecular weight and hence were able to hold their shape for significant periods of time.

The development of rubber and rubber elasticity theory can be traced through several stages. Perhaps the first scientific investigation of rubber was by Gough in 1805 (13). Working with unvulcanized rubber, Gough reached three conclusions of far-reaching thermodynamic impact:

1. A strip of rubber warms on stretching and cools on being allowed to contract. (This experiment can easily be confirmed by a student using a rubber band. The rubber is brought into contact with the lips and stretched rapidly, constituting an adiabatic extension. The warming is easily perceived by the temperature-sensitive lips.)

2. Under conditions of constant load, the stretched length decreases on heating and increases on cooling. Thus it has more retractive strength at higher temperatures. This is the opposite of that observed for most other materials.

3. On stretching a strip of rubber and putting it in cold water, the rubber loses some of its retractile power, and its relative density increases. On warming, however, the rubber regains its original shape. In the light of present-day knowledge, this last set of experiments involved the phenomenon known as strain-induced crystallization, since unvulcanized natural rubber crystallizes easily under these conditions.

In 1844, Goodyear vulcanized rubber by heating it with sulfur (14). In modern terminology, he cross-linked the rubber. (Other terms meaning cross-linking include "tanning" of leather, "drying" of oil-based paints, and "curing" of inks.) Vulcanization introduced dimensional stability, reduced creep and flow, and permitted the manufacture of a wide range of rubber articles, where before only limited uses, such as waterproofing, were available (15). (The "MacIntosh" raincoat of that day consisted of a sandwich of two layers of fabric held together by a layer of unvulcanized natural rubber.)

Using the newly vulcanized materials, Gough's line of research was continued by Kelvin (16). He tested the newly established second law of thermodynamics with rubber and calculated temperature changes for adiabatic stretching. The early history of rubber research has been widely reviewed (17, 18).

All of the above, of course, was accomplished without an understanding of the molecular structure of polymers or of rubber in particular. Beginning in 1920, Staudinger developed his theory of the long-chain structure of polymers (19, 20). [Interestingly, Staudinger's view was repeatedly challenged, many investigators tenaciously adhering to ring formulas or colloid structures held together by partial valences (21).] See Appendix 5.1.

9.3 RUBBER NETWORK STRUCTURE

Once the macromolecular hypothesis of Staudinger was accepted, a basic understanding of the molecular structure was possible. Before cross-linking, rubber (natural rubber in those days) consists of linear chains of high molecular weight. With no molecular bonds between the chains, the polymer may flow under stress if it is above T_g.

The original method of cross-linking rubber, via sulfur vulcanization, results in many reactions. One such may be written

$$2 \sim CH_2-\underset{\underset{\displaystyle H_3C}{|}}{C}=CH-CH_2 \sim + \text{ sulfur}$$

$$\rightarrow \sim CH_2-\underset{\underset{\displaystyle |}{\overset{\displaystyle H_3C}{|}}}{\overset{\displaystyle |}{C}}-\underset{\overset{\displaystyle S-S-R}{|}}{CH}-CH_2 \sim \qquad (9.1a)$$

where R represents other rubber chains.

Two other methods of cross-linking polymers must be mentioned here. One is radiation cross-linking, with an electron beam or gamma irradiation. Using polyethylene as an example,

$$2(\sim CH_2-CH_2-CH_2-CH_2\sim)$$

$$\xrightarrow{h\nu} \begin{array}{c} \sim CH_2-CH-CH_2-CH_2\sim \\ | \\ \sim CH_2-CH-CH_2-CH_2\sim \end{array} +H_2 \qquad (9.2)$$

Another method involves the use of a multifunctional monomer in the simultaneous polymerization and cross-linking of polymers. Taking poly(ethyl acrylate) as an example, with divinyl benzene as cross-linker,

$$(9.3)$$

where the upper and lower reactions take place independently in time.

After cross-linking, flow of one molecule past another (viscoelastic behavior) is suppressed. Excluding minor impurities, an object such as a rubber band can be considered as one huge molecule. (It fulfills the two basic requirements of the definition of a molecule: (a) every atom is covalently bonded to every other atom, and (b) it is the smallest unit of matter with the characteristic properties of rubber bands.)

The structure of a cross-linked polymer may be idealized (Figure 9.3). The primary chains are cross-linked at many points along their length. For materials such as rubber bands, tires, and gaskets, the primary chains may have molecular weights of the order of 1×10^5 g/mol and be cross-linked (randomly) every $5-10 \times 10^3$ g/mol along the chain, producing 10–20 cross-links per primary molecule. It is convenient to define the average molecular weight between cross-links as M_c and to call chain portions bound at both ends by cross-link junctions active network chain segments.

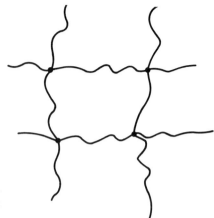

Figure 9.3 Idealized structure of a cross-linked polymer. Wavy lines, polymer chains; circles, cross-links.

In the most general sense, an elastomer may be defined as an amorphous, cross-linked polymer above its glass transition temperature (see Section 8.12). The two terms "rubber" and "elastomer" mean nearly the same thing. The term rubber comes from the "rubbing out" action of an eraser. Originally, of course, rubber was natural rubber, cis-polyisoprene. The term elastomer is more general and refers to the elastic-bearing properties of the materials.

9.4   RUBBER ELASTICITY CONCEPTS

The first relationships between macroscopic sample deformation, chain extension, and entropy reduction were expressed by Guth and Mark (22) and by Kuhn (23, 24) (see Section 5.3). Mark and Kuhn proposed the model of a random coil polymer chain (Figure 9.4) which forms an active network chain segment in the cross-linked polymer. When the sample was stretched, the chain has extended in proportion, now called an affine deformation. When the sample is relaxed, the chain has an average end-to-end distance, r_0 (Figure 9.4), which increases to r when the sample is stretched. (Obviously, if the sample is compressed or otherwise deformed, other chain dimensional changes will occur.)

Through the research of Guth and James (25–29), Treloar (30), Wall (31), and Flory (32), the quantitative relations between chain extension and entropy reduction were clarified. In brief, the number of conformations that a polymer chain can assume in space were calculated. As the chain is extended, the number of such conformations diminishes. (A fully extended chain, in the shape of a rod, has only one conformation, and its conformational entropy is zero.)

The idea was developed, in accordance with the second law of thermodynamics, that the retractive stress of an elastomer arises through the reduction

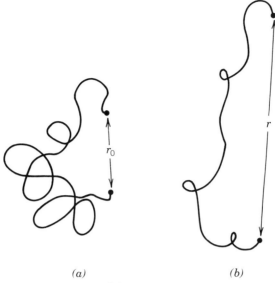

(a)   *(b)*

Figure 9.4 A network chain segment: (*a*) relaxed, with a random coil conformation, and (*b*) extended, owing to an external stress.

of entropy rather than through changes in enthalpy. Thus long-chain molecules, capable of reasonably free rotation about their backbone, and joined together in a continuous, monolithic network are required for rubber elasticity.

In brief, the basic equation relating the retractive stress, σ, of an elastomer in simple extension to its extension ratio, α, is given by

$$\sigma = nRT\left(\alpha - \frac{1}{\alpha^2}\right) \tag{9.4}$$

where the original length, L_0, is increased to L ($\alpha = L/L_0$), and RT is the gas constant[†] times the absolute temperature. The quantity n represents the number of active network chain segments per unit volume (33–36). The quantity n equals ρ/M_c, where ρ is the density and M_c the molecular weight between cross-links.

It will be observed that equation (9.4) is nonlinear; that is, the Hookean simple proportionality between stress and strain does not hold.

Equation (9.4) is compared to experiment in Figure 9.5 (37, 38). The theoretical value of M_c was chosen for the best fit at low extensions. The

[†]A convenient value of R for calculation purposes is 8.31×10^7 dynes · cm/mol · K. The stress then has units of dynes/cm^2.

Theory works at low extensions
Hooke's law doesn't work

Figure 9.5 Stress – strain behavior of lightly cross-linked natural rubber at 50°C. Curve (a), experimental. Theoretical is equation (9.4). Curve (c) illustrates the reversible nature of the extension up to $\alpha = 5.5$. At higher elongations, curve (b), hysteresis effects become important. The theoretical curve has been fitted to the experimental data in the region of small extensions, with $nRT = 0.39$ N / mm^2 (37, 38).

0% extension   100% extension

sharp upturn of the experimental data above $\alpha = 7$ is due to the limited extensibility of the chains themselves,[†] which can be explained in part by more advanced theories (36).

In the following sections, the principal equations of the theory of rubber elasticity are derived, emphasizing the relationships between molecular chain characteristics, stress, and strain.

9.5 THERMODYNAMIC EQUATION OF STATE

As a first approach to the equation of state for rubber elasticity, we analyze the problem via classical thermodynamics. The Helmholtz free energy, F, is given by

$$F = U - TS \tag{9.5}$$

where U is the internal energy and S is the entropy.

[†]Part of the effect is usually strain-induced crystallinity also, for natural rubber and *cis*-polybutadiene.

The retractive force, f, exerted by the elastomer depends on the change in free energy with length:

$$f = \left(\frac{\partial F}{\partial L}\right)_{T,V} = \left(\frac{\partial U}{\partial L}\right)_{T,V} - T\left(\frac{\partial S}{\partial L}\right)_{T,V} \qquad (9.6)$$

For an elastomer, Poisson's ratio is nearly 0.5, so the extension is nearly isovolume. When the experiment is done isothermally, the analysis becomes significantly simplified, noting the subscripts to equation (9.6).

According to the statistical thermodynamic approach to be developed below, each conformation that a network chain segment may take is equally probable. The number of such conformations depends on the end-to-end distance, r, of the chain, reaching a rather sharp maximum at r_0. The retractive force of an elastomer is developed by the thermal motions of the chains, statistically driven toward their most probable end-to-end distance, r_0.

The changes in numbers of chain conformations can be expressed as an entropic effect. Thus, for an ideal elastomer, $(\partial U/\partial L)_{T,V} = 0$.

By contrast, most other materials develop internal energy-driven retractive forces. For example, on extension the iron atoms in a steel bar are forced farther apart than normal, calling into play energy well effects and concomitant increased atomic attractive forces. Such a model assumes the opposite effect, $(\partial S/\partial L)_{T,V} = 0$.

As derived by Wall (39), there is a perfect differential mathematical relationship between the entropy and the retractive force:

$$-\left(\frac{\partial S}{\partial L}\right)_{T,V} = \left(\frac{\partial f}{\partial T}\right)_{L,V} \qquad (9.7)$$

Equation (9.6) can then be expressed as

$$f = \left(\frac{\partial U}{\partial L}\right)_{T,V} + T\left(\frac{\partial f}{\partial T}\right)_{L,V} \qquad (9.8)$$

which is sometimes called the thermodynamic equation of state for rubber elasticity.

Equation (9.8) can be analyzed with the aid of a construction due to Flory (18b) (see Figure 9.6), which illustrates a general experimental curve of force versus temperature. This line is extended back to 0 K. For an ideal elastomer, the quantity $(\partial U/\partial L)_{T,V}$ is zero, and the entropic portion (tangent) goes through the origin. Of course, the experimental line is straight in the ideal case, the slope being proportional to $-(\partial S/\partial L)_{T,V}$ or $(\partial f/\partial T)_{L,V}$.

The first term on the right of equation (9.8) expresses the energetic portion of the retractive force, f_e, and the second term on the right expresses

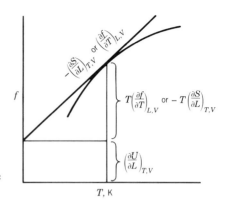

Figure 9.6   An analysis of the thermodynamic equations of state for rubber elasticity (18b).

the entropic portion of the force, f_s. Thus

$$f = f_e + f_s \tag{9.9}$$

Equations (9.8) and (9.9) call for stress–temperature (isometric) experiments (40).[†] While a detailed analysis of such experiments will be presented below in Section 9.10, Figure 9.7 (40) shows the results of such an isometric study. The quantity f_s accounts for more than 90% of the stress, whereas f_e hovers near zero. The turndown of f_e above 300% elongation may be due to incipient crystallization.

Incidentally, equation (9.4) fits Figure 9.7 much better than it does Figure 9.5. There are two reasons: (a) Figure 9.7 represents a much closer approach to an equilibrium stress–strain curve, and (b) the much higher level of sulfur used in the vulcanization (8% vs. 2%) reduces the quantity of crystallization at high elongations.

9.6   EQUATION OF STATE FOR GASES

The equation of state of rubber elasticity will now be calculated via statistical thermodynamics, rather than the classical thermodynamics of Section 9.5. Statistical thermodynamics makes use of the probability of finding an atom, segment, or molecule in any one place as a means of computing the entropy. Thus tremendous insight is obtained into the molecular processes of entropic phenomena, although classical thermodynamics illustrates energetic phenomena adequately.

However, most students not broadly exposed to statistical thermodynamics find such calculations difficult to follow at first. For this reason, we first

[†]Of course, the term "stress" refers to the force per unit initial cross section (see Section 8.1). Much of the early literature talks about force but measures stress.

Figure 9.7 The total retractive force and its entropic, f_s, and energetic components, f_e, as a function of elongation. Natural rubber vulcanized with 8% sulfur; values at 20° C (40).

derive the ideal gas law via the very same principles that are employed in calculating the stress–elongation relationships in rubber elasticity.

Consider a gas of ν molecules in an original volume of V_0 (18c). Let us calculate the probability Ω that all the gas molecules will move spontaneously to a smaller volume, V (see Figure 9.8).

The probability of finding one molecule in the volume V is given by

$$p_1 = V/V_0 \tag{9.10}$$

Neglecting the volume actually occupied by the molecules (an ideal gas assumption), the probability of finding two molecules in the volume V is given by

$$p_2 = (V/V_0)^2 \tag{9.11}$$

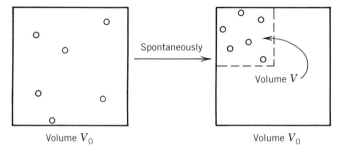

Figure 9.8   Ideal gas molecules spontaneously moving from a volume V_0 to a volume V.

and the probability of finding all the molecules (spontaneously) in the volume V is given by

$$\Omega = (V/V_0)^\nu \tag{9.12}$$

The change in entropy, ΔS, is given by the Boltzmann relation[†]

$$\Delta S = +k \ln \Omega \tag{9.13}$$

which yields

$$\Delta S = k\nu \ln(V/V_0) \tag{9.14}$$

The pressure of the gas is given by

$$P = -\left(\frac{\partial F}{\partial V}\right)_T = -\left(\frac{\partial U}{\partial V}\right)_T + T\left(\frac{\partial S}{\partial V}\right)_T \tag{9.15}$$

where F is the Helmholtz free energy. For a perfect gas, $(\partial U/\partial V)_T = 0$, and

$$P = T\left(\frac{\partial S}{\partial V}\right)_T = \frac{k\nu T}{V} \tag{9.16}$$

If moles instead of molecules are considered, k becomes R and ν becomes n, yielding the familiar

$$PV = nRT \tag{9.17}$$

[†]The Boltzmann hypothesis *assumes* rather than derives a logarithmic relationship between the probability of a state (or the number of such states, the inverse) and the entropy. Boltzmann's constant, k, is defined as the constant of proportionality.

Table 9.1   Corresponding concepts in ideal gases and ideal elastomers

$PV = nRT$	$G = nRT$
Entropy calculated from probabilities of finding n molecules in a given volume	Entropy calculated from probability of finding end-to-end distance r at r_0
Probability of the gas volume spontaneously decreasing $(\partial U/\partial V)_T = 0$; internal energy assumed zero	Probability of an elastomer strip spontaneously elongating $(\partial U/\partial L)_{T,V} = 0$; internal energy assumed constant
Molar volume of gas assumed zero	Elastomer assumed incompressible (molar volume is constant)
Pressure P given by $-(\partial F/\partial V)_T$	Retractive force f given by $-(\partial F/\partial L)_{T,V}$

If the internal energy is not required to be zero, more complex equations arise. For example, van der Waal's law per mole gives

$$P = -\frac{a}{V^2} + \frac{nRT}{V-b} \tag{9.18}$$

where the first term on the right indicates an energetic (attractive force) term and the second term on the right is the entropic term corrected for the molar volume of the gas (40).

Returning to Figure 9.8, if any large number of molecules are involved (say, 1 mol), the probability of the gas molecules spontaneously moving to a much smaller volume is very small. Of course, if they are arbitrarily so moved, a pressure P is required to keep them there.

In an analogous strip of rubber, the corresponding situation would be the spontaneous elongation of the strip (a rubber band stretching itself). On a molecular level, such motions are spontaneous, but because of the low probabilities involved, are but momentary. Again, the phenomenon is possible but unlikely. Instead of a pressure P to hold the gas in the volume V, a stress σ (force per unit area) will be required to keep the elastomer stretched from L_0 to L ($\alpha = L/L_0$).

In both problems, to the first approximation, the internal energy component can be assumed to be zero. Table 9.1 compares the concepts of an ideal gas with those of an ideal elastomer, to be developed below.

9.7   STATISTICAL THERMODYNAMICS OF RUBBER ELASTICITY

It is useful to consider again the freely jointed chain (Section 5.3.1). In this case, the root-mean-square end-to-end distance is given by $(\overline{r_f^2})^{1/2} = lx^{1/2}$. In a real random coil, with fixed bond angles, the quantity $(\overline{r_f^2})^{1/2}$ is larger but still obeys the $x^{1/2}$ relationship. For a given value of x, however, the root-mean-square end-to-end distance can vary widely, from zero, where the ends touch, to xl, the length of the equivalent rod. The probability of finding particular values of r underlies the following subsections.

9.7.1 The Equation of State for a Single Chain

It is convenient to divide the derivation of equations such as (9.4) into two parts. First, the equation of state for a single chain in space is derived. Then we show how a network of such chains behaves.

It is convenient to start again with the general equation for the Helmholtz free energy, equation (9.5):

$$F = U - TS$$

This can be rewritten in statistical thermodynamic notation:

$$F = \text{constant} - kT \ln \Omega(r, T) \tag{9.19}$$

where the quantity $\Omega(r, T)$ [see equations (9.12) and (9.13)] now refers to the probability that a polymer molecule with end-to-end distance r at temperature T will adopt a given conformation.[†]

From a quantitative point of view, at each particular end-to-end distance, all possible conformations of the chain need to be counted, holding the ends fixed in space. Then the sum of all such conformations as the end-to-end distance is varied needs to be calculated. (Later, the sum of all conformations of a distribution of molecular weights is considered.)

The retractive force is given by

$$f = \left(\frac{\partial F}{\partial r}\right)_{T,V} = -kT\left(\frac{\partial \ln \Omega(r, T)}{\partial r}\right)_{T,V} \tag{9.20}$$

As before, the quantity U, assumed to be constant (or zero), drops out of the calculation, leaving only the entropic contribution. In this case, for a single chain, the quantity f for force must be used. The cross section of the individual chain, necessary for a determination of the stress, remains undefined.

A particular direction in space is selected first. Then, using vector notation, the probability that r lies between r and $r + dr$ in that direction is

---

[†]The term "conformation" refers to those arrangements of a molecule that can be attained by rotating about single bonds. Configurations refer to tacticity, steric arrangements, cis and trans, and so on; see Chapter 2.

given by

$$W(r)\, dr = \frac{\Omega(r,T)\, dr}{\int_0^\infty \Omega(r,T)\, dr} \tag{9.21}$$

where the denominator serves as a normalizing factor. Of course, the integral does not need to extend to infinity; in reality different conformations only go to the fully extended, rodlike chain.

Removing the directional restrictions on r,

$$W(r)\, dr = \frac{\Omega(r,T)\, dr}{\int_0^\infty \Omega(r,T)\, dr} 4\pi r^2 \tag{9.22}$$

A spherical shell between r and $r + dr$ is generated (see Figure 9.9), depicted in Cartesian coordinates.

Rearranging equation (9.22) and taking logarithms,

$$\ln \Omega(r,T) = \ln W(r) + \ln \int_0^\infty \Omega(r,T)\, dr - \ln 4\pi - \ln r^2 \tag{9.23}$$

Differentiating with respect to r,

$$\left(\frac{\partial \ln \Omega(r,T)}{\partial r}\right)_{T,V} = \left(\frac{\partial \ln W(r)}{\partial r}\right)_{T,V} - \frac{2}{r} \tag{9.24}$$

since $\int_0^\infty \Omega(r,T)\, dr$ is independent of r.

Figure 9.9 A spherical shell at a distance r (inner surface) and $r + dr$ (outer surface), defining all conformations in space having that range of r.

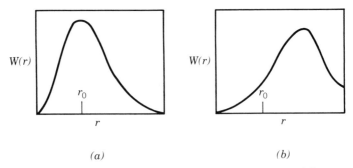

Figure 9.10 The radial distribution function $W(r)$ as a function of r for (a) a relaxed chain, where the most probable end-to-end distance is r_0, and (b) for a chain under an extensive force, f.

The quantity $W(r)$ can be expressed as a Gaussian distribution (18):

$$W(r) = \left(\frac{\beta}{\pi^{1/2}}\right)^3 e^{-\beta^2 r^2} 4\pi r^2 \tag{9.25}$$

In molecular terms, $\beta^2 = 3/(2\overline{r_0^2})$, where $\overline{r_0^2}$ represents the average of the squares of the relaxed end-to-end distances.

The quantity $W(r)$ is shown as a function of r in Figure 9.10. The radial distribution function $W(r)$ is shown for a relaxed chain and a chain extended by a force f.

Equation (9.24) can now be expressed:

$$\left(\frac{\partial \ln \Omega(r, T)}{\partial r}\right)_{T,V} = -2\beta^2 r \tag{9.26}$$

Substituting equation (9.26) into equation (9.20),

$$f = \frac{3kTr}{\overline{r_0^2}} \tag{9.27}$$

and the equation of state for a single chain is obtained. The force appears to be zero at $r = 0$, because of the spherical shell geometry assumed (Figure 9.9). Again, the quantity r_0 is the isotropic end-to-end distance for a free chain in space. This approximates the end-to-end distance expected both in θ-solvents and in the bulk state, for linear chains.

9.7.2 The Equation of State for a Macroscopic Network

Equation (9.27) expresses the retractive force of a single chain on extension. Since the retractive force is proportional to the quantity r, the chain behaves as a Hookean spring on extension. The problem now is to link a large

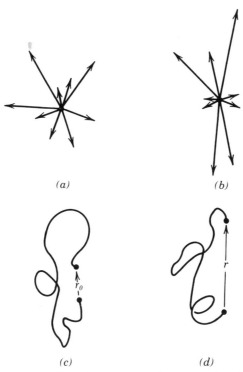

(a)   (b)

(c)   (d)

Figure 9.11 Illustration of an affine deformation, end-to-end vectors drawn from a central point. (a) End-to-end vectors have spherical symmetry. (b) After extending the macroscopic sample, end-to-end vectors have ellipsoidal symmetry. Parts (c) and (d) illustrate the corresponding effects on a single chain.

number, n, of these chains together per unit volume to form a macroscopic network.

The assumption of affine deformation is required; that is, the junction points between chains move on deformation as if they were embedded in an elastic continuum (Figure 9.11). The mean square length of a chain in the strained state is given by

$$\overline{r^2} = \tfrac{1}{3}\left(\alpha_x^2 + \alpha_y^2 + \alpha_z^2\right)\overline{r_i^2} \tag{9.28}$$

The alphas represent the fractional change in shape in the three directions, equal to $(L/L_0)_i$, where $i = x$, y, or z.

The quantity $\overline{r_i^2}$ represents the isotropic, unstrained end-to-end distance in the network. The two quantities $\overline{r_i^2}$ and $\overline{r_0^2}$ [see equation (9.27)] bear exact comparison. They represent the same chain in the network and uncross-linked states, respectively. Under many circumstances, the quantity $\overline{r_i^2}/\overline{r_0^2}$ approximately equals unity. In fact, the simpler derivations of the equation of state for rubber elasticity do not treat this quantity, implicitly assuming it to be unity (36). However, deviations from unity may be caused by swelling, cross-linking while in tension, changes in temperature, and so on and play an important role in the development of modern theory.

The difference between r_i and r_0 may be illustrated by way of an example. Assume a cube of dimensions $1 \times 1 \times 1$ cm. Its volume, of course, is 1 cm³. If it is swelled to 10 cm³ volume, then each linear dimension in the sample is increased by $(10)^{1/3}$, the new length of the sides, in this instance. Assuming an affine deformation, the end-to-end distance of the chains will also increase by $(10)^{1/3}$; that is, $r_i = (10)^{1/3} r_0$. The value of r_0 does not change, because it is the end-to-end distance of the equivalent free chain. The value of r_i is determined by the distances between the cross-link sites binding the chain. Again, neither r_i nor r_0 represents the end-to-end distance of the whole primary chain but rather the end-to-end distance between cross-link junctions.

For the work done by the n network chains,

$$-W = \Delta F_{el} \tag{9.29}$$

where ΔF_{el} represents the change in the Helmholtz free energy due to elastic deformation. Returning to equation (9.27), for n chains,

$$\Delta F_{el} = \frac{3nRT}{r_0^2} \int_{(r_i^2)^{1/2}}^{(\overline{r^2})^{1/2}} r\, dr \tag{9.30}$$

Integrating and substituting equation (9.28) yields

$$\Delta F_{el} = \frac{nRT}{2} \frac{\overline{r_i^2}}{\overline{r_0^2}} \left(\alpha_x^2 + \alpha_y^2 + \alpha_z^2 - \alpha_{x_0}^2 - \alpha_{y_0}^2 - \alpha_{z_0}^2 \right) \tag{9.31}$$

By definition, $\alpha_{x_0}^2 = \alpha_{y_0}^2 = \alpha_{z_0}^2 = 1$, since these terms deal with the unstrained state.

Since it is assumed that there is no volume change on deformation (Poisson's ratio very nearly equals 0.5—i.e., an incompressible solid),

$$\alpha_x \alpha_y \alpha_z = 1 \tag{9.32}$$

If α_x is taken simply as α, then $\alpha_z = \alpha_y = 1/\alpha^{1/2}$, at constant volume.

After making the above substitutions, the work on elongation equation may be written

$$-W = \Delta F_{el} = \frac{nRT}{2} \frac{\overline{r_i^2}}{\overline{r_0^2}} \left(\alpha^2 + \frac{2}{\alpha} - 3 \right) \tag{9.33}$$

The stress is given by

$$\sigma = \left(\frac{\partial F}{\partial \alpha}\right)_{T,V} = nRT\frac{\overline{r_i^2}}{r_0^2}\left(\alpha - \frac{1}{\alpha^2}\right) \tag{9.34}$$

which is the equation of state for rubber elasticity. Equation (9.34) bears comparison with equation (9.4). The quantity $\overline{r_i^2}/r_0^2$ is known as the "front factor."

The quantity n in the above represents the number of active network chains per unit volume, sometimes called the network or cross-link density. The number of cross-links per unit volume is also of interest. For a tetrafunctional cross-link (see Figure 9.3), the number of cross-links is one-half the number of chains. (See Section 9.9.2.)

Several other relationships may be derived immediately. Young's modulus can be written

$$E = L\left(\frac{\partial \sigma}{\partial L}\right)_{T,V} \tag{9.35}$$

which yields

$$E = nRT\frac{\overline{r_i^2}}{r_0^2}(2\alpha^2 + 1/\alpha) \cong 3n\frac{\overline{r_i^2}}{r_0^2}RT \tag{9.36}$$

for small strains. This is the engineering modulus, which utilizes the actual cross-section at α, rather than the relaxed value (i.e., $\alpha = 1$). At low extensions, the two moduli are nearly identical. The shear modulus may be written

$$G = E/2(1 + \nu) \tag{9.37}$$

Poisson's ratio, ν, for rubber is approximately 0.5 (incompressibility assumption), so that

$$G = n\frac{\overline{r_i^2}}{r_0^2}RT \tag{9.38}$$

Then, to a good approximation,

$$\sigma = G(\alpha - 1/\alpha^2) \tag{9.39}$$

thus defining the work to stretch, the stress–strain relationships, and the modulus of an ideal elastomer. As equations (9.34) and (9.39) illustrate, the stress–strain relationships are non-Hookean; that is, the strain is not proportional to the stress. Again, these equations yield curves of the type illustrated in Figures 9.5 and 9.7.

Other moduli are occasionally used to characterize polymeric materials. Appendix 9.1 describes the use of the ball indentation method to characterize the cross-link density of gelatin. In general, this method can be used for sheet rubber and other large sample methods.

9.7.3   An Example of Rubber Elasticity Calculations

Suppose an elastomer of 0.1 cm × 0.1 cm × 10 cm is stretched to 25 cm length at 35°C, a stress of 2×10^7 dynes/cm^2 being required. What is the concentration of active network chain segments?

Use equation (9.34), assuming that $\overline{r_i^2} = \overline{r_0^2}$.

$$n = \frac{\sigma}{RT\left(\alpha - 1/\alpha^2\right)}$$

$$= \frac{2 \times 10^7 \text{ dynes/cm}^2}{8.31 \times 10^7 (\text{dynes} \cdot \text{cm/mol} \cdot \text{K}) \times 308 \text{ K} \times \left(2.5 - 1/2.5^2\right)}$$

noting that $\alpha = L/L_0 = 25/10 = 2.5$.

$$n = 3.34 \times 10^{-4} \text{ mol/cm}^3$$

Note that the rubber band in Appendix 9.2 has $n = 1.9 \times 10^{-4}$ mol/cm^3.

9.7.4   Comparison to Metal Springs

As described above, the elasticity in elastomers arises through the entropic straightening and recoiling of the polymer chains. This is substantially an isovolume phenomenon, Poisson's ratio being close to 0.5.

By contrast, the elasticity in metals arises through increases in the distances between atoms, removing the atoms from their equilibrium positions in the energy well. This is an enthalpic effect. The volume of such materials increases substantially on stretching, Poisson's ratio being about 0.35 for many metals.

Thus elasticity in solid materials can arise through two general mechanisms: reduction in entropy for elastomers and increases in internal energy for most other materials. As described in Section 9.2, a strip of rubber warms on stretching; a metal spring cools on stretching.

9.8   THE "CARNOT CYCLE" FOR ELASTOMERS

In elementary thermodynamics, the Carnot cycle illustrates the production of useful work by a gas in a heat engine. This section outlines the corresponding thermodynamic concepts for an elastomer and illustrates a demonstration experiment.

The conservation of energy for a system may be written

$$dU = V\,dp + T\,dS + \sigma\,dL + \cdots \qquad (9.40)$$

where the internal energy, U, is equated to as many variables as exist in the system. For an ideal gas (Section 9.6), P–V–T variables are selected. The corresponding variables for an ideal elastomer are σ–L–T [see equation (9.34)]. Since Poisson's ratio is nearly 0.5 for elastomers, the volume is substantially constant on elongation.

By carrying a gas, elastomer, or any material through the appropriate closed loop with a high- and low-temperature portion, they may be made to perform work proportional to the area enclosed by the loop. A system undergoing such a cycle is called a heat engine.

9.8.1   The Carnot Cycle for a Gas

In the Carnot heat engine, a gas is subjected to two isothermal steps, which alternate with two adiabatic steps, all of which are reversible (see Figure 9.12) (41). Briefly, the gas undergoes a reversible adiabatic compression from state 1 to state 2. The temperature is increased from T_1 to T_2. During this step, the surroundings do work $|\omega_{12}|$ on the gas. The absolute signs are used because conventions require that the signs on some of the algebraic quantities herein be negative.

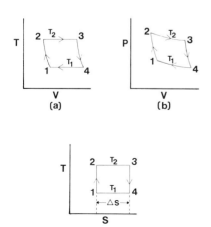

Figure 9.12   Carnot cycle for a gas (41).

Next, the gas undergoes a reversible isothermal expansion from state 2 to state 3. While expanding, the gas does work $|\omega_{23}|$ on the surroundings while absorbing heat $|q_2|$. Then there follows a reversible adiabatic expansion of the gas from state 3 to state 4, the temperature dropping from T_2 to T_1. During this step, the gas does work $|\omega_{34}|$ on the surroundings.

Lastly, there is an isothermal compression of the gas from state 4 to state 1 at T_1. Work $|\omega_{41}|$ is performed on the gas, and heat $|q_1|$ flows from the gas to the surroundings.

9.8.2 The Carnot Cycle for an Elastomer

For an elastomer, the rubber goes through a series of stress–length steps, two adiabatically and two isothermally, as in the Carnot cycle (see Figure 9.13) (42). Beginning at length L_1 and temperature T^I, a stress, σ, is applied stretching the elastomer adiabatically to L_2. The elastomer heats up to T^{II}. The quantity σ is related to the length by the nonlinear equation

$$\sigma = nRT\left[\frac{L}{L_0} - \left(\frac{L_0}{L}\right)^2\right] \tag{9.41}$$

[see equation (9.34)]. In this step work is done on the elastomer.

Figure 9.13 Thermal cycle for an elastomer (42).

Table 9.2   Heat and work equations for a Carnot cycle and a rubber thermal cycle

Gas	Elastomer
$\Delta U_{12} = \int_{V_1}^{V_2} P\,dV$	$\Delta U_{12} = -\int_{L_1}^{L_2} \sigma\,dL$
$\Delta U_{23} = 0 = T_2\,\Delta S + \int_{V_2}^{V_3} P\,dV$	$\Delta U_{23} = 0 = T^{II}\,\Delta S + \int_{L_2}^{L_3} \sigma\,dL$
$\Delta U_{34} = \int_{V_3}^{V_4} P\,dV$	$\Delta U_{34} = \int_{L_3}^{L_4} \sigma\,dL$
$\Delta U_{41} = -T_1\,\Delta S - \int_{V_4}^{V_1} P\,dV$	$\Delta U_{41} = -T^{I}\,\Delta S - \int_{L_4}^{L_1} \sigma\,dL$
Governing equations for entropy:	
$\Delta S = \dfrac{q}{T}$	$\Delta S = nR\left[\dfrac{L}{L_0} - \left(\dfrac{L_0}{L}\right)^2\right]$

At T^{II}, the elastomer is allowed to contract isothermally to L_3. It absorbs heat from its surroundings in this step and does work. As the length decreases, its entropy increases by ΔS (see Figure 9.13c). The elastomer then is allowed to contract adiabatically to L_4, doing work, and its temperature falls to T^{I} again. The length of the sample is then increased isothermally from L_4 to L_1, work being done on the sample, and heat is given off to its surroundings. This step completes the cycle.

The thermodynamic equations for each step of the gas and elastomer bear comparison (see Table 9.2). For the elastomer, again, stress replaces pressure, both with units of dynes/cm^2, and length replaces volume.

An increase in the volume of the gas, however, corresponds to a decrease in the length of a stretched elastomer. It is important to note that at no time does the elastomer come to its rest length, L_0. Interestingly, the corresponding "rest volume" of a gas is infinitely large.

9.8.3   Work and Efficiency

The equations governing the work done during the two cycles may also be compared. For a gas,

$$\omega_g = -\oint P\,dV \tag{9.42}$$

For an elastomer,

$$\omega_e = -\oint \sigma\,dL \tag{9.43}$$

In both cases, the cyclic integral measures the area enclosed by the four steps in Figures 9.12 and 9.13.

The efficiencies, $\bar{\eta}$, of the two systems may also be compared. For a gas,

$$\bar{\eta}_g = \frac{q_1 + q_2}{q_2} \tag{9.44}$$

where q_1 and q_2 are the heat absorbed and released (opposite signs), as above. For the elastomer,

$$\bar{\eta}_e = \frac{\oint \sigma \, dL}{Q_{II}} = \frac{(T^{II} - T^{I}) \, \Delta S}{Q_{II}} = \frac{Q_I + Q_{II}}{Q_{II}} \tag{9.45}$$

or in a different form,

$$\bar{\eta}_e = \frac{T^{II} - T^{I}}{T^{II}} \tag{9.46}$$

where Q_I and Q_{II} are the amounts of heat released to the low-temperature reservoir (T^{I}) and absorbed from the high-temperature reservoir (T^{II}), respectively.

While the entropy change is zero for either system during the reversible adiabatic steps (see Figures 9.12c and 9.13c), it must be emphasized that the entropy change is greater than zero for an irreversible adiabatic process. An example for an elastomer is "letting go" of a stretched rubber band.

9.8.4 An Elastomer Thermal Cycle Demonstration

The elastomer thermal cycle is demonstrated in Figure 9.14 (42). A bicycle wheel is mounted on a stand, with a source of heat on one side only. Stretched rubber bands replace the spokes. On heating, the stress that the stretched rubber bands exert is increased, so that the center of gravity of the

Figure 9.14 A thermally rotated wheel, employing an elastomer as the working substance (42).

wheel is displaced toward 9 o'clock in the drawing. The wheel then rotates counterclockwise (43).

Each of the steps in Figure 9.13 may be traced in Figure 9.14, although none of the steps in Figure 9.13 are purely isothermal or adiabatic, and then of course they are not strictly reversible. Steps 1 to 2 in Figure 9.13 occur at 6 o'clock in Figure 9.14, where there is a (near) adiabatic length increase due to gravity. At 3 o'clock, at T^{II}, heat is absorbed (nearly) isothermally, and the length decreases, doing work. At 12 o'clock, corresponding to steps 3 to 4, there is an adiabatic length decrease due to gravity. Lastly, at 9 o'clock, steps 4 to 1, there is a (nearly) isothermal length increase, and heat is given off to the surroundings at T^I, and work is done on the elastomer.

9.9 CONTINUUM THEORIES OF RUBBER ELASTICITY

9.9.1 The Mooney–Rivlin Equation

The statistical theory of rubber elasticity is based on the concepts of random chain motion and the restraining power of cross-links; that is, it is a molecular theory. Amazingly, similar equations can be derived strictly from phenomenological approaches, considering the elastomer as a continuum. The best known such equation is the Mooney–Rivlin equation (44–47),

$$\sigma = 2C_1\left(\alpha - \frac{1}{\alpha^2}\right) + 2C_2\left(1 - \frac{1}{\alpha^3}\right) \tag{9.47}$$

which is sometimes written in the algebraically identical form,

$$\sigma = \left(2C_1 + \frac{2C_2}{\alpha}\right)\left(\alpha - \frac{1}{\alpha^2}\right) \tag{9.48}$$

Equations (9.47) and (9.48) appear to be corrections to equation (9.34), with an additional term being added. However, they are derived from quite different principles. $\approx G$

According to equation (9.34), the quantity $\sigma/(\alpha - 1/\alpha^2)$ should be a constant. Equation (9.48), on the other hand, predicts that this quantity depends on α:

$$\frac{\sigma}{(\alpha - 1/\alpha^2)} = 2C_1 + \frac{2C_2}{\alpha} \tag{9.49}$$

Plots of $\sigma/(\alpha - 1/\alpha^2)$ versus $1/\alpha$ are found to be linear, especially at low elongation (see Figure 9.15) (48). The intercept on the $\alpha^{-1} = 0$ axis yields $2C_1$, and the slope yields $2C_2$. The value of $2C_1$ varies from 2 to 6 kg/cm^2,

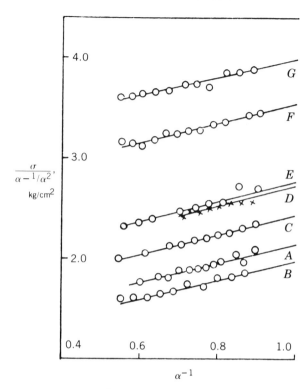

Figure 9.15 Plot of $\sigma / (\alpha - 1 / \alpha^2)$ versus α^{-1} for a range of natural rubber vulcanizates. Sulfur content increases from 3 to 4%, with time of vulcanization and other quantities as variables (48).

$$Slope : 2C_2$$
$$Intercept : 2C_1$$

but the value of $2C_2$, interestingly, remains constant near 2 kg/cm². Appendix 9.2 describes a demonstration experiment that illustrates both rubber elasticity [see equation (9.34)] and the nonideality expressed by equation (9.49).

On swelling, the value of $2C_2$ drops rapidly (see Figure 9.16) (48), reaching a value of zero near v_2 (volume fraction of polymer) equal to 0.2. This same dependence is observed for the same polymer in different solvents, different levels of cross-linking for the same polymer, or (as shown) different polymers entirely (48).

The interpretation of the constants $2C_1$ and $2C_2$ has absorbed much time; the results are inconclusive (36). It is tempting but generally considered incorrect to equate $2C_1$ and $nRT\left(\overline{r_i^2}/r_0^2\right)$. The original derivation of Mooney (39) shows that $2C_2$ has to be finite, but it does not indicate its value relative to $2C_1$. According to Flory (35), the ratio $2C_2/2C_1$ is related to the looseness with which the cross-links are embedded within the structure. Trifunctional

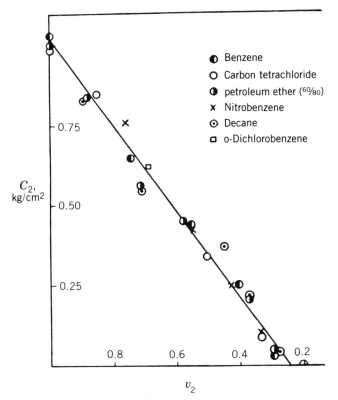

Figure 9.16 Dependence of C_2 on v_2 for synthetic rubber vulcanizates (48). Polymers: O, butadiene – styrene, (95 / 5); ◑, butadiene – styrene, (90 / 10); ◐, butadiene – styrene, (85 / 15); ●, butadiene – styrene, (75 / 25); ⊙, butadiene – styrene, (70 / 30); X, butadiene – acrylonitrile, (75 / 25).

cross-links have larger values of $2C_2/2C_1$ than tetrafunctional cross-links, for example (49).

As indicated above, $2C_2$ decreases with the degree of swelling. Furthermore, Gee (50) showed that during stress relaxation, swelling increased the rate of approach to equilibrium. Ciferri and Flory (51) showed that $2C_2$ is markedly reduced by swelling and deswelling the sample at each elongation. The samples, actually measured dry, had $2C_2$ values about half as large after the swelling–deswelling operation as those measured before. These results suggest that the magnitude of $2C_2$ is caused by nonequilibrium phenomena. Gumbrell et al. (48) stated it in terms of the reduced numbers of conformations available in the dry state versus the swollen state.

Other possible explanations include non-Gaussian chain or network statistics (see Section 9.10.6) and internal energy effects (36). The latter, bearing on the front factor, will be treated in Section 9.10.

9.9.2 Generalized Strain – Energy Functions

Following the work of Mooney, more generalized theories of the stress–strain relationships in elastomers were sought. The central problem was how to calculate the work, W, stored in the body as strain energy.

Rivlin (47) considered the most general form that such strain–energy functions could assume. As basic assumptions, he took the elastomer to be incompressible and isotropic in the unstrained state. Symmetry conditions required that the three principal extension ratios, α_1, α_2, and α_3, depend only on even powers of the α's. In three dimensions (see Figure 9.17), the simplest functions that satisfy these requirements are

$$I_1 = \alpha_1^2 + \alpha_2^2 + \alpha_3^2 \tag{9.50}$$

$$I_2 = \alpha_1^2\alpha_2^2 + \alpha_2^2\alpha_3^2 + \alpha_3^2\alpha_1^2 \tag{9.51}$$

$$I_3 = \alpha_1^2\alpha_2^2\alpha_3^2 \tag{9.52}$$

where I_1, I_2, and I_3 are termed strain invariants.

The third strain invariant is equal to the square of the volume change,

$$I_3 = \left(\frac{V}{V_0}\right)^2 = 1 \tag{9.53}$$

which under the assumption of incompressibility equals unity. Alternate formulations have been proposed by Valanis and Landel (52) and by Ogden (53), which have been reviewed by Treloar (36).

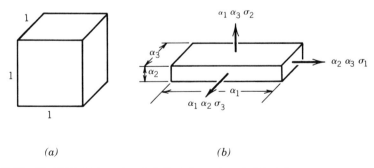

(a)   *(b)*

Figure 9.17 An elastomeric cube. (a) Undeformed and (b) deformed states, showing principal stresses and strains.

Consider the deformation of a cube (Figure 9.17). The work that is stored in the body as strain energy can be written (54).

$$W(\alpha) = \int \sigma_1 \, d\alpha_1 + \int \sigma_2 \, d\alpha_2 + \int \sigma_3 \, d\alpha_3 \qquad (9.54)$$

where the σ's are the stresses.

The work, in a more general form, can be expressed as a power series (47):

$$W = \sum_{i, j, k = 0}^{\infty} C_{ijk}(I_1 - 3)^i (I_2 - 3)^j (I_3 - 1)^k \qquad (9.55)$$

Equation (9.55) is written so that the strain energy term in question vanishes at zero strain.

For the lowest member of the series, $i = 1$, $j = 0$, and $k = 0$,

$$W = C_{100}(I_1 - 3) \qquad \text{—Stretching only.} \qquad (9.56)$$
$$\text{,in one direction}$$

which is functionally identical to the free energy of deformation expressed in equation (9.31). For the case of uniaxial extension,

$$\alpha_1 = \alpha \qquad (9.57)$$

and noting equation (9.32) and equation (9.53),

$$\alpha_2 = \alpha_3 = \left(\frac{1}{\alpha} \right)^{1/2}$$

Equation (9.56) can now be written

$$W = C_{100}\left(\alpha^2 + \frac{2}{\alpha} \right) \qquad (9.58)$$

and the stress can be written [see equation (9.54)]

$$\sigma = \frac{\partial W}{\partial \alpha} = 2C_{100}\left(\alpha - \frac{1}{\alpha^2} \right) \qquad (9.59)$$

This equation is readily identified with equation (9.34), suggesting (for this case only)

$$2C_{100} = nRT \frac{\overline{r_i^2}}{\overline{r_0^2}} \qquad (9.60)$$

On retention of an additional term in equation (9.55), with $i = 0$, $j = 1$, and $k = 0$,

$$W = C_{100}(I_1 - 3) + C_{010}(I_2 - 3) \tag{9.61}$$

which leads directly to the Mooney–Rivlin equation, equation (9.48).

Interestingly, if we retain one more term, C_{200}, an equation of the form (54)

$$\sigma = \left(C + \frac{C'}{\alpha} + C''\alpha^2\right)\left(\alpha - \frac{1}{\alpha^2}\right) \tag{9.62}$$

can be written, where

$$C = 2(C_{100} - 6C_{200}) \tag{9.63}$$

$$C' = 2(4C_{200} + C_{010}) \tag{9.64}$$

and

$$C'' = 4C_{200} \tag{9.65}$$

Equation (9.62), with two additional terms over the statistical theory of rubber elasticity, fits the data quite well (see Figure 9.18) (55).

Because no particular molecular model was assumed, theoretical values cannot be assigned to C, C', and C'', nor can any molecular mechanisms be assigned. These phenomenological equations of state, however, accurately express the form of the experimental stress–strain data.

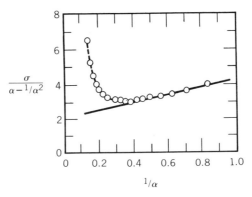

Figure 9.18 Mooney – Rivlin plot for sulfur-vulcanized natural rubber. Solid line, equation (9.48); dotted line, equation (9.62).

9.10 SOME REFINEMENTS TO RUBBER ELASTICITY

The statistical theory of rubber elasticity has undergone significant and continuous refinement, resulting in a series of correction terms. These are sometimes omitted and sometimes included in scientific and engineering research, as the need for them arises. In this section, we briefly consider some of these.

9.10.1 The Inverse Langevin Function

The Gaussian statistics leading to equation (9.34) are valid only for relatively small strains—that is, under conditions where the contour length of the chain is much more than its end-to-end distance. In the region of high strains, where the ratio of the two parameters approaches $\frac{1}{3}$ to $\frac{1}{2}$, this limit is exceeded.

Kuhn and Grün (56) derived a distribution function based on the inverse Langevin function. The Langevin function itself can be written

$$L(x) = \coth x - \frac{1}{x} \tag{9.66}$$

and was first applied to magnetic problems. In this case,

$$L(\beta) = \frac{r}{n'l} \tag{9.67}$$

where n' is the number of links of length l. (It must be pointed out that the quantity n' in this case need not be identical with the number of mers in the chains.) Thus the quantity $n'l$ represents a measure of the contour length of the chain, and $r/n'l$ is the fractional chain extension. Of course, for the inverse Langevin function of interest here,

$$L^{-1}\left(\frac{r}{n'l}\right) = \beta \tag{9.68}$$

The stress of an elastomer obeying inverse Langevin statistics can be written (36, 54)

$$\sigma = nRT(n')^{1/2}\left[L^{-1}\left(\frac{\alpha}{(n')^{1/2}}\right) - \alpha^{-3/2}L^{-1}\left(\frac{1}{\alpha^{1/2}(n')^{1/2}}\right)\right] \tag{9.69}$$

At intermediate values of α (and hence of $r/n'l$), equation (9.69) predicts a sharp upturn in the stress at α's greater than 4, as observed in experiments. Because of the complexity of equation (9.69), the Gaussian-based (9.34) is preferred where possible.

9.10.2 Cross-link Functionality

In order to form a network, at least some of the mers need to have a functionality greater than 2; that is, more than two chain portions must emanate from those mers. In the structure depicted in Figure 9.3, the functionality of each cross-link is 4. When divinyl benzene or sulfur is used as a cross-linker, the functionality will indeed be 4.

Suppose, however, that glycerol is used as the cross-linker in the synthesis of a polyester. Then the functionality of the cross-link site will be 3. Use of trimethylol propane trimethacrylate or pentaerythritol tetramethacrylate results in functionalities of 6 and 8, respectively (57).

Duiser and Staverman (58) and Graessley (59) have shown that the front factor depends on the functionality of the network. Representing the network functionality as f^*, equation (9.34) can be written

$$\sigma = \left(\frac{f^* - 2}{f^*}\right) nRT \frac{\overline{r_i^2}}{r_0^2} \left(\alpha - \frac{1}{\alpha^2}\right) \tag{9.70}$$

For tetrafunctional cross-links, defined as four chain segments emanating from each cross-link site [the same type as obtained with the use of divinyl benzene (see Figure 9.3)], $f^* = 4$, equation (9.70) predicts one-half the stress that equation (9.34) predicts.

Another way of writing the correction for cross-link functionality is (35, 60)

$$\sigma = (n - \mu) RT \frac{\overline{r_i^2}}{r_0^2} \left(\alpha - \frac{1}{\alpha^2}\right) \tag{9.71}$$

where n and μ are the number densities of elastically active strands and junctions. A junction is elastically active if at least three paths leading away from it are independently attached to the network. A strand, meaning a polymer chain segment, is elastically active if it is bound at each end by elastically active junctions (61). Equation (9.71) also predicts a front-factor correction of $\frac{1}{2}$ for a tetrafunctional network, since there are half as many cross-links as there are chain segments.

9.10.3 Network Defects

There are two major types of network defects: (a) the formation of inactive rings or loops, where the two ends of the chain segment are connected to the same cross-link junction, and (b) loose, dangling chain ends, attached to the network by only one end (62–65) (see Figure 9.19).

Both of these defects tend to decrease the retractive stress, because they are not part of the network. The equation in use to correct for dangling ends

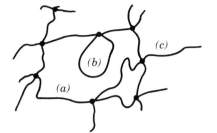

Figure 9.19 Network structure and defects: (a) elastically active chain, (b) loop, and (c) dangling chain end.

may be written

$$\sigma = nRT\left(1 - \frac{2M_c}{M}\right)\frac{\overline{r_i^2}}{\overline{r_0^2}}\left(\alpha - \frac{1}{\alpha^2}\right)$$ (9.72)

where M_c is the molecular weight between cross-links and M is the primary-chain molecular weight. Where $M \gg M_c$, the correction becomes negligible.

9.10.4 Volume Changes and Swelling

If a polymer network is swollen with a "solvent" (it does not dissolve), equation (9.32) may be rewritten,

$$\alpha_x\alpha_y\alpha_z = \frac{1}{v_2}$$ (9.73)

where v_2 is the volume fraction of polymer in the swollen material. Of course, v_2 is less than unity, and $\alpha_x\alpha_y\alpha_z$ is larger than unity, as is commonly experienced.

The detailed effect has two parts:

1. Effect on the front factor, $\overline{r_i^2}/\overline{r_0^2}$. The quantity $\overline{r_i^2}$ increases with the volume V to the two-thirds power, since r_i itself is a linear quantity. Of course, $\overline{r_0^2}$ remains constant. Thus

$$\left(\frac{\overline{r_i^2}}{\overline{r_0^2}}\right)_s = \left(\frac{V}{V_0}\right)^{2/3}\left(\frac{\overline{r_i^2}}{\overline{r_0^2}}\right) = \frac{1}{v_2^{2/3}}\left(\frac{\overline{r_i^2}}{\overline{r_0^2}}\right)$$ (9.74)

where the subscript s refers to the swollen state and where V_0 is the volume of the unswollen polymer.

2. Effect on the number of network chain segments concentration, n. The quantity n decreases with volume:

$$\left(\frac{V_0}{V}\right)n = n_s \tag{9.75}$$

where n_s is the chain segment concentration in the swollen state.

$$\left(\frac{V_0}{V}\right)n = v_2 n \tag{9.76}$$

Incorporating the right-hand sides of equations (9.74) and (9.76) into equation (9.34) leads to an equation of the form (66, 67)

$$\sigma = nRTv_2^{1/3} \frac{\overline{r_i^2}}{r_0^2}\left(\alpha - \frac{1}{\alpha^2}\right) \tag{9.77}$$

The stress, defined as force per unit actual cross-section, is decreased by $v_2^{1/3}$, since the number of chains occupying a given volume has decreased.

When volume change caused by deformation alone is considered (usually less than 1%), the equation of state can be written (68–71)

$$\sigma = nRT\left(\frac{V_0}{V}\right)^{2/3} \frac{\overline{r_i^2}}{r_0^2}\left(\alpha - \frac{1}{\alpha^2}\frac{V}{V_0}\right) \tag{9.78}$$

9.10.5   Physical Cross-links

9.10.5.1   Trapped Entanglements

So far, the discussion has been restricted to ordinary covalent cross-links. There are, however, several types of physical cross-links, which are permanent loops or entanglements existing in the network structure. (They may slide, however, yielding a mode of stress relaxation also.)

Three types of trapped entanglements are shown in Figure 9.20 (72–77). They each portray the same phenomenon, but with increasing rigor of definition.

Early works referred to the chemical and physical cross-links in a simple manner,

$$\sigma = (n_c + n_p)RT\frac{\overline{r_i^2}}{r_0^2}\left(\alpha - \frac{1}{\alpha^2}\right) \tag{9.79}$$

(a)   *(b)*   *(c)*

Figure 9.20 Three types of trapped entanglements: (*a*) The Bueche trap (72 – 77), (*b*) the Ferry trap (75), and (*c*) the Langley trap (76). The black circles are chemical cross-link sites. After Ferry (77).

where n_c and n_p are the concentration of chains bound by chemical and physical cross-links, respectively. It had been established early, for example, that the retractive stress was higher than expected by nearly a constant amount; indeed, for short relaxation times even linear polymers above T_g behaved as if they had some type of cross-linking (78, 79) (see also Section 8.2).

Figure 9.21 illustrates the rubbery plateau (see Section 8.2) for a dynamic mechanical study of polystyrene as a function of frequency. The plateau

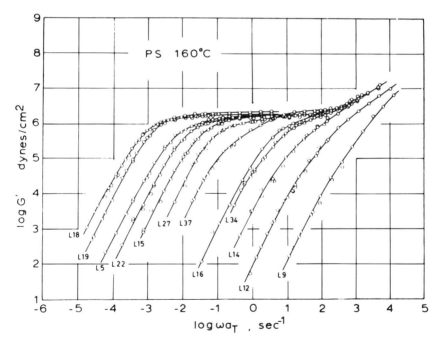

Figure 9.21 Shear storage modulus versus frequency for narrow molecular-weight polystyrenes at 160° C. Molecular weights range from $M_w = 8900$ g / mol (L9) to $M_w = 581,000$ g / mol (L18) (79).

shear modulus, near 3×10^6 dynes/cm^2, corresponds to a number of active network chains of near 1×10^{-4} mol/cm^3, nearly independent of the molecular weight of the polymer.

A more recent approach uses the concept of the potential entanglements that have been trapped by the cross-linking process. Langley (76) defines the quantity T_e as the fraction (or probability) that an entanglement is trapped in this manner. The quantity T_e can be expressed as a function of the gel fraction, W_g, which is easily measured:

$$T_e = \left[2 - W_g - 2W_g \left(\ln \frac{1}{1 - W_g} \right)^{-1} \right]^2 \tag{9.80}$$

For a perfect network, $W_g = 1$ and $T_e = 1$. (It should be noted, however, that this equation fails for low gel contents.)

Two theories, developed by Flory (80) and Scanlan (81), yield the calculation of chemical cross-links (57, 58). For Flory's theory, the total number of effective cross-links is

$$n_{\text{tot}} = n_c W_g T_e^{1/2} + n_e T_e \tag{9.81}$$

where n_e is the concentration of potential entanglement strands. For $W_g = 1$, equation (9.81) reduces to $n_c + n_p$ [see equation (9.79)].

With the Scanlan criterion,

$$n_{\text{tot}} = \tfrac{1}{2} n_c T_e^{1/2} \left(3W_g - T_e^{1/2} \right) + n_e T_e \tag{9.82}$$

which also reduces to $n_c + n_p$ for $W_g = 1$. The value of these relationships, of course, is that for real networks, $W_g < 1$, and the way is open to evaluate n_c and n_e. Some calculations are shown in Figure 9.22, which illustrate that

Figure 9.22 Contributions of chemical cross-links and trapped entanglements to the total cross-link level, for polystyrene, $M_n = 5 \times 10^5$ g/mol, $M_w = 1 \times 10^6$ g/mol, based on equation (9.81) (82).

the effective (permanent) physical cross-links start out at zero when the system is linear and increase rapidly to a plateau level (82).

9.10.5.2 The Phantom Network

It must be remarked that considerable controversy exists over the existence of physical cross-links (34, 35, 78, 83). A theory has been proposed by Flory (34, 35) using mathematics of a simplified network, called the "phantom network."

This model consists of a network of Gaussian chains connected in any arbitrary manner. The physical effect of the chains is assumed to be confined exclusively to the forces they exert on the junctions to which they are attached. For a Gaussian chain, this force is proportional to the end-to-end distance [see equation (9.27)]. The properties of the chains as material bodies that occupy space are dismissed along with the integrity of their chemical structures. Thus they may transect one another freely and are not restricted to neighboring chains, through which they may move without restraint. In the ideal case, physical cross-links are few in number in this theory.

In Gaussian phantom networks, (a) the mean positions of the cross-link junctions are determined by the macroscopic dimensions, (b) displacements of these mean positions are affine to the macroscopic strain, and (c) fluctuations of the junctions from these mean positions are Gaussian, and the magnitude of the fluctuations is invariant with strain.

Thus the phantom network differs from earlier theories, which assumed the junctions were firmly embedded in their surroundings. Under strain, they experience displacements like occlusions in a homogeneous isotropic medium, but the motions on deformation are also affine.

For a perfect phantom network of functionality f^*, the front factor contains the term $(f^* - 2)/f^*$, leading to equation (9.70), which considers chemical cross-links of arbitrary functionality, but no physical cross-links.

Flory (34) argues that the presence or absence of the term $(f^* - 2)/f^*$ may depend on the magnitude of the strain. At small deformations the displacement of junctions may conform more nearly to the older assumptions (i.e., affine in the macroscopic sense), and hence the term $(f^* - 2)/f^*$ might not appear for real chains that do have entanglements.

9.10.6 Small-Angle Neutron Scattering

In the above text, the chains were assumed to deform in an affine manner when the networks were swelled or stretched. Until recently, there was no way to approach this problem experimentally. With the advent of small-angle neutron scattering (SANS), the conformation of the chains in the bulk state could be investigated (see Section 5.2.2.1).

On extension or swelling, the classical theory of rubber elasticity (Section 9.7) predicts several specific points with regard to both cross-link site positions and chain conformations:

1. The distances between pairs of cross-linking points are governed by a Gaussian distribution. The end-to-end distance, r, of these pairs follows the form $\exp(-K^2 r^2/6)$, predicting a steadily decreasing scattering intensity with increasing K^2.

2. On swelling, according to the affine deformation principle, the mean pair separation distance between chain ends, h, goes as $h_0 Q^{1/3}$, where Q is the swelling ratio. (The quantity h, measured by scattering techniques, arises irrespective of whether the chains are in fact connected to both of the ends in question; h is closely related to r.)

3. Gaussian theory dictates that h (as well as r and R_g) is proportional to $M_c^{0.5}$, where M_c represents the molecular weight between cross-links.

4. On extension, the cross-link sites also separate according to the affine principle; see point 2.

5. The polymer chain radii of gyration increase as $Q^{1/3}$ during swelling. This is because swelling is a volume property and R_g is a linear property.

6. On stretching, the parallel and perpendicular radii of gyration, $R_{\parallel}$ and $R_{\perp}$, respectively, depend on the undeformed radii of gyration according to

$$R_{\parallel}^2 = \alpha^2 R_0^2 \tag{9.83}$$

$$R_{\perp}^2 = \alpha^{-1} R_0^2 \tag{9.84}$$

for the chain affine case, where α is the extension ratio. It should be noted that when SANS studies are done on the stretched elastomers, the scattering pattern yields greater dimensions in one direction than the other, because the chains are anisotropic (see Section 5.2) (84–97).

First, the question arose whether polymer chains in a network had the same conformation as in the melt before cross-linking. Beltzung et al. (84) prepared well-defined poly(dimethyl siloxane) (PDMS) chains containing Si—H linkages in the α and ω positions. Blends of PDMS(H) and PDMS(D) were prepared, where H and D, of course, represent the protonated and deuterated analogues. These blends were end-linked by tetrafunctional or hexafunctional cross-linkers under stoichiometric conditions.

Neutron scattering was carried out on both the PDMS melts and the corresponding networks. The principal results were that (a) the Gaussian character of the network chains in the undeformed state was confirmed, and (b) the chain dimensions were not changed by the cross-linking process.

Benoit et al. (85–89) prepared two types of tagged polystyrene networks: (a) "A" networks containing labeled (deuterated) cross-link sites. This permitted a characterization of the spatial distribution of the cross-link points.

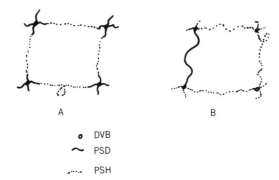

A   B

o   DVB

~   PSD

.....   PSH

Figure 9.23 Schematic representation of labeled polystyrene networks. (*A*) Cross-linking points labeled. (*B*) Random labeled chains added (87).

(b) "B" networks containing a few percent of perdeuterated polystyrene chains (see Figure 9.23) (85). Cross-linking utilized divinyl benzene, DVB.

Benoit et al. (85) studied these polystyrene networks as is, swollen in several solvents, and stretched. For the latter, stretching was done above T_g, followed by cooling in the stretched state. They found a maximum in the angular scattering curves of the A networks, contradicting point 1 above. However, they found that the quantity h increased as $Q^{1/3}$, thus confirming point 2.

The quantity h was found to be proportional to $M_c^{0.5}$ both in the dry state and in the swollen state, confirming point 3 above. On extension, $h_{\parallel}$ and $h_{\perp}$ values followed the expected affine deformation, thus confirming point 4.

The B network, on the other hand (85), appeared to deviate significantly from the affine, contradicting point 5. As illustrated in Figure 9.24 (85), the chain radius of gyration increased on swelling far less than predicted by the affine deformation mechanism. The values of R_g were also less than predicted by the "end-to-end pulling mechanism" (88), which accentuates the

Figure 9.24 Variation of the radius of gyration of B-type networks of different functionalities. Dotted line, theoretical behavior for affine deformation; dashed line, theoretical behavior for the end-to-end pulling mechanism (87), for polystyrene.

extension of the end portions of the chains rather than the central section. One might imagine that entanglements prevent the motion of the central portions of long chains. Another possible explanation (see below) is that the chain's cross-link junction points rearrange to yield the system with the lowest free energy. This rearrangement minimizes the actual extension of the chain.

In the above studies, one experimental conclusion may warrant reevaluation, the one bearing on point 1. In addition to working with very narrow molecular weight distributions (anionic polymerization), Benoit et al. (85) worked with high-functionality cross-link junctions: up to 10 DVB mers were added to each living polymer end. In the case of block copolymers (see below), domain spacing regularity increases with the number of "arms" in the order diblock, triblock, and starblock. It appears that Benoit et al. (85) have made a high-functionality, star-shaped cross-link site, which would tend to appear at more regularly spaced intervals than, for example, randomly placed tetrafunctional cross-links. In fact, a more recent theoretical examination of the network structure shows that such a maximum should exist. It is caused by the volume of the chains preventing or reducing close contact of the cross-links with one another, and it is called the correlation hole effect (87).

More recent SANS experiments on stretched networks were performed by Hinkley et al. (90) and by Clough et al. (91, 92). Hinkley et al. (90) prepared blends of polybutadiene and polybutadiene-d_6. Both polymers were made by the "living polymer" technique, end-capped with ethylene oxide, and water-washed to yield the dihydroxy liquid prepolymer. Uniform networks were prepared by reacting the prepolymers with stoichiometric amounts of triphenyl methane triisocyanate. The value of using polybutadiene over polystyrene, of course, is that the networks are elastomeric at ambient temperatures.

These networks (90) were extended up to $\alpha = 1.6$ and characterized by SANS. Owing to large experimental error, no definitive conclusion could be reached, although the data fit the junction affine model better than either the chain affine model or the phantom network model (Section 9.10.5.2). This experiment tends to support point 6.

Random types of cross-linking are of special interest for real systems. Clough et al. (91, 92) blended anionically polymerized polystyrene with PS-d_8 and cross-linked the mutual solution with ^{60}Co γ-radiation. Bars of the cross-linked polystyrene were elongated at 145°C and cooled. Specimens were cut in both the longitudinal and transverse directions and characterized by SANS. The quantity

$$R = \frac{R_g \text{ (stretched)}}{R_g \text{ (unstretched)}} \qquad (9.85)$$

was plotted versus α, as illustrated in Figure 9.25 (91, 92). The transverse

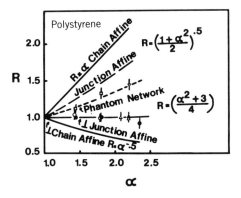

Figure 9.25 The ratio $R = R_g$ (stretched) / R_g (unstretched) as a function of sample elongation (91): □, longitudinal; ▽, longitudinal; ○, transverse, anisotropic; ●, transverse, "end-on"; △, transverse, anisotropic; ▲, transverse, "end-on."

measurements are divided into "anisotropic" and "end-on." The anisotropic measurements refer to the case where the beam was perpendicular to the direction of orientation (92), and the end-on measurements refer to the case where the beam was parallel to the stretch direction. In neither of these cases was affine chain deformation followed (point 6). Both of these experiments, however, yielded identical results within experimental error, confirming important macromolecular hypotheses.

The data for the end-on systems achieved the same quality in a quarter of the experimental time required for the anisotropic systems, because the scattering was symmetric about the beam and thus could be averaged about the azimuthal angle (72). Again, this represents the extreme need to conserve beam time, which is very expensive.

In Figure 9.25 the quantity $R = [(\alpha^2 + 3)/4]^{0.5}$ was obtained from the dependence predicted for a tetrafunctional phantom network (92). The quantity $R = [(\alpha^2 + 1)/2]^{0.5}$ represents the affine junction case (point 4 above).

The constant value of $R_\perp$ up to $\alpha = 2$ suggests that the chains are deforming far less than the junctions, supporting point 4. These results follow neither the $R = \alpha$ (parallel) nor the $R = \alpha^{-0.5}$ (perpendicular) prediction of point 6, but rather support Benoit et al. (85) that affine chain behavior is not followed.

In a series of theoretical papers, Ullman (94–96) reexamined the phantom network theory of rubber elasticity, especially in the light of the new SANS experiments. He developed a semiempirical equation for expressing the lower than expected chain deformation on extension:

$$\lambda^{*2} = \lambda^2(1 - \alpha') + \alpha' \qquad (9.86)$$

as the basis of a network unfolding model. The quantity λ^* is defined as the

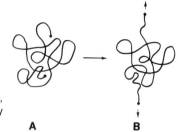

Figure 9.26 Model of the end-pulling mechanism, showing how $R_{\parallel}$ increases, while $R_{\perp}$ remains nearly constant: (A) relaxed and (B) stretched.

A   **B**

ensemble average of the deformation of junction pairs connected by a single submolecule in the network, and λ is the corresponding quantity calculated for the phantom network model. The quantity α' expresses the fractional deviation from ideality. The phantom network corresponds to $\alpha' = 0$. If $\alpha' = 1$, the chain does not deform at all upon network stretching or swelling. From the data of Clough et al. (91, 92), Ullman (94, 95) concluded that α' was in the range of 0.36–0.53.

In the above, Hinkley et al. (90), Clough et al. (91, 92), and Benoit et al. (85, 86) utilized end-linked networks. Ullman (94, 95) delineated the differences between the two types of network. He pointed out that randomly cross-linked chains deform to a greater extent than end-linked chains, that sensitivity to network functionality is much greater for end-linked chains, and that for high cross-linking levels, the randomly cross-linked chain approaches the macroscopic deformation of the sample. Ullman (96) recently reviewed these and other SANS experiments on the deformation of polymer networks.

Recently, Hadziioannou et al. (97) prepared amorphous polystyrene with extrusion ratios up to 10, using a solid-state coextrusion technique. Their polystyrene had a molecular weight of 5×10^5 g/mol. The anisotropy of the R_g values agreed with those predicted on the basis of a chain affine model, in accord with point 6 above.

While general conclusions appear to be premature, it appears that the cross-link sites rearrange themselves during deformation to achieve their lowest free-energy states; thus the chains deform less than the affine mechanism predicts. A modified end-pulling mechanism is also possible. A possible molecular mechanism, which results in minimal changes in $R_{\perp}$, is illustrated in Figure 9.26 (98). The debate over the exact molecular mechanism of deformation is sure to continue.

9.11   INTERNAL ENERGY EFFECTS

9.11.1   Thermoelastic Behavior of Rubber

In Section 9.5, some of the basic classical thermodynamic relationships for rubber elasticity were examined. Now the classical and statistical formulations are combined (99, 100).

Rearranging equation (9.8),

$$f_e = f - T\left(\frac{\partial f}{\partial T}\right)_{L,V} \tag{9.87}$$

Dividing through by f and rearranging,

$$\frac{f_e}{f} = 1 - \left(\frac{\partial \ln f}{\partial \ln T}\right)_{L,V} \tag{9.88}$$

Rewriting equation (9.78) in terms of force, and substituting equation (9.38),

$$f = GA_0\left(\alpha - \frac{V}{V_0}\frac{1}{\alpha^2}\right) \tag{9.89}$$

where A_0 is the initial cross-sectional area; substituting equation (9.89) into the right-hand side of equation (9.88) and carrying out the partial derivative,

$$\left(\frac{\partial \ln f}{\partial \ln T}\right)_{L,V} = \frac{d \ln G}{d \ln T} + \frac{\beta T}{3} \tag{9.90}$$

where β is the isobaric coefficient of bulk thermal expansion, $(1/V)(\partial V/\partial T)_{L,P}$.

Substituting equation (9.90) into equation (9.88),

$$\frac{f_e}{f} = 1 - \frac{d \ln G}{d \ln T} - \frac{\beta T}{3} \tag{9.91}$$

Shen and Blatz (99) show that f_e/f may take several nearly equivalent forms, depending on the assumptions (see Table 9.3).

Table 9.3   Summary of thermoelastic equations

Number	Theory	Equation
I	Thermodynamic (constant volume)	$f_e/f = 1 - (T/f)(\partial f/\partial T)_{V,L}$
II	Thermodynamic (constant volume)	$f_e/f = 1 - (T/f)(\partial f/\partial T)_{P,L} - (\beta T/f\kappa)(\partial f/\partial P)_{T,L}$
III	Infinitesimal	$f_e/f = 1 - (T/f)(\partial f/\partial T)_{P,L} - (\beta\lambda T3f)(\partial f/\partial\lambda)_{P,T}$
IV	Statistical	$f_e/f = 1 - (T/f)(\partial f/\partial T)_{P,L} - [\beta T/(\alpha^3 - 1)]$
V	Neo-Hookean	$f_e/f = 1 - T(d \ln G/dT) - (\beta T/3)$
Va	Neo-Hookean	$f_e/f = (m - 1)(\beta T/3)$

Returning to equation (9.38), and differentiating the natural logarithm of the network end-to-end distance with respect to the natural logarithm of the temperature,

$$\frac{d \ln \overline{r_0^2}}{d \ln T} = 1 - \frac{d \ln G}{d \ln T} - \frac{\beta T}{3} \qquad (9.92)$$

Noting that the right-hand sides of equation (9.91) and (9.92) are identical,

$$\frac{f_e}{f} = \frac{d \ln \overline{r_0^2}}{d \ln T} = \frac{1}{T} \frac{d \ln \overline{r_0^2}}{dT} \qquad (9.93)$$

which expresses the fractional force due to internal energy considerations in terms of the temperature coefficient of the free chains end-to-end distance.

Values of f_e/f are usually derived by applying the above equations to force–temperature data of the type presented in Figure 9.27 (101). These data, carefully taken after extensive relaxation at elevated temperatures, are

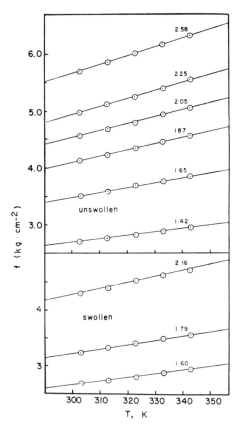

Figure 9.27 Force – temperature relationships for natural rubber. Extension ratios, α, are indicated by the numbers associated with the lines (101).

Table 9.4   Thermoelastic behavior of various polymers

Polymer	f_e/f	Reference
Natural rubber	0.12	(a)
trans-Polyisoprene	0.17	(b)
cis-Polybutadiene	0.10	(c)
Polyethylene	−0.42	(d)
Poly(ethyl acrylate)	−0.16	(e)
Poly(dimethyl siloxane)	0.15	(e)

References: (a) G. Allen, M. J. Kirkham, J. Padget, and C. Price, *Trans. Faraday Soc.*, **67**, 1278(1971). (b) J. A. Barrie and J. Standen, *Polymer*, **8**, 97(1967). (c) M. Shen, T. Y. Chem, E. H. Cirlin, and H. M. Gebhard, in *Polymer Networks, Structure, and Mechanical Properties*, A. J. Chompff and S. Newman, Eds., Plenum Press, New York, 1978. (d) A. Ciferri, C. A. J. Hoeve, and P. J. Flory, *J. Am. Chem. Soc.*, **83**, 1015(1961). (e) L. H. Sperling and A. V. Tobolsky, *J. Macromol. Chem.*, **1**, 799(1966).

reversible within experimental error; that is, the same result is obtained whether the temperature is being lowered (usually first) or raised.

Some values of f_e/f are shown in Table 9.4. For most simple elastomers, f_e/f is a small fraction, near ± 0.20 or less. This indicates that some 80% or more of the retractive force is entropic in nature, as illustrated from early data in Figure 9.7.

These same values, of course, lead to temperature coefficients of polymer chain expansion [equation (9.93)], the subject of the following subsection.

9.11.2   Effects in Dilute Solution: A Confirmation

Starting from the intrinsic viscosity, $[\eta]$, the molecular weight equation (see Section 3.8.3)

$$[\eta] = \Phi\left(\frac{\overline{r_0^2}}{M}\right)^{3/2} M^a \tag{9.94}$$

which may be rewritten

$$\ln[\eta] = \ln \Phi M^{a-3/2} + \tfrac{3}{2} \ln \overline{r_0^2} \tag{9.95}$$

yields, after taking the temperature derivative,

$$\frac{d \ln \overline{r_0^2}}{dT} = \frac{2}{3}\frac{d \ln[\eta]}{dT} \tag{9.96}$$

The right-hand side of equation (9.96) is easily measured, using a series of

Flory θ-solvents, for example (102), or from the temperature coefficient of an athermal solvent, provided r_0 is known from a θ-solvent at some temperature (103).

Amazingly, values of $d \ln \overline{r_0^2}/dT$ from equations (9.93) and (9.96) agree quite well. For example, for poly(dimethyl siloxane), $d \ln \overline{r_0^2}/dT$ is calculated to be 0.43×10^{-3} by thermoelastic measurements (71), whereas intrinsic viscosity measurements yield 0.33×10^{-3} (103) (see also Table 9.4).

One further calculation can be performed. Assume that the temperature dependence of $\overline{r_0^2}$ goes as follows (71):

$$\overline{r_0^2} = A e^{\varepsilon'/RT} \qquad (9.97)$$

where ε' represents the trans–gauche or similar energy differences, and A is a constant. Then

$$\ln \overline{r_0^2} = \ln A + \frac{\varepsilon'}{RT} \qquad (9.98)$$

and

$$\frac{d \ln[\eta]}{dT} = -\frac{3}{2} \frac{\varepsilon'}{RT^2} \qquad (9.99)$$

Of course, from thermoelastic calculations,

$$\frac{f_e}{f} = -\frac{\varepsilon'}{RT} \qquad (9.100)$$

From equations (9.100), a value of -28 cal/mol was calculated for poly(dimethyl siloxane).

9.12   THE FLORY – REHNER EQUATION

The equilibrium swelling theory of Flory and Rehner (67) treats simple polymer networks in the presence of small molecules. The theory considers forces arising from three sources:

1. The entropy change caused by mixing polymer and solvent. The entropy change from this source is positive and favors swelling.

2. The entropy change caused by reduction in numbers of possible chain conformations on swelling. The entropy change from this source is negative and opposes swelling.

3. The heat of mixing of polymer and solvent, which may be positive, negative, or zero. Usually, it is slightly positive, opposing mixing.

The Flory–Rehner equation may be written

$$-\left[\ln(1 - v_2) + v_2 + \chi_1 v_2^2\right] = V_1 n\left[v_2^{1/3} - \frac{v_2}{2}\right] \tag{9.101}$$

where v_2 is the volume fraction of polymer in the swollen mass, V_1 is the molar volume of the solvent, and χ_1 is the Flory–Huggins polymer–solvent dimensionless interaction term. Appendix 9.3 describes the application of the Flory–Rehner theory. This theory, of course, is also related to the thermodynamics of solutions (see Section 3.2). As a rubber elasticity phenomenon, it is an extension in three dimensions.

The value of equation (9.101) here lies in its complementary determination of the quantity n [see equation (9.4) for simplicity]. Both equations (9.4) and (9.101) determine the number of elastically active chains per unit volume (containing, implicitly, corrections for front factor changes). By measuring the equilibrium swelling behavior of an elastomer (χ_1 values are known for many polymer–solvent pairs), its modulus may be predicted. Vice versa, by measuring its modulus, the swelling behavior in any solvent may be predicted.

Generally, values from modulus determinations are somewhat higher, because physical cross-links tend to count more in the generally less relaxed mechanical measurements than in the closer-to-equilibrium swelling data. However, agreement is usually within a factor of 2, providing significant interplay between swelling and modulus calculations (104–106).

Simple elastomers may swell a factor of 4 or 5 or so, leading to a quantitative determination of n. However, several factors need to be considered before the final numerical result is accepted:

1. The front-factor, not explicitly stated in the Flory–Rehner equation, may be significantly different from unity (107).

2. While step polymerization methods lead to more or less statistical networks and good agreement with theory, addition polymerization and vulcanization nonuniformities lead to networks that may swell as much as 20% less than theoretically predicted (108, 109).

9.13   CROSS-LINKS IN GELS

Let us consider a polymeric network that contains solvent, usually called a polymeric gel. There are several types of gels. A previously cross-linked polymer subsequently swollen in a solvent follows the Flory–Rehner equation (Section 9.12). If the network was formed in the solvent so that the chains are relaxed, the Flory–Rehner equation will not be followed, but rubber elasticity theory can still be used to count the active network segments.

Gels may be prepared using either chemical or physical cross-links. Physical crosslinks in gels may involve dipole–dipole interactions, traces of crystallinity, multiple helices, and so on and thus vary greatly with the number and strength of the bonds. The number of physical crosslinks present in a given system depends on time, pressure, and temperature. Many such gels are thermoreversible; that is, the bonds break at elevated temperature and reform at lower temperatures.

A common thermoreversible gel is gelatin in water. Gelatin is made by hydrolytic degradation, boiling collagen in water. Collagen forms a major constituent of the skin and bones of animals. The gelatin made from it forms the basis for many foods and desserts. (The latter is usually made by dissolving gelatin plus sugar in hot water, then refrigerating to cause gelation.)

Thermoreversible gels may be bonded at single points, called point cross-links; junction zones, where the chains interact over a portion of their length; or in the form of fringe micelles—see Figure 9.28 (110–112). Gelatin has the junction zone type of cross-links, where the chains form multiple helices. Natural collagen is in the form of a triple helix. In hot aqueous solution, the denatured protein forms random coils. The helix–coil transition is at about 40°C at about 0.5% concentration, as commonly used in foods. Under these conditions, gelatin forms either double or triple helices, a subject of current research (110). The critical nucleation length is about 20–50 peptide (mer) units, the most important interactions being among proline, hydroxyproline, and glycine.

Gels can be made from a number of polymers, including poly[acrylamide–*stat*–(acrylic acid)], poly(vinyl acetate), poly(dimethyl siloxane), and polyiso-

Figure 9.28 Types of thermoreversible cross-links: (a) point cross-links, (b) junction zones, and (c) fringed micelles.

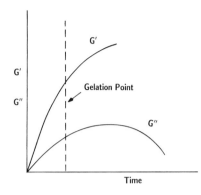

Figure 9.29 Schematic of dynamic viscoelastic measurements on curing. Gelation occurs just before the maximum in G''; see dotted line.

cyanurates. The poly[acrylamide–*stat*–(acrylic acid)] hydrogels can be made to swell up to 20,000 times, v/v. Thermorevesible gels can be prepared in organic solvents from polyethylene and *i*-polystyrene, both of which crystallize on cooling but go back in solution on heating.

Dynamic mechanical analysis of the gels during formation shows that G' increases steadily throughout the reaction; see Figure 9.29 (110–112). In the region just after gelation, G' does not increase proportionately with the number of cross-links, because of the large number of dangling chain ends and cycles. However, G'' goes through a maximum just after reaching the gel stage. Thus the loss term depends significantly on nonactive network segments. Fully formed gels may be highly elastic, as illustrated for gelatin gels in Appendix 9.1.

Another example of gel formation involves the globular proteins. Egg white is a typical example. On cooking, it goes from a sol to a gel irreversibly, becoming denatured. Current models (112) show the hydrophobic groups partially flipping out in the range of 75–80°C, making contact with similar groups on other globules. Above 85°C, the $-S-S-$ intramolecular cross-links in cysteine become labile, partially interchanging to form intermolecular crosslinks. As is well known, hard-boiled egg white is highly elastic. The white appearance is caused by the nonuniform distribution of protein and water, causing intense light-scattering.

Gels sometimes undergo the phenomenon known as syneresis, where the solvent is exuded from the gel. Two types of syneresis are distinguishable (113): (a) The χ-type, where the polymer phase separates from the solvent due to poor thermodynamics of mixing. Spinodal decomposition is common in such circumstances. Such gels may be turbid in appearance. (b) The *n*-type, which exudes solvent because of increasing cross-link density. The polymeric gel still forms one phase with the solvent, but its equilibrium swelling level decreases. Such gels remain clear. In both cases, various amounts of fluid surround the gel.

9.14 GELS AND GELATION

Most of the above cross-linked polymers were considered in the dry state, although the Flory–Rehner theory (Section 9.12) made use of equilibrium swollen gels in the evaluation of the cross-link density. Generally, a polymeric gel is defined as a system consisting of a polymer network swollen with solvent. It must be understood that the solvent is dissolved in the polymer, not the other way around.

Polymeric gels may be categorized into two major classes (114): thermoreversible gels and permanent gels. The thermoreversible gels undergo a transition from a solid-like form to a liquid-like form at a certain characteristic temperature. The links between the polymeric chains are transient in nature and support a stable polymeric network only below a characteristic "melting" point.

The permanent gels consist of solvent-logged covalently bonded polymer networks. One family of such networks is formed by crosslinking preexisting polymer chains, such as by vulcanization. Another family makes use of simultaneous polymerization and cross-linking. Some of the more important types of gels are delineated in Table 9.5. some specific examples include Jello® (comprised of approximately 3% of collagen-derived protein gelatin plus colored, flavored, and sweetened water), vitreous humor that fills the interior of the eye, membranes (both natural and synthetic), and soft-contact lenses (115).

Taking gelatin as an example, the thermal naturation and denaturation transition occurs near 40°C in water for native soluble collagen, and near 25°C for partially renatured gelatin. After setting, the initial growth rate of the modulus is nearly third order in gelatin concentration. The mechanism of gelation thus is thought to involve slow association of three rapidly formed single-helix segments (Figure 9.30) (116). a demonstration experiment measuring the modulus of gelatin as a function of concentration is given in Appendix 9.1.

The gelatin-type of triple helix may be considered a type of crystal; indeed, the denaturing and renaturing of these gels is a first-order transition. How-

Table 9.5   Types of gels

Class	Bonding Mode
Thermoreversible	Hydrogen bonds
	Microcrystals
	Entwined helical structures
	Specific interactions
Permanent Gels	Vulcanization
	Multifunctional monomers
	(a) Chain polymerizable
	(b) Stepwise polymerizable

Collagen-fold cross-link

Figure 9.30   Formation of triple-helix cross-links in gelatin gels (116).

ever, it has been modeled as a nucleated, one-dimensional crystallization, as opposed to the ordinary three-dimensional crystallization of bulk meterials.

9.15 EFFECT OF STRAIN ON THE MELTING TEMPERATURE

Most elastomers are amorphous in use. Indeed, significant crystallinity deprives the polymer of its rubbery behavior. However, some elastomers crystallize during strains such as extension. The most important of these are cis-polybutadiene, cis-polyisoprene, and cis-polychloroprene. Crystallization on extension can be responsible for a rapid upturn in the stress–strain curves at high elongation; see Figure 9.5.

Such crystallization can result in significant engineering advantage. Consider, for example, the wear mechanisms in automotive tires. It turns out that abrasion is the most important mode of loss of tread. Tiny shreds of rubber are torn loose, hanging at one end. In contact with the road, these shreds are strained at each revolution of the tire, gradually tearing off more rubber. If the rubber crystallizes during extension, it becomes self-reinforcing when it is needed most, thus slowing the failure process. When the strain is released, the crystallites melt, returning the rubber to its amorphous state reversibly.

The elevation in melting temperature can be expressed (117) as

$$\frac{1}{T_f} = \frac{1}{T_f^0} - \frac{R}{2N_u \, \Delta H_f}\left(\alpha^2 + \frac{2}{\alpha} - 3\right) \qquad (9.102)$$

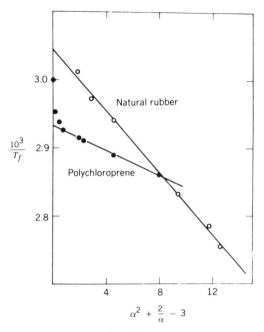

Figure 9.31   Fit of equation (9.102) to melting point elevation (114).

where T_f and T_f^0 are the thermodynamic melting temperatures for the strained and unstrained polymers, respectively, N_u is the number of repeating units per network chain, and ΔH_f is the heat of fusion per mer. The student will note the relationship to equations (6.38)–(6.43), albeit that these equations treat melting point depressions, and the present case treats melting point elevation. Also note equation (9.33).

Equation (9.102) was derived from the assumption that the system can be treated as being composed of two independent phases, T_f obtained directly by equating the chemical potentials of the strained amorphous and crystalline phases. The function of the extension ratio, given in equation (9.33) yields the dependence of the change in free energy on elongation.

Figure 9.31 (117) illustrates the application of the equation to natural rubber and polychloroprene (see Section 9.15 for structures). The equation fits the data for natural rubber somewhat better than for polychloroprene, although both polymers are fit for reasonable extensions. It must be pointed out that while the pure polymers both crystallize somewhat above room temperature, ordinary vulcanization and compounding lower the melting temperatures significantly, shifting the data in Figure 9.31 upward. Typical numerical values for equation (9.102) are as follows:

Polymer	ΔH_f (cal/mol)	N_u
Natural rubber	1000	73
Polychloroprene	2000	37

9.16   ELASTOMERS IN CURRENT USE

The foregoing sections outline the theory of rubber elasticity. This section describes the classes of elastomers in current use. While many of these materials exhibit low modulus, high elongation, and rapid recovery from deformation and obey the theory of rubber elasticity, some materials deviate significantly, have limited extensibility or poorly defined rubbery plateaus, but are considered elastomers.

9.16.1   Classes of Elastomers

9.16.1.1   Diene Types

The diene elastomers are based on polymers prepared from butadiene, isoprene, their derivatives and copolymers. The oldest elastomer, natural rubber (polyisoprene), is in this class (see Section 9.2). Polybutadiene, poly-chloroprene, styrene–butadiene rubber (SBR), and acrylonitrile–butadiene rubber (NBR) are also in this class.

The general polymerization scheme may be written

$$n CH_2 {=} \overset{\overset{\displaystyle X}{|}}{C}{-}CH{=}CH_2 \; \rightarrow \; {\left(\!CH_2{-}\overset{\overset{\displaystyle X}{|}}{C}{=}CH{-}CH_2\!\right)}_n \quad (9.103)$$

where X— may be H—, CH_3—, Cl—, and so on (see Table 9.6).

The diene double bond in equation (9.103) may be either cis or trans. The cis products all have lower glass transition temperatures and/or reduced crystallinity, and they make superior elastomers. A random copolymer of butadiene and styrene is polymerized to form SBR (styrene–butadiene rubber). This copolymer forms the basis for tire rubber (see below). The trans materials, such as the balata and gutta percha polyisoprenes (12), are highly crystalline and make excellent materials such as golf ball covers.

Natural rubber is widely used in truck and aircraft tires, which require heavy duty. They are self-reinforcing because the rubber crystallizes when stretched.

9.16.1.2   Saturated Elastomers

The polyacrylates exemplify these materials:

$$n CH_2 {=} \underset{\underset{\displaystyle O{-}X}{\overset{\displaystyle |}{\underset{\displaystyle |}{C{=}O}}}}{CH} \quad \rightarrow \; {\left(\!CH_2{-}\underset{\underset{\displaystyle O{-}X}{\overset{\displaystyle |}{\underset{\displaystyle |}{C{=}O}}}}{CH}\!\right)}_n \quad (9.104)$$

where X— may be CH_3—, CH_3CH_2—, and so on (see Table 9.6). Ethyl and butyl are the two most important derivatives, with glass transition tempera-

Table 9.6   Structures of elastomeric materials

Name	Structure
A. Diene elastomers	$\begin{array}{c} X \\ | \\ +CH_2-C=CH-CH_2 \, +_n \end{array}$
Polybutadiene	X— = H—
Polyisoprene	X— = CH_3—
Polychloroprene	X— = Cl—
B. Acrylics	$\begin{array}{c} +CH_2-CH+_n \\ | \\ O=C-O-X \end{array}$
Poly(ethyl acrylate)	X— = CH_3CH_2—
C. EPDM[a]	$+CH_2-CH_2+_n +CH_2-\overset{\displaystyle CH_3}{\underset{|}{CH}}+_m$
D. Thermoplastic elastomers	ABA
Poly(styrene–*block*–butadiene–*block*–styrene)	A = polystyrene B = polybutadiene
Segmented polyurethanes	$+AB+_n$ A = polyether (soft block)
	B = aromatic urethane (hard block)
	A = poly(butylene oxide)
	B = poly(terephthalic acid–ethylene glycol)
E. Inorganic elastomers	
Silicone rubber	$\begin{array}{c} CH_3 \\ | \\ +Si-O+_n \\ | \\ CH_3 \end{array}$
Polyphosphazines	$+N=P+_n$ with R, R′

[a]Ethylene–propylene diene monomer.

tures in the range of $-22°C$ and $-50°C$, respectively. The main advantage of the saturated elastomers is resistance to oxygen, water, and ultraviolet light, which attack the diene elastomers in outdoor conditions (118).

An important saturated elastomer is based on a random copolymer of ethylene and propylene, EPDM, or ethylene–propylene–diene monomer. The diene is often a bicyclic compound introduced at the 2% level to provide cross-linking sites:

(9.105)

Dicyclopentadiene   5-Ethylidene-
2-norbornene

It must be pointed out that although both polyethylene and polypropylene are crystalline polymers, the random polymer at midrange compositions is totally amorphous. These materials, with a glass transition of $-50°C$, make especially good elastomers for toughening polypropylene plastics (119–121).

9.16.1.3 Thermoplastic Elastomers

These new materials contain physical cross-links rather than chemical cross-links. A physical cross-link can be defined as a noncovalent bond that is stable under one condition but not under another. Thermal stability is the most important case. These materials behave like cross-linked elastomers at ambient temperatures but as linear polymers at elevated temperatures, having reversible properties as the temperature is raised or lowered.

The most important method of introducing physical crosslinks is through block copolymer formation (122–125). At least three blocks are required. The simplest structure contains two hard blocks (with a T_g or T_f above ambient temperature) and a soft block (with a low T_g) in the middle (see Figure 4.15). The soft block is amorphous and above T_g under application temperatures, and the hard block is glassy or crystalline.

The thermoplastic elastomers depend on phase separation of one block from the other (see Chapter 4), which in turn depends on the very low entropy gained on mixing the blocks. The elastomeric phase must form the continuous phase to produce rubbery properties; thus the center block has a higher molecular weight than the two end blocks combined.

Examples of the thermoplastic elastomers include polystyrene–*block*–polybutadiene–*block*–polystyrene, SBS, or the saturated center block counterpart, SEBS. In the latter, the EB stands for ethylene–butylene, where a combination of 1,2 and 1,4 copolymerization of butadiene on hydrogenation presents the appearance of a random copolymer of ethylene and butylene (see Table 9.6).

When the polymer illustrated in Figure 4.15 is of the SBS or SEBS type, it is sold under the trademark Kraton.®

Important applications of the triblock and its cousins, the starblock copolymer thermoplastic elastomers, include the rubber soles of running shoes and sneakers (125, 126) and hot melt adhesives. In the former application, sliding friction generated heat momentarily turns the elastomer into an adhesive (see Section 8.2), reducing slips and falls. When sliding stops, the sole surface cools again, regenerating the rubber.

The so-called segmented polyurethanes form thermoplastic elastomers of the $+AB+_n$ type, where A is usually a polyether such as poly(ethylene oxide) and B contains aromatic urethane groups (127, 128). These polyurethanes make excellent elastic fibers that stretch about 30% and are widely used in undergarments under the trade names Spandex® and Lycra®.

A newer type of $(AB)_n$ block copolymer, known as the poly(ether–ester) elastomers and sold as Hytrel®, contains alternating blocks of poly(butylene oxide) and butylene terephthalate as the soft and hard blocks, respectively

Table 9.7 Poly(ether–ester) material characterization (130)

Hard Block (%)	H33	H50	H57	H63	H76	H84
4GT content, wt%	33	50	57	63	76	84
Av 4GT[a] block length, L	2.64	4.95	6.43	8.14	14.8	24.26
DSC						
T_g, °C	−68	−59	−55	−51	−33	−9
T_f, °C	163	189	196	200	209	214
ΔH_f, J/g	16	33	41	48	58	61
Crystallinity %	11.5	22.9	28.6	33.3	40.7	42.8
Young's modulus, E_y, MPa	36.8	102	—	132	328	375
Rheovibron[b]						
E'' peak, °C[c]	−63	−58	−53	−48	−30	−4
tan δ peak, °C[c]	−51	−41	−34	−27	10	30

[a]Tetramethylene terephthalate.
[b]Compression-molded samples.
[c]β Relaxation.

(129–131):

Usually, $m = 1$ or 2, and $n = 40$–60; thus the soft segment is much longer than the hard segment.

Typical sample compositions for the poly(ether–ester)–AB–block copolymer structures are shown in Table 9.7 (130). The hard blocks crystallize in this case.

9.16.1.4 Inorganic Elastomers

The major commercial inorganic elastomer is poly(dimethyl siloxane), known widely as silicone rubber (see Table 9.6). This specialty elastomer has the lowest known glass transition temperature, $T_g = -130$°C (132); it also serves as a high-temperature elastomer. A common application of this elastomer is as a caulking material. It cross-links on exposure to air.

Another covalently bonded inorganic elastomer class is the polyphosphazenes (132, 133),

$$\left(N{=}P\right)_n \quad \substack{R \\ | \\ | \\ R'}$$

(9.107)

In elastomeric compositions R and R' are mixed substituent fluoroalkoxy groups. The current technological applications depend on the oil resistance and nonflammability of these elastomers; low T_g's are also important. Gaskets, fuel lines, and O-rings are made from this class of elastomer.

9.16.2 Reinforcing Fillers and Other Additives

Natural rubber has a certain degree of self-reinforcement, since it crystallizes on elongation (134). The thermoplastic elastomers also gain by the presence of hard blocks (135). However, nearly all elastomeric materials have some type of reinforcing filler, usually finely divided carbon black or silicas (122).

These reinforcing fillers, with dimensions of the order of 100–200 Å, form a variety of physical and chemical bonds with the polymer chains. Tensile and tear strength are increased, and the modulus is raised. The reinforcement can be understood through chain slippage mechanisms (see Figure 9.32) (122). The filler permits local chain segment motion but restricts actual flow.

The theory of rubber elasticity can be modified to incorporate the effects of reinforcement (136). The internal energy changes suggested by chain slippage and rearrangement in the filler–polymer complex can be represented by a new front-factor term, X/X_0, which multiplies the right-hand side of equation (9.34):

$$\sigma = nRT \left(\frac{\overline{r_i^2}}{\overline{r_0^2}} \right) \left(\frac{X}{X_0} \right) \left(\alpha - \frac{1}{\alpha^2} \right) \tag{9.108}$$

The quantity X/X_0 represents the increase in the energetic portion of the free-energy contribution.

The value of X/X_0 was shown (136) to be given by

$$\frac{X}{X_0} = T^{f_e/f} \tag{9.109}$$

where f_e/f attains values of 0.4–0.6 in fully reinforced systems. See equations (9.9) and (9.87). This leads to an increase in the retractive stress, and hence the modulus, of a factor of about 10-fold. The field of carbon-black- and silica-reinforced rubber has been reviewed (137–140).

While it is beyond the scope of this book to treat all the other components of commercial elastomers, a few must be mentioned. Frequently, an extender is added. This is a high-molecular-weight oil that reduces the melt viscosity of the polymer while in the linear state, easing processing, and lowers the price of the final product.

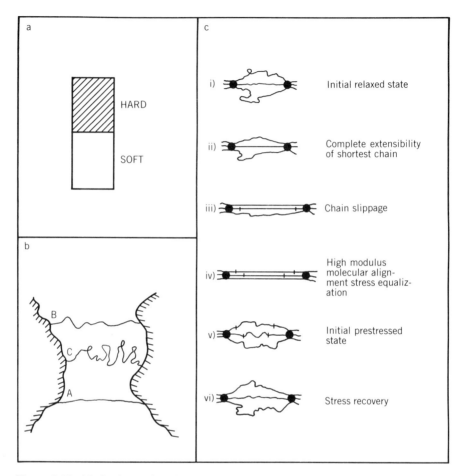

Figure 9.32 Mechanisms of reinforcement of elastomers by carbon black: (*a*) Takayanagi model approach; (*b*) on stretching, some chains (A) will become taut before others (B, C); (*c*) chain slippage on filler surface maintains polymer–filler bonds. From J. A. Harwood, L. Mullins, and A. R. Payne, *J. IRI*, **1**, 17(**1967**).

Antioxidants or antiozonites are added, especially to diene elastomers, to slow attack on the double bond. Ultraviolet srceens are used for outdoor applications (118).

The major application of carbon-black-reinforced elastomers is in the manufacture of automotive tires. Table 9.8 illustrates two recipes, one involving natural rubber (mainly for trucks and aircraft) and one involving synthetic rubber (automobiles). The synthetic-rubber recipes usually contain two or more elastomers that are blended together with the other ingredients and then covulcanized.

Table 9.8   Typical tire tread recipes (12)

Ingredient	phr[a] Natural Rubber	phr[a] SBR Synthetic	Function
Smoked sheet	100	—	Elastomer
Styrene–butadiene/oil masterbatch	—	103.1	Elastomer–extender masterbatch
Cis-polybutadiene	—	25	Special purpose elastomer
Oil-soluble sulfonic acid	2.0	5.0	Processing aid
Stearic acid	2.5	2.0	Accelerator–activator
Zinc oxide	3.5	3.0	Accelerator–activator
Phenyl-β-naphthylamine	2.0	2.0	Antioxidant
Substituted N,N'-p-phenylenediamine	4.0	4.0	Antiozonant
Microcrystalline wax	1.0	1.0	Processing aid and finish
Mixed process oil	5.0	7.0	Softener
HAF carbon black	50	—	Reinforcing filler
ISAF carbon black	—	65	Reinforcing filler
Sulfur	2.5	2.8	Vulcanizing agent
Substituted benzothiazole-2-sulfonamide	0.5	1.5	Accelerator
N-nitrosodiphenylamine	0.5	—	Retarder
Total weight	173.5	220.4	
Specific gravity	1.12	1.13	

[a]Parts per hundred parts of rubber, by weight.

9.17   SUMMARY OF RUBBER ELASTICITY BEHAVIOR

When an amorphous cross-linked polymer above T_g is deformed and re-leased, it "snaps back" with rubbery characteristics. The dependence of the stress necessary to deform the elastomer depends on the crosslink density, elongation, and temperature in a way defined by statistical thermodynamics.

The theory of rubber elasticity explains the relationships between stress and deformation in terms of numbers of active network chains and temperature but cannot correctly predict the behavior on extension. The Mooney–Rivlin equation is able to do the latter but not the former. While neither theory covers all aspects of rubber deformation, the theory of rubber elasticity is more satisfying because of its basis in molecular structure.

The theory of rubber elasticity is one of the oldest theories in polymer science and has played a central role in its development. Now, one of the key assumptions of rubber elasticity theory, the concept of affine deformation, is under question in the light of small-angle neutron scattering experiments. The SANS experiments suggest a nonaffine chain deformation via an end-pulling mechanism.

If these new results stand the test of time, they may help answer one of the important challenges to the Mark–Flory random coil model (see Section 5.3). In this challenge, the considerable discrepancy between rubber elasticity theory and experiment is blamed on the presumed nonrandom coiling of the polymer chain in space. These discrepancies may, however, lie rather in the realm of the mode of chain disentanglement on extension.

An understanding of how an elastomer works, however, has led to many new materials and new types of elastomers. A leading new type of elastomer is based on physical cross-links rather than chemical cross-links. The new materials are known as thermoplastic elastomers.

REFERENCES

1. D. Adolf, J. E. Martin, and J. P. Wilcoxon, *Macromolecules*, **23**, 527(1990).
2. D. Adolf and J. E. Martin, *Macromolecules*, **23**, 3700(1990).
3. J. E. Martin, D. Adolf, and J. P. Wilcoxon, *Phys. Rev. A*, **39**, 1325(1989).
4. P. G. de Gennes, *J. Phys. Lett.*, **37**, 61(1976).
5. D. J. Stauffer, *J. Chem. Soc. Faraday Trans. 2*, **72**, 1354(1976).
6. D. Stauffer, A. Coniglio, and M. Adam, *Adv. Poly. Sci.*, **44**, 103(1982). [*Polymer Networks*, K. Dusek, Ed., Springer-Verlag, New York, 1982).]
7. B. Gutte and R. B. Merrifield, *J. Am Chem. Soc.*, **91**, 501(1969).
8. F. Rodriguez, *Principles of Polymer Systems*, 3rd ed., Hemisphere, New York, 1988, Chap. 14.
9. K. Dusek, in *Advances in Polymerization. 3*, R. N. Haward, Ed., Applied Science Publishers, London, 1982.
10. K. Dusek and W. J. MacKnight, in *Cross-Linked Polymers: Chemistry, Properties, and Applications*, R. A. Dickie, S. S. Labana, and R. S. Bauer, Eds., ACS Symposium Series 367, ACS Books, Washington, DC, 1988.
11. H. F. Mark, *Giant Molecules*, Time Inc., New York, 1966, p. 124.
12. H. J. Stern, *Rubber: Natural and Synthetic*, 2nd ed., Palmerton, New York, Chap. 1.
13. J. Gough, *Mem. Lit. Phil. Soc. Manchester*, **1**, 28(1805).
14. C. Goodyear, U.S. Patent 3, 633(1844).
15. J. L. White, *Rubber Industry*, August 1974, p. 148.
16. Lord Kelvin (W. Thompson), *J. Pure Appl. Math.*, **1**, 57(1857).
17. C. Price, *Proc. R. Soc. Lond.*, **A351**, 331(1976).
18. P. J. Flory, *Principles of Polymer Chemistry*, Cornell University Press, Ithaca, NY, 1953, (a) pp. 434–440, (b) p. 442, (c) p. 464.
19. H. Staudinger, *Ber. Otsch. Chem. Ges.*, **53**, 1073(1920).

20. H. Staudinger, *Die hochmolekularen organische Verbindungen*, Verlag von Julius Springer, Berlin, 1932.

21. Reference 18, Chap. 1.

22. E. Guth and H. Mark, *Monatsh. Chem.* **65**, 93(1934).

23. W. Kuhn, *Angew. Chem.*, **51**, 640(1938).

24. W. Kuhn, *J. Polym. Sci.*, **1**, 380(1946).

25. E. Guth and H. M. James, *Ind. Eng. Chem.*, **33**, 624(1941).

26. E. Guth and H. M. James, *J. Chem. Phys.*, **11**, 455(143).

27. H. M. James, *J. Chem. Phys.*, **15**, 651(1947).

28. H. M. James and E. Guth, *J. Chem. Phys.*, **15**, 669(1947).

29. H. M. James and E. Guth, *J. Polym. Sci.*, **4**, 153(1949).

30. L. R. G. Treloar, *Trans. Faraday Soc.*, **39**, 36, 241(1943).

31. F. T. Wall, *J. Chem. Phys.*, **11**, 527(1943).

32. P. J. Flory, *Trans. Faraday Soc.*, **57**, 829(1961).

33. G. Allen, *Proc. R. Soc. Lond.*, **A351**, 381(1976).

34. P. J. Flory, *Polymer*, **20**, 1317(1979).

35. P. J. Florym, *Proc. R. Soc. Lond.*, **A351**, 351(1976).

36. L. R. G. Treloar, *The Physics of Rubber Elasticity*, 3rd ed., Clarendon Press, Oxford, 1975.

37. L. R. G. Treloar, *Trans. Faraday Soc.*, **40**, 49(1944).

38. L. R. G. Treloar, *Trans. Faraday Soc.*, **42**, 83(1946).

39. F. T. Wall, *Chemical Thermodynamics*, 2nd ed., Freeman, San Francisco, 1965, p. 314.

40. R. L. Anthony, R. H. Caston, and E. Guth, *J. Phys. Chem.*, **46**, 826(1942).

41. R. A. Alberty and F. Daniels, *Physical Chemistry*, 5th ed., Wiley, New York, 1979, Chap. 2.

42. E. Pines, K. L. Wun, and W. Prins, *J. Chem. Ed.*, **50**, 753(1973).

43. K. J. Mysels, *Introduction to Colloid Chemistry*, Wiley-Interscience, New York, 1965.

44. M. Mooney, *J. Appl. Phys.*, **11**, 582(1940).

45. M. Mooney, *J. Appl. Phys.*, **19**, 434(1948).

46. R. S. Rivlin, *Trans. R. Soc.* (*Lond.*), **A240**, 459, 491, 509(1948).

47. R. S. Rivlin, *Trans. R. Soc.* (*Lond.*), **A241**, 379(1948).

48. S. M. Gumbrell, L. Mullins, and R. S. Rivlin, *Trans. Faraday Soc.*, **49**, 1495(1953).

49. J. E. Mark, R. R. Rahalkar, and J. L. Sullivan, *J. Chem. Phys.*, **70** 1794(1979).

50. G. Gee, *Trans. Faraday Soc.*, **42**, 585(1946).

51. A. Ciferri and P. J. Flory, *J. Appl. Phys.*, **30**, 1498(1959).

52. K. C. Valanis and R. F. Landel, *J. Appl. Phys.*, **38**, 2997(1967).

53. R. W. Ogden, *Proc. R. Soc.*, **A326**, 565(1972).

54. M. Shen, in *Science and Technology of Rubber*, F. R. Eirich, Ed., Academic Press, Orlando, 1978.

55. Y. Sato, *Rep. Prog. Polym. Phys. Jpn.*, **9**, 369(1969).

56. W. Kuhn and F. Grün, *Kolloid Z.*, **101**, 248(1942).

57. J. K. Yeo, L. H. Sperling, and D. A. Thomas, *J. Appl. Polym. Sci.*, **26**, 3977(1981).

58. J. A. Duiser and J. A. Staverman, in *Physics of Noncrystalline Solids*, J. A. Prins, Ed., North-Holland, Amsterdam, 1965.

59. W. W. Graessley, *Macromolecules*, **8**, 186(1975).

60. K. J. Smith, Jr., and R. J. Gaylord, *J. Polym. Sci. Polym. Phys. Ed.*, **13**, 2069(1975).

61. D. S. Pearson and W. W. Graessley, *Macromolecules*, **13**, 1001(1980).

62. P. J. Flory, *Ind. Eng. Chem.*, **38**, 417(1946).

63. L. Mullins and A. G. Thomas, *J. Polym. Sci.*, **43** 13(1960).

64. A. V. Tobolsky, D. J. Metz, and R. B. Mesrobian, *J. Am. Chem. Soc.*, **72**, 1942(1950).

65. J. Scanlan, *J. Polym. Sci.*, **43**, 501(1960).

66. H. M. James and E. Guth, *J. Chem. Phys.*, **11**, 455(1943).

67. P. J. Flory and J. Rehner, *J. Chem. Phys.*, **11**, 521(1943).

68. H. M. James and E. Guth, *J. Polym. Sci.*, **4**, 153(1949).

69. P. J. Flory, *Trans. Faraday Soc.*, **57**, 829(1961).

70. A. V. Tobolsky and M. C. Shen, *J. Appl. Phys.*, **37**, 1952(1966).

71. L. H. Sperling and A. V. Tobolsky, *J. Macromol. Chem.*, **1**, 799(1966).

72. A. M. Bueche, *J. Polym. Sci.*, **19**, 297(1956).

73. L. Mullins, *J. Appl. Polym. Sci.*, **2**, 1(1959).

74. G. Kraus, *J. Appl. Polym. Sci.*, **7**, 1257(1963).

75. R. G. Mancke, R. A. Dickie, and J. O. Ferry, *J. Polym. Sci.*, **A-2**(6), 1783(1968).

76. N. R. Langley, *Macromolecules*, **1**, 348(1968).

77. J. D. Ferry, *Viscoelastic Properties of Polymers*, 3rd ed., Wiley, New York, 1980, pp. 408–411.

78. O. Kramer, *Polymer*, **20**, 1336(1979).

79. S. Onogi, T. Masuda, and K. Kitagawa, *Macromolecules*, **3**, 111(1970).

80. P. J. Flory, *Chem. Rev.*, **35**, 51(1944).

81. J. Scanlan, *J. Polym. Sci.*, **43**, 501(1960).

82. N. R. Langley and K. E. Polmanter, *J. Polym. Sci. Polym. Phys. Ed.*, **12**, 1023(1974).

83. J. D. Ferry, *Polymer*, **20**, 1343(1979).

84. M. Beltzung, C. Picot, P. Rempp, and J. Hertz, *Macromolecules*, **15**, 1594(1982).

85. H. Benoit, D. Decker, R. Duplessix, C. Picot, P. Remp, J. P. Cotton, B. Farnoux, G. Jannick, and R. Ober, *J. Polym. Sci. Polym. Phys. Ed.*, **14**, 2119(1976).

86. H. Benoit, R. Duplessix, R. Ober, M. Daoud, J. P. Cotton, B. Farnoux, and G. Jannick, *Macromolecules*, **8** 451(1975).

87. L. H. Sperling, *Polym. Eng. Sci.*, **24**, 1(1984).

88. L. H. Sperling, A. M. Fernandez, and G. D. Wignall, in *Characterization of Highly Cross-linked Polymers*, S. S. Labana and R. A. Dickie, Eds., ACS Symposium Series No. 243, American Chemical Society, Washington, 1984.

89. C. Picot, R. Duplessix, D. Decker, H. Benoit, F. Boue, J. P. Cotton, and P. Pincus, *Macromolecules*, **10**, 436(1977).

90. J. A. Hinkley, C. C. Han, B. Mozer, and H. Yu, *Macromolecules*, **11**, 837(1978).

91. S. B. Clough, A. Maconnachie, and G. Allen, *Macromolecules*, **13**, 774(1980).

92. S. B. Clough, private communication, December 29, 1982.

93. D. S. Pearson, *Macromolecules*, **10**, 696(1977).

94. R. Ullman, *Macromolecules*, **15**, 1395(1982).

95. R. Ullman, *Macromolecules*, **15**, 582(1982).

96. R. Ullman, in *Elastomers and Rubber Elasticity*, J. E. Mark and J. Lal, Eds., ACS Symposium Series No. 193, American Chemical Society, Washington, DC, 1982, Chap. 13.

97. G. Hadziiannou, L. H. Wang, R. S. Stein, and R. S. Porter, *Macromolecules*, **15**, 880(1982).

98. F. Boue, M. Nierlich, G. Jannick, and R. C. Ball, *J. Physiol.* (*Paris*), **43**, 137(1982).

99. M. Shen and P. J. Blatz, *J. Appl. Physiol.*, **39**, 4937(1968).

100. M. Shen, *Macromolecules*, **2**, 358(1969).

101. A. Ciferri, *Macromol. Chem.*, **43**, 152(1961).

102. P. J. Flory, L. Mandelkern, J. B. Kinsinger, and W. B. Schultz, *J. Am. Chem. Soc.*, **74**, 3364(1952).

103. H. Ciferri, *Trans. Faraday Soc.*, **57**, 853(1961).

104. K. Dusek, Ed., *Polymer Networks*, Springer-Verlag, New York, 1982.

105. J. E. Mark, *J. Chem. Ed.*, **58**(11), 898(1981).

106. J. E. Mark and J. Lal, Eds., *Elastomers and Rubber Elasticity*, American Chemical Society, Washington, DC, 1982.

107. A. V. Galanti and L. H. Sperling, *Polym. Eng. Sci.*, **10**, 177(1970).

108. B. A. Rozenberg, presented at *Networks 91*, Moscow, April 1991.

109. B. A. Rozenberg, *Epoxy Resins and Composites II*, Advances in Polymer Science Vol. 75, Springer-Verlag, Berlin, 1986.

110. A. H. Clark and S. B. Ross-Murphy, *Adv. Polym. Sci.*, **83**, 57(1987).

111. S. B. Ross-Murphy, *Polymer* (A. Keller Retirement Symposium issue), to be published 1992.

112. S. B. Ross-Murphy, presented at *Networks 91*, Moscow, April 1991.

113. L. Z Rogovina and V. G. Vasil'yev, *Networks 91*, Moscow, April 1991.

114. S. M. Aharoni and S. F. Edwards, *Macromolecules*, **22**, 3361(1989).

115. T. Tanaka, in *Encyclopedia of Polymer Science and Engineering*, 2nd ed., Vol., 7, Wiley, New York, 1987, p. 514.

116. P. I. Rose, in *Encyclopedia of Polymer Science and Engineering*, 2nd ed., Vol. 7, Wiley, New York, 1987, p. 488.

117. W. R. Krigbaum, J. V. Dawkins, G. H. Via, and Y. I. Balta, *J. Polym. Sci. A-2*, **4**, 475(1966).

118. F. Rodriguez, *Principles of Polymer Systems*, 2nd ed., McGraw-Hill, New York, 1982.

119. J. Karger-Kocis, A. Kallo, A. Szafner, G. Bodor, and Z. Senyei, *Polymer*, **20**, 37(1979).

120. D. W. Bartlett, J. W. Barlow, and D. R. Paul, *J. Appl. Polym. Sci.*, **27**, 2351(1982).

121. E. Martuscelli, M. Pracela, M. Avella, R. Greco, and G. Ragosta, *Makromol. Chem.*, **181**, 957(1980).

122. J. A. Manson and L. H. Sperling, *Polymer Blends and Composites*, Plenum, New York, 1976.

123. J. E. McGrath, *J. Chem. Ed.*, **58**(11), 914(1981).

124. A. Noshay and J. E. McGrath, *Block Copolymers—Overview and Critical Survey*, Academic Press, Orlando, 1977.

125. B. M. Walker, *Handbook of Thermoplastic Elastomers*, Van Nostrand–Reinhold, New York, 1979.

126. G. Holden, in *Recent Advances in Polymer Blends, Grafts, and Blocks*, L. H. Sperling, Ed., Plenum, New York, 1974.

127. E. Pechhold, G. Pruckmayr, and I. M. Robinson, *Rubber Chem. Technol.*, **53**, 1032(1980).

128. J. Blackwell and K. H. Gardner, *Polymer*, **20**, 13(1979).

129. L. L. Zhu, G. Wegner, and U. Bandara, *Makromol. Chem.*, **182**, 3639(1981).

130. A. Lilaonitkul and S. L. Cooper, *Rubber Chem. Technol.*, **50**, 1(1977).

131. P. C. Mody, G. L. Wilkes, and K. B. Wagener, *J. Appl. Polym. Sci.*, **26**, 2853(1981).

132. H. R. Allcock and F. W. Lampe, *Contemporary Polymer Chemistry*, Prentice-Hall, Englewood Cliffs, NJ, 1981, Chap. 7.

133. D. P. Tate, *J. Polym. Sci. Polym. Symp.*, **48**, 33(1974).

134. H. W. Greensmith, L. Mullins, and A. G. Thomas, in *The Chemistry and Physics of Rubber-like Substances*, L. B. Bateman, Ed., Maclaren and Sons, London, 1963, Chap. 19.

135. M. Morton and J. C. Healy, *Appl. Polym. Symp.*, **7**, 155(1968).

136. A. V. Galanti and L. H. Sperling, *Polym. Eng. Sci.*, **10**, 177(1970).

137. G. Kraus, *Rubber Chem. Technol.*, **51**, 297(1978).

138. B. B. Boonstra, *Polymer*, **20**, 691(1979).

139. Z. Rigbi, *Adv. Polym. Sci.*, **36**, 21(1980).

140. K. E. Polmanteer and C. W. Lentz, *Rubber Chem. Technol.*, **48**, 795(1975).

GENERAL READING

M. F. Bukhina, *Technical Physics of Elastomers*, Chemi, Moscow, 1984.

W. Burchard and S. B. Ross-Murphy, *Physical Networks: Polymers and Gels*, Elsevier, New York, 1990.

K. Dusek, Ed., *Polymer Networks*, Springer-Verlag, New York, 1982.

U. Eisele, *Introduction to Polymer Physics*, Springer-Verlag, New Yor, 1990.

R. N. Haward, Ed., *Developments in Polymerization-3, Network Formation and Cyclization in Polymer Reactions*, Applied Science Publishers, London, 1982.

O. Kramer, Ed., *Biological and Synthetic Polymer Networks*, Elsevier, London, 1988.

J. E. Mark and B. Erman, *Rubberlike Elasticity—A Molecular Primer*, Wiley-Interscience, New York, 1988.

M. Morton, Ed., *Rubber Technology*, 3rd ed., Van Nostrand–Reinhold, New York, 1987.

J. P. Queslel and J. E. Mark, *Encyclopedia of Polymer Science and Engineering*, 2nd ed., Vol. 5, Wiley, New York, 1986 p. 365.

L. H. Sperling, *Interpenetrating Polymer Networks and Related Materials*, Plenum, New York, 1981.

STUDY PROBLEMS

1. Why does a rubber band snap back when stretched and released? An explanation including both thermodynamic and molecular aspects is required. (Equations/diagrams/figures and as few words as possible will be appreciated.)

2. A strip of elastomer 1 cm × 1 cm × 10 cm is stretched to 25-cm length at 25°C, a stress of 1.5×10^7 dynes/cm^2 being required.

   (a) Assuming a tetrafunctional cross-linking mode, how many moles of network chains are there per cubic centimeter?

   (b) What stress is required to stretch the sample to only 15 cm, at 25°C?

   (c) What stress is required to stretch the sample to 25 cm length at 100°C?

3. The theory of rubber elasticity and the theory of ideal gas dynamics show that the two equations, $G = nRT$ and $PV = n'RT$, share certain common thermodynamic ideas. What are they?

4. Write the chemical structure for polybutadiene, and show its vulcanization reaction with sulfur.

5. For a swollen elastomer, the equation of state can be written

$$\sigma = nRTv_2^{1/3}\frac{\overline{r_i^2}}{r_0^2}\left(\alpha - \frac{1}{\alpha^2}\right)$$

Explain, qualitatively and very briefly, where the term $v_2^{1/3}$ originates. A derivation is not required.

6. Read any paper, 1990 or more recent, on rubber elasticity and write a 200-word report on it *in your own words*. (Give the reference!) Key

figures, tables, or equations may be photocopied. How is the science or engineering of elastomers advanced beyond this book?

7. Show how equations (9.81) and (9.82) reduce to $n = n_c + n_p$ for $W_g = 1$. What do you get when $W_g = 0$?

8. Recent experimental evidence using SANS instrumentation suggests that the ends of a network segment deform affinely, yet the chain itself barely extends in the direction of the stress and contracts in the transverse direction even less. Develop a model to explain the results, and comment on how you think the theory of rubber elasticity ought to be modified to accommodate the new finding.

9. We just had Halloween. Do you believe in "phantom networks"? Why?

10. A sample of vulcanized natural rubber, cis-polyisoprene, swells to five times its volume in toluene. What is Young's modulus of the unswollen elastomer at 25°C? (The interaction parameter, χ_1, is 0.39 for the system cis-polyisoprene–toluene.) [*Hint:* See equation (9.101); the molar volume of toluene is 106.3 cm^3/mol.]

11. Young's modulus for an elastomer at 25°C is 3×10^7 dynes/cm^2. What is its shear modulus? What is the retractive stress if a sample 1 cm × 1 cm × 10 cm is stretched to 25 cm length at 100°C?

12. A sample of elastomer, cross-linked at room temperature, is swollen afterward to 10 times its original volume. Then it is stretched. What value of the "front factor" should be used in the calculation of the stress?

13. Two identical 10-cm rubber bands, A and B, are tied together at their ends and stretched to a total of 40 cm length and held in that position. Rubber band A is at 25°C, and rubber band B is at 150°C. How far from the B end is the knot?

14. In the rubber heat engine described in Figure 9.14, the wheel is heated and turns accordingly. Does this experiment have the equivalent of all four steps illustrated in Figure 9.13a? [*Hint:* Don't forget gravity.]

15. A rubber ball is dropped from a height of 1 yard and bounces back 18 in. Assuming a perfectly elastic floor, approximately how much did the ball heat up? The heat capacity, C_p, of SBR rubber is about 1.83 kJ kg$^{-1} \cdot$ K^{-1}.

16. According to recent papers published in the *J. Theor. Hypothet. Polym. Sci.*, wheels go round, plastics break, and balls don't bounce. In one paper of recent vintage, two identical rubber bands, A and B, were dissolved in identical baths A' and B'. The solvent de-cross-linked the rubber bands but otherwise was a simple solvent. Rubber band A was dropped in unstretched. Rubber band B was rapidly stretched from its

initial length of 10 cm to 25 cm and instantly placed in the bath while held stretched by a holder of no physical properties during the solution process. Right after the solution process was completed, the poor investigator found he had mixed up the baths. Quick, how would you help him identify the baths? What basic and simple experiment would you perform? Assuming you found the difference, how does this difference change, algebraically, with the extent of stretch of rubber band B? If the two baths are identical, and no difference should exist, write a brief paragraph giving the correct reasons for full credit.

17. A sample of rubber was mixed with sulfur and other curing agents, molded into a sheet, and heated briefly in the relaxed state. A total of n_1 cross-links were introduced. Then the sheet was stretched α times its original length, and the heating continued, introducing n_2 new cross-links. The sample was then released and cooled. In terms of α, to what extent will the sample remain stretched?

APPENDIX 9.1   GELATIN AS A PHYSICALLY CROSS-LINKED ELASTOMER[†]

Introduction

Ordinary gelatin is made from the skins of animals by a partial hydrolysis of their collagen, an important type of protein (A1, A2). At home, a crude type of gelatin can be prepared from the broth of cooked meats and fowl; this material also frequently gels on cooling.

When dissolved in hot water, the gelatin protein has a random coil type of conformation. On cooling, a conformational change takes place to a partial helical arrangement. At the same time, intermolecular hydrogen bonds form, probably involving the N—H linkage. On long standing, such gels may also crystallize locally. The bonds that form in gelatin are known not to be permanent, but rather they relax in the time frame of 10^3–10^6 sec (A3–A5). The amount of bonding also decreases as the temperature is raised. The purpose of this appendix is to demonstrate the counting of these bonds via modulus measurements.

Theory

By observing the depth of indentation of a sphere into the surface of gelatin, "indentation" modulus is easily determined. The indentation modulus yields its close relative, Young's modulus. The cross-link density and thus the number of hydrogen bonds (simple physical cross-links) are readily determined by treating the gelatin as a hydrogen-bonded elastomer.

[†]Reproduced in part from the G. V. Henderson, D. O. Cambell, V. Kuzmicz, and L. H. Sperling, *J. Chem. Ed.*, **62**, 269(1985).

Young's modulus may be determined by indentation using the Hertz (A6) equation:

$$E = \frac{3(1 - \nu^2)F}{4h^{3/2}r^{1/2}}$$ (A9.1.1)

where F represents the force of sphere against the gelatin surface = mg (dynes), h represents the depth of indentation of sphere (cm), r is the radius of sphere (cm), g represents the gravity constant, and ν is Poisson's ratio.

The ball indentation experiment is the scientific analogue of pressing on an object with one's thumb to determine hardness. The less the indentation, the higher the modulus.

Young's modulus is related to the cross-link density through rubber elasticity theory, see equation (9.36):

$$E = 3nRT$$ (A9.1.2)

Assuming a tetrafunctional cross-linking mode (four chain segments emanating from the lcous of the hydrogen bond):

$$E = 6C_x RT$$ (A9.1.3)

where n represents the number of active chain segments in network and C_x is the cross-link density (moles of cross-links per unit volume). For this experiment, the gelatin was at 278.0 K, the temperature of the refrigerator employed.

Experimental

Time: About 30 min, the gelatin prepared previously.

Principles Illustrated:

1. Helix formation and physical cross-linking in gelatin.
2. Rubber elasticity in elastomers.
3. Physical behavior of proteins.

Equipment and Supplies:

Five 150 × 75 mm Pyrex® crystallizing dishes or soup dishes
Five 2-cup packets of flavored Jello® brand gelatin (8 g protein per packet)
Eighteen 2-cup packets of unflavored Knox® brand gelatin (6 g protein per packet)
One metric ruler
One steel bearing (1.5-in. diameter and 0.226 kg—or any similar spherical object)
One lab bench
One knife

Table A9.1.1 Gelatin concentrations

Dish	1	2	3	4	5
Concentration[a]	3.0	2.0	1.0	0.75	0.50
Jello[b]	1	1	1	1	1
Gelatin[c]	8	5	2	1.25	0.5

[a]Concentration = number of times the normal gelatin concentration (each dish contains 600 ml of water).
[b]Jello = number of 2-cup packets of Jello® brand black raspberry fla-vored gelatin.
[c]Gelatin = number of 2-cup packets of Knox® brand unflavored gelatin.

eye level   ruler   Gelatin cross section

DEPTH OF INDENTATION.

$$h(cm) = d - h^*$$

Figure A9.1.1 Schematic of experiment, measuring the indentations of the heavy ball in the gelatin.

Five different concentrations (see Table A9.1.1) of gelatin were prepared, each in 600 ml of water, and allowed to set overnight in a refrigerator at 5.0°C. Then indentation measurements were made by placing the steel bearing in the center of the gelatin samples and measuring the depth of indentation, h (see Figure A9.1.1). As it is difficult to see through the gelatin to observe this depth, it is desirable to measure the height of the bearing from the level surface of the gelatin and subtract this quantity from the diameter of the bearing (Figure A9.1.1).

The measured depth of indentation, the radius of the bearing, and the force due to the bearing are algebraically substituted into equation (A9.1.1). This value of Young's modulus is substituted into equation A9.1.3 to yield hydrogen bond cross-link density.

Results

A plot of E as a function of gelatin concentration (Figure A9.1.2) demon-strates a linear increase in Young's modulus at low concentrations. The slight upward curvature at high concentrations is caused by the increasing efficiency of the network. However, the line should go through the origin.

Figure A9.1.2 Modulus of gelatin samples versus concentration. By comparison, a rubber band has a Young's modulus of about 1×10^6 pascals (1 pascal = 10 dynes/cm^2).

Table A9.1.2   Gelatin indentations yield bond numbers

Concentration	h (cm)	C_x (mol/cm^3)	N
0.5	1.50	3.5×10^{-7}	1.20
0.75	1.30	4.4×10^{-7}	1.00
1.0	1.20	5.0×10^{-7}	1.00
2.0	0.80	9.1×10^{-7}	0.86
3.0	0.40	2.6×10^{-6}	1.70

Key: h, indentation measured at gelatin temperature of 5°C; C_x, number of hydrogen bonds; N, number of bonds per molecule.

Physical cross-link concentrations were determined using equation (A9.1.3), and the results are shown in Table A9.1.2. Assuming a molecular weight of about 65,000 g/mol for the gelatin, there is about 1.2 physical bonds per molecule (see Table A9.1.2).

Using gelatin as a model cross-linked elastomer, its rubber elasticity can also be demonstrated by a simple stretching experiment. Thin slices of the more concentrated gelatin samples were cut and stretched by hand. On release from stretches up to about 50%, the sample snaps back, illustrating the rubberlike elasticity of these materials. At greater elongation the sample breaks, however. The material is weak because the gelatin protein chains are much diluted with water.

Discussion

For rubbery materials, Young's modulus is related to the number of cross-links in the system. In this case, the cross-links are of a physical nature, caused by hydrogen bonding. Measurement of the modulus via ball indentation techniques allows a rapid, inexpensive method of counting these bonds. Table A9.1.2 shows that the number of these bonds is of the order of 10^{-7} mol/cm^3. The number of these bonds also was shown to increase linearly with concentration, except at the highest concentrations.

Table A9.1.2 also demonstrates that at each gelatin concentration, the number of bonds per gelatin molecule is relatively constant. This number, of course, is the number of bonds taking part in three-dimensional network formation. Not all the gelatin chains are bound in a true tetrafunctionally cross-linked network. Many dangling chain ends exist at these low concentrations, and the network must be very imperfect.

The gelation molecule is basically composed of short α-helical segments in the form of a triple helix with numerous intramolecular bonds at room temperature, see section 9.13. The α-helical segments are interrupted by proline and hydroxy proline functional groups. These groups disrupt the helical structure, yielding intervening portions of chain that behave like random coils, and which may be relatively free to develop intermolecular bonds. The subject has been reviewed recently by Djabourov (A7) and Mel'nichenko, et al. (A8).

In this experiment, the concentration of sugar was kept constant so as to minimize its effect on the modulus. In concluding, it must be pointed out that if sanitary measures are maintained, the final product may be eaten at the end of the experiment. If gelation five times normal or higher is included in the study, the student should be prepared for his or her jaws springing open after biting down!

REFERENCES

A1. A. Veis, *Macromolecular Chemistry of Gelatin*, Academic Press, Orlando, 1964.

A2. E. M. Marks, in *Encyclopedia of Chemical Technology*, Kirk–Othmer, Interscience, New York, 1966, Vol. 10, p. 499.

A3. J. L. Laurent, P. A. Janmey, and J. D. Ferry, *J. Rheol.*, **24**, 87(1980).

A4. M. Miller, J. D. Ferry, F. W. Schremp, and J. E. Eldridge, *J. Phys. Colloid Chem.*, **55**, 1387(1951).

A5. J. D. Ferry, *Viscoelastic Properties of Polymers*, 3rd ed., Wiley, New York, 1980, pp. 529–539.

A6. L. H. Sperling, *Interpenetrating Polymer Networks and Related Materials*, Plenum Press, New York, 1981, p. 177.

A7. M. Djabourov, *Contemp. Phys.*, *29*(3) 273(1988).

A8. Yu. Mel'nichenko, Yu. P. Gomza, V. V. Shilov, and S. I. Osipov, *Polym. Intern.* (*Brit. Polym. J.*), *25(3)*, 153(1991).

Figure B9.1.1   Simple rubber–elastic behavior of a rubber band under increasing load.

APPENDIX 9.2   ELASTIC BEHAVIOR OF A RUBBER BAND[†]

Stretching a rubber band makes a good demonstration of the stress–strain relationships of cross-linked elastomers. The time required is about 30 min. The equipment includes a large rubber band (Star® band size 107, E. Faber, Inc., Wilkes-Barre, PA, is suitable), a set of weights up to 25 kg, and a meter stick. Also required are hooks to attach the weights and a high place from which to hang the rubber band.

First, the rubber band is measured, both in length and cross section, and the hooks are weighed. Increasing weight is hung from the rubber band, its length being recorded at each step. When it nears its breaking length, caution is advised.

A plot of stress (using initial cross-sectional area) as a function of α, Figure B9.1.1, demonstrates the nonlinearity of the stress–strain relationship. Initial values of the slope of the curve yield Young's modulus, E. The sharp upturn of the experimental curve at high elongations is due to the limited

[†]Reproduced in part from A. J. Etzel, S. J. Goldstein, H. J. Panabaker, D. G. Fradkin, and L. H. Sperling, *J. Chem. Ed.*, **63**, 731(1986).

extensibility of the chains themselves. The number of active network chains per unit volume can be calculated from equation (9.34) as 1.9×10^2 mol/m^3.

A Mooney–Rivlin plot according to equation (9.49) yields a curve that rapidly increases for values of $1/\alpha$ greater than 0.25; see Figure 9.18. The constants $2C_1$ and $2C_2$ are calculated from the intercept and slope, respectively. Values of 2.3×10^5 Pa and 2.8×10^5 Pa were obtained, respectively.

APPENDIX 9.3   DETERMINATION OF THE CROSS-LINK DENSITY OF RUBBER BY SWELLING TO EQUILIBRIUM[†]

The present experiment is based on the rapid swelling of elastomers by organic solvents. Application of the Flory–Rehner equation yields the number of active network chain segments per unit volume, a measure of the extent of vulcanization (C1, C2).

Experiment

Time: About 1 hr

Level: Physical Chemistry

Principles Illustrated:

1. The cross-linked nature of rubber
2. Diffusion of a solvent into a solid

Equipment and Supplies:

One large rubber band

One 600-ml beaker (containing 300 ml toluene)

One ruler or yardstick

One long tweezers to remove swollen rubber band

Paper towels to blot wet swollen rubber band

One clock or watch

One laboratory bench

First, cut the rubber band in one place to make a long rubber strip. Measure and record its length in the relaxed state. Place in the 600-ml beaker with toluene, making sure the rubber band is completely covered. Remove after 5–10 min. Blot dry. **Caution: toluene is toxic and can be absorbed through the skin**. Again, measure and record length. Repeat for about 1 hr. Optional: Cover and store overnight. Measure the length of the band the next day.

[†]Reprinted in part from L. H. Sperling and T. C. Michael, *J. Chem. Ed.*, **59**, 651(1982).

Expected Results

The rubber band swells to about twice its original length, but then it remains stable. (Note that swelling to twice its length means a volume increase of about a factor of 8.) Also, *note that the swollen rubber band is much weaker than the dry material and may break if not treated gently.*

Chemically, most rubber bands and similar materials are composed of a random copolymer of butadiene and styrene, written poly(butadiene–*stat*–styrene), meaning that the placement of the monomer units is statistical along the chain length. Usually, this product is made via emulsion polymerization.

SWELLING OF A RUBBER BAND WITH TIME

Length, cm	Time, min
16.5	0
24.0	14
26.0	25
27.0	36
28.0	70

Typical results are shown in the table. Over a period of 70 min, the length of the rubber band increased from 16.5 to 28.0 cm, for a volume increase of about 4.9. This is sufficiently obvious to be seen at the back of an ordinary classroom. The rubber band would continue to swell slowly for some hours, or even days, but for the purposes of demonstrations and classroom calculations, the swelling can be considered nearly complete.

Calculations

For the system poly(butadiene–*stat*–styrene) and toluene, χ_1 is 0.39. Assuming additivity of volumes v_2 is found from the swelling data to be 0.205. The quantity V_1 is 106.3 cm^3/mol for toluene. Algebraic substitution into equation (9.101) yields n equal to 1.55×10^{-4} mol/cm^3. (Compare result with Appendix 9.2.)

Extra Credit

Two experiments (or demonstrations) can be done easily for extra credit.

1. Obtain some unvulcanized rubber. Most tire and chemical companies can supply this.[†] Put a piece of this material into toluene overnight and observe the results. It should dissolve to form a uniform solution.

[†]Commercial source: Firestone Synthetic Rubber and Latex Co. Trade name: Stereon®.

2. The quantity n can be used also to predict Young's modulus (the stiffness) of the rubber band. The equation is

$$E = 3nRT \qquad (A9.2.1)$$

where E represents Young's modulus, and R in these units is 8.31×10^7 dynes $\cdot$ cm/mol $\cdot$ K. For the present experiment, E is calculated to be 1.1×10^7 dynes/cm^2, typical of such rubbery products.

REFERENCES

C1. P. J. Flory, and J. Rehner, *J. Chem. Phys.*, **11**, 521(1943).

C2. P. J. Flory, *Principles of Polymer Chemistry*, Cornell University Press, Ithaca, NY, 1953.

10

POLYMER VISCOELASTICITY AND RHEOLOGY

The study of polymer viscoelasticity treats the interrelationships among elasticity, flow, and molecular motion. In reality, no liquid exhibits pure Newtonian viscosity, and no solid exhibits pure elastic behavior, although it is convenient to assume so for some simple problems. Rather, all motion of real bodies includes some elements of both flow and elasticity. Because of the long-chain nature of polymeric materials, their viscoelastic characteristics come to the forefront. This is especially true when the times for molecular motion are of the same order of magnitude as an imposed mechanical motion.

Chapters 8 and 9 have introduced the concepts of the glass transition and rubber elasticity. In particular, Section 8.2 outlined the five regions of viscoelasticity, and Section 8.6.1.2 derived the WLF equation. This chapter treats the subjects of stress relaxation and creep, the time–temperature superposition principle, and melt flow.

10.1 STRESS RELAXATION AND CREEP

In a stress relaxation experiment, the sample is rapidly stretched to the required length, and the stress is recorded as a function of time. The length of the sample remains constant, so there is no macroscopic movement of the body during the experiment. Usually the temperature remains constant also (1).

Creep experiments are conducted in the inverse manner. A constant stress is applied to a sample, and the dimensions are recorded as a function of time. Of course, these experiments can be generalized to include shear motions, compression, and so on.

Section 8.1.6 defined the modulus of a material as a measure of its stiffness, and compliance as a measure of its softness. Under conditions far from a transition, $E \cong 1/J$. Frequently, stress relaxation experiments are reported as the time-dependent modulus, $E(t)$, whereas creep experiments are reported as the time-dependent compliance, $J(t)$.

10.1.1 Molecular Bases of Stress Relaxation and Creep

While the exact molecular causes of stress relaxation and creep are varied, they can be grouped into five general categories (2):

1. *Chain scission.* Oxidative degradation and hydrolysis are the primary causes.

The reduction in stress caused by chain scission can be illustrated by a model where three chains are bearing a load:

$$(10.1)$$

2. *Bond interchange.* While this is not a degradation in the sense that the molecular weight is decreased, chain portions changing partners cause a release of stress. Examples of stress relaxation by bond interchange include polyesters and polysiloxanes. Equation (10.2) provides a simple example (3).

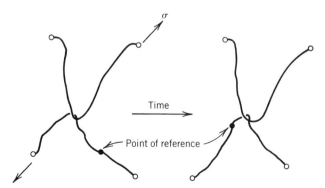

Figure 10.1 An illustration of reversible motion in a trapped entanglement under stress. The marked point of reference moves with time. When the stress is released, entropic forces return the chains to near their original positions.

Bond interchange is going on constantly in polysiloxanes, with or without stress. In the presence of a stress, however, the statistical rearrangements tend to reform the chains so that the stress is reduced.

3. *Viscous flow.* Caused by linear chains slipping past one another, this mechanism is responsible for viscous flow in pipes and elongational flow under stress. An example is the pulling out of Silly Putty.®

4. *Thirion relaxation* (4). This is a reversible relaxation of the physical cross-links or trapped entanglements in elastomeric networks. Figure 10.1 illustrates the motions involved. Usually, an elastomeric network will relax about 5% by this mechanism, most of it in a few seconds. It must be emphasized that the chains are in constant motion of the reptation type (5) (see Section 5.4).

$$
\begin{array}{c}
\text{[structure with } \sigma = \text{constant]}
\end{array}
\qquad (10.2)
$$

interchange along dotted line →

[structure with $\sigma = 0$]

5. *Molecular relaxation, especially near T_g.* This topic was the major subject of Chapter 8, pointing out that near T_g the chains relax at about the same rate as the time frame of the experiment. If the chains are under stress during the experiment, the motions will tend to relieve the stress.

It must be emphasized that more than one of the above relaxation modes may be operative during any real experiment.

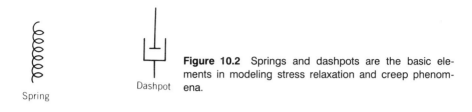

Figure 10.2 Springs and dashpots are the basic elements in modeling stress relaxation and creep phenomena.

10.1.2 Models for Analyzing Stress Relaxation and Creep

To permit a mathematical analysis of the creep and relaxation phenomenon, spring and dashpot elements are frequently used (see Figure 10.2). A spring behaves exactly like a metal spring, stretching instantly under stress and holding that stress indefinitely. A dashpot is full of a purely viscous fluid. Under stress, the plunger moves through the fluid at a rate proportional to the stress. On removing the stress, there is no recovery. Both elements may be deformed indefinitely.

10.1.2.1 The Maxwell and Kelvin Elements

The springs and dashpots can be put together to develop mathematically amenable models of viscoelastic behavior. Figure 10.3 illustrates the simplest two such arrangements, the Maxwell and the Kelvin (sometimes called the Voigt) elements. While the spring and the dashpot are in series in the Maxwell element, they are in parallel in the Kelvin element. In such arrangements, it is convenient to assign moduli E to the various springs, and viscosities η to the dashpots.

In the Maxwell element, both the spring and the dashpot are subjected to the same stress but are permitted independent strains. The inverse is true for the Kelvin element, which is equivalent to saying that the horizontal connect-

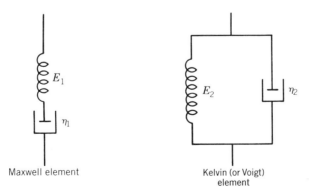

Figure 10.3 The Maxwell and Kelvin (or Voigt) elements, representing simple series and parallel arrays of springs and dashpots.

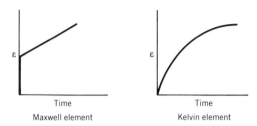

Figure 10.4 Creep behavior of the Maxwell and Kelvin elements. The Maxwell element exhibits viscous flow throughout the time of deformation, whereas the Kelvin element reaches an asymptotic limit to deformation.

ing portions on the right-hand side of Figure 10.3 are constrained to remain parallel.

As examples of the behavior of combinations of springs and dashpots, the Maxwell and Kelvin elements will be subjected to creep experiments. In such an experiment, a stress, σ, is applied to the ends of the elements, and the strain, ε, is recorded as a function of time. The results are illustrated in Figure 10.4.

On application of the stress to the Maxwell element, the spring instantly responds, as illustrated by the vertical line in Figure 10.4. The height of the line is given by $\varepsilon = \sigma/E$. The spring term remains extended, as the dashpot gradually pulls out, yielding the slanted upward line. This model illustrates elasticity plus flow.

The spring and the dashpot of the Kelvin element undergo concerted motions, since the top and bottom bars (see Figure 10.3) are constrained to remain parallel. The dashpot responds slowly to the stress, bearing all of it initially and gradually transferring it to the spring as the latter becomes extended.

The rate of strain of the dashpot is given by

$$\frac{d\varepsilon}{dt} = \frac{\sigma}{\eta} \tag{10.3}$$

When the spring bears all the stress, both the spring and the dashpot stop deforming together, and creep stops. Thus, at long times, the Kelvin element exhibits the asymptotic behavior illustrated in Figure 10.4. In more complex arrangements of springs and dashpots, the Kelvin element contributes a retarded elastic effect.

10.1.2.2 The Four-Element Model

While a few problems in viscoelasticity can be solved with the Maxwell or Kelvin elements alone, more often they are used together or in other combinations. Figure 10.5 illustrates the combination of the Maxwell element

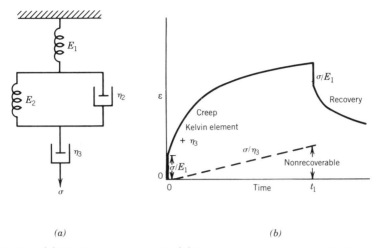

Figure 10.5 (*a*) The four-element model. (*b*) Creep behavior as predicted by this model. At t_1, the stress is relaxed, and the model makes a partial recovery.

and the Kelvin element in series, known as the four-element model. It is the simplest model that exhibits all the essential features of viscoelasticity.

On the application of a stress, σ, the model (Figure 10.5*a*) undergoes an elastic deformation, followed by creep (Figure 10.5*b*). The deformation due to η_3, true flow, is nonrecoverable. Thus, on removal of the stress, the model undergoes a partial recovery.

The four-element model exhibits some familiar behavior patterns. Consider the effects of stretching a rubber band around a book. Initially, E_1 stretching takes place. As time passes, $E_2 + \eta_2 + \eta_3$ relaxations take place. On removing the rubber band at a later time, the remaining E_1 recovers. Usually, the rubber band circle is larger than it was initially. This permanent stretch is due to η_3. Although less obvious, the Kelvin element motions can also be observed by measuring the rubber band dimensions immediately after removal and again at a later time.

The quantities E and η of the models shown above are not, of course, real-life values of modulus and viscosity. However, as shown below, they can be used in numerous calculations to provide excellent predictions or understanding of viscoelastic creep and stress relaxation. It must be emphasized that E and η themselves can be governed by theoretical equations. For example, if the polymer is above T_g, the theory of rubber elasticity can be used. Likewise, the WLF equation can be used to represent that portion of the deformation due to viscous flow, or for the viscous portion of the Kelvin element.

More complex arrangements of elements are often used, especially if multiple relaxations are involved or if accurate representations of engineering data are required. The Maxwell–Weichert model consists of a very large (or infinite) number of Maxwell elements in parallel (2). The generalized Voigt–Kelvin model places a number of Kelvin elements in series. In each of these models, a spring or a dashpot may be placed alone, indicating elastic or viscous contributions.

10.1.2.3 The Takayanagi Models

Most polymer blends, blocks, grafts, and interpenetrating polymer networks are phase-separated. Frequently, one phase is elastomeric, and the other is plastic. The mechanical behavior of such a system can be represented by the Takayanagi models (6). Instead of the arrays of springs and dashpots, arrays of rubbery (R) and plastic (P) phases are indicated (see Figure 10.6) (7). The quantities λ and φ or their indicated multiplications indicate volume fractions of the materials.

As with springs and dashpots, the Takayanagi models may also be expressed analytically. For parallel model Figure 10.6a, the horizontal bars connecting the two elements must remain parallel and horizontal, yielding an

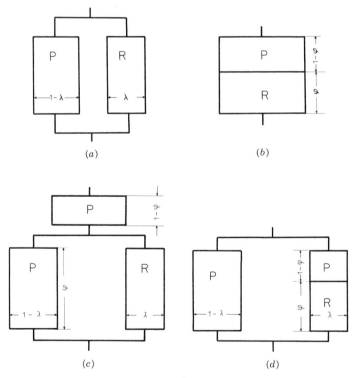

Figure 10.6 The Takayanagi models for two-phase systems: (a) an isostrain model; (b) isostress model; (c, d) combinations. The area of each diagram is proportional to a volume fraction of the phase.

isostrain condition ($\varepsilon_P = \varepsilon_R$). Then

$$\sigma = \sigma_R + \sigma_P \tag{10.4}$$

$$\sigma_i = \varepsilon E_i; \quad i = P, R \tag{10.5}$$

$$E = (1 - \lambda)E_P + \lambda E_R \tag{10.6}$$

Figure 10.6a represents an upper bound model, meaning that the modulus predicted is the highest achievable.

For example, take a 50/50 blend of a plastic and an elastomer. Typical Young's moduli are $E_P = 3 \times 10^{10}$ dynes/cm^2, and $E_R = 2 \times 10^7$ dynes/cm^2. The quantity λ equals 0.5, and equation (10.6) yields 1.5×10^{10} dynes/cm^2. To an excellent approximation, this is half the modulus of the plastic.

For the series model Figure 10.6b, an isostress strain condition exists. The strains are additive,

$$\varepsilon = \varepsilon_P + \varepsilon_R \tag{10.7}$$

yielding a Young's modulus of

$$E = \left(\frac{\varphi}{E_R} + \frac{1 - \varphi}{E_P} \right)^{-1} \tag{10.8}$$

This is a lower bound modulus, meaning that equation (10.8) yields the lowest modulus any material may have. If the above example is used, with $\varphi = 0.5$, equation (10.8) yields $E = 4 \times 10^7$ dynes/cm^2, or approximately twice the elastomer value.

Surprisingly, it is possible to make two different 50/50 blends of plastics and elastomers and obtain very nearly both of the results predicted by the above examples. This depends on the morphology, and particularly on phase continuity; see Figure 4.3. When the plastic phase is continuous in space and the elastomer dispersed (Figure 10.6a), the material will be stiff, and substantially exhibit upper bound behavior. On the other hand, when the elastomer is continuous and the plastic discontinuous (Figure 10.6b), a much softer material results and a lower bound situation prevails. The rubber phase is much more obviously discontinuous in the models represented in Figures 10.6c and 10.6d.

While the above examples used Young's modulus, many other parameters may be substituted. These include other moduli, rheological functions such as creep, stress relaxation, melt viscosity, and rubber elasticity. Each element may itself be expressed by temperature-, time-, or frequency-dependent quantities. It must also be noted that these models find application in composite problems as well. For example, a composite of continuous fibers in a plastic matrix can be described by Figure 10.6a if deformed in the direction of the fibers, and by Figure 10.6b if deformed in the transverse direction.

Figures 10.6c and 10.6d represent a higher level of sophistication, allowing for more precise calculations involving continuous and discontinuous phases. For Figures 10.6c and 10.6d, respectively, Young's moduli are given by

$$E = \left[\frac{\varphi}{\lambda E_R + (1 - \lambda) E_P} + \frac{1 - \varphi}{E_P} \right]^{-1} \tag{10.9}$$

$$E = \lambda \left(\frac{\varphi}{E_R} + \frac{1 - \varphi}{E_P} \right) + (1 - \lambda) E_P \tag{10.10}$$

For these latter models, the volume fraction of each phase is given by a function of λ multiplied by a function of φ. The volume fraction of the upper right-hand portion of Figure 10.6d is given by $\lambda(1 - \varphi)$.

10.2 RELAXATION AND RETARDATION TIMES

The various models were invented explicitly to provide a method of mathematical analysis of polymeric viscoelastic behavior. The Maxwell element expresses a combination of Hooke's and Newton's laws. For the spring,

$$\sigma = E\varepsilon \tag{10.11}$$

the time dependence of the strain may be expressed as

$$\frac{d\varepsilon}{dt} = \frac{1}{E} \frac{d\sigma}{dt} \tag{10.12}$$

The time dependence of the strain on the dashpot is given by

$$\frac{d\varepsilon}{dt} = \frac{\sigma}{\eta} \tag{10.13}$$

Since the Maxwell model has a spring and dashpot in series, the strain on the model is the sum of the strains of its components:

$$\frac{d\varepsilon}{dt} = \frac{1}{E} \frac{d\sigma}{dt} + \frac{\sigma}{\eta} \tag{10.14}$$

10.2.1 The Relaxation Time

For a Maxwell element, the relaxation time is defined by

$$\tau_1 = \frac{\eta}{E} \tag{10.15}$$

The viscosity, η, has the units of dynes $\cdot$ sec/cm^2, and the modulus, E, has the units of dynes/cm^2, so τ_1 has the units of time. Thus τ_1 relates modulus and viscosity.

On a molecular scale, the relaxation time of a polymer indicates the order of magnitude of time required for a certain proportion of the polymer chains to relax—that is, to respond to the external stress by thermal motion. It should be noted that the chains are in constant thermal motion whether there is an external stress or not. The stress tends to be relieved, however, when the chains happen to move in the right direction, degrade, and so on.

Alternatively, τ_1 can be a measure of the time required for a chemical reaction to take place. Common reactions that can be measured in this way include bond interchange, degradation, hydrolysis, and oxidation. Combining equations (10.14) and (10.15) leads to

$$\frac{d\varepsilon}{dt} = \frac{1}{E}\frac{d\sigma}{dt} + \frac{\sigma}{\tau_1 E} \tag{10.16}$$

where the first term on the right is important for short-time changes in strain, and the second term on the right controls the longer-time changes.

The stress relaxation experiment requires that $d\varepsilon/dt = 0$; that is, the length does not change, and the strain is constant. Integrating equation (10.16) under these conditions leads to

$$\sigma = \varepsilon_0 E e^{-(t/\tau_1)} \tag{10.17}$$

or

$$\sigma = \sigma_0 e^{-(t/\tau_1)} \tag{10.18}$$

Equations (10.17) and (10.18) predict a straight-line relationship between $\ln \sigma$ or $\log \sigma$ and linear time if a single mechanism controls the relaxation process. If experiments other than simple elongation are done (i.e., relaxation in shear), the appropriate modulus replaces E in equation (10.17).

10.2.2 Applications of Relaxation Times to Chemical Reactions

10.2.2.1 Chemical Stress Relaxation

As indicated in Section 10.1.1, stress relaxation can be caused by either physical or chemical phenomena. Examples will be given of each.

Figure 10.7 (8) illustrates the stress relaxation of a poly(dimethyl siloxane) network, silicone rubber, in the presence of dry nitrogen. The reduced stress,

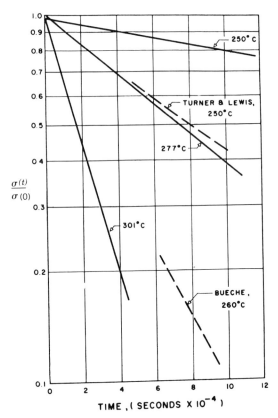

Figure 10.7 Stress relaxation of silicone rubber, poly(dimethyl siloxane). The rate of stress relaxation increases with temperature, but the lines remain straight (8).

$\sigma(t)/\sigma(0)$, is plotted, so that under the initial conditions its value is always unity. Since the theory of rubber elasticity holds (Chapter 9), what is really measured is the fractional decrease in effective network chain segments. The bond interchange reaction of equation (10.2) provides the chemical basis of the process. While the rate of the relaxation increases with temperature, the lines remain straight, suggesting that equation (10.2) can be treated as the sole reaction of importance.

The relaxation times may be estimated from the time necessary for $\sigma(t)/\sigma(0)$ to drop to $1/e = 0.368$. The results are shown in Table 10.1 (8).

Appendix 10.1 derives a relationship between the relaxation time and the energy of activation, ΔE_{act},

$$\tau_1 = \text{constant} \times e^{\Delta E_{act}/RT} \tag{10.19}$$

Table 10.1   Chemical stress relaxation times for silicone rubber (8)

Temperature, °C	$\tau_1 \times 10^{-4}$, sec
250	48
277	10.5
301	2.45

A plot of $\ln \tau_1$ versus $1/T$ yields $\Delta E_{act}/R$ for the slope. An apparent energy of activation of 35 kcal/mol was estimated from such a plot. Thus a purely chemical quantity can be deduced from a mechanical experiment.

Stress relaxation experiments were used to determine the mechanism of degradation in synthetic polymers (9) during World War II, when these materials were first being made. Tobolsky later described the results of these famous experiments (2):

It was found that in the temperature range of 100 to 150°C, these vulcanized rubbers showed a fairly rapid decay to zero stress at constant extension. Since in principle a cross-linked rubber network in the rubbery range of behavior should show little stress relaxation, and certainly no decay to zero stress, the phenomenon was attributed to a chemical rupture of the rubber network. This rupture was specifically ascribed to the effect of molecular oxygen since under conditions of *very low* oxygen pressures ($< 10^{-4}$ atm) the stress–relaxation rate was markedly diminished. However at moderately low oxygen pressures the rate of chemical stress relaxation was the same as at atmospheric conditions. This result parallels the very long established fact that in the liquid phase the rate of reaction of hydrocarbons with oxygen is independent of oxygen pressure down to fairly low pressures.

10.2.2.2 Procedure X

Stress relaxation can also be used to separate and identify two or more reactions causing relaxation, provided the rates are sufficiently different. Consider reactions a and b going on simultaneously,

$$\sigma(t) = \sigma_a(0)e^{-t/\tau_{1a}} + \sigma_b(0)e^{-t/\tau_{1b}} \qquad (10.20)$$

If the two relaxation times, chemical or physical, are sufficiently different, two straight lines may be obtained by algebraic analysis. Tobolsky named this method of analysis "procedure X" (2).

10.2.2.3 Continuous and Intermittent Stress Relaxation

Another "trick" to separate two reactions involves continuous and intermittent stress relaxation measurements. Figure 10.8 illustrates the separation of degradation and cross-linking in cis-1,4-polybutadiene. In an intermittent stress relaxation experiment, the sample is maintained in a relaxed, unstretched condition at a constant temperature. At suitably spaced time intervals the rubber is rapidly stretched to a fixed elongation, and the stress is measured. Then the sample is returned to its unstretched length. Of course, in the continuous stress relaxation experiment, the strain is maintained continuously.

At 130°C, the temperature of the experiment, oxidative scission and cross-linking are both going on all the time. The reactions happen, however, in the condition of the network at the time. This means that the continuous stress relaxation experiment measures only the degradation step, because the new cross-links form in the stretched chains; that is, the second network develops in the extended state, at equilibrium. There is no significant change in conformational entropy on formation of these cross-links, and the change in stress is near zero.

On the other hand, if the oxidative cross-links form when the sample is unstretched, then they can be measured afterward by using the theory of

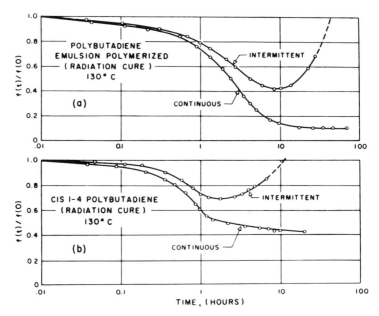

Figure 10.8 Continuous and intermittent stress relaxation at 130°C on radiation cured cis-1,4-polybutadiene in the presence of air. The experiment separates oxidative cross-linking and chain scission. The quantity $f(t)$ is the time-dependent retractive force. From A. V. Tobolsky, *Properties and Structure of Polymers*, Wiley, New York, 1960.

rubber elasticity. The intermittent experiment measures the total number of active chain segments at any given instant of time. As illustrated in Figure 10.8 the cross-linking reaction predominates after 10 h, and the sample actually gets harder. This last is often observed at home, where old rubber materials tend to stiffen up.

10.2.3   The Retardation Time

As the relaxation time, τ_1, is defined for the Maxwell elements, the retardation time, τ_2, is defined for the Kelvin element. The equation for the Kelvin element under stress can be written

$$\sigma = \eta \frac{d\varepsilon}{dt} + E\varepsilon \tag{10.21}$$

Under conditions of constant stress, equation (10.21) can be integrated to

$$\varepsilon = \frac{\sigma}{E}(1 - e^{-(E/\eta)t}) \tag{10.22}$$

which can be rewritten in terms of the retardation time,

$$\varepsilon = \frac{\sigma}{E}(1 - e^{-t/\tau_2}) \tag{10.23}$$

The equation for the four-element model (Figure 10.5a) may now be written for the condition of constant stress:

$$\varepsilon = \varepsilon_1 + \varepsilon_2 + \varepsilon_3 \tag{10.24}$$

$$\varepsilon = \frac{\sigma}{E_1} + \frac{\sigma}{E_2}(1 - e^{-t/\tau_2}) + \frac{\sigma}{\eta_3}t \tag{10.25}$$

The first term on the right of equation (10.25) represents an elastic term, the second term expresses the viscoelastic effect, and the third term expresses the viscous effect.

It is of interest to compare the retardation time with the relaxation time. The retardation time is the time required for E_2 and η_2 in the Kelvin element to deform to $1 - 1/e$, or 63.21% of the total expected creep. The relaxation time is the time required for E_1 and η_3 to stress relax to $1/e$ or 0.368 of σ_0, at constant strain. Both τ_1 and τ_2, to a first approximation, yield a measure of the time frame to complete about half of the indicated phenomenon, chemical or physical. A classroom demonstration experiment showing the determination of the constants in the four-element model is shown in Appendix 10.2.

10.2.4 Dynamic Mechanical Behavior of Springs and Dashpots

In addition to stress relaxation and creep, springs and dashpots can model the loss and storage characteristics of polymers undergoing cyclic motions. Since a principal application of such modeling is for noise and vibration damping analysis (see Section 8.12) where the motions are of a shearing nature, the equations for shear are emphasized below. The viscoelastic motions of a Maxwell element in shear may be written for an angular frequency ω (rad/sec) (10):

$$J(t) = J + \frac{t}{\eta} \tag{10.26}$$

$$G(t) = Ge^{-t/\tau_1} \tag{10.27}$$

$$G'(\omega) = \frac{G\omega^2\tau_1^2}{1 + \omega^2\tau_1^2} \tag{10.28}$$

$$G''(\omega) = \frac{G\omega\tau_1}{1 + \omega^2\tau_1^2} \tag{10.29}$$

$$J'(\omega) = J \tag{10.30}$$

$$J''(\omega) = \frac{J}{\omega\tau_1} = \frac{1}{\omega\eta_1} \tag{10.31}$$

$$\tan\delta = \frac{1}{\omega\tau_1} \tag{10.32}$$

The scaling of time in relaxation processes is achieved by means of the "Deborah number," which is defined as

$$D_e = \tau_e/t \tag{10.33}$$

where t is a characteristic time of the deformation process being observed and τ_e is a characteristic time of the polymer. For a Hookian elastic solid, τ_e is infinite, and for a Newtonian viscous liquid, τ_e is zero; see Section 10.2.1. For polymer melts, τ_e is often of the order of a few seconds. Of course, τ_e can be the relaxation time, or the retardation time of the polymer. The Deborah number is widely used in engineering as a single-number approximation to the several molecular phenomena involved in the deformation and flow of many materials, including polymers.

The loss tangent is seen to be a maximum when $\tau_1 = 1/\omega$—that is, when the time required for one cycle of the experiment equals the relaxation time. Of course, $G = 1/J$ far from a transition (Chapter 8).

The corresponding quantities for the Kelvin element are as follows (10).

$$J(t) = J(1 - e^{-t/\tau_2}) \tag{10.34}$$

$$G(t) = G \tag{10.35}$$

$$G'(\omega) = G \tag{10.36}$$

$$G''(\omega) = G\omega\tau_2 = \omega\eta_2 \tag{10.37}$$

$$\eta'(\omega) = \eta \tag{10.38}$$

$$J'(\omega) = \frac{J}{1 + \omega^2\tau_2^2} \tag{10.39}$$

$$J''(\omega) = \frac{J\omega\tau_2}{1 + \omega^2\tau_2^2} \tag{10.40}$$

$$\tan \delta = \omega\tau_2 \tag{10.41}$$

If springs are equated to capacities and resistances to dashpots, the storage and dissipative units are seen to correspond to time-dependent electrical behavior. However, the topology is backward; that is, series electrical connections correspond to parallel mechanical connections (10).

10.2.5 Molecular Relaxation Processes

The preceding sections describe the mechanical behavior of a polymeric sample in terms of creep and stress relaxation. Both elastomeric and plastic materials can be modeled by combinations of springs and dashpots. However, these are only models. Ultimately, stress relaxation and creep derive from molecular origins, and it is in this area that more recent studies have been concentrated (11–15).

The most important method of characterizing the molecular relaxation processes has been through the use of small-angle neutron scattering (see Figure 10.9) (16). Boue et al. (15) investigated linear polystyrene of medium and high molecular weight. In each case, blends of deuterated and protonated polystyrenes were prepared, and the samples stretched up to an α value of 3 at temperatures above T_g in the range of 113–134°C. The samples were held for various periods of time at that extension, the stress being recorded (15), and then the samples were quickly quenched to the glassy state at room temperature. Then SANS measurements were made in both the perpendicular and parallel directions.

Figure 10.10 (15) shows the changes in the transverse radius of gyration with time, presented in the form of a master curve (see next section). On an absolute scale, R_g for the isotropic sample was 280 Å, and R_g for the samples undergoing relaxation ranged from 161 to 210 Å. Boue et al. (15)

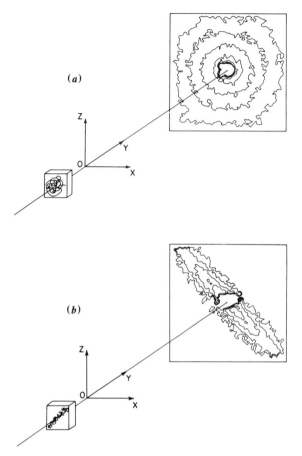

Figure 10.9 Geometry of the SANS experiment and resulting intensity contour plots: (a) unoriented amorphous polymer and (b) oriented polymer (16).

point out that immediately after stretching their sample, the radius of gyration showed affine deformation. This can be interpreted as indicating that the experiment did not fail to capture any major coil relaxation process.

Figure 10.10 (15) also applies the molecular diffusion theories of de Gennes (17), Doi and Edwards (18), and Daoudi (19), which assume that the major mode of molecular relaxation is by reptation. In this way, the chains move back and forth within a hypothetical tube. Relaxation occurs by the chain disengaging itself from the tube, only at the ends, in a backward-and-forward reptation.

Two characteristic times are employed: (a) T_d, defined as the longest of the relaxation times for defect equilibration, proportional to M^2, and (b) T_r, defined as the renewal time for chain conformation, proportional to M^3. This latter time is the time to form a new isotropic tube. For polystyrene of

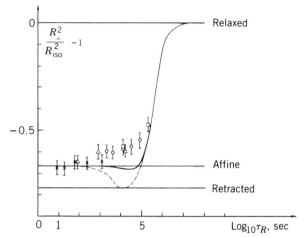

Figure 10.10 Molecular relaxation of polystyrene at different relaxation times and temperatures, as followed by SANS data taken in the transverse direction (15). Data are reduced to 117°C (just above T_g) via the time–temperature superposition principle.

$M_w = 650,000$ g/mol at 117°C, they found

$$3 \times 10^3 \text{ sec} \leq T_d \leq 4 \times 10^4 \text{ sec} \qquad (10.42)$$

$$6 \times 10^5 \text{ sec} \leq T_r \leq 1 \times 10^7 \text{ sec} \qquad (10.43)$$

The data in Figure 10.10 are in good agreement with the value of T_r. The authors state that more data are required to determine T_d decisively. The reader is referred to Section 5.4.2 and Appendix 5.2.

By way of contrast, Maconnachie et al. (12), using polystyrene of $M_w = 144,000$ g/mol, found that most of the relaxation was complete in about 5 min at 120°C. However, the chains never quite returned to their initial dimensions, both $R_{\parallel}$ and $R_{\perp}$ remaining slightly higher than the unstretched polymer dimensions even after 1×10^4 sec. The deformation appeared to behave as if the affine deformation theory held only for distances separating effective cross-links.

For better comparisons, these data are in the rubbery flow portion of the spectrum. The SANS data in Figure 10.10 correspond to the earlier stress relaxation studies by Tobolsky (20) on almost identical polystyrenes.

More recently, Wool (21) described three relaxation times. Referring to Figure 10.11,

$$T_0 \sim M_e^2 \qquad (10.44)$$

which is constant for a given polymer. The quantity M_e refers to the segment molecular weight involved in movement about the diameter of the equivalent

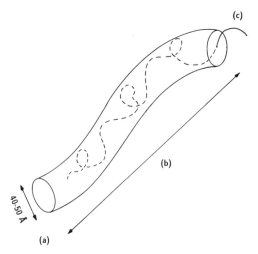

Figure 10.11 Tube model illustrating three relaxation times, each with a specific motion: (a) across the tube, (b) along the tube, and (c) in and out of the tube.

tube. The Rouse relaxation time is given by

$$T_R \sim T_0^2 \tag{10.45}$$

which expresses the time for one end of the chain to respond to or communicate with the other end. If one end of the chain is stretched, this is the characteristic time for the other end to retract. Reptation is given by

$$T_r \sim T_0 M^3 \tag{10.46}$$

similar to the de Gennes theory, where M is the molecular weight of the whole chain.

Figure 10.12 (22) illustrates stress relaxation in poly(methyl methacrylate) resulting from molecular relaxation processes. Here, the logarithm of the relaxation Young's modulus is plotted against the logarithm of time. At low temperatures, the polymer is glassy, and only slow relaxation is observed. As the glass transition temperature is approached (106°C at 10 sec), the relaxation rate increases, reaching a maximum just above the classical glass transition temperature range. Then the rate of relaxation decreases as the rubbery plateau is approached. This is exemplified by the data at 125 and 135°C in the range of 0.01–1 hr. At higher temperatures or longer times, the polymer begins to flow.

Figure 10.12 Stress–relaxation of poly(methyl methacrylate) with $M_v = 3.6 \times 10^6$ g / mol (20).

The data in Figure 10.12 thus show two distinct relaxation phenomena: first, chain portions corresponding to 10–50 carbon atoms are relaxing, which corresponds to the glass–rubber transition; then, at higher temperatures, whole chains are able to slide past one another.

10.2.6 Physical Aging in the Glassy State

Whereas the molecular motion in the rubbery and liquid states involves 10–50 carbon atoms, molecular motion in the glassy state is restricted to vibrations, rotations, and motions by relatively short segments of the chains.

The extent of molecular motion depends on the free volume. In the glassy state, the free volume depends on the thermal history of the polymer. When a sample is cooled from the melt to some temperature below T_g and held at constant temperature, its volume will decrease (see Figure 8.18). Because of the lower free volume, the rate of stress relaxation, creep, and related properties will decrease (23–27). This phenomenon is sometimes called physical aging (23, 26), although the sample ages in the sense not of degradation or oxidation but rather of an approach to the equilibrium state in the glass.

The effect of physical aging can be illustrated through a programmed series of creep studies. First, the sample is heated to a temperature T_0, about 10–15°C above T_g. A period of 10–20 min suffices to establish thermodynamic equilibrium. Then the sample is quenched to a temperature T_1 below T_g, and kept at this temperature. At a certain elapsed time after the quench, a creep experiment is started. The sample is subjected to a constant stress, σ_0; the resulting strain, ε, is determined as a function of time. The sample is then allowed to undergo creep recovery. Each creep period is short in comparison with the previous aging time as well as the last recovery period preceding it. The student should note the difference between the two time scales: t_e is the aging time, beginning at the time of quenching, whereas t is the creep time, which begins at the moment of loading for each run.

Figure 10.13 (23) illustrates typical creep compliance results for poly(vinyl chloride), which has a glass transition temperature of about 80°C. The most important conclusion obtained from such studies is that the rate of creep slows down as the sample ages. Noting the first and last curves in Figure

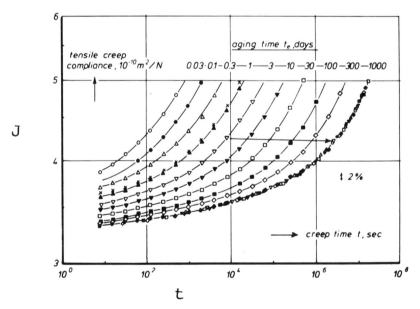

Figure 10.13 Small-strain tensile creep curves for glassy poly(vinyl chloride) quenched from 90°C (about 10°C above T_g) to 40°C and stored at that temperature for 4 years. The different curves were measured for various times t_e elapsed after the quench. The master curve gives the result of a superposition by shifts that were almost horizontal; the shifting direction is shown by the arrow. The crosses refer to another sample quenched in the same way, but only measured for creep at t_e of 1 day (23).

10.13, the polymer reaches a tensile creep compliance of 5×10^{-10} m^2/N after about 10^3 sec after being aged for 0.03 days, but after an aging period of 1000 days, 10^7 sec are required to reach the same tensile creep compliance level.

The problem of determining the theoretical behavior of polymers in the glassy rates is treated by Curro et al. (28, 29). The time dependence of the volume in the glassy state is accounted for by allowing the fraction of unoccupied volume sites to depend on time. This permits the application of the Doolittle equation to predict the shift in viscoelastic relaxation times,

$$\tilde{D} = \tilde{D}_r \exp\left[-B\left(f^{-1} - f_r^{-1} \right) \right] \tag{10.47}$$

where $\tilde{D}$ is the diffusion constant for holes, and the subscript r represents the reference state, taken for convenience at the glass transition temperature, and f is the fractional free volume; see Section 8.6.1.2.

10.3   THE TIME–TEMPERATURE SUPERPOSITION PRINCIPLE

10.3.1   The Master Curve

As indicated above, relaxation and creep occur by molecular diffusional motions which becomes more rapid as the temperature is increased. Temperature is a measure of molecular motion. At higher temperatures, time moves faster for the molecules. The WLF equation, derived in Section 8.6.1.2, expresses a logarithmic relationship between time and temperature. Building on these ideas, the time–temperature superposition principle states that with viscoelastic materials, time and temperature are equivalent to the extent that data at one temperature can be superimposed on data at another temperature by shifting the curves along the log time axis (2).

The importance of these ideas becomes clear when one considers that data can be obtained conveniently over only a narrow time scale, say from 1 sec to 10^5 sec (see Figure 10.12). The time–temperature superposition principle allows an estimation of the relaxation modulus and other properties over many decades of time.

Figure 10.14 (30–32) illustrates the time–temperature superposition principle using polyisobutylene data. The reference temperature of the master curve is 25°C. The reference temperature is the temperature to which all the data are converted by shifting the curves to overlap the original 25°C curve. Other equivalent curves can be made at other temperatures. The shift factor shown in the inset corresponds to the WLF shift factor, especially at the lower temperatures. Thus the quantitative shift of the data in the range T_g to $T_g + 50°C$ is governed by the WLF equation, and this equation can be used both to check the data and to estimate the shift where data are missing.

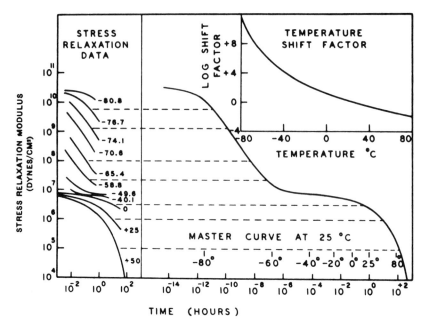

Figure 10.14 The making of a master curve, illustrated with polyisobutylene data. The classical T_g at 10 sec of this polymer is $-70°C$ (30–32).

Most interestingly, the master curve shown in Figure 10.14 looks much like the modulus temperature curves (see Figures 1.6 and 8.2). This likeness derives, of course, from the equivalence of log time with temperature.

In multicomponent polymeric systems such as polymer blends or blocks, each phase stress relaxes independently (33–35). Thus each phase will show a glass–rubber transition relaxation. While each phase follows the simple superposition rules illustrated above, combining them in a single equation must take into account the continuity of each phase in space. Attempts to do so have been made using the Takayanagi models (35), but the results are not simple.

10.3.2 The Reduced Frequency Nomograph

As illustrated in Figure 10.14, the master curve allows the extrapolation of data over broad temperature and time ranges. Similar master curves can be constructed with frequency as the variable, instead of time. More elegant still is the reduced frequency nomograph, which permits both reduced frequency and temperature simultaneously; see Figure 10.15 (36).

The reduced frequency nomograph (36, 37) is constructed as follows. First, the storage and loss modulus (or tan δ) are plotted versus reduced frequency,

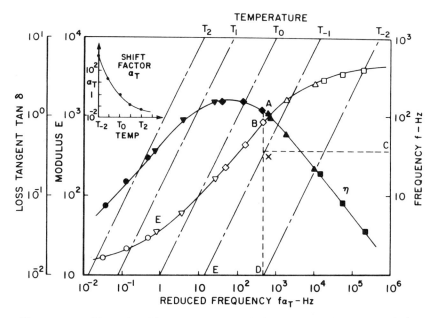

Figure 10.15   The reduced frequency nomograph. Here the quantity $\eta \equiv \tan\delta$ (36).

making use of the reduced variables shift factor, a_T. The reduced frequency is defined as $f_j a_{T_i}$, where f_j is the frequency and a_{T_i} is the value of a_T at temperature T_i. Then an auxiliary frequency scale is constructed as the ordinate on the right side of the graph; see Figure 10.15. The values of $f_j a_{T_i}$ and f form a set of corresponding oblique lines representing temperature. Weissman and Chartoff (36) point out that computer software has been developed to speed the generation of such nomographs.

Simulated data are used in Figure 10.15 for ease of instruction. Assume that the value of E' and $\tan\delta$ at T_{-1} and some frequency f are needed, illustrated by point C in Figure 10.15. The intersection of the horizontal line CX, with $f_j a_{T_i}$ (point X) defines a value of fa_T at point D of approximately 4×10^2. From this value of fa_T it follows from the plots of E' and $\tan\delta$ that the quantity E' equals 10^{-3} N/m², point B, and $\tan\delta$ equals 1.2, point A.

A critical point of general information in Figure 10.15 is that, at high enough frequencies, the storage modulus increases into the glassy range. At low enough frequencies, any linear (or lightly cross-linked polymer) will flow (or be rubbery). As the frequency increases beyond the capability of the polymer chains to respond, the polymer glassifies. Since the reduced frequency is usually plotted with increasing values to the right, this figure appears backward to figures where log time is the x-axis variable. While stress relaxation studies employ time as the variable, dynamic experiments employ frequency.

Table 10.2   WLF parameters

Polymer	C_1	C_2	T_g, K
Polyisobutylene	16.6	104	202
Natural rubber (Hevea)	16.7	53.6	200
Polyurethane elastomer	15.6	32.6	238
Polystyrene	14.5	50.4	373
Poly(ethyl methacrylate)	17.6	65.5	335
"Universal constants"	17.4	51.6	

Source: J. J. Aklonis and W. J. MacKnight, *Introduction to Polymer Viscoelasticity*, Wiley-Interscience, New York, 1983, Table 3-2, p. 48.

Applications of the reduced frequency nomograph include sound and vibration damping (see Section 8.12), an interdisciplinary field involving polymer scientists and mechanical engineers, as well as other disciplines. Frequencies of interest range from about 0.01 Hz (tall buildings) to 10^5 Hz, in the ultrasonic range.

10.4   POLYMER MELT VISCOSITY

10.4.1   The WLF Constants

In regions 4 and 5 of the modulus–temperature curve, linear polymers are capable of flow if they are subjected to a shear stress. While the melt viscosity, η, of low-molecular-weight substances may be Newtonian [see equation (8.3)], the flow behavior of polymers always contains some elements of viscoelasticity. In Section 8.6.1, the WLF equation was derived:

$$\log\left(\frac{\eta}{\eta_{T_g}}\right) = \frac{-C_1(T - T_g)}{C_2 + (T - T_g)} \tag{10.48}$$

where C_1 and C_2 are constants, and η_{T_g} is the melt viscosity at the glass transition temperature. If data are not available on the polymer of interest, values of $C_1 = 17.44$ and $C_2 = 51.6$ may be used. However, these constants vary significantly from polymer to polymer because of differences in the free volumes, expansion coefficients, and so on. Selected values of C_1 and C_2 are tabulated in Table 10.2 (38–40). Frequently, $\eta_{T_g} \cong 1 \times 10^{13}$ poises.

10.4.2   The Molecular-Weight Dependence of the Melt Viscosity

10.4.2.1   Critical Entanglement Chain Length

Viscoelasticity in polymers ultimately relates back to a few basic molecular characteristics involving the rates of chain molecular motion and chain entanglement. The increasing ability of chains to slip past one another as the temperature is increased governs the temperature dependence of the melt viscosity. One embodiment of this concept is the WLF equation, above. The increased resistance to flow caused by entanglement governs the molecular-weight dependence.

Basic parameters in discussing the molecular-weight dependence of the melt viscosity are the degree of polymerization (DP), which represents the number of monomer units linked together, and the number of atoms along the polymer chain's backbone (Z). For styrenics, acrylics and vinyl polymers, $Z = 2DP$, and for diene polymers, $Z = 4DP$. The point is that melt viscosity characteristics depend more on the number of backbone atoms than on the side foliage. It was also found very early that the melt viscosity depends on the weight-average degree of polymerization.

The molecular-weight dependence of the melt viscosity exhibits two distinct regions (41), depending on whether the chains are long enough to be significantly entangled (see Figure 10.16). A critical entanglement chain length, $Z_{c,w}$ is defined as the weight-average number of chain atoms in the polymer molecules to cause intermolecular entanglement. Below $Z_{c,w}$ the

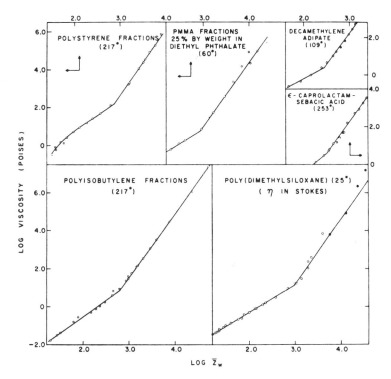

Figure 10.16   A plot of log η versus log Z_w for several polymers of importance. Below about $Z_w = 600$, the melt viscosity goes as the first power of Z_w; above 600, the melt viscosity depends on the 3.4 power of Z_w (41).

melt viscosity is given by

$$\eta = K_L Z_w^{1.0} \tag{10.49}$$

and above $Z_{c,w}$ the melt viscosity is given by

$$\eta = K_H Z_w^{3.4} \tag{10.50}$$

where K_L and K_H are constants for low and high degrees of polymerization.

The 1.0 power dependence in equation (10.49) represents the simple increase in viscosity as the chains get longer. The dependence of the viscosity on the 3.4 power of the degree of polymerization as shown in equation (10.50) arises from entanglement and diffusion considerations. The algebraic relationship was derived by de Gennes (42), using scaling concepts. According to the reptation model (see Section 5.4), there is a maximum relaxation time of the chains, τ_{max}, which is the time required for complete renewal of the tube that constrains the chain. That is, τ_{max} represents the time necessary for the polymer chain to diffuse out of its original tube and assume a new conformation.

10.4.2.2 An Example Calculation

You are the engineer in charge of a polystyrene drinking cup manufacturing unit. Normally, you process the material at 160°C, where the viscosity is 1.5×10^3 poises at $Z_w = 800$. Today, your polystyrene has $Z_w = 950$. What changes in processing temperature will bring the viscosity down to 1.5×10^3 again?

T_g for polystyrene is 100°C. The melt viscosity relationships are given by equations (10.50) and (8.48):

$$\eta = K_H Z_w^{3.4}$$

$$\log\left(\frac{\eta}{\eta_{T_g}}\right) = -\frac{17.44(T - T_g)}{51.6 + (T - T_g)}$$

Solving for K_H,

$$K_H = 1.5 \times 10^3 / (800)^{3.4} = 2.02 \times 10^{-7}$$

At 160°C, the new viscosity is

$$\eta = 2.02 \times 10^{-7} \times (950)^{3.4} = 2.69 \times 10^3 \text{ P}.$$

Solving the WLF equation for η_{T_g}:

$$\log\left(\frac{2.69 \times 10^3}{\eta_{T_g}}\right) = -\frac{17.44(433 - 373)}{51.6 + (433 - 373)}$$

$$\eta_{T_g} = 6.39 \times 10^{12} \text{ P.}$$

Now the new temperature must be calculated, using the WLF equation:

$$\log\left(\frac{1.5 \times 10^3}{6.39 \times 10^{12}}\right) = -\frac{17.44(T - 373)}{51.6 + (T - 373)}$$

$$T = 436.6 \ K \quad \text{or} \quad 163.6° \ C$$

Thus the temperature should be raised 3.6°C, to return to the same melt viscosity. It is often important in operating polymer melt machinery to work at the same melt viscosity. These two equations, used in tandem, can be used to solve many problems involving simultaneous changes in temperature and molecular weight.

The quantity Z_w may cross 600, necessitating a change from equation (10.49) to equation (10.50) or vice versa. The required new constant, (K_L or K_H) can be obtained by noting that the viscosity for both equations is the same at Z_w equals 600, see Figure 10.16.

10.4.2.3   The Tube Diffusion Coefficient

Under a steady force, f, the chain moves with a velocity, v, in the tube. Then the "tube mobility" of the chain, μ_{tube}, is given by

$$\mu_{\text{tube}} = \frac{v}{f} \tag{10.51}$$

If long-range backflow effects are assumed negligible, then the friction force v/μ_{tube} is essentially proportional to the number of atoms in the chain,

$$\mu_{\text{tube}} = \frac{\mu_1}{Z} \tag{10.52}$$

where μ_1 is independent of Z. Similarly, the tube diffusion coefficient, D_{tube}, is related to μ_{tube} and Z through an Einstein relationship $D_{\text{tube}} = \mu_1 T/Z$

$$D_{\text{tube}} = \frac{D_1}{Z} \tag{10.53}$$

Since the time necessary to diffuse a certain distance depends on the square of the distance, the time necessary to completely renew the tube depends on

Table 10.3   Selected entanglement chain lengths (41)

Polymer	$Z_{c,w}$	Reference
Polyisobutylene	610	(a, b)
Polystyrene	730	(c, d)
Poly(dimethyl siloxane)	950	(e, f)
Poly(decamethylene sebacate)	290	(g)

References: (a) T. G. Fox and P. J. Flory, *J. Am. Chem. Soc.,* **70**, 2384(1948). (b) T. G. Fox and P. J. Flory, *J. Phys. Chem.,* **55**, 221(1951). (c) T. G. Fox and P. J. Flory, *J. Appl. Phys.,* **21**, 581(1950). (d) T. G. Fox and P. J. Flory, *J. Polym. Sci.,* **14**, 315(1954). (e) A. J. Barry, *J. Appl. Phys.,* **17**, 1020(1946). (f) M. J. Hunter, E. L. Warrick, J. F. Hyde, and C. C. Currie, *J. Am. Chem. Soc.,* **68**, 2284(1946). (g) P. J. Flory, *J. Am. Chem. Soc.,* **62**, 1057(1940).

the length of the tube, L, squared. Then

$$\tau_{max} \cong \frac{L^2}{D_{tube}} \qquad (10.54)$$

and

$$\tau_{max} \cong \frac{ZL^2}{D_1} \qquad (10.55)$$

since L is proportional to Z,

$$\tau_{max} = \tau_1 Z^3 \qquad (10.56)$$

The viscosity of the system is given by equation (10.15), $\eta = \tau E$. According to the reptation model, the modulus E depends on the distance between obstacles and does not depend on the chain length. Therefore

$$\eta \propto Z^3 \qquad (10.57)$$

Equation (10.57) should be compared with equation (10.50). While the power dependence is not quite correct in this simple derivation, it illustrates the principal molecular-weight dependence of the viscosity.

Returning to the critical entanglement chain length, this value was found to be approximately 600 chain atoms (see Table 10.3) (41). Values of $Z_{c,w}$ are easily estimated from the break in Figure 10.16. Since the average molecular weight of a mer is about 100, then $Z_{c,w}$ corresponds to about 30,000 g/mol.[†] Interestingly, this molecular weight corresponds to the low end of the useful mechanical properties of many polymers. Polymers below this molecular weight usually exhibit inferior strengths, as illustrated in Figure 1.4. Thus, for

[†]The actual molecular weight *between* the entanglements, M_e, is half this value, or about 15,000 g/mol.

mechanical applications, the definition of what chain length constitutes a polymer is sometimes taken as a molecular weight of about these values or higher.

10.5 POLYMER RHEOLOGY

Rheology is the study of the deformation and flow of matter. As such, the reader will recognize that a significant fraction of this book already involves rheological concepts. Some important areas not yet considered include shear rate dependence of flow and the effect of normal stress differences (43, 44).

The range of melt viscosities ordinarily encountered in materials is given in Table 10.4 (43). Polymer latexes and suspensions are aqueous dispersions with viscosities dependent on solid content and additives. Polymer solutions may be much more viscous, depending on the concentration, molecular weight, and temperature.

The temperature dependence of the viscosity is most easily expressed according to the Arrhenius relationship:

$$\eta = Ae^{-B/T} \tag{10.58}$$

where T represents the absolute temperature and A and B are constants of the liquid. The Arrhenius equation may easily be shown to be an approximation of the WLF equation far above the glass transition temperature.

The rate of energy dissipation per unit volume of the sheared polymer may be expressed as either the product of the shear stress and the shear rate or, equivalently, the product of the viscosity and the square of the shear rate. The rate of heat generated during viscous flow may be significant. Thus the heat generated may actually reduce the viscosity of the material. From an engineering point of view, these quantities provide an important measure of the power necessary to maintain a given flow rate.

Table 10.4 Viscosities of some common materials (43)

Composition	Viscosity Pa · s	Consistency
Air	10^{-5}	gaseous
Water	10^{-3}	fluid
Polymer latexes	10^{-2}	fluid
Olive oil	10^{-1}	liquid
Glycerine	10^{0}	liqiud
Golden Syrup	10^{2}	thick liquid
Polymer melts	10^{2}–10^{6}	toffee-like
Pitch	10^{9}	stiff
Plastics	10^{12}	glassy
Glass	10^{21}	rigid

All real liquids have both viscous and elastic components, although one or the other may predominate. For example, water behaves as a nearly perfect viscous medium, while a rubber band is a nearly perfect elastomer. A polymer solution of, for example, polyacrylamide in water may exhibit various ranges of viscoelasticity, depending on the concentration and temperature. According to Weissenberg (45), when an elastic liquid is subjected to simple shear flow, there are two forces to be considered:

1. The shear stress, characteristic of ordinary viscosity.
2. A normal force, observed as a pull along the lines of flow.

The several aspects of polymer rheology are developed below.

10.5.1 Shear Dependence of Viscosity

Any liquid showing a deviation from Newtonian behavior is considered non-Newtonian. As soon as reliable viscometers became available, workers found departures from Newtonian behavior for polymer solutions, dispersions, and melts. In the vast majority of cases, the viscosity decreases with increasing shear rate, giving rise to what is often called "shear-thinning," also called dilatancy.

An example for xanthan gum solutions is given in Figure 10.17 (46). Note especially the wide range of shear rates obtainable. Xanthan gum is a

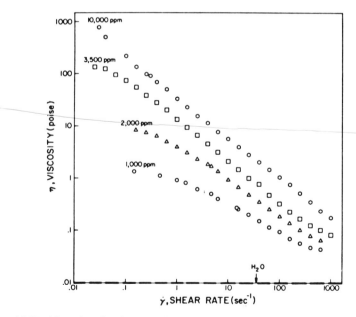

Figure 10.17 Viscosity of polymer melts and solutions usually decreases with increasing shear rate (46).

Table 10.5   Typical power-law parameters of a selection of well-known materials for a particular range of shear rates (44)

Material	$K_2 (Pa \cdot s^n)$	n	Shear rate range (s^{-1})
Ball-point pen ink	10	0.85	10^0–10^3
Fabric conditioner	10	0.6	10^0–10^2
Polymer melt	10000	0.6	10^2–10^4
Molten chocolate	50	0.5	10^{-1}–10
Synovial fluid	0.5	0.4	10^{-1}–10^2
Toothpaste	300	0.3	10^0–10^3
Skin cream	250	0.1	10^0–10^2
Lubricating grease	1000	0.1	10^{-1}–10^2

high-molecular-weight polysaccharide (MW = 7.6×10^6 g/mol in Figure 10.17), a biopolymer used in food applications, oil recovery, and textile printing. Less than 1% of xanthan increases water's viscosity by a factor of 10^5 at low shear rates, yet only by a factor of 10 at high shear rates. The viscosity reduction is caused by the reduced number of chain entanglements as the chains orient along the lines of flow. Note that water has 0.01 poise (1×10^{-3} Pa · sec), shown as the base line in Figure 10.17.

The general case of shear rate dependence includes a limit of constant viscosity at very low shear rates, and a lower constant viscosity at the limit of very high shear rates. While this is not often observed with polymeric materials, a suggestion of both limits can be seen in Figure 10.17. The reason is that, even at the lowest shear rates, the chains begin to orient, and entanglements slip easier. At the limit of very high shear rates, polymers degrade. An extremely useful relationship is the well known "power-law" model,

$$\eta = K_2 \dot{\gamma}^{n-1} \qquad (10.59)$$

where $\dot{\gamma}$ is the shear rate, n is the power-law index, and K_2 is called the "consistency." When n equals zero, a form of Newton's law is generated; see Section 8.2. The quantity $\dot{\gamma}$ is identical to ds/dt, equation (8.3), with units of reciprocal seconds. Typical values of n and K_2 for a number of materials are given in Table 10.5 (44). Many practical materials, such as skin creams and inks, contain polymers.

10.5.2   Normal Stress Differences

Consider a small plane surface of area ΔA drawn in a deforming medium; see Figure 10.18. The material flows or deforms through ΔA. For analytic purposes, the vector direction of motion is divided among x, y, and z. The direction of motion is not necessarily normal to the plane of ΔA.

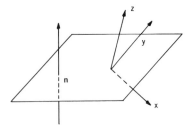

Figure 10.18 Arbitrary infinitesimal plane of area ΔA; n is the direction perpendicular to the plane. The quantities x, y, and z are orthogonal coordinates.

The stress components may be written f_{nx}, f_{ny}, and f_{nz}, the first index referring to the orientation of plane surface, and the second to the direction of the stress. Then the quantity f_{xx} means that a stress perpendicular to the plane ΔA is under consideration. There are three quantities that are rheologically relevant (47)—the shear stress, f_{yx}, and the first and second normal stress differences, N_1 and N_2, respectively:

$$N_1 = f_{xx} - f_{yy} \qquad (10.60)$$

$$N_2 = f_{yy} - f_{zz} \qquad (10.61)$$

Both N_1 and N_2 are shear rate dependent. From a physical point of view, the generation of unequal normal stress components arises from the anisotropic structure of a polymer fluid undergoing flow; that is, the polymer becomes oriented. The largest of the three normal stress components is always f_{xx}, the component in the direction of flow. However, since the molecular structures undergoing flow are anisotropic, the forces are also anisotropic. The deformed chains often have the rough shape of ellipsoids, which have their major axis tilted toward the direction of flow. Thus the restoring force is greater in this direction than in the two orthogonal directions. It is these restoring forces that give rise to the normal forces (44).

The observable consequences of the normal forces are quite dramatic. Perhaps the best known is the so-called Weissenberg effect; see Figure 10.19 (44). It is produced when a rotating rod is placed into a vessel containing a viscoelastic fluid such as a concentrated polymer solution. Whereas a Newtonian liquid would be forced toward the rim of the vessel by inertia, the viscoelastic fluid moves *toward* the rod, climbing it. The chains are extended yet remain partly entangled, which causes a rubber elasticity type of retractive force. This "hoop stress" around the rod forms the basis for the Weissenberg effect.

Another phenomenon, seen in fiber and plastic extruding operations, is "die swell"; see Figure 10.20 (48). Here a viscoelastic fluid is oriented in a tube during flow. When it is extruded, it flows from the exit of the tube and retracts. This results in swelling to a much greater diameter than that of the hole or slit. Die swell becomes much greater when the polymer chains are

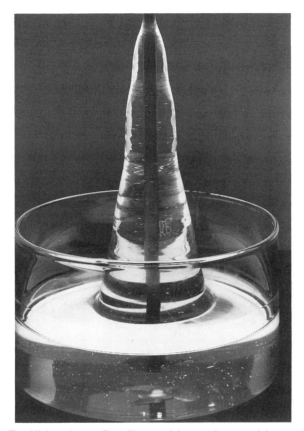

Figure 10.19 The Weissenberg effect illustrated by rotating a rod in a solution of poly-isobutylene in polybutene (44).

long enough to contain significant numbers of entanglements, being a direct consequence of the elastic energy stored by the polymer in the tube.

10.5.3 Dynamic Viscosity

Periodic or oscillatory experiments provide a powerful rheological analogue to dynamic mechanical experiments. The theoretical relations are, in fact, closely interlocked. The complex viscosity, η^*, is defined as (10)

$$\eta^* = \eta' - i\eta'' \qquad (10.62)$$

The complex viscosity can be determined from dynamic mechanical data,

$$\eta^* = \frac{\left[(G')^2 + (G'')^2\right]^{1/2}}{\omega} \qquad (10.63)$$

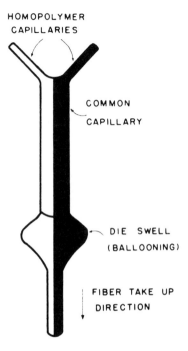

HOMOPOLYMER
CAPILLARIES

COMMON
CAPILLARY

DIE SWELL
(BALLOONING)

FIBER TAKE UP
DIRECTION

Figure 10.20 Bicomponent fibers are often used in the manufacture of rugs, because a mismatch in the expansion coefficients causes curling on cooling. However, the second normal forces must match, so that the die swell on both sides is the same (48).

where ω is the angular frequency. The individual components are given by

$$\eta' = G''/\omega \qquad (10.64)$$

and

$$\eta'' = G'/\omega \qquad (10.65)$$

Thus the phase relations are the opposite of those for G^*. The in-phase, or real component, η' for a viscoelastic liquid approaches the steady-state flow viscosity as the frequency approaches zero.

10.5.4 Instruments and Experiments

There are several different types of viscometers, of varying complexity, good for specific purposes and/or ranges of viscosity (49, 50). The major classes are described below.

Falling-Ball Viscometers. This is the simplest type of viscometer, useful for determining relatively low viscosities. In one simple case, a graduated cylinder is filled with the fluid of interest (perhaps a concentrated polymer solution), and a steel ball is dropped in. The time to fall a given distance is recorded. According to Stokes, the terminal velocity is given by

$$\bar{u} = \frac{2}{9} \frac{R^2}{\eta} g(\rho - \rho_0) \tag{10.66}$$

where R is the sphere radius, ρ and ρ_0 are the densities of the sphere and the medium, respectively, and g is the gravitational constant. Within its range of operation, it is inexpensive, easily used, and the viscosity is absolute.

Capillary Viscometers. These instruments are used for intrinsic viscosities and also for more viscous melts, solutions, and dispersions. The viscosity is given by the Hagen–Poiseuille expression,

$$\eta = \frac{\pi r^4 \Delta p \, t}{8VL} \tag{10.67}$$

where r is the capillary radius, Δp represents the pressure drop, and V is the volume of liquid that flows through the capillary of length L in time t. For a given viscometer with similar liquids,

$$\eta = Kt \quad \text{or} \quad \bar{\nu} = \eta/\rho_0 = K't \tag{10.68}$$

where $\bar{\nu}$ is the kinematic viscosity.

Rotational Viscometers. These relatively complex instruments can be used in the steady state or in an oscillatory, dynamic mode. Some are useful up to the glassy state of the polymer. The working mechanism, in all cases, is one part that moves past another. Designs include concentric cylinders (cup and bob), cone-and-plate, parallel-plate, and disk, paddle, or rotor in a cylinder.

The most important device is the cone-and-plate viscometer; see Figure 10.21 (44). The advantage of the cone-and-plate geometry is that the shear rate is very nearly the same everywhere in the fluid, provided the gap angle,

Figure 10.21 The cone-and-plate viscometer, showing the cone on top, a rotating plate, and couple attached to the cone. The inset shows a form of truncation employed in many instruments (44).

θ_0, is small. The shear rate in the fluid is given by

$$\dot{\gamma} = \Omega_1/\theta_0 \qquad (10.69)$$

where Ω_1 represents the angular velocity of the rotating platter. The viscosity is given by

$$\eta = \frac{3f\theta_0}{2\pi a^3 \Omega_1} \qquad (10.70)$$

where the stress f is measured by the couple C on the cone, and a is the radius of the cone-and-plate. An important factor in deciding on a viscometer, of course, is the viscosity of the fluid and the information desired.

10.5.5   General Definitions and Terms

Several shear-rate-dependent phenomena are illustrated in Figure 10.22. A Bingham plastic has a linear relationship between shear stress and shear strain but does not go through the origin. It must be realized that this is an idealization. Pseudoplastic materials, which include most polymer solutions and melts, exhibit shear thinning. Dilatant materials shear thicken.

There are several terms that apply to time-dependent behavior. Thixotropic fluids possess a structure that breaks down as a function of time and shear rate. Thus the viscosity is lowered. A famous example of this reversible phenomenon is the ubiquitous catsup bottle, yielding its contents only after sharp blows. The opposite effect, although rarely observed, is called anti-thixotropic, or rheopectic behavior, where materials "set up" as a function of time and shear rate; that is, the viscosity increases or the material gels.

Viscoelastic fluids exhibit elastic recovery from deformations that occur during flow. The Weissenberg effect and die swell have been discussed

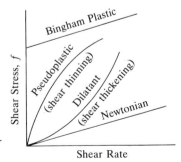

Figure 10.22   Several   shear-rate-dependent   rheological phenomena.

Table 10.6   Flow models and equations (50)

Model	Equation
Newtonian	$f = \eta \dot{\gamma}$
Bingham plastic	$f - f_0 = \eta \dot{\gamma}$
Power law	$f = k \dot{\gamma}^n$
Power law with yield value	$f - f_0 = k \dot{\gamma}^n$
Casson fluid	$f^{1/2} - f_0^{1/2} = \eta_\infty^{1/2} \dot{\gamma}^{1/2}$
Williamson	$\eta - \eta_\infty = \dfrac{(\eta_0 - \eta_\infty)}{1 + f/f_m}$
Cross	$\eta - \eta_\infty = \dfrac{(\eta_0 - \eta_\infty)}{1 + \alpha \dot{\gamma}^n}$

previously. Another very simple example is the stirring of concentrated polymer solutions rapidly, then stopping. Elastic effects make the fluid move backward for a time.

Several flow equations are summarized in Table 10.6 (50) for various models. Here, f_0 is the yield stress, η_0 the viscosity at low shear rates and η_∞ at high shear rates, and α and n are constants.

10.6   OVERVIEW OF VISCOELASTICITY AND RHEOLOGY

This chapter has illustrated how stress relaxation, creep, and rheology in polymers depend on the rate of molecular motion of the chains and on the presence of entanglements. It must be remembered that all macroscopic deformations of matter depend ultimately on molecular motion. In the case of high polymers, the chain's radius of gyration is changed during initial deformation or flow. Thermal motions tend to return the polymer to its initial conformation, thus raising its entropy. Thus there is a direct relationship between the mechanical or viscous behavior of polymeric materials and their molecular behavior.

REFERENCES

1. J. J. Aklonis, *J. Chem. Ed.*, **58**, 893(1981).
2. A. V. Tobolsky, *Properties and Structure of Polymers*, Wiley, New York, 1960.
3. L. H. Sperling, S. L. Cooper, and A. V. Tobolsky, *J. Appl. Polym. Sci.*, **10**, 1735(1966).
4. P. Thirion and R. Chasset, *Proceedings of the 4th Rubber Technology Conference*, London, 1962, p. 338.
5. P. G. de Gennes, *J. Chem. Phys.*, **55**, 572(1971).

6. M. Takayanagi, *Mem. Fac. Eng. Kyushu Univ.*, **23**, 11(1963).

7. J. A. Manson and L. H. Sperling, *Polymer Blends and Composites*, Plenum, New York, 1976, Chap. 2.

8. T. C. P. Lee, L. H. Sperling, and A. V. Tobolsky, *J. Appl. Polym. Sci.*, **10**, 1831(1966).

9. A. V. Tobolsky, I. B. Prettyman, and J. H. Dillon, *J. Appl. Phys.*, **15**, 380(1944).

10. J. D. Ferry, *Viscoelastic Properties of Polymers*, 3rd ed., Wiley, New York, 1980.

11. C. Picot, R. Duplessix, D. Decker, H. Benoît, F. Boue, J. P. Cotton, M. Daoud, B. Farnoux, G. Jannick, M. Nierloch, A. J. deVries, and P. Pincus, *Macromolecules*, **10**, 957(1977).

12. A. Maconnachie, G. Allen, and R. W. Richards, *Polymer*, **22**, 1157(1981).

13. F. Boue and G. Jannink, *J. Phys. Colloq.*, **39**, C2-183(1978).

14. F. Boue, M. Nierlich, G. Jannink, and R. C. Ball, *J. Phys. Lett. (Paris)*, **43**, 593(1982).

15. F. Boue, M. Nierlich, G. Jannink, and R. C. Ball, *J. Phys. (Paris)*, **43**, 137(1982).

16. G. Hadziiannou, L. H. Wang, R. S. Stein, and R. S. Porter, *Macromolecules*, **15**, 880(1982).

17. P. G. de Gennes, *J. Chem. Phys.*, **55**, 572(1971).

18. M. Doi and J. F. Edwards, *J. Chem. Soc. Faraday Trans.*, **74**, 1789, 1802, 1818(1978).

19. S. Daoudi, *J. Phys.*, **38**, 731(1977).

20. A. V. Tobolsky, *J. Polym. Sci. Lett.*, **2**, 103(1964).

21. R. P. Wool, seminar, Lehigh University, October 1988.

22. J. R. McLaughlin and A. V. Tobolsky, *J. Colloid Sci.*, **7**, 555(1952).

23. L. C. E. Struik, *Physical Aging in Amorphous Polymers and Other Materials*, Elsevier, New York, 1978.

24. H. C. Booij and J. H. M. Palmen, *Polym. Eng. Sci.*, **18**, 781(1978).

25. S. Matsuoka, H. E. Bair, S. S. Bearder, H. E. Kern, and J. T. Ryan, *Polym. Eng. Sci.*, **18**, 1073(1978).

26. F. H. J. Maurer, J. H. M. Palmen, and H. C. Booij, *Polym. Mater. Sci. Eng. Prepr.*, **51**, 614(1984).

27. F. H. J. Maurer, in *Rheology, Vol. 3: Applications*, G. Astarita, G. Marrucci, and L. Nicolais, Eds., Plenum, New York, 1980.

28. J. G. Curro, R. R. Lagasse, and R. Simha, *J. Appl. Phys.*, **52**, 5892(1981).

29. J. G. Curro, R. R. Lagasse, and R. Simha, *Macromolecules*, **15**, 1621(1982).

30. E. Castiff and A. V. Tobolsky, *J. Colloid Sci.*, **10**, 375(1955).

31. E. Castiff and A. V. Tobolsky, *J. Polym. Sci.*, **19**, 111(1956).

32. L. E. Nielsen, *Mechanical Properties of Polymers*, Reinhold, New York, 1962.

33. T. Horino, Y. Ogawa, T. Soen, and H. Kawai, *J. Appl. Polym. Sci.*, **9**, 2261(1965).

34. R. E. Cohen and N. W. Tschoegl, *Int. J. Polym. Mater.*, **2**, 205(1973).

35. D. Kaplan and N. W. Tschoegl, *Polym. Eng. Sci.*, **14**, 43(1974).

36. P. T. Weissman and R. P. Chartoff, in *Sound and Vibration Damping with Polymers*, R. D. Corsaro and L. H. Sperling, Eds., ACS Symposium Series No. 424, American Chemical Society, Washington, DC, 1991.

37. D. I. G. Jones, *Shock Vibration Bull.*, **48**(2), 13(1978).

38. M. L. Williams, R. F. Landel, and J. D. Ferry, *J. Am. Chem. Soc.*, **77**, 3701(1955).

39. J. D. Ferry, *Viscoelastic Properties of Polymers*, 3rd ed., Wiley, New York, 1980, Chap. 11.

40. J. J. Aklonis and W. J. MacKnight, *Introduction to Polymer Viscoelasticity*, 2nd ed., Wiley, New York, 1983, Chap. 3.

41. T. G. Fox, S. Gratch, and S. Loshaek, in *Rheology*, F. R. Eirich, Ed., Academic Press, Orlando, 1956, Vol. 1, Chap. 12.

42. P. G. de Gennes, *Scaling Concepts in Polymer Physics*, Cornell University Press, Ithaca, NY, 1979, Chap. 8.

43. F. N. Cogswell, *Polymer Melt Rheology*, G. Godwin, Ed., Wiley, New York, 1981.

44. H. A. Barnes, J. F. Hutton, and K. Walters, *An Introduction to Rheology*, Elsevier, New York, 1989.

45. K. Weissenberg, *Nature*, **159**, 310(1947).

46. P. J. Whitcomb and C. W. Macosko, *J. Rheol.*, **22**, 493(1978).

47. J. Meissner, R. W. Garbella, and J. Hostettler, *J. Rheol.*, **33**, 843(1989).

48. J. A. Manson and L. H. Sperling, *Polymer Blends and Composites*, Plenum, New York, 1976.

49. *Chemical Engineer's Handbook*, 5th ed., J. H. Perry, Ed., McGraw-Hill, New York, 1973.

50. C. K. Schoff, in *Encyclopedia of Polymer Science and Engineering*, Vol. 14, Wiley, New York, 1988, p. 454.

GENERAL READING

J. J. Aklonis and W. J. MacKnight, *Introduction to Polymer Viscoelasticity*, 2nd ed., Wiley, New York, 1983.

H. A. Barnes, J. F. Hutton, and K. Walters, *An Introduction to Rheology*, Elsevier, Amsterdam, 1989.

J. D. Ferry, *Viscoelastic Properties of Polymers*, 3rd ed., Wiley, New York, 1980.

L. E. Nielsen, *Polymer Rheology*, Dekker, New York, 1977.

L. C. E. Struik, *Physical Aging in Amorphous Polymers and Other Materials*, Elsevier, Amsterdam, 1978.

A. V. Tobolsky, *Properties and Structure of Polymers*, Wiley, New York, 1960.

N. W. Tschoegl, *The Theory of Linear Viscoelastic Behavior*, Academic Press, Orlando, 1981.

I. M. Ward, *Mechanical Properties of Solid Polymers*, 2nd ed., Wiley, New York, 1983.

STUDY PROBLEMS

1. Draw the creep and creep recovery curves for the three-element model consisting of a Kelvin element and a spring in series.

2. If the modulus of the cis-1,4-polybutadiene in Figure 10.8 was 2.5×10^7 dynes/cm^2, plot the total number of remaining active chain segments from the original network as a function of time. Also, plot the number of active chain segments formed by the oxidative cross-linking.

3. At 200°C, how long would it take for the silicone rubber in Figure 10.7 to relax 50%?

4. Derive equations to express Young's modulus as a function of rubber and plastic composition using the Takayanagi models (c) and (d) in Figure 10.6.

5. Draw stress relaxation curves for the four-element model.

6. Derive an equation for the creep recovery of a Kelvin element, beginning after a creep experiment extending it to t_1, a later time.

7. A poly(methyl methacrylate) bridge is to be placed across a river in the tropics, where the average temperature is 40°C. The bridge is a simple platform 100 ft long and 5 ft thick and 10 ft wide. The bridge will fail when creep slopes the sides of the bridge more than a 30° angle, so cars get stuck in the middle. How long will the bridge last? [*Hint:* As a simple approximation, the simple beam, center-loaded bending distance Y is given by $Y = fL^3/4a^3bE$, where a is the thickness, b is the width, f is the force, L is the length, and E is Young's modulus. See Figure 10.12.]

8. Prepare a master curve at 110°C for poly(methyl methacrylate), using the data in Figure 10.12.

9. The melt viscosity of a fraction of natural rubber is 2×10^4 poises at 240 K. What is the melt viscosity of this fraction at 250 K?

10. A new polymer with a mer weight of 211 g/mol and five atoms in the chain was found to have a weight-average molecular weight of 300,000 g/mol. Its melt viscosity is 1500 poises. What is the viscosity of the polymer if its molecular weight is doubled?

11. A polymer with a Z_w of 200 was found to have a melt viscosity of 100 poises. What is the viscosity of this polymer when $Z_w = 800$?

12. A polymer with a T_g of 110°C and a Z_w of 400 was found to have a melt viscosity of 5000 poises at 160°C. What is its melt viscosity at 140°C when $Z_w = 900$? [*Hint:* Combine the WLF equation with the DP dependence.]

13. The three-element springs and dashpot model shown is subject to a creep experiment. Show how the length (or strain) increases with time. At time $= t$, the stress is removed. Show how the sample recovers.

14. A new polymer follows the Kelvin model. The quantity η obeys the WLF equation, and E obeys rubber elasticity theory. The glass transition temperature of the polymer is 5°C, where it has a viscosity of 1×10^{13} poises. The concentration of active chain segments is 1×10^{-4} mol/cm³. The temperature of the experiment is 30°C.

(a) How does this polymer creep with time under a stress of 1×10^7 dynes/cm²? A plot of strain versus time is required.

(b) Briefly discuss two ways to slow down or reduce the rate of creep in part (a).

15. A certain extruder for plastics was found to work best at a melt viscosity of 2×10^4 poises. The polymer of choice had this viscosity at 145°C when its DP_w was 700. This polymer has a T_g of 75°C. Because of a polymerization kinetics miscalculation by someone who did not take polymer science, today's polymer has a DP_w of 500. At what temperature should the extruder be run so that the viscosity will remain at optimum conditions?

16. In the spinning of rayon fibers, the viscosity of the viscose solution must be carefully controlled. In a falling ball experiment on a viscose solution, it took a 2 mm lead ball 324 sec to fall 10 cm. The density of the viscose was 1.2 g/cm³. What is its viscosity?

17. A cone-and-plate experiment was performed on a polymer exhibiting first and second normal forces. Will the cone and plate be forced together, pushed apart, pushed sideways, or what? Discuss your reasoning.

APPENDIX 10.1 ENERGY OF ACTIVATION FROM CHEMICAL STRESS RELAXATION TIMES

From first-order chemical kinetics,

$$\frac{-dc_A}{dt} = kC_A \tag{A10.1.1}$$

where C_A is the concentration of species A. Then on integration,

$$-\ln C_A = kt + \text{constant} \tag{A10.1.2}$$

$$C_A = C_{A_0}e^{-kt} \tag{A10.1.3}$$

Note that k has the units of inverse time. If $k = 1/\tau_1$, an immediate relationship with equation (10.18) is noted.

From the Arrhenius equation,

$$k = \frac{RT}{Nh}e^{\Delta S/R}e^{-\Delta H/RT} \tag{A10.1.4}$$

For the present purposes,

$$k = se^{-\Delta E_{\text{act}}/RT} \tag{A10.1.5}$$

which may be rewritten

$$\tau_1 = s^{-1}e^{\Delta E_{\text{act}}/RT} \tag{A10.1.6}$$

where ΔE_{act} is the activation energy of the process. Then

$$\ln \tau_1 = \text{constant}^{-1} + \Delta E_{\text{act}}/RT \tag{A10.1.7}$$

A plot of $\ln \tau_1$ versus $1/T$ yields $\Delta E_{\text{act}}/R$ as the slope.

APPENDIX 10.2 VISCOELASTICITY OF CHEESE[†]

Cheese is composed of protein with the general structure

$$\left(\begin{matrix} & R & O \\ & | & \| \\ \text{--NH--CH--C} \end{matrix} \right)_{n} \tag{A10.2.1}$$

Segments containing 10–50 amino acids or mers are capable of long-range, coordinated molecular motions when above their glass transition temperature (the present case). These coordinated molecular motions are the cause of the observed viscoelastic behavior.

[†]Reprinted in part from Y. S. Chang, J. S. Guo, Y. P. Lee, and L. H. Sperling, *J. Chem. Ed.*, **63**, 1077(1986).

Equipment and Supplies:

one rectangular Plexiglas® or wooden box, slightly tilted back

one 1-pound block processed cheese (Velveeta® or similar), at room temperature

one clock (stopwatch is fine)

one meter stick

one small, flat plate (insert between cheese and weight)

one flat, hard surface (most desks are suitable)

Weights of about 1 kg (rectangular weights of the same cross-sectional area as the cheese are fine)

The cheese is placed inside the box, and the original height of the cheese is recorded. Then the plate and weight are placed on top of the cheese. Its new height is recorded immediately and each succeeding 5 min thereafter. This experiment measures creep in compression.

Replace the weight and deformed cheese with another weight and new cheese, and repeat.

Results

The initial dimensions of the 1-pound block of cheese were 4 cm × 6 cm × 15 cm. Weights of 500 and 700 g were employed. The resulting creep curves are illustrated in Figure A10.2.1. An immediate elastic compression is noted, a measure of the modulus E_1. This is followed by a curved line of

Figure A10.2.1 Experimental creep curves for Velveeta® cheese under two loads.

Figure A10.2.2   Resolution of the experimental data according to the four-element model. Curve 1 represents flow according to η_3. Curve 2 illustrates the behavior of the middle element, which is really a Kelvin element. The strain at zero time up to the interception of curve 1 represents the instantaneous deformation according to the spring E_1.

strain versus time, and a straight-line portion lasting at least 2 h; see Figure A10.2.1.

The data were analyzed in Figure A10.2.2 according to the four-element model. First, at zero time, the strain yields E_1. The straight-line portion at long times, separated as curve 1, yields η_3. Then, by subtraction, curve 2 was obtained. By simple curve fittings E_2, η_2, and τ_2 can be determined. The retardation time for Velveeta cheese was found to be about 7–9 min (see Table A10.2.1).

The value of E_1, near 5×10^5 dynes/cm^2, places it in the fourth region of viscoelasticity, rubbery flow.

Table A2.10.1   Characteristics of Velveeta cheese at room temperature according to the four-element model

	Applied Weight	
Variable	500 g	700 g
σ (dynes/cm^2)	2.00×10^4	2.88×10^4
E_1 (dynes/cm^2)	4.88×10^5	5.18×10^5
E_2 (dynes/cm^2)	2.82×10^5	4.24×10^5
η_2 (dynes · sec/cm^2)	1.52×10^8	1.78×10^8
η_3 (dynes · sec/cm^2)	1.00×10^9	1.21×10^9
τ_2 (min)	9	7

11

MECHANICAL BEHAVIOR
OF POLYMERS

Up until not too long ago, problems of fatigue, fracture, and failure in polymers were treated empirically. The old adage, "a chain is as strong as its weakest link," prevailed. To an increasing extent, fortunately, ideas of thermodynamics and viscoelasticity are now being used to explain what happens when polymers undergo fracture.

Fracture in polymers presents two important aspects. On the one hand, a knowledge of the strengths (and weaknesses) of polymers is of enormous engineering importance. Second, it is becoming increasingly popular as a field of fundamental research, combining the skill of the theoretician with that of the experimenter.

11.1 AN ENERGY BALANCE FOR DEFORMATION AND FRACTURE

11.1.1 The First Law of Thermodynamics

Consider a closed system, into which an increment of energy, δU_1, is transferred. While the nature of the energy can and should remain general, for the most part δU_1 is intended to consist of mechanical work; see Figure 11.1 (1). Inside the closed system, δU_1 is divided into three types of energy: δU_2, the change in irreversibly dissipated energy; δU_3, the change in stored or potential energy; and δU_4, the change in the kinetic energy.

The quantity δU_2 represents that energy dissipated by plastic or viscous flow being converted to heat. Two subcases must be considered—adiabatic systems and isothermal systems. In the adiabatic case, there is a temperature rise in the system. In the isothermal case, the thermal energy is all transferred to the surroundings across the closed system's boundary.

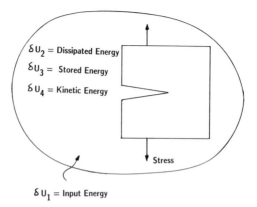

Figure 11.1 Mechanical behavior of a closed system, with energy input, causing crack growth (1).

The quantity δU_3 indicates the stored elastic energy in the system (note the springs in Chapter 10) but also includes stored thermal energy. The quantity δU_4 indicates changes in energy associated with changes in linear or angular velocity.

For the isothermal case where the stored energy is solely elastic, the net input of energy can be written

$$\delta U_1 - \delta U_2 \tag{11.1}$$

The first law of thermodynamics yields the conservation of energy,

$$\delta U_1 - \delta U_2 = \delta U_3 + \delta U_4 \tag{11.2}$$

Now consider that the body undergoes a displacement δu. Then

$$\frac{\delta U_1}{\delta u} - \frac{\delta U_2}{\delta u} = \frac{\delta U_3}{\delta u} + \frac{\delta U_4}{\delta u} \tag{11.3}$$

An incremental increase in energy in the system may be written

$$\delta U = \sigma \, \delta u \tag{11.4}$$

where σ represents the stress in the direction of the deformation or displacement u.

In the simplest case, that of rigid body dynamics, U_3 and U_2 are both zero. Then

$$\frac{\delta U_1}{\delta u} = \frac{\delta U_4}{\delta u} \tag{11.5}$$

11.1.2   Relations for Springs and Dashpots

By way of review, if a mass M moves with a velocity v,

$$U_4 = \left(\tfrac{1}{2}\right)Mv^2 \tag{11.6}$$

The derivative of U_4 with respect to u yields the acceleration, a:

$$\frac{dU_4}{du} = Ma \tag{11.7}$$

Since the external force f equals dU_1/du [see equation (11.5)], Newton's second law of motion is obtained:

$$f = Ma \tag{11.8}$$

The springs and dashpots of Chapter 10 clearly have both a thermodynamic and a mechanical basis. For a spring of modulus E attached to a mass M, the stored energy on stretching is given by

$$U_3 = \tfrac{1}{2}E\varepsilon^2 \tag{11.9}$$

where ε is the strain, and as a function of stretch,

$$\frac{dU_3}{du} = E\varepsilon \tag{11.10}$$

The dashpot, of course, always dissipates its energy, according to its viscosity,

$$\frac{dU_2}{du} = \eta \frac{d\varepsilon}{dt} \tag{11.11}$$

Substitution of these quantities into equation (11.3) yields the equation of forced vibration of a viscously damped system:

$$\sigma = E\varepsilon + \eta \frac{d\varepsilon}{dt} + M\frac{d^2\varepsilon}{dt^2} \tag{11.12}$$

This is the motion of a Maxwell element, including both mass and acceleration concepts.

The important points here are that the mechanical deformation of a body, including that of a polymer, is governed by simple thermodynamic laws applied to a mechanical system. Two additional quantities will be defined before leaving this section.

Consider the case of the body fracturing, where the area of the crack increases by δA. Then the energy changes can be written

$$\frac{\delta U_1}{\delta A} - \frac{\delta U_2}{\delta A} = \frac{\delta U_3}{\delta A} + \frac{\delta U_4}{\delta A} \qquad (11.13)$$

The term $\delta U_2/\delta A$ is called the fracture resistance, $\overline{R}$:

$$\overline{R} = \frac{\partial U_2}{\partial A} \qquad (11.14)$$

It indicates the energy dissipated in propagating a fracture over an increment of crack area δA. This is the work required to create new surfaces. Clearly, a large value of $\overline{R}$ indicates a tougher material.

As will be amplified below, contributions to $\overline{R}$ arise from the energy actually required to create the new surface (a quantity related to the surface tension), the orientation of chains near the surface, the breaking of chains that spanned the cracking region, and rubber elasticity energy storage effects.

A closely related quantity is the strain energy release rate, $\mathscr{G}$,

$$\mathscr{G} = \frac{dU_1}{dA} - \frac{dU_3}{dA} \qquad (11.15)$$

If $\mathscr{G} > \overline{R}$, then the system is unstable, and the crack velocity increases. These terms define the conditions of crack growth and stability, and energy storage within the body (1).

11.2 DEFORMATION AND FRACTURE IN POLYMERS

The previous section described the thermodynamics of deformation and fracture in terms of the energy required to elongate and break the sample. This section describes two major experiments to evaluate the deformation and fracture energy: stress–strain and impact resistance. In a stress–strain experiment, the sample is elongated until it breaks. The stress is recorded as a function of extension. Stress–strain studies are usually relatively slow, of the order of inches per minute. Impact strength measures the material's resistance to a sharp blow and by definition constitutes a much faster experiment (2). In both stress–strain and impact studies, energy is absorbed within the sample by viscoelastic deformation of the polymer chains, and finally by the creation of new surface areas (3). Energy may be absorbed by shear yielding, crazing, or cracking.

It is important to distinguish between a craze and a crack. When a stress is applied to a polymer, the first deformation involves shear flow of the polymer molecules past one another if it is above T_g, or bond bending and stretching

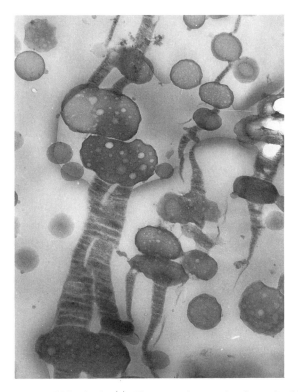

Figure 11.2 Crazing in ABS plastic (4). Electron micrograph of an ultrathin section cut parallel to the stress-whitened surface. The dark portions have been stained with OsO_4. The white plastic phase is poly(styrene – *stat* – acrylonitrile), and the stained phase is poly(butadiene – *stat* – acrylonitrile).

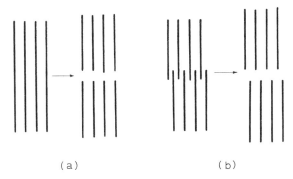

(a)   (b)

Figure 11.3 Schematic of fracture mechanisms in polymeric materials: (a) bond breakage and (b) chain slippage (3).

for glassy polymers. Eventually, a crack will begin to form, presumably at a flaw of some kind, and then propagate at high speed, causing catastrophic failure. A craze is not an open fissure but is spanned top to bottom by fibrils that are composed of highly oriented polymer chains (see Figure 11.2) (4). Although the volume of the sample is increased by the void space, these fibrils hold the material together.

A third mechanism of energy absorption is called shear yielding. In shear yielding, oriented regions are formed at 45° angles to the stress. Shear bands may be seen as birefringent entities; no void space is produced.

Another fundamental point relates to the actual mechanism(s) of fracture. Consider a sample composed of long polymer chains capable of viscoelastic motion. Does the crack grow through the polymer by breaking the chains, by viscoelastic flow of one chain past another, or by some combination of these factors? See Figure 11.3 (3). This will be considered further in Section 11.5.

11.2.1 Stress–Strain Behavior of Polymers

11.2.1.1 General Behavior

Three important types of stress–strain curves are illustrated in Figure 11.4. The brittle plastic stress–strain curve is linear up to fracture at about 1–2% elongation. Typical stresses at break are of the order of 10,000 psi, or about 6×10^7 Pa. Ordinary polystyrene behaves this way.

An example of a tough plastic is polyethylene, which is semicrystalline, with the amorphous portions above T_g. Its Young's modulus, given by the initial slope of the stress–strain curve, is somewhat lower than that of the brittle, glassy plastic (see Figure 11.4). Typically, this class of polymer exhibits a yield point, followed by extensive elongation at almost constant stress. This is called the plastic flow region and is clearly a region of nonlinear viscoelasticity. Stresses in the range of $2–5 \times 10^7$ Pa are commonly exhibited in this

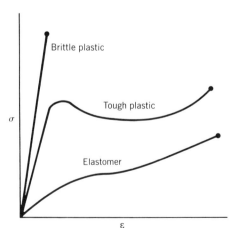

Figure 11.4 Stress – strain behavior of three types of polymeric materials.

range. Extension at constant stress in the plastic flow region is often referred to as cold drawing (see Section 11.4). Finally, the polymer strain hardens, then ruptures.

The third type of stress–strain curve is that exhibited by elastomers. The equation of state for rubber elasticity governs here, with its peculiar nonlinear curve. The student will remember that $\varepsilon = \alpha - 1$.

Elongation to break for both the tough plastic and the elastomer may be of the order of several hundred percent and is indicated by the dot at the end of the curve. For crystallizing elastomers such as natural rubber, the curve swings upward rather sharply at the point of crystallization, and tensile strengths of $2\text{--}5 \times 10^3$ psi are common. Noncrystallizing elastomers such as SBR have much lower tensile strengths, often below 1000 psi. However, with the addition of reinforcing fillers such as finely divided carbon black, the tensile strength is much increased. Of course, this last material is widely used for automobile tires, one of the toughest materials known.

Toughness as such is measured by the area under the stress–strain curve. This area has the units of energy per unit volume and is the work expended in deforming the material (see Section 11.1). The deformation may be elastic, and recoverable, or permanent (irreversible deformation). Elastic energy is stored in the sample in terms of energy per unit volume.

There are several terms in use that describe the "strength" of a polymer. The tensile strength describes the stress to break the material, usually in a simple test as in an Instron, for example. The tensile strength is certainly important, but in engineering practice a polymer is rarely stressed so greatly that it breaks immediately. The toughness of the polymer is frequently a more useful parameter.

Because of the development of crazes within the strained material, which are microscopic voids, the volume of the sample may increase, sometimes by several percent. This volume increase should not be equated with volume increases related to Poisson's ratio.

11.2.1.2 Time and Temperature Effects

The viscoelastic characteristics of stress–strain behavior of glassy amorphous polymers can be approached through considering that the stress at any time t is the sum of all the little stresses, $\delta\sigma$, each of which is the result of many incremental relaxing stresses each started at a progressively different time, t'. Each $\delta\sigma$ produces an incremental strain, $\delta\varepsilon$, and

$$\delta\sigma(t) = E(t')\delta\varepsilon \tag{11.16}$$

Thus the polymer can be considered as undergoing stress relaxation during the actual deformation process. On dividing the time interval from 0 to t into equal time increments, δx, we find the total stress, $\sigma(t)$, given by the convolution integral (5),

$$\sigma(t) = \int_0^t E(t - x) \frac{d\varepsilon}{dx} \, dx \qquad (11.17)$$

While the stress relaxation studies in Chapter 10 emphasized the presence of a single relaxation time, or at most a few relaxation times, real plastics undergoing extensive deformation display a broad distribution of relaxation times. Furthermore, mechanical strain in tension reduces individual relaxation times significantly, because the free volume in the system is increased. Of course, if the sample is compressed, the opposite effect is noted.

Figures 11.5 and 11.6 illustrate the effects of time and temperature (5). Both sets of data show yield stresses, followed by strains nearly independent of the stress. Significantly, the data taken at different temperatures are substantially congruent with the data taken at different strain rates, as should be expected from the time–temperature superposition principle.

Figure 11.5   Stress – strain curves for poly(vinyl chloride) at various strain rates. Note the well-defined yield point at about 5% strain. (1 psi $= 6.9 \times 10^3$ N / m²) (5).

Figure 11.6 For poly(vinyl chloride) in the glassy region, Young's modulus stays substantially constant as the temperature is lowered from 23 to −60°C, but the stress needed to stretch the sample beyond the yield point increases (5).

11.2.1.3 Semicrystalline Polymers

There are two cases to be considered in the stress–strain relationships of semicrystalline plastics. If the amorphous portion is rubbery, then the plastic will tend to have a lower modulus, and the extension to break will be very large. If the amorphous portion is glassy, however, then the effect will be much more like that of the glassy amorphous polymers. Orientation of semicrystalline polymers is also much more important than for the amorphous polymers. A special case involves fibers, where tensile strength is a direct function of the orientation of the chains in the fiber direction.

Typical data for low density polyethylene are shown in Figure 11.7 (5). At the lowest temperature shown, −60°C, the polymer is just above its glass transition temperature, −80°C. As the temperature is reduced, the modulus increases, and a yield point is developed at about 5% strain. At higher elongations, polyethylene exhibits cold drawing; see Section 11.2.2.

Some semicrystalline polymers such as poly(ethylene terephthalate) and polycarbonates based on bisphenol A are used commercially at very low molecular weights, down to $M_v = 13,000$ g/mol. Under these conditions, the polymer may be extremely brittle, with low tensile strength, because it is

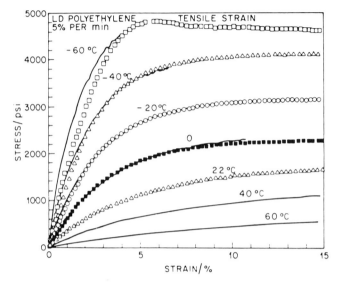

Figure 11.7 Low-density polyethylene's Young's modulus increases greatly as the temperature is lowered. Fit curves were generated theoretically using a scaling rule for crystalline polymers (5).

below its critical entanglement molecular weight. The chains do not form Gaussian coils at such low molecular weights, but rather are stiff, semicoils. Such polymers are most often used as matrix resins with glass fibers, useful because of their very low melt viscosities.

Polycarbonates and other polymers are considered by some to be toughened by the presence of low-temperature secondary transitions, such as the β-transition. The theory is that the flexibility imparted to the chain by the secondary chain motion contributes to the chain's ability to respond to stress.

11.2.1.4 Liquid Crystalline Polymers

Fibers made from polymers in the liquid crystalline state exhibit both very high moduli and very high tensile strengths. As described in Chapter 7, these fibers are actually very highly crystalline, being made up of rigid-rod polymer chains, and hence highly aligned. Table 11.1 compares the mechanical behavior of several plastics and fibers. As noted previously, polystyrene is a brittle plastic, with a high modulus, but breaking at only 2% elongation. Polyethylene is softer, but much more extensible. Ultradrawn polyethylene fibers are made from very high molecular weight polymer by drawing the material 50–100 times its original spun length, thus introducing a very high orientation. Note that ultradrawn polyethylene has a modulus 70 times higher than the ordinary plastic.

Table 11.1 Mechanical behavior of fibers

Polymer	Modulus, GPa	Tensile Strength, GPa	Elongation to Break, %	T_f, °C	Reference
Polystyrene	4	0.08	2	100, T_g	(a)
Polyethylene	1	0.05	50	140	—
Polyethylene, ultradrawn	172	3.5	5	149	(a)
Rayon	3	0.3	20	180 decomp.	
Nylon	3.8	0.8	25	250	(b)
Poly(1,4-benzamide)	100	3.0	6	500 decomp.	(b)

Note: $1 \text{ GPa} = 10^9 \text{ N/m}^2 = 10^{10} \text{ dynes/cm}^2 = 1.1 \times 10^4 \text{ kg/cm}^2 = 1.45 \times 10^5 \text{ lb/in.}^2$.

References: (a) S. Kavesh and D. C. Prevorsek, US Pat. No. 4,413,110 (1983), basis for Spectra." (b) J. R. Schaefgen, T. I. Bair, J. W. Ballou, S. L. Kwolek, P. W. Morgan, M. Panar, and J. Zimmerman, both in *Ultra-High Modulus Polymers*, A. Ciferri and I. M. Ward, Eds., Applied Science, London, 1979.

The liquid crystalline-forming polymer, poly(1,4-benzamide), has a modulus and tensile strength still higher than the ultradrawn polyethylene. The properties of poly(1,4-benzamide) are compared with the well-known fibers nylon and rayon. The latter are much softer but are more extensible.

The fibers made from poly(1,4-benzamide) and other liquid crystalline forming polymers are used in the manufacture of bullet-proof jackets. Here, large and rapid energy-absorbing capabilities are essential.

The ultradrawn polyethylenes are being researched as the simplest fiber possible, with every carbon atom contributing to its load-bearing characteristics. Even though it is clearly not a rigid-rod polymer, through ultradrawing virtually all the crystalline chain folds are eliminated, and the finished material mimics those actually made from rigid rods.

11.2.2 Cold Drawing in Crystalline Polymers

The tough plastic in Figure 11.4 is shown to have a yield point, followed by a region of cold drawing at almost constant stress. There are two basic causes for this phenomenon. First, for rubber-toughened amorphous plastics, the region of cold drawing is where extensive orientation of the chains takes place, accompanied by significant viscoelastic flow.

Second, for semicrystalline polymers with amorphous portions above T_g, cold drawing rearrangement of the chains takes place in a characteristic, complex manner, beginning with necking. A neck is a narrowing down of a portion of the stressed material to a smaller cross section. This is a form of shear yielding. The neck grows, at the expense of the material at either end, eventually consuming the entire specimen.

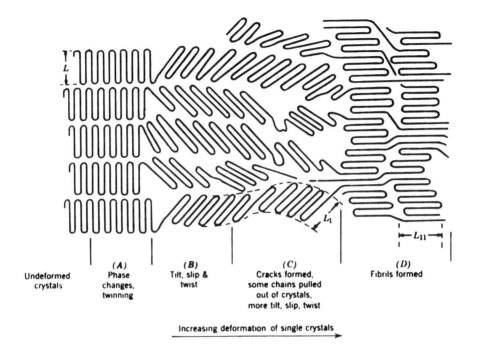

	(A)	(B)	(C)	(D)
	Undeformed crystals	Phase changes, twinning	Tilt, slip & twist	Cracks formed, some chains pulled out of crystals, more tilt, slip, twist	Fibrils formed

Increasing deformation of single crystals

Figure 11.8 Proposed mechanism of reorientation of crystalline structures during necking (6).

In the region of the neck, a very extensive reorganization of the polymer is taking place. Spherulites are broken up, and the polymer becomes oriented in the direction of stretch (see Figure 11.8) (6). The number of chain folds decreases, and the number of tie molecules between the new fibrils is increased. The crystallization is usually enhanced by the chain alignment. At the end of the reorganization, a much longer, thinner, and stronger fiber or film is formed.

This phenomenon is easily demonstrated at home or in the classroom. A strip of polyethylene film, perhaps with print on it, is satisfactory. The strip is slowly stretched between the hands at room temperature, and the neck will form after several percent elongation. Such a material can then be elongated 2–5 times its original length. Draw ratios can be significantly higher under more scientific conditions.

The molecular reorganization illustrated in Figure 11.8 has attracted the attention of the theoreticians (7, 8). Does the polymer actually melt under stress, and then reorganize, or do the crystals themselves rotate? At this point the evidence seems to be divided between the two possibilities.

It must be pointed out that drawing operations are critically important in the manufacture of fibers such as nylon and rayon. In the case of nylon, the polymer is spun in the melt state. The fiber is then cooled until it crystallizes, and finally stretched with a draw ratio of 4–8.

The case of rayon is slightly complicated because actually an alkaline solution of sodium cellulose xanthate is spun. The nascent fiber is spun into an acid bath, which removes the xanthate groups, precipitating the polymer. As it precipitates, it crystallizes. The fiber is then stretched wet, approximately four to seven times its original length. In both cases, the fibers are highly oriented, as shown by x-ray studies. The high degree of orientation contributes to the high strength of such fibers.

11.2.3 The Brittle–Ductile Transition

As illustrated in Figure 11.9 (8a) crazing involves the formation of an intricate network of fibrils connecting the upper and lower surfaces of the craze. The craze itself forms at right angles to the applied stress. Shear yielding, on the other hand, involves molecular slip, usually at 45° to the applied stress, assuming uniaxial stress–strain relationships. If the polymer forms crazes, it usually fails after 1 or 2% extension, and is said to be brittle. If it undergoes yielding, the polymer usually can be stretched at least 10 or 20% and is said to be ductile.

What criteria determine which mechanism will prevail? Polymers sometimes change from brittle to ductile on raising the temperature. Similarly, if the polymer is tested under compression, of under hydrostatic pressure, it tends to fail by shear yielding. Sternstein and Ongchin (9) studied the yield criteria for plastic deformation of glassy polymers in general stress fields, because the fracture resistance of a polymer is determined by its ability to develop a yield zone in the region of a crack tip, where it is usually in a state of triaxial stress. Sternstein and Ongchin used the so-called von Mises criterion (10–13),

$$(\sigma_1 - \sigma_2)^2 + (\sigma_2 - \sigma_3)^2 + (\sigma_3 - \sigma_1)^2 = 6C^2 \qquad (11.18)$$

where σ_1, σ_2, and σ_3 are the triaxial stresses. For polymers, the quantity C is pressure, temperature, and strain rate dependent. For pressure, a linear dependence has been used. If the left-hand side exceeds $6C^2$, then yield occurs, according to the von Mises criterion.

Sternstein and Ongchin calculated the biaxial stress envelopes for poly(methyl methacrylate); Figure 11.10 (9) shows separately the envelopes for craze initiation and shear yielding predicted from equation (11.18). In the first quadrant of stress space, the crazing envelope is everywhere inside the

(a)

(b)

Figure 11.9 (a) Transmission electron micrograph of a craze in polystyrene. Courtesy of R. P. Kambour, General Electric Co. Note fibrillar structures spanning the craze. (b) Section of polystyrene deformed at 60°C, viewed between crossed polarizers. These shear bands started to form at a compressive strain of about 4%. P. B. Bowden and S. Raha, *Philos. Mag.*, **22**, 463(1970).

shear yielding envelope, which implies that all combinations of tensile biaxial stress produce crazes prior to shear yielding. When the sample is under compression, third quadrant yielding will always take place. In the second and fourth quadrants, the two envelopes intersect each other. Generally, the failure mechanism that takes place is the one requiring the lower stress. The 45° line running through the second and fourth quadrants represents pure shear deformation, marking the boundary between hydrostatic compression

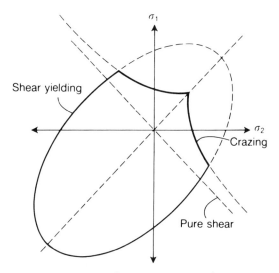

Figure 11.10 Failure envelope for poly(methyl methacrylate) under biaxial stress at room temperature; $\sigma_3 = 0$. Note intersection of crazing and shear yielding envelopes, heavy continuous line (9).

and hydrostatic tension. Crazing does not take place below this line because the pressure component of the stress tends to reduce rather than increase the volume.

While it can be argued that the intersection points in the second and fourth quadrants of Figure 11.10 constitute a brittle–ductile transition, both yielding mechanisms being coexistent. The brittle–ductile "transition" is a transition in the sense of which mode is predisposed to happen first in a given stress field.

11.2.4 Impact Strength

11.2.4.1 Instrumentation

Resistance to impact loading may be measured by using specially designed universal testing machines, which permit very high rates of loading, or, more commonly, by using one of two types of instruments: the Izod and the Charpy impact test machines. Each of these strikes the specimen with a calibrated pendulum. The Charpy instrument is illustrated in Figure 11.11 (2). In both cases, the sample is often notched to provide a standardized weak point for the initiation of fracture. In the Charpy experiment, the specimen is supported on both ends and struck in the middle. The notch is on the side away

Figure 11.11 Charpy-type impact instrument. *Inset:* Charpy test piece showing notch and point of striker (2).

from the striker. In the Izod experiment, on the other hand, the sample is supported at one end only, cantilever style. The notch is placed on the same side as the striker.

The interpretation of the results may be complicated even if the experiment is one of the simplest. The striker is released from an elevated position. On striking the sample, part of the momentum of the striker goes into the

Table 11.2   Izod impact strengths of several plastics[a,b](3)

Polymer	Impact Strength
Polystyrene	0.25–0.4
High-impact polystyrene (Hips)	0.5–4
ABS plastics	1–8
Epoxy resin (no filler)	0.2–1.0
Epoxy resin (glass-fiber-filled)	10–30
Cellulose acetate	0.4–5.2
Poly(methyl methacrylate)	0.3–0.5
Phenol–formaldehyde plastics	0.20–0.36
Poly(vinyl chloride)	0.4–1
High-impact poly(vinyl chloride)	10–30

[a]*Modern Plastics Encyclopedia* (1973); Lannon (1967).
[b]ASTM test D256 was followed, and the values reported have the units ft · lb/in. of notch.

creation of new surface area. The broken portion(s) of the sample also leaves with a certain momentum, which must be accounted for during analysis. For certain kinds of standardized width samples, the impact strength is reported in terms of foot-pounds per inch (of notch). When multiplied by the width, the work required to create a unit surface is determined. Typical values for Izod impact strengths are summarized in Table 11.2 (3). Ideally, impact strength ought to measure the work required to create new surfaces, especially for brittle, glassy plastics. However, viscoelastic effects cause the numerical results to be much larger (3). The student will note that two surfaces are created in each fracture.

11.2.4.2   Toughening by Rubber Addition

The impact resistance of glassy plastics may be increased by the addition of small quantities of rubber in the form of a polymer blend or graft copolymer (3, 4, 14–18). The rubber promotes crazing in the material, which absorbs the energy locally. This may be seen in everyday life as the white spot that results after a plastic has been hit accidently. This is called stress whitening.

The characteristics of the rubber additive are of critical importance in determining the toughness of the final product. Several important factors include the size of the rubber droplets, the phase structure within the rubber particles (see Figures 4.14 and 11.2), grafting of the rubber to the plastic, and the glass transition of the rubber.

The rubber droplets must be at least as large as the crack they are trying to stop. This puts the minimum size at several hundred angstroms, and a maximum size at about 3000–5000 Å.

Table 11.3   Impact resistance of ABS polymers at 30°C (16)

Sample Number	Composition of Rubber Component		T_g of Rubber Component, °C	Charpy Impact Strength, kg · cm/cm^2
	BD^a	ST^b		
1	35	65	40	0.75
2	55	45	−20	18
3	65	35	−35	30
4	100	0	−85	40

^{a}Polybutadiene.
^{b}Poly(styrene–*stat*–acrylonitrile).

The glass transition effect arises from a consideration of the speed of crack growth, which in turn depends on the velocity of sound in the material (3). In a plastic matrix having Young's modulus equal to 3×10^{10} dynes/cm^2, the velocity of a crack under impact conditions is about 620 m/sec. If the crack radius is about 1000 Å, an equivalent cyclic deformation frequency of about 10^9 Hz can be calculated. If the glass transition temperature increases at about 6–7°C per decade of frequency, the effective glass transition temperature of the rubbery phase is increased about 60°C above values measured at low frequencies, about 10^{-1} Hz. This calculation, while grossly oversimplified, suggests that the T_g of the elastomer phase must be about 60°C below the test temperature, which correlates well with the experimental evidence (see Table 11.3) (16). Thus, if the impact experiments are done at about 20°C, the glass transition of the rubber must be below about $-40°C$ in order to attain significant improvement in impact resistance (19).

One of the ways of obtaining the greatest overall toughness in a plastic is by combining shear yielding, crazing, and cracking in the proper order to absorb the highest total energy. Usually, it is desirable to have the sample yield in shear first. This absorbs energy without serious damage to the plastic. Then crazes should be encouraged to form within the shear-banded areas. The shear bands tend to stop the propagation of the crazes, limiting their growth. Only lastly does the "molecular engineer" want an open crack to form, because its propagation leads to failure.

A certain amount of viscoelastic molecular motion is required for shear yielding to occur first. The rubber domains actually cause crazes to multiply. This is a valuable occurrence, because as the crazes multiply they absorb energy locally in great quantities.

For example, polystyrene homopolymer is brittle, and under impact loading it tends to crack. High-impact polystyrene, HiPS, the graft copolymer of polystyrene with 5–10% polybutadiene, crazes first. A blend of this graft copolymer with poly(2,6-dimethyl-1,4-phenylene oxide), PPO, undergoes shear yielding first, then crazes. The ABS plastics craze, but with some shear yielding.

Similarly, homopolymers and copolymers with a flexible bond in their backbone tend to be tougher than those without. Examples include the polycarbonates and poly(methyl methacrylate) containing small amounts of acrylate comonomers. Flexible side chains seem not to be as effective, probably because the mechanical energy is not absorbed where it is needed.

Semicrystalline plastics such as nylon 610 and polyethylene are particularly tough. In this case, the amorphous portions are well above T_g, the $-CH_2-$ groups being highly flexible. The crystalline portions are hard and act to hold the entire material together.

11.2.5 Mechanical Behavior of Elastomers

11.2.5.1 *Stress–Strain Behavior*

In Chapter 9, the theory of rubber elasticity was developed. Young's modulus was given as $E = 3nRT$. Indeed, the modulus, a direct measure of the stiffness, increases with cross-link density. Ordinary cross-linked networks have a distribution of active chain lengths. Assuming a random cross-linking process, then $(M_c)_w/(M_c)_n$ should be about 2.

Mark and co-workers (20–24) examined the influence of the distribution of M_c, using end-linked poly(dimethyl siloxane) networks. Networks were formed from polymers having various ratios of low-molecular-weight component (660 g/mol) and high-molecular-weight component (21.3×10^3 g/mol), both number averages. The stress–strain behavior of the bimodal networks is illustrated in Figure 11.12 (20). A maximum in the stress to break as well as the toughness (as measured by the area under the curves) was attained at about 95 mol% of short chains. However, because of the difference in chain length between the two species, this is actually close to a 50/50 weight ratio.

At high elongations, the curves in Figure 11.12 turn upward owing to the limited extensibility of the short network chains. The elongations at rupture increase with decreasing mol% of short chains, causing the energy to rupture and the stress to break to go through maxima as the composition is changed.

Since most real networks contain a random distribution of active chain lengths, it is interesting to speculate that the distribution might bring about increased toughness.

Another feature of some elastomeric networks is their capability of crystallization. This too brings about an anomalous increase in the modulus at high elongations (23). Natural rubber has a melting temperature near room temperature in the relaxed state. (The melting temperature is slightly depressed by the vulcanization process.) At high elongations, the melting temperature climbs, so that at $\alpha = 4.5$, T_f is about 75°C.

Strain-induced crystallization also has a pronounced reinforcing effect within the network, and this increases the ultimate strength and maximum extensibility (24). The capability of crystallizing with extension while remaining elastomeric at low elongations provides an important self-reinforcement

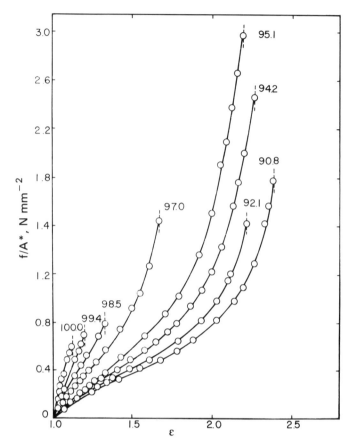

Figure 11.12 Stress–strain behavior of bimodal poly(dimethyl siloxane) networks. Each curve is labeled with the mol% of the short chains. The area under each curve represents the energy required for rupture (20).

mode for natural rubber. The only other elastomer with these properties is cis-polybutadiene. These materials are the elastomers of choice for many heavy-duty applications.

11.2.5.2 Viscoelastic Rupture of Elastomers

Failure in highly stretched elastomers is by no means instantaneous (25–27). A common home example involves putting a rubber band around some objects and setting them aside. Some later time, the rubber band may be seen to be broken. Interestingly, the time frame of the failure can be predicted through the application of ordinary principles of viscoelasticity.

In the laboratory, highly stretched elastomers undergo a period of smooth relaxation followed by sudden failure. The time frame of failure decreases with increasing initial stretch (see Figure 11.13) (25). The dashed line

Figure 11.13 Stress relaxation of *cross* – poly(butadiene – *stat* – styrene) at 1.7°C. Elongations range from 350 to 525%. Solid points indicate rupture. The dashed line gives the ultimate stress as a function of log time (25).

illustrates the locus of failure. The data in Figure 11.13 are for one temperature. When the experiment is repeated at different temperatures, a family of such curves will be generated, as shown in Figure 11.14 (25). The quantity σ_b represents the stress to break. The quantity T_0/T multiplying the stress to break applies the theory of rubber elasticity, reducing the stress to the temperature of interest.

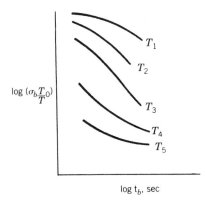

Figure 11.14 The locus of failure points obtained at different temperatures (25).

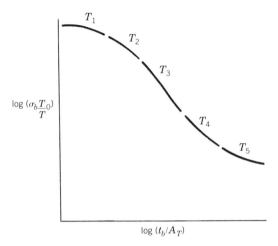

Figure 11.15   The master curve composed of the data in Figure 11.14.

Through the application of the time–temperature superposition principle (Section 10.3), these curves can be shifted to yield a master curve at a particular temperature (see Figure 11.15) (25). Note that the quantity A_T appearing in Figure 11.15 is the shift factor appearing in the WLF equation (Section 8.6.12).

It is customary to record the strain to break, ε_b, as well as the stress to break. When log ε_b is plotted against $\log(\sigma_b T_0/T)$, a special curve called the failure envelope is generated (see Figure 11.16) (25). The failure envelope is widely used to predict the stability of elastomers under stress. If a specimen falls in region A of Figure 11.16, it will not fail. On the other hand, a specimen in region B is subject to eventual failure.

Elastomers exposed to wear and tear, such as automotive tires, are always reinforced. The usual reinforcing substance is either finely divided silicas or carbon blacks (see Section 9.16.2). Tires contain up to 60% of the latter (see Table 9.8), raising their tensile strength several times over. It must be pointed out that the major cause of wear in tires is abrasion. Tiny shreads of rubber

Figure 11.16   The failure envelope prepared from the master curve in Figure 11.15, obtained by plotting the reduced stress to break versus the strain to break, on a log–log scale. The specimen will not fail in region A, but will in region B. The arrow indicates the direction of lower temperatures or higher strain rates, the latter for dynamic experiments.

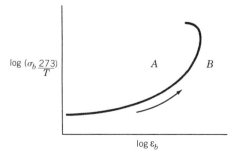

are torn loose and stretched on the road, gradually ripping them off (28). While tear strength and tensile strength are not the same, they are closely related.

11.2.6 Fiber-Reinforced Plastics

In the manufacture of engineering materials based on plastics, much emphasis has been placed on high modulus and great strength. The practice of polyblending small quantities of rubber into plastics has already been mentioned as a way to improve impact resistance and strain to break. The addition of carbon black to rubber increases tensile strength and abrasion resistance.

A third way to toughen plastics is to add fibers, to make composite materials (3). Fibers in use today include glass, boron, graphite, or other polymers. These may be chopped or added as continuous filaments. With continuous filaments, the fibers carry most of the mechanical load, while the polymeric matrix serves to transfer stresses to the load-bearing fibers and to protect them against damage (29–32). Thus a controlled amount of adhesion between the fibers and the matrix is required.

11.2.6.1 Matrix Materials

Polymers used for matrix materials are usually thermosets, with epoxies being the most common (33). Others are polyesters and phenolics. Sometimes thermoplastics are used, principally polyimides, polysulfones, and polyetheretherketone (PEEK). The matrix has several roles in polymer–fiber composites. These include maintaining desired fiber spacing, transmitting shear loads between layers of fiber, and reducing the tendency to transmit stress concentration from broken fibers to intact fibers.

The role of the fiber is to support all main loads and limit deformations. Thus the fibers increase strength, stiffness, and toughness, and decrease corrosion, creep, and fatigue. Figure 11.17 (34) illustrates a scanning electron micrograph of the fracture surface of an injection molded PEEK composite reinforced with 30 wt% of short carbon fibers. The sample was static loaded, with crack growth being from lower right to upper left. Minimum pull-out of fibers indicates good adhesion between fiber and matrix.

Because of their light weight and great strength and stiffness, fiber–polymer composites are among the strongest materials known, especially on specific modulus and strength basis (the property is divided by the density). Table 11.4 (33) illustrates the increase in properties that can be achieved. Thus the modulus is increased up to 80 times, and the tensile strength about 20 times on the addition of various fibers. On the other hand, the densities are less than double.

Figure 11.18 (33) illustrates the specific modulus and strength of these materials, compared to common metals. It is especially important to note

Figure 11.17 Fracture surface of short carbon fiber – PEEK composite (34).

Table 11.4 Comparative fiber and unidirectional composite properties (33)

Fiber/Composite	Elastic Modulus, GPa[a]	Tensile Strength, GPa[a]	Density, g/cm³	Specific Stiffness, MJ/kg	Specific Strength, MJ/kg
Epoxy matrix resin	3.5	0.09	1.20	—	—
E-glass fiber	72.4	2.4	2.54	28.5	0.95
Epoxy composite	45	1.1	2.1	21.4	0.52
S-glass fiber	85.5	4.5	2.49	34.3	1.8
Epoxy composite	55	2.0	2.0	27.5	1.0
Boron fiber	400	3.5	2.45	163	1.43
Epoxy composite	207	1.6	2.1	99	0.76
High-strength graphite fiber	253	4.5	1.8	140	2.5
Epoxy composite	145	2.3	1.6	90.6	1.42
High-modulus graphite fiber	520	2.4	1.85	281	1.3
Epoxy composite	290	1.0	1.63	178	0.61
Aramid[b] fiber	124	3.6	1.44	86	2.5
Epoxy composite	80	2.0	1.38	58	1.45

[a]To convert GPa to psi, multiply by 145,000.
[b]Aromatic polyamide.

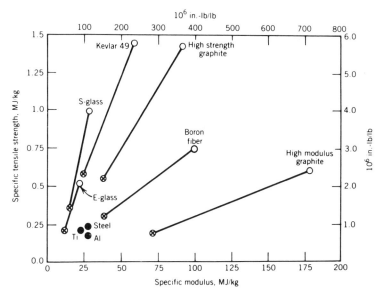

Figure 11.18 Specific strength versus specific modulus for fiber-reinforced epoxy materials: ○, unidirectional composite; ⊗, quasi-isotropic material. Properties are compared to some common metallic materials (●). To convert MJ / kg to 10^6 in.· lb / lb, multiply by 4.023 (33).

that for many applications, particularly aerospace, weight is the important quantity, not volume of material.

11.2.6.2 Silane Interfacial Bonding Agents

A critically important parameter involves the bonding of the fiber to the matrix. If the fiber is not well bonded to the matrix, it will pull out under load. If it is too well bonded, no energy can be absorbed by the debonding process. Ideally, the matrix should fracture first, followed by the debonding of the fiber–matrix interface, followed by the fracture of the fiber itself for maximum absorption of energy; see Figure 11.17. For glass fibers particularly, the extent of bonding is controlled by the use of silanes as interfacial bonding agents. The silane molecule has one end that will react with the glass surface, and the other end with the matrix. For an epoxy–glass system, a silane such as γ-aminopropyltriethoxysilane is used,

$$NH_2-CH_2-CH_2-CH_2-Si{\overset{\displaystyle O-C_2H_5}{\underset{\displaystyle O-C_2H_5}{-O-C_2H_5}}} \qquad (11.19)$$

The amino group reacts with the oxirane of the epoxy, and the silane bonds to the glass as illustrated in Figure 11.19 (35).

$$(CH_3O)_3Si-R \xrightarrow{H_2O} \begin{array}{c} R \\ | \\ HO-Si-OH \\ | \\ OH \end{array}$$

+

Figure 11.19 Illustration of the process by which silane coupling agents are adsorbed onto silica surfaces (35). The group R may be a vinyl-bearing moiety or may contain another reactive group that will bond it to the polymeric matrix.

Coil-Like Polymer (Matrix)

Rigid Rod Polymer (Reinforcement)

Mixing Process

Planar Isotropic Molecular Composite

Orientation Process

Uniaxially Oriented Molecular Composite

Ordered PBT

Reinforcing Molecules

Random PBT

Matrix

Molecular Composite
• Unidirectional Orientation

Figure 11.20 The molecular composites concept utilizes a rod-shaped polymer dispersed in a random-coil-shaped polymer. When the rod-shaped chains are oriented, reinforced fibers can be made (36).

11.2.6.3 Molecular Composites

Among the newest experimental materials are the so-called molecular composites, which make use of rod-shaped polymer chains dispersed or dissolved in a random coil matrix; see Figure 11.20 (36). Since the extent of reinforcement increases with the ratio of length to width of the fiber (called the aspect ratio), ideally very strong materials should result. A major experimental problem, however, has been a proper dispersal of the rod-shaped molecules in the matrix, caused by the very low entropy of mixing two kinds of polymer molecules; see Section 4.3. Many scientists and engineers think that a rapid coagulation followed by orientation, such as used in fiber technology, may yield a break-through for the anticipated excellent properties.

11.3 CRACK GROWTH

11.3.1 The Griffith Equation

As a material is strained, energy is stored internally by chain extension, bond bending, or bond stretching modes (37–40). This energy will be dissipated if bond breakage or viscoelastic flow occurs. The first analysis of the balance between the energy applied and the energy released in bond breakage as a crack propagates was due to Griffith (37). He showed that when the release of strain energy per unit area of the crack surface exceeds the energy required to break the bonds associated with the unit area of surface, the intrinsic surface energy being designated as γ_s, a crack would propagate. Several designs of specimens containing various cracks, or notches, are shown in Figure 11.21 (40).

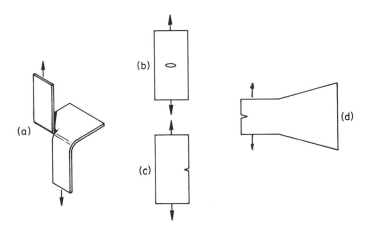

Figure 11.21 Important specimen shapes for the determination of the characteristic energy and stress intensity parameters for fracture mechanics (40): (a) Trouser-leg design, often used with elastomers. (b) Center-notched panel. (c) Edge-notched panel. (d) Cantilever-beam type. Types (b) – (d) are used for plastics.

When σ is the gross applied stress, E is Young's modulus, and a is half the crack length, Griffith showed that the critical point is defined by the relation

$$\sigma = \left(\frac{2E\gamma_s}{\pi a}\right)^{1/2} \tag{11.20}$$

This equation was specifically derived for the center-notched panel (Figure 11.21b). Equation (11.20) works reasonably well for inorganic glasses, because viscoelastic motions are almost nonexistent. However, experimentally determined values for the surface energy for polymers are 10^2–10^3 times values calculated on the basis of bond breakage alone (38, 39). The increase is largely due to viscoelastic flow, which absorbs large amounts of energy.

11.3.2 The Stress Intensity Factor

A more general equation was derived by Orowan (41), who replaced γ_s with the term $\gamma_s + \gamma_p$, where γ_p accounts for the energy involved in plastic deformation. Irwin (42) considered the fracture of solids from a thermodynamic point of view and arrived at the equation

$$\sigma = \left(\frac{E\mathscr{G}}{\pi a}\right)^{1/2} \tag{11.21}$$

where $\mathscr{G}$ represents the strain energy release rate, $\partial U/\partial a$ (see Section 11.1). Then,

$$\mathscr{G} = 2(\gamma_s + \gamma_p) \tag{11.22}$$

A key development in fracture mechanics was the analysis of the stress field at a crack tip (43). If σ_{ys} is the material's yield strength and r_y the crack-tip plastic zone radius (see Figure 11.22),

$$r_y = \frac{1}{2\pi}\frac{K^2}{\sigma_{ys}^2} \tag{11.23}$$

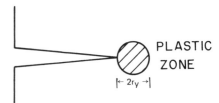

Figure 11.22 Schematic illustration of the plastic zone at the crack tip (43).

where the stress intensity factor, K, is given by

$$K^2 = E\mathscr{G} \tag{11.24}$$

for the case of plane stress, or $K^2 = E\mathscr{G}/(1 - \nu^2)$ for plane strain, where ν is Poisson's ratio (43).

For the simple case represented by Figure 11.21b the stress intensity factor is given by

$$K = \sigma(\pi a)^{1/2} \tag{11.25}$$

At the critical conditions of crack spreading, the critical stress intensity factor, K_c, is given by

$$K_c = \sigma_c(\pi a)^{1/2} \tag{11.25a}$$

Similarly, the work done per unit area of new crack surface is given by $\mathscr{G}_c$. It must be remembered that a crack will spread only if the total energy of the system is decreased. A more general case may be written,

$$K_c = Y\sigma a^{1/2} \tag{11.26}$$

where Y represents a geometrical factor including the sample width.

The stress intensity factor forms the central part of a great deal of the fracture and fatigue literature. Once the stress intensity factor is known, it is possible to determine the maximum value that would cause failure. This critical value, K_c, is known as the fracture toughness of the material. Under plane strain (tensile) test conditions, the fracture toughness is given the special designation K_{1c} (44). The corresponding notation for shear strain is K_{2c}, and the notation for tearing (antiplane) test conditions is K_{3c}. These tests refer to conditions where the crack is pulled open, where the portions of the sample above and below the crack are pulled in opposite directions parallel to the crack, and where portions of the sample above and below the crack are pulled in opposite directions normal to the direction of the crack, respectively.

11.3.3 A Worked Example

A center-notched panel of polystyrene at 25°C has a crack of 1.0 cm width; see Figure 11.21b. The critical stress intensity factor, K_c, was determined by other experiments to be 6×10^8 dynes $\cdot$ cm$^{1/2}$/cm^2. Will the sample fail under a stress of 1×10^8 dynes/cm^2?

From equation (11.25), we have

$$K = \sigma(\pi a)^{1/2}$$

$$K = 1 \times 10^8 \; (\text{dynes/cm}^2)(3.14 \times 0.5 \text{ cm})^{1/2}$$

since a is half the width of the crack.

$$K = 1.25 \times 10^8 \text{ dynes} \cdot \text{cm}^{1/2}/\text{cm}^2$$

Since K is smaller than K_c, the sample will not fail.

What is the critical strain energy release rate, $\mathscr{G}_c$? From equation (11.24),

$$\mathscr{G}_c = K_c^2/E$$

Young's modulus for polystyrene at room temperature is approximately 3×10^{10} dynes/cm^2; see Chapter 8.

$$\mathscr{G}_c = \left(6 \times 10^8 \text{ dynes} \cdot \text{cm}^{1/2}/\text{cm}^2\right)^2/3 \times 10^{10} \text{ dynes}/\text{cm}^2$$

$$= 1.2 \times 10^7 \text{ dynes}/\text{cm} \quad \text{or} \quad 1.2 \text{ ergs}/\text{cm}^2$$

Note the units of work per unit area for $\mathscr{G}_c$.

11.4 CYCLIC DEFORMATIONS

In cyclic deformation, stress and strain patterns are repeated over and over again, perhaps tens of thousands of times. While the stresses are usually well below the ultimate failure stress measured in simple extension, cracks may grow and the sample may fail. This experiment simulates the mechanical vibrations that engineering plastics are subjected to in service; indeed, such plastics are frequently more likely to fail because of repeated stressing than because of a single excessive deformation or hard blow.

Material failure because of cyclic stressing is called fatigue. In terms of molecular processes, fatigue failure involves two principal mechanisms: (a) the sample may get hot owing to adiabatic heating (see Section 11.1)—this is especially true at the crack tip, and (b) growth of the crack itself (45, 46). The number of cycles to failure depends on the stress level, as illustrated in Figures 11.23 (46) and 11.24 (47). In some cases a so-called endurance limit is observed below which fatigue failure does not occur in realistic lifetimes of experiments.

The fatigue phenomenon is not an all-or-nothing situation. First, it is assumed by many investigators that all materials are flawed as made, even if the flaws are too small to be ordinarily noticed. On repeated stressing, these tiny flaws initiate cracks. In the next stage, these cracks propagate slowly, perhaps growing a bit with every cycle, or sometimes intermittently. Finally, the material fails, the crack growing through the sample in one stress cycle.

Figure 11.23 Fatigue lifetimes versus stress level for nylon 6, poly(methyl methacrylate), and polytetrafluoroethylene (46). This plot is known as an S – N plot, for stress and number of cycles to failure.

11.4.1 Mechanisms of Crack Growth

Hertzberg and Manson (43) have identified several important molecular aspects of the fatigue process:

1. High molecular weights and narrow moleclar-weight distributions are generally more fatigue resistant.
2. Chemical changes such as bond breakage are to be avoided or minimized.
3. Elastic, anelastic, and viscoelastic deformation of the chains is desired, as these motions absorb energy and tend to prevent crack growth, provided the sample does not heat up excessively.
4. Morphological changes such as drawing, orientation, and crystallization also absorb energy and are desired.
5. Samples undergoing cyclic stressing at frequencies and temperatures near the glass transition temperature or secondary transitions will tend to heat up more than otherwise, causing softening or degradation.
6. Adiabatic heating caused by such hysteretic effects is clearly also undesirable.
7. Inhomogeneous deformations, such as crazing and shear banding, absorb energy and are desired. It is especially desired to have shear banding occur before crazing; that is, the former mechanism should have a lower free energy of initiation.

Of course, several of these factors may be operative simultaneously.

Figure 11.24 The relationship between ΔK and fatigue crack growth rates for poly(methyl methacrylate), PMMA; polysulfone, PSF; polystyrene, PS; poly(vinyl chloride), PVC; poly(2,6-dimethyl-1,4-phenylene oxide), PPO; polycarbonate, PC; nylon 66; and poly(vinylidene fluoride), PVF$_2$ (47).

11.4.2 Fatigue Crack Propagation

In many fatigue crack propagation experiments, a notch is deliberately introduced in the specimen at a convenient location (48). The rate of growth of this crack and the morphology of the crack surface outline areas of current interest.

For a crack of length a, the fatigue crack growth rate is da/dn, where n represents the number of cycles. The stress intensity factor range is given by

$$\Delta K = K_{max} - K_{min} \tag{11.27}$$

where K_{max} and K_{min} are the maximum and minimum values of the stress

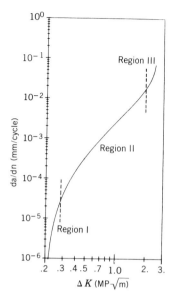

Figure 11.25 Linear polystyrene, $M_n = 160,000$ g / mol, narrow molecular-weight distribution. Region I, substantially nil crack growth. Region II, fatigue crack propagation proceeds, with thousands of cycles required to produce failure. Region III, failure after only a few cycles (49).

intensity factor during the cyclic loading. The crack growth rate is related to the stress intensity factor range by the equation (Paris' law)

$$\frac{da}{dn} = A \, \Delta K^m \tag{11.28}$$

where A and m are parameters dependent on material variables, mean stress, environment, and frequency.

While the rate of crack growth increases with crack length, equation (11.28) implies a straight line, which is frequently observed over much of the range in ΔK (see Figure 11.25) (49).

The actual rates of crack growth depend on several factors. First is the temperature of the experiment. For low frequencies, the temperature of the crack tip may approximate the ambient conditions. At higher frequencies, the crack tip may get quite hot. This is important because viscoelastic motions depend strongly on the temperature and the frequency of the loading. Either the glass transition or secondary transitions such as the β transition may be involved (48), but the latter is usually the more important because most engineering plastics are well below their glass transition temperatures. The frequency sensitivity, defined as the multiple by which the fatigue crack propagation rate changes per decade of frequency, goes through a maximum at the temperature where the β transition segmental jump frequency is comparable to the test machine loading frequency (48).

Second, the rates of crack growth depend on the value of ΔK. If ΔK is low enough, the rate of crack growth may be essentially zero. If ΔK is high enough, one cycle may produce fracture. This results in a sigmoidal character to the fatigue crack growth curve (see Figure 11.25) (49). The data in Figure 11.24 are essentially in the middle range.

Third, fatigue propagation depends inversely on the weight-average molecular weight (49). Below a critical value of M_w, the sample is brittle. At higher molecular weight ranges, the fatigue crack propagation rate decreases. Indeed, higher molecular weight samples are stronger in a variety of experiments.

11.5   MOLECULAR ASPECTS OF FRACTURE AND HEALING IN POLYMERS

From a kinetics point of view, formation of new surfaces, fracture, and subsequent crack healing have a certain symmetry, with chains variously being pulled out or diffusing into the fracture surface. The symmetry is far from complete, however, because of chain breakage, the chemical and physical state of the surfaces at the time of crack healing, and so on (50–57).

In general, the two surfaces being "healed" are from different sources. Examples of this include molding operations, film formation from latexes, and adhesion. This section examines the micromechanisms involved in fracture and healing in polymeric materials. The effect of chemical bonding and adhesion between a polymer and a substrate surface is considered in Section 11.6.

11.5.1   Interdiffusion at a Polymer–Polymer Interface

When two otherwise identical polymer surfaces are brought into juxtaposition at $T > T_g$ and annealed, a healing process may take place. As a result, the interface gradually disappears. Basic additional requirements are that the polymer be linear (or branched) and amorphous. The major mechanism involves the interdiffusion of polymer chain segments across the interface. For polymer chain interdiffusion, the reptation theory of de Gennes (51) and Edwards (52) is employed. In this model, the chain is confined to a tube, as described in Chapter 5. The equations of motion of this chain in the bulk state are described in Section 10.4.2. In the following, a linear monodisperse molecular weight polymer whose chain ends are uniformly distributed in the bulk melt material is considered. Assuming instantaneous wetting of the two surfaces, Wool and co-workers (53–56) and Tirrell (57) analyzed the molecular aspects of the interface as interdiffusion proceeds. Wool developed the "minor chain model"; see Figure 11.26 (54). At time equals zero, molecular contact is achieved at the interface. The chains begin to reptate via their chain ends, which are shown initially as dots randomly distributed throughout the polymer. As time proceeds, only that portion of the chain that has escaped its original conformation (i.e., the minor chain) can contribute to

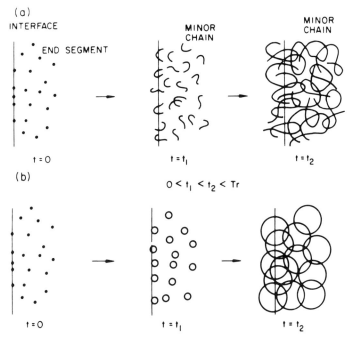

Figure 11.26 The interdiffusion process at a polymer – polymer interface (54): (*a*) appearance of the minor chains and (*b*) the minor chain spherical envelopes.

interdiffusion across the interface. Wool does not consider the exact location of each segment, but only the spherical envelope to which it belongs; see Figure 11.26.

11.5.2 Fracture Micromechanisms at a Polymer–Polymer Interface

While chain scission is important in the deformation and fracture of many polymers, this micromechanism normally consumes only a small fraction of the fracture energy. However, its appearance limits the extent of viscoelastic energy dissipation, the major energy consuming factor in most polymer fractures. Thus, for tougher materials, chain scission should be delayed or avoided in favor of molecular relaxation.

In the relaxed amorphous state (Chapter 5), polymer chains exist in the form of random coils. The initial effect of an external stress is to elongate those coil segments, which reside in the fracture plane; see Figure 11.27 (50). Compared to chain scission that consumes only approximately 83 kcal/mol per $C — C$ bond broken, this process consumes much energy. When the chains reach their ultimate extension, sufficient stress on individual bonds may be generated for chain breakage to occur. Alternately, the chains may slip out of one side or the other of the new surface as it is formed, a process

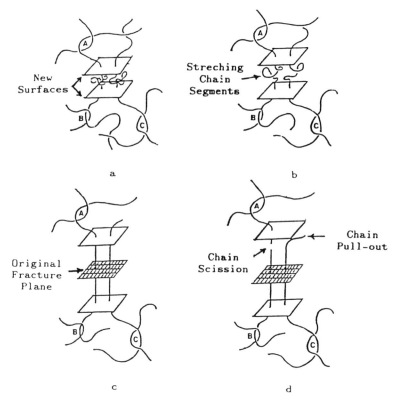

Figure 11.27 Micromechanisms of fracture: (*a*) new surfaces appear under stress; (*b*) chain segments elongate; (*c*) maximum elongation; and (*d*) chain scission or pull-out (50).

that also consumes much energy, or a combination of both may occur. Molecular weight, polydispersity, temperature, and stress rates all affect the contribution of each to the overall deformation process.

11.5.3 Scaling Laws for a Polymer–Polymer Interface

Kim and Wool (54) derived a number of functions that describe the molecular state at the interface as a function of time. For example, the number of random-coil chains intersecting unit area of the interface as a function of contact time, t, and molecular weight, M, can be written

$$n(t) \sim t^{1/4}M^{-5/4} \qquad (t < \tau_r) \tag{11.29}$$

$$n_\infty \sim M^{-1/2} \qquad (t > \tau_r) \tag{11.30}$$

where τ_r is the relaxation time. Note that the static or virgin state solution at

Table 11.5   Molecular aspects of interdiffusion at a polymer – polymer interface (56)

Molecular Aspect	Symbol	Dynamic Relation $H(t)$	Static Relation H_∞	r	s
General property	$H(t)$	$t^{r/4}M^{-s/4}$	$M^{(3r-s)/4}$	r	s
Number of chains	$n(t)$	$t^{1/4}M^{-5/4}$	$M^{-1/2}$	1	5
Number of bridge	$p(t)$	$t^{1/2}M^{-3/2}$	M^0	2	6
Average monomer depth	$X(t)$	$t^{1/4}M^{-1/4}$	$M^{1/2}$	1	1
Total monomer depth	$X_0(t)$	$t^{1/2}M^{-3/2}$	M^0	2	6
Average contour length	$\ell(t)$	$t^{1/2}M^{-1/2}$	M	2	2
Total contour length, number of monomers, N	$L_0(t)$	$t^{3/4}M^{-7/4}$	$M^{1/2}$	3	7
Average bridge length	$\ell_p(t)$	$t^{1/4}M^{-1/4}$	$M^{1/2}$	1	1
Center of mass depth[a]	X_{cm}	$t^{1/2}M^{-1}$	$M^{1/2}$	2	5
Diffusion front length	N_f	$t^{1/2}M^{-3/2}$	M^0	2	6

[a]This equation applies to chains in the bulk and does not apply to chains whose center of mass is within a radius of gyration of the surface.

$t > \tau_r$ is obtained by substituting $t = \tau_r \sim M^3$ in the dynamic solution. This condition constitutes the state when the crack is completely healed, and the sample behaves as if the crack never existed. A number of analytical aspects of interdiffusion at polymer–polymer interfaces are shown in Table 11.5 (56). The general relations may be written

$$H(t) = H_\infty (t/\tau_r)^{r/4} \tag{11.31}$$

$$H_\infty = M^{(3r-s)/4} \tag{11.32}$$

Values for r and s for individual properties are shown in Table 11.5 (56). Related fractal geometry (58–60) aspects are treated in Section 11.5.5.

11.5.4   Tensile Strength Dependence on Interdiffusion Distance

One of the major conclusions of the Wool theory (53, 54) is that mechanical strength buildup during interface healing should have a one-fourth power time dependence. Note that the time dependence of equation (11.29) and others are $t^{1/4}$ rather than $t^{1/2}$ as expected for usual atomic diffusion processes.

Yoo et al. (61, 62) confirmed this dependence using uniform polystyrene latex films. (See Section 5.4.4.) The process of film formation from a latex can be divided into three stages: (a) evaporation of the water containing the latex, (b) coalescence and deformation of the latex particles, and (c) interdiffusion

Figure 11.28 Wool plot, showing tensile strength buildup from polystyrene latex (61).

of the polymer chains into adjacent latex particles.[†] The value of using latex particles was that they could be made very small (275 Å radius), thus presenting a relatively large surface area. Through small-angle neutron scattering studies using deuterated particles mixed into the samples, the interdiffusion depth as a function of annealing time was determined.

Figure 11.28 (61) shows the tensile strength buildup as a function of time to the one-fourth power. There is an apparent induction period, followed by a portion that agrees with Wool's theory. At long times, the fully developed tensile strength remains nearly constant. Concomitant small-angle neutron scattering experiments showed that the level-off conditions correspond to an interdiffusion distance of 110–120 Å, which corresponds roughly to the radius of gyration of the polystyrene chains. In other words, at full tensile strength, the lead interdiffusing chains were half in the old latex particle and half in the new. Thus no further increase in tensile strength could be expected, because the interface between the latex particles was substantially healed.

11.5.5 The Fractal Nature of the Diffusing Species

Fractal geometry achieved fame with its solution of the "coastline problem" (58): How long is the coastline of Britain, measured with smaller and smaller steps? Fractal geometry is now being used to examine a wide range of chemical and physical problems. In Euclidean geometry, a line is expressed as distance to the first power, and area as distance to the second power. However, certain irregular structures (such as coastlines or, in the present case, interdiffusion boundaries) are better expressed as fractional powers between one and two. An important case is that the structures are "self-simi-

---

[†]These are the stages of film formation in ordinary latex paint, such as used on buildings. In this case, the glass transition temperature of the polymer is usually just below room temperature to facilitate interdiffusion. The latex dispersion contains filler to increase modulus and decrease scratch resistance, as well as colorants and so on.

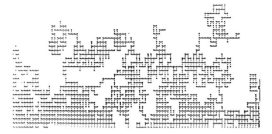

Figure 11.29 Computer simulation of a polymer interface (53, 60).

lar"; that is, as one measures them by finer and finer scales, one finds the same geometrical patterns.

When viewed in two or three dimensions on the molecular scale, the diffusion field has a very rough nature. Sapoval et al. (59) considered two-dimensional lattice diffusion, with D_f being the fractal dimension of the diffusion front. A computer simulation of a fractal diffusion front is shown in Figure 11.29 (60). In two dimensions, $D_f = \frac{7}{4}$. This is perceived as being significantly more complex than a single line, but not quite entirely two dimensional. Interface roughness and molecular connectivity have interesting implications for both the mechanical and electrical property development of interfaces. For example, roughness promotes mechanical interlocking, improving bonding. However, its fractal characteristic reduces electrical conductance as well as signal speed.

11.6 THE BEHAVIOR OF ADHESIVES

11.6.1 Forms of Adhesives

Adhesives and glues were known to antiquity. Early materials were based on proteins from boiled down fish, bones, albumin, and so on, making gelatinous water solutions. Many animal glues form bonds stronger than the wood they bond together in furniture construction. However, such materials were restricted to indoor uses, for the polymers would rapidly dissolve in rainwater. Later, rubber dissolved in solvents formed the so-called rubber cement or glues, still in service.

Adhesives may be used to bond two different structures together to form a composite, as in polymer–polymer welding. The problem of adhesion also extends to the bonding of an engineering polymer directly to a substrate.

In Chapter 8, an adhesive was defined as a linear amorphous polymer above its glass transition temperature. While this definition holds in its simplest form for pressure-sensitive tapes, many real adhesives are more complex in application. For example, the adhesive may be applied as a monomer. This is true for the epoxy glues and for the "instantaneous" adhesives based on cyanoacrylates, such as Super Glue.5®. Similarly, the adhesive may be applied as a prepolymer (low-molecular-weight reactive polymeric species), such as is the case for urethane types. Other kinds of adhesives include block copolymers, suspensions, and latexes.

After application, the adhesive may remain linear or become cross-linked. It may be either above or below its glass transition temperature. While individual cases are designed for the various special applications, each shares the stage where a more or less soluble polymer above T_g is present.

There are several types of adhesive bonds.

1. Mechanical adhesion is defined as when the adhesive flows around the substrate surface roughness. An interlocking action takes place, like the pieces of a puzzle.

2. Specific adhesion is defined as the case where secondary bonds, such as hydrogen bonds, between the adhesive and the substrate are formed.

3. A special case of specific adhesion is when there are primary chemical bonds present. For example, graft or block copolymers may bond the different phases of a multicomponent polymeric material together. Direct bonding to substrates is also encouraged in many systems.

4. There are several cases of special attachments that are known. The most important of these is surface interpenetration between a polymer network and another material, which depends on diffusion. This second material may, of course, also be a polymer network. In the case of interpenetration, the rates of interdiffusion are important.

5. There are various non-specific forces, such as the van der Waals forces, which contribute to bonding.

It must be emphasized that several of the above may be active simultaneously in real systems.

11.6.2 Strength of Adhesion with Cross-linking

The science of adhesion recognizes two types of failure. Adhesive failure is when the bond between the adhesive and the adherend breaks on stressing. Cohesive failure is when the failure takes place within the material, either the substrate or the adhesive. Many cases are known where the adhesive bond between two substances is stronger than the substance itself. For example, when two pieces of wood are glued together and the joint is fractured, it may be the wood that fails and not the adhesive.

Gent and co-workers (63–67) and Lake and Thomas (68) examined the molecular aspects of both cohesive and adhesive failure in elastomeric networks. One of the most interesting results is that the work of fracture per unit area in cohesive failure of elastomers, G_c, is about 30–100 J/m², which compares with only a few J/m^2 expected for the theoretical value of a plane of $C - C$ covalent bonds (see Appendix 11.1, part d).

According to Lake and Thomas (68), this difference has been attributed to the polymeric nature of the molecular chains comprising the network; many bonds must be stressed in order to break any one bond. They concluded that the work of fracture is given by

$$G_c = KM_c^{1/2} \tag{11.33}$$

where K is a constant involving the density of the polymer and aspects of bond strength. As defined in Chapter 9, M_c is the molecular weight of an active network chain segment. This equation predicts that lower cross-link densities will produce stronger materials. It also must be remarked that G_c is closely related to the quantity $\bar{R}$ of Section 11.1.

The adhesive strength of elastomers was examined by Chang and Gent (65). Two partially cross-linked layers of elastomer were pressed together, and the gelation reaction was then taken to completion. By varying the extent of initial gelation, the degree of chemical interlinking could be varied systematically. The strength of adhesion between the layers increased linearly as the amount of interlinking between the two layers was increased. This was true both when the elastomers were identical (65), a special case of cohesive failure, and when the elastomers were different (66), more obviously a case of adhesive failure. Gent and co-workers also concluded that equation (11.33) was valid, as the extent of cross-linking was varied.

Equation (11.33) has been found to be valid for a number of cross-linked plastics such as epoxies as well as elastomers. Large values of M_c contribute to increased viscoelastic motions, and hence greater toughness.

However, densely cross-linked plastics are sometimes deliberately used. Reasons tend to be nonmechanical: increased solvent and swelling resistance, high temperature resistance, and reduced flammability are cited. The last arises because the polymer chars rather than degrades with volatile (and flammable) components.

11.6.3 Adhesion in Plastics

The reversible work required to create a unit of surface area in a substance is called the surface tension,

$$\bar{\gamma} = \left(\frac{\partial G}{\partial A} \right)_{T,P,n} \tag{11.34}$$

where $\bar{\gamma}$ represents the surface tension. The surface tension is also widely known as the surface free energy, with units of $ergs/cm^2$, which is sometimes simplified to dyn/cm, see Section 12.1. Selected values of the surface tension are shown in Table 11.6 (69). The very low values of $\bar{\gamma}$ for polytetrafluo-

Table 11.6 Surface tension of polymers (69)

Polymer	Surface Tension, dynes/cm		$-(d\bar{\gamma}/dT)$, dynes/cm · deg	References
	20°C	140°C		
Polyethylene	35.7	28.8	0.057	(a)
Polystyrene	40.7	32.1	0.072	(b)
Polytetrafluoro-ethylene	23.9	16.9	0.058	(c)
Poly(methyl methacrylate)	41.1	32.0	0.076	(b)
Poly(dimethyl siloxane)	19.8	14.0	0.048	(d)
Polycarbonate	42.9	35.1	0.060	(e)

References: (a) S. Wu, *J. Colloid Interface Sci.*, **31**, 153(1969). (b) S. Wu, *J. Phys. Chem.*, **74**, 632(1970). (c) S. Wu, *J. Macromol. Sci.*, **C10**, 1(1974). (d) S. Wu, *J. Polym. Sci.*, **C34**, 19(1971). (e) *Polymer Handbook*, 3rd ed., VI 420, J. Brandrup and E. H. Immergut, Eds. Wiley, New York, 1987.

roethylene are responsible for its "nonstick" application in frying pans. Low values of the surface tension mean low values of adhesion, in general.

The reversible work per unit area formed required to separate the interface between two bulk phases 1 and 2 from their equilibrium positions to infinity is the work of adhesion,

$$W_a = \bar{\gamma}_1 + \bar{\gamma}_2 - \bar{\gamma}_{12} \qquad (11.35)$$

where $\bar{\gamma}_{12}$ represents the interfacial tension between phases 1 and 2 (69).

As an adhesive bond is stressed, a stress concentration will appear around unwetted, and hence unbonded, interfacial defects. These defects act as flaws. Since real surfaces are never smooth, many microscopic unwetted voids usually appear at the interface. The driving force for the wetting of these interfacial voids is known as the spreading coefficient,

$$\lambda_{12} = \bar{\gamma}_2 - \bar{\gamma}_1 - \bar{\gamma}_{12} \qquad (11.36)$$

By definition, phase 1 has a lower surface tension than phase 2. The quantity λ_{12} can be either positive or negative, as illustrated in Table 11.7 (69) for polymers in the melt state. Shear strength increases with the spreading coefficient.

Of course, important chemical ways of increasing the bond strength between two polymers in an interface increase the adhesion between them. Grafting together the two polymers has already been mentioned. Block copolymers of the two components or miscible with them also serve to anchor

Table 11.7   Selected values of shear strength and spreading coefficient (69)

Polymer Pair		Shear Strength, psi	$\bar{\gamma}_{12}$, dynes/cm at 140°C	λ_{12}, ergs/cm² at 140°C	W_a, ergs/cm² at 140°C
Polymer 1	Polymer 2				
Polyethylene	Poly(methyl methacrylate)	750	9.7	−6.5	51.1
Poly(n-butyl methacrylate)	Poly(vinyl acetate)	1400	2.9	+1.6	49.8
Poly(dimethyl siloxane)	Polychloroprene	1850	6.5	+12	40.8

the two polymers together. Each block is dissolved in the surface of its respective species.

Chemical bonding can also be achieved through the use of coupling agents. Silanes are widely used because they react with many surfaces, forming strong bonds (70, 71). Although there are many silanes, a typical structure is vinyltriethoxysilane, $CH_2 = CHSi(OCH_2CH_3)_3$. In this case the vinyl group is polymerizable, and the silane portion can react with other materials, such as glass. Because of the need to bond to a variety of surfaces, commercial paints and adhesives usually contain several different functional groups.

11.7 SUMMARY

Deformation, fracture, and fatigue in polymeric materials are clearly established as being controlled by viscoelastic quantities as well as the thermodynamics of energy storage and the energy required to create new surfaces. There are several ways of modifying polymers so that they can withstand greater stresses. The most important include the incorporation of finely divided fillers, which reinforce polymers, particularly elastomers; the use of fiber composites, which bear the load; and the blending in of a rubbery phase to toughen an otherwise brittle plastic.

Under some well-defined conditions, both plastics and elastomers may be used indefinitely without significant fear of failure. This is specified by the failure envelope for elastomers and by the endurance limit in plastics. The glass transition temperature and the secondary transitions are important because their associated molecular motions absorb energy, heating the sample. Thus both the temperature and frequency of the experiment are important.

Two major mechanisms prevail in determining the fracture resistance of polymeric materials (72). One primary contribution comes from the energy required to extend polymer chains, or some subsection thereof to the point of rupture. This can be divided into two subparts: (a) extensional energy is required to deform a polymer chain to a fully extended conformation, and (b) each of the main-chain bonds along this fully extended chain will store its rupture energy before the chain breaks. This is a consequence of the shape of the energy well.

The second major contribution arises from viscous dissipation. Viscoelastic loss associated with chain deformation in regions around the crack tip is significant in determining fracture resistance in polymers.

Keeping in mind that thermodynamic conservation of energy holds everywhere, the energy to propagate a crack, supplied by an outside source, is consumed by the deformation of the chains, their breaking, and slipping. New surface formation also consumes energy, as well as a host of other mechanisms. If they work well, the polymer is strong or tough. If not, the polymer is weak or brittle, and easy to break.

In concluding, it must be emphasized that research in the areas of fracture and fatigue in polymers is in its infancy. Indeed, research in the whole area of polymers is developing rapidly and at an increasing rate.

Although natural polymers were used since biblical times and before for clothing, building, and transportation [note that Noah's Ark was painted with pitch, a polymeric material (73)], it is only since Staudinger's macromolecular hypothesis in the 1920s that significant understanding has begun to be achieved.

REFERENCES

1. J. G. Williams, *Fracture Mechanics of Polymers*, Ellis Horwood, Chichester, England, 1984, Chap. 1.

2. "Methods of Test for Impact Resistance of Plastics and Electrical Insulating Materials," ASTMD 256-54T, ASTM Standards, Part 6, p. 182 (1955). Also ASTM D-256-56, ASTM Standards 1956. American Society for Testing Materials, Baltimore, 1955 and 1956.

3. J. A. Manson and L. H. Sperling, *Polymer Blends and Composites*, Plenum, New York, 1976, Chap. I.

4. M. Matsuo, *Jpn. Plastics*, **2**, 6(1968).

5. S. Matsuoka, in *Failure of Plastics*, W. Brostow and R. D. Corneliussen, Eds., Hanser Publishers, New York, 1986.

6. A. Peterlin, *J. Polym. Sci.*, **9C**, 61(1965).

7. J. D. Hoffman, *Polymer*, **24**, 3(1983).

8. J. D. Hoffman, *Polymer*, **23**, 656 (1982).

8a. P. B. Bowden and S. Raha, *Philos. Mag.*, **22**, 463(1970).

9. S. S. Sternstein and L. Ongchin, *Polym. Prepr.*, **10(2)**, 1117 (1969).

10. J. C. Bauwens, *J. Polym. Sci. A-2*, **5**, 1145(1967).

11. J. C. Bauwens, *J. Polym. Sci. A-2*, **8**, 893(1970).

12. A. J. Kinlock and R. J. Young, *Fracture Behavior of Polymers*, Applied Science Publishers, London, 1983, p. 115.

13. N. G. McCrum, C. P. Buckley, and C. P. Bucknall, *Principles of Polymer Engineering*, Oxford University Press, New York, 1988, p. 117.

14. C. B. Bucknall and D. G. Street, *Soc. Chem. Ind. Monogr.*, **26**, 272(1967).

15. D. R. Paul and L. H. Sperling, Eds., *"Multicomponent Polymer Materials,"* ACS Adv. in Chem. Ser. No. 211, American Chemical Society, Washington, DC, 1986.

16. M. Matsuo, *Polym. Eng. Sci.*, **9**, 206(1969).

17. K. Kato, *Jpn. Plastics*, **2**, 6(1968).

18. L. H. Sperling, *J. Polym. Sci. Polym. Symp.*, **60**, 175(1977).

19. C. B. Bucknall and R. R. Smith, *Polymer*, **6**, 437(1965).

20. M. Y. Tang and J. E. Mark, *Macromolecules*, **17**, 2616(1984).

21. J. G. Curro and J. E. Mark, *J. Chem. Phys.*, **80**, 4521(1984).

22. M. A. Llorente, A. L. Andrady, and J.E. Mark, *J. Polym. Sci. Polym. Phys. Ed.*, **19**, 621(1981).

23. J. E. Mark, *Polym. Eng. Sci.*, **19**, 254(1979).

24. J. E. Mark, *Polym. Eng. Sci.*, **19**, 409(1979).

25. K. W. Scott, *Polym. Eng. Sci.*, **7**, 158(1967).

26. T. L. Smith and P. J. Stedry, *J. Appl. Phys.*, **31**, 1892(1960).

27. T. L. Smith, *J. Appl. Phys.*, **35**, 27(1964).

28. J. C. Halpin and F. Bueche, *J. Appl. Phys.*, **35**, 3142(1964).

29. L. J. Broutman and R. H. Krock, *Modern Composite Materials*, 8 Vols., Academic Press, Orlando, 1974.

30. L. Nicholais, *Polym. Eng. Sci.*, **15**, 137(1975).

31. S. Tsai, J. C. Halpin, and N. J. Pagano, Eds., *Composite Materials Workshop*, Technomic, Stamford, CT, 1969.

32 J. C. Halpin, *Polym. Eng. Sci.*, **15**, 132(1975).

33. F. P. Gerstle, Jr., *Encyclopedia of Polymer Science and Engineering*, Vol. 3, J. I. Kroschwitz, Ed., Wiley, New York, 1985, p. 776.

34. K. Friedrich, R. Walter, H. Voss, and J. Karger-Kocsis, *Composites*, **17**, 205(1986).

35. J. C. Seferis and L. Nicolais, in *The Role of the Polymeric Matrix in the Processing and Structural Properties of Composite Materials*, J. L. Koenig and C. Chiang, Eds., Plenum, New York, 1983.

36. D. R. Ulrich, *Polymer*, **28**, 533(1987).

37. A. A. Griffith, *Philos. Trans. R. Soc.*, **A221**, 163(1921).

38. J. P. Berry, *J. Polym. Sci.*, **50**, 107(1961).

39. B. Rosen, Ed., *Fracture Processes in Polymeric Solids, Phenomena and Theory*, Interscience, New York, 1964.

40. R. S. Rivlin and A. G. Thomas, *J. Polym. Sci.*, **10**, 291(1953).

41. E. Orowan, *Phys. Soc. Rep. Prog. Phys.*, **12**, 186(1948).

42. G. R. Irwin, in *Fracture, Handbuch der Physik*, Vol. VI, S. Flugge, Ed., Springer, Berlin, 1958, p. 551.

43. R. W. Hertzberg and J. A. Manson, *Fatigue of Engineering Plastics*, Academic Press, Orlando, 1980, Chap. 3.

44. ASTM Standard E399-78, "Test for Plane-Strain Fracture Touchness in Metallic Materials," Part 10, American Society for Testing Materials, Philadelphia, 1979.

45. J. A. Manson and R. W. Hertzberg, *CRC Crit. Rev. Macromol. Sci.*, **1**, 433(1973).

46. N. M. Riddell, G. P. Koo, and J. L. O'Tool, *Polym. Eng. Sci.*, **6**, 363(1966).

47. R. W. Hertzberg, J. A. Manson and M. D. Skibo, *Polym. Eng. Sci.*, **15**, 252(1975).

48. R. W. Hertzberg, J. A. Manson, and M. D. Skibo, *Polymer*, **19**, 359(1978).

49. J. Michel, R. W. Hertzberg, and J. A. Manson, *Polymer*, **25**, 1657(1984).

50. N. Mohammadi, J. N. Yoo, A. Klein, and L. H. Sperling, submitted *J. Polym. Sci. Polym. Phys. Ed.*, (1991).

51. P. G. de Gennes, *J. Chem. Phys.*, **55**, 572(1971).

52. S. F. Edwards, *Proc. Phys. Soc. London*, **92**, 9(1967).

53. R. P. Wool, B. L. Yuan, and O. J. McGarel, *Polym. Eng. Sci.*, **29**, 1340(1989).

54. Y. H. Kim and R. P. Wool, *Macromolecules*, **16**, 1115(1983).

55. R. P. Wool, *Rubber Chem. Technol.*, **57**, 307(1984).

56. H. Zhang and R. P. Wool, *Macromolecules*, **22**, 3018(1989).

57. M. Tirrell, *Rubber Chem. Technol.* **57**, 523(1984).

58. B. B. Mandelbrot, *The Fractal Geometry of Nature*, Freeman, New York, 1983.

59. B. Sapoval, M. Rosso, and J. F. Gouyet, *J. Phys. Lett.* (*Paris*), **46**, 149(1985).

60. R. P. Wool and J. M. Long, *Polym. Prepr.*, **31**(**2**), 558(1990).

61. J. N. Yoo, L. H. Sperling, C. J. Glinka, and A. Klein, *Macromolecules*, **23**, 3962(1990).

62. L. H. Sperling, A. Klein, J. N. Yoo, K. D. Kim, and N. Mohammadi, *Polym. Adv. Technol.*, **1**, 263(1990).

63. A. N. Gent and R. H. Tobias, *J. Polym. Sci. Polym. Phys. Ed.*, **22**, 1483(1984).

64. A. N. Gent and R. H. Tobias, *J. Polym. Sci. Polym. Phys. Ed.*, **20**, 2051(1982).

65. R. J. Chang and A. N. Gent, *J. Polym. Sci. Polym. Phys. Ed.*, **19**, 1619(1981).

66. R. J. Chang and A. N. Gent, *J. Polym. Sci. Polym. Phys. Ed.*, **19**, 1635(1981).

67. A. Ahagon and A. N. Gent, *J. Polym. Sci. Polym. Phys. Ed.*, **13**, 1903(1975).

68. G. J. Lake and A. G. Thomas, *Proc. R. Soc. Lond. Ser. A*, **A300**, 108(1967).

69. S. Wu, *Polymer Interface and Adhesion*, Dekker, New York, 1982.

70. E. P. Plueddemann and W. T. Collins, in *Adhesion Science and Technology*, Vol. 9A, L. H. Lee, Ed., Plenum, New York, 1975.

71. E. P. Plueddemann, *J. Paint Technol.*, **42**(550), 600(1970).

72. K. A. Mazch, M. A. Samus, C. A. Smith, and G. Rossi, *Macromolecules*, **24**, 2766(1991).

73. The Bible, *Genesis* 6 : 14.

GENERAL READING

E. H. Andrews, *Fracture in Polymers*, American Elsevier, New York, 1968.

W. Brostow and R. D. Corneliussen, Eds., *Failure of Plastics*, Hanser Publishers, New York, 1986.

C. B. Bucknall, *Toughened Plastics*, Applied Science, London, 1977.

A. Ciferri and I. M. Ward, *Ultra-High Modulus Polymers*, Applied Science, London, 1979.

W. J. Feast and H. S. Munro, Eds., *Polymer Surfaces and Interfaces*, Wiley, New York, 1987.

J. E. Gordon, *The New Science of Strong Materials*, 2nd ed., Penguin, Harmondsworth, England, 1976.

R. W. Hertzberg and J. A. Manson, *Fatigue of Engineering Plastics*, Academic Press, Orlando, 1980.

H. H. Kausch, *Polymer Fracture*, 2nd ed., Springer-Verlag, Berlin, 1987.

A. J. Kinloch and R. J Young, *Fracture Behavior of Polymers*, Applied Science, New York, 1983.

J. A. Manson and L. H. Sperling, *Polymer Blends and Composites*, Plenum, New York, 1976.

L. E. Nielsen, *Mechanical Properties of Polymers and Composites*, Vols. I and II, Dekker, New York, 1974.

A. Noshay and J. E. McGrath, *Block Copolymers—Overview and Critical Survey*, Academic Press, Orlando, 1977.

M. Panar and B. N. Epstein, *Science*, **226** (4675), 642(1984).

D. R. Paul and S. Newman, Eds., *Polymer Blends*, Vols. I and II, Academic Press, Orlando, 1978.

L. H. Sperling, *Interpenetrating Polymer Networks and Related Materials*, Plenum, New York, 1981.

L. G. Williams, *Fracture Mechanics of Polymers*, Horwood, Chichester, England, 1984.

A. E. Woodard, *Atlas of Polymer Morphology*, Hanser Publishers, New York, 1989.

S. Wu, *Polymer Interface and Adhesion*, Marcel Dekker, New York, 1982.

STUDY PROBLEMS

1. An engineering plastic is stretched until it breaks. Briefly, discuss all the possible places that the energy might have gone.

2. Define the following terms: impact resistance; tensile strength; failure envelope; fatigue crack propagation; craze.

3. Theoretically, polyethylene can be perfectly oriented and crystallized 100%. At perfect orientation,

   (a) What is the theoretical work to break of this material?

   (b) What is its theoretical tensile strength?

   (c) What is its theoretical modulus in the direction of orientation?

   [*Hints:* See Appendix 11.1. Consider the strength of the C $-$ C bond and the shape of the energy well.]

4. Read any scientific or engineering paper in the field of mechanical behavior of polymers written in the past year, and show how it advances the understanding of polymer science beyond when this book was finished.

5. What is the approximate shear strength that a bond between polyethylene and polystyrene will have? (The work of adhesion for this system is estimated to be 55 ergs/cm^2 at 140°C.)

6. You are designing new plastics and elastomers for space and planetary exploration. Some of the planets to be visited are very cold, and some of them have very dense atmospheres.

   (a) What will the effect of hydrostatic pressure and temperature be on the relaxation behavior of polystyrene? Of cis-polyisoprene?.

   (b) What will be the effects of hydrostatic pressure and temperature on the stress–strain behavior of these polymers?

7. A sample of poly(methyl methacrylate) at room temperature has a center notched crack of 2.0 cm in diameter. If K_c is 0.80 MPa $\cdot$ m$^{1/2}$, what is the critical stress to propagate the crack?

8. A sample of a new plastic 100 cm long, 1.0 cm in width, and 0.1 cm in thickness was subjected to a force of 1×10^6 dynes. If it is stretched to 100.1 cm in length, what is its Young's modulus?

9. Note in Figure 11.5 the 5% per minute strain rate. How much work would it take to break a $1 \times 1 \times 10$ cm^3 strip of this poly(vinyl chloride) under these conditions?

APPENDIX 11.1   CALCULATION OF MECHANICAL BEHAVIOR[†]

The following problem illustrates the determination of mechanical properties from molecular structures.

An elastomer based on polybutadiene, with a density of 0.93 g/cm^3 and with a cross section of 2 mm^2, was stretched to break, with the strain being

[†]D. Fradkin and L. H. Sperling, polymer class calculations.

seven times its original length (15 cm) and the force being 10 pounds. The following treats several mechanical problems at the molecular level.

a. What is the stress to break?

$$\sigma_b = \frac{F}{A} = \left(\frac{10 \text{ lb}}{2 \text{ mm}^2}\right)\left(\frac{10 \text{ mm}}{\text{cm}}\right)^2\left(2.54 \frac{\text{cm}}{\text{in.}}\right)^2$$

$$\sigma_b = 3225.8 \text{ lb/in}^2.$$

$$= 2.22 \times 10^8 \text{ dynes/cm}^2$$

b. What is the true stress to break (based on the final cross section)? First, determine the final area:

$$V_0 = V_f$$

$$L_0 A_0 = L_f A_f$$

$$\varepsilon_b = 7 = \alpha - 1$$

$$\alpha = 8 = \frac{L_f}{L_0} = \frac{L_f}{15 \text{ cm}}$$

$$L_f = 120 \text{ cm}$$

$$(15 \text{ cm})(0.02 \text{ cm}^2) = (120 \text{ cm})(A_f)$$

$$A_f = 0.0025 \text{ cm}^2$$

$$\sigma_b = \left(\frac{10 \text{ lb}}{0.0025 \text{ cm}^2}\right)\left(2.54 \frac{\text{cm}}{\text{in.}}\right)^2$$

$$\sigma_b = 2.58 \times 10^4 \text{ lb/in.}^2$$

$$= 1.78 \times 10^9 \text{ dynes/cm}^2$$

c. What is the work to break? First, calculate the work of stretching W:

$$W = f \times d$$

$$W = A \int \sigma \times L$$

$$\sigma = nRT\left(\alpha - \frac{1}{\alpha^2}\right); \qquad \alpha = \frac{L}{L_0}$$

This is based on a per-cm^2 area cross section.

$$W = A_0 \int_{15}^{120} nRT\left[\frac{L}{L_0} - \left(\frac{L_0}{L}\right)^2\right] dL$$

Determine n:

$$n = \frac{\sigma}{RT\left(\alpha - \dfrac{1}{\alpha^2}\right)}$$

$$n = \frac{2.22 \times 10^8 \text{ dynes/cm}^2}{\left(8.31 \times 10^7 \dfrac{\text{dynes} \cdot \text{cm}}{\text{mol} \cdot \text{K}}\right)(298 \text{ K})\left(8 - \dfrac{1}{8^2}\right)}$$

$$n = 1.12 \times 10^{-3} \text{ mol/cm}^3$$

Then,

$$W = A_0 nRT \int_{15}^{120} \left[\frac{L}{L8d0} - \left(\frac{L_0}{L}\right)^2 \right] dL$$

$$W = A_0 nRT \left[\frac{L^2}{2L_0} + \frac{L_0^2}{L} \right]\Big|_{15}^{120}$$

$$W = \left(1.12 \times 10^{-3} \frac{\text{mol}}{\text{cm}^3}\right)(0.02 \text{ cm}^2)\left(8.31 \times 10^7 \frac{\text{dynes} \cdot \text{cm}}{\text{mol} \cdot \text{K}}\right)(298 \text{ K})$$

$$\times \left[\frac{1}{2(15 \text{ cm})}(120^2 \text{ cm}^2 - 15^2 \text{ cm}^2) + (15 \text{ cm})^2\left(\frac{1}{120 \text{ cm}} - \frac{1}{15 \text{ cm}}\right) \right]$$

$$W = 2.55 \times 10^8 \text{ dynes} \cdot \text{cm} = 2.55 \times 10^8 \text{ ergs}$$

This result is high because rubber elasticity theory is not obeyed when $\alpha = 8$.

d. What is the theoretical work to break the chemical bonds? Consider $-CH_2-$ of MW = 14 g/mol

$$\rho = 0.93 \text{ g/cm}^3$$

$$N_A = 6.02 \times 10^{23} \text{ molecules/mol}$$

$$\left(14 \frac{\text{g}}{\text{mol}}\right)\left(\frac{1 \text{ mol}}{6.02 \times 10^{23} \text{ molecules}}\right)\left(\frac{1 \text{ cm}^3}{0.93 \text{ g}}\right) = 2.50 \times 10^{-23} \frac{\text{cm}^3}{\text{molecule}}$$

Compute the area occupied by 1 molecule ($-CH_2-$ group) (see figure above):

$$(2.50 \times 10^{-23} \text{ cm}^3)^{2/3} = 8.55 \times 10^{-16} \text{ cm}^2/\text{molecule}, \quad \text{or 3 Å on a side}$$

Work to break, utilizing the energy well (see figure below):

$$= \left(\frac{1 \text{ molecule}}{8.55 \times 10^{-16} \text{ cm}^2} \right) \left(\frac{1 \text{ mol}}{6.02 \times 10^{23} \text{ molecule}} \right) \left(83 \frac{\text{kcal}}{\text{mol}} \right)$$

The last being the energy to break the $C-C$ bond.

$$W_{\text{bond}} = \left(1.61 \times 10^{-7} \frac{\text{kcal}}{\text{cm}^2} \right) \left(4.19 \times 10^{10} \frac{\text{ergs}}{\text{kcal}} \right)$$

$$= 6.75 \times 10^3 \text{ ergs/cm}^2$$

This is the work to break the bonds only. Then, add the work to stretch, from part c, to obtain the total work to break.

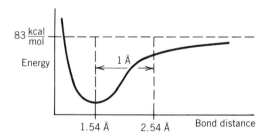

e. What is the theoretical stress to break? (Tensile strength.) Work = force × distance; 1-Å distance assumed.

$$83 \text{ kcal/mol} = F \times 1 \times 10^{-8} \text{ cm}$$

$$F = \left(83 \frac{\text{kcal}}{\text{mol}} \right) \left(\frac{1}{1 \times 10^{-8} \text{ cm}} \right) \left(4.19 \times 10^{10} \frac{\text{ergs}}{\text{kcal}} \right) = 3.47 \times 10^{20} \frac{\text{ergs}}{\text{cm} \cdot \text{mol}}$$

$$\left(3.47 \times 10^{20} \frac{\text{ergs}}{\text{cm} \cdot \text{mol}} \right) \left(\frac{1 \text{ mol}}{6.02 \times 10^{23} \text{ molecules}} \right)$$

$$= 5.76 \times 10^{-4} \frac{\text{erg}}{\text{molecule} \cdot \text{cm}}$$

$$\left(5.76 \times 10^{-4} \frac{\text{erg}}{\text{cm} \cdot \text{molecule}} \right) \left(\frac{1 \text{ molecule}}{8.55 \times 10^{-16} \text{ cm}^2} \right)$$

$$\sigma_b = 6.74 \times 10^{11} \text{ ergs/cm}^3 = 6.74 \times 10^{11} \text{ dynes/cm}^2$$

Or, about $300 \times$ the actual experimental value, 1.78×10^9 dynes/cm^2, part b.

Kevlar's tensile strength is

$$400{,}000 \text{ psi} \cong 2 \times 10^{10} \text{ dynes/cm}^2$$

at around 5% of theoretical. Thus, the actual tensile strengths of real polymeric materials, and indeed of all engineering materials, including metals and ceramics, are far short of their ultimate potential. Therefore, the student should not despair at there being no important professional work left to be done!

12

MODERN TOPICS

The preceding chapters provide an introduction to physical polymer science. However, there are many more topics of great interest not yet covered. While a few topics to be discussed have been known for some time, most of this chapter describes relatively new developments, often still in their infancy.

12.1 POLYMER SURFACES AND INTERFACES

The surface of any material is different from the interior or bulk of the material. Differences can be chemical or physical, or both. An extremely well-known example is rust on iron. In polymers, surface oxidation can also exist. Special effects on surfaces include contamination, plasticization, grafting, and a host of other chemical modifications or changes, some wanted and some adventitious.

There are numerous physical changes as well. The orientation of the polymer at the surface is almost always different from the interior; the polymer chains may be lying flat, oriented in the surface plane, or, if some special group is attracted to the surface, the orientation may be normal to the plane of the surface.

To say that a polymer has a surface implies that the surface is an interface to another material. Of course, the surface may be exposed to a vacuum. Otherwise, the surface forms an interface with a gas, liquid, or solid. The interface may be wetted by the second substance, it may be bonded to it by any of a variety of adhesion mechanisms, or there may be interdiffusion.

In general, people talk about the surface region of a material as being 50–100 Å thick. There are two aspects. First, the depth of the surface of interest depends on the nature of what is being studied. Perhaps only the first

Table 12.1 Surface-analytical methods used in biomedical studies (1, 2)

Technique	Advantages	Disadvantages
Electron spectroscopy for chemical analysis (ESCA), also called XPS	Sampling depth is relevant to surface interactions ($\sim$ 10–100 Å) All elements detected (except H and He) High sensitivity Simple sample preparation Well-established theory Nondestructive Many levels of information can be obtained Hydrated samples can be observed	Expensive Artifact-prone High vacuum required
Contact angle methods	Inexpensive Rapid Provides information related to surface energetics Hydrated samples can be observed	Artifact-prone: swelling of substrate, leaching and extraction, penetration, surface reorientation, roughness, operator bias Liquid purity is of critical importance Only surface energetics information is obtained Surface chemistry information is ambiguous

Technique	Advantages	Disadvantages
Scanning electron microscopy (SEM)	Rapid Visually rich Three-dimensional perspective	Differences in secondary electron emission can be confused with topographic features Great care must be exercised to use quantitatively Specimen preparation and drying artifacts can occur
Attenuated total reflection (ATR) infrared	Readily available Spectra are rich in chemical information Orientation information can be obtained	Analysis to 1-μm depth is studying the bulk and surface Determination is difficult Some samples are not amenable to ATR analysis
Secondary ion mass spectrometry (SIMS)	Looks at a highly surface-localized region ($\sim$ 10 Å) Rich in chemical information Depth-profiling (destructive) is possible	Expensive Artifact-prone Theory, particularly for polymers, is not developed Interpretation is complex
Auger electron spectroscopy (AES) or scanning Auger microprobe (SAM)	Rapid Good spatial resolution ($2.5 \times 10^{-3}\ \mu m^2$) All elements detected	Destructive to organics Limited chemical information Quantitative analysis is difficult

10 Å are of interest. Second, the measurable thickness depends on the instrument being used or on the limitations of instruments so far invented.

Some polymer interface problems have already been addressed in this book; for example, polymer blending and miscibility (Sections 4.3 and 11.2.4), adhesion (Section 11.6), and film formation from latexes (Section 5.4.4). This section centers on methods of characterizing polymer surfaces and interfaces.

12.1.1 Instrumental Methods of Characterization

Table 12.1 (1, 2) summarizes several of the more important methods of instrumental analysis of polymer surfaces and interfaces. Electron spectroscopy for chemical analysis, ESCA, also known as x-ray photoelectron spectroscopy, XPS, is based on the observation that electrons are emitted by atoms under x-ray irradiation. The energy of the emitted electrons yields the binding energy of the electron to the particular atom, which is the primary information. While the energies measured are characteristic of the elements, they are sensitive to the electronic environment of the atom, which, of course, yields information about surrounding atoms and molecules. The exciting radiation is a monoenergetic beam of soft x-rays.[†] Energies of interest range from 0–1500 eV. Since the electrons are emitted with low energy, they are easily scattered unless produced very close to the surface (3). Thus the sampling depth is 10–100 Å. ESCA can be used to study the oxidation of polymers, surface fluorination (used to improve the chemical resistance of polyethylene and other polymers), and copolymer composition, which may differ from the interior.

Secondary ion mass spectrometry, SIMS, uses a noble gas ion beam, usually Ar^+ or Xe^+, which is directed onto the sample surface (4). On collision with the surface material, molecular fragments consisting of one or more atoms are sputtered into a vacuum. These secondary ions are directed to a mass spectrometer to provide, in sequential experiments, the positive and negative ion mass spectra, respectively. For example, Figure 12.1 (4) illustrates the SIMS spectra of poly(ethylene terephthalate). The protonated repeat unit is seen at 193 D (daltons, atomic mass units, g/mol), and peaks at lower masses represent fragmentation with dominant peaks at 149 D ($HOOC—C_6H_4—CO^+$), 104 D ($C_6H_4CO^+$), and 77 D ($C_6H_5^+$). SIMS can be used for microanalysis of polymer surfaces. In one case (4), the experiment was used to detect the presence of dimethyl siloxane contamination on the surface of the polymer.

Auger electron spectroscopy, AES, has an atom in an M^+ excited state emit another electron to form an M^{2+} lower lying state. Surface examination is limited to 3–6 Å. In this process, an electron beam gun fires electrons at a low angle of incidence, which increases the interaction with surface atoms. The energy and number of emitted Auger electrons are detected by an

---

[†]Three to 30 Å in wave length.

Figure 12.1 Positive SIMS of poly(ethylene terephthalate) using 4-keV Ar$^+$ (film sample), 10^3 counts / sec fsd (4).

energy analyzer and counter. The method is used for measuring surface oxidation, surface strain, and a variety of lubrication problems.

Contact angle refers to the angle a drop of liquid makes where it meets the surface, Table 12.1. If it wets the surface, it will spread out, making a shallow angle. If it does not wet the surface, it may "bead up," the angle being larger than 90°.

12.1.2 Surface and Interfacial Tension

The free energy of formation of a surface is always positive, since work is required to create a new surface. This increases the total free energy of the system. The work, W, required to create a unit surface area, dA, is given by

$$W = \bar{\gamma} \, dA \qquad (12.1)$$

which is another formulation of equation (11.34). The surface tension, $\bar{\gamma}$, has cgs units of ergs/cm^2 or SI units of J/m^2. Since ergs, or work generally, can be expressed as force times distance, customary units are dynes/cm. Thus surfaces may be treated using either the concept of surface tension or the equivalent concept of surface free energy (5). Surface tension is the earlier of the two terms, going back to ideas that the surface of a liquid has some kind of contractile "skin." The term surface free energy implies that work is required to form more surface, involving the bringing up of more molecules from the interior of the substance, associating it more readily with conventional thermodynamic language. Thus the surface tension is actually the free

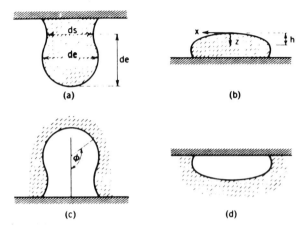

Figure 12.2   The shape of drops and bubbles can be used to calculate the surface tension of polymer and other substances (5): (a) hanging drop, (b) sessile drop, (c) hanging bubble, and (d) sessile bubble.

energy per unit area. At constant volume, the entropy of surface formation per unit area is given by

$$\Delta S^s = \frac{-d\bar{\gamma}}{dT} \tag{12.2}$$

Methods of measuring the surface tension include the pendant drop method, the sessile drop method, and others (5–7). These methods depend on the shape of a drop of the polymer or bubble in it, and depend on the sum of surface tension and gravitational forces; see Figure 12.2 (5).

Some numerical values of $\bar{\gamma}$ and $d\bar{\gamma}/dT$ were shown in Table 11.6. Positive values of $-d\bar{\gamma}/dT$ mean an increase in the entropy on the formation of the surface. This may be associated with the highly irregular structure of the surface, with molecules or segments literally protruding into the air, or (as below) into another phase. This density gradient may be described in terms of fractals; see Section 11.5.5.

Surface tension is usually defined as involving a single phase in equilibrium with its vapor. In the case of a polymer, where the vapor pressure is substantially zero, the polymer is considered in contact with air.

When the polymer contacts a condensed phase, it is common to talk about the interfacial tension of the system. A special case of interest involves two polymers in contact. Interfacial tensions are shown in Table 12.2 (6) for a few selected polymer pairs. The low surface tension of poly(dimethyl siloxane), relative to almost all other polymers, favors spreading. In order for liquid b to spread over liquid a, it is necessary that the spreading coefficient,

$$\lambda_{a/b} = \bar{\gamma}_b - \left(\bar{\gamma}_a + \bar{\gamma}_{ab}\right) \tag{12.3}$$

Table 12.2   Interfacial tension between select polymer pairs (6)

Polymers	Interfacial Tension 20°C erg/cm^2	100°C erg/cm^2	$-(d\gamma/dT)$ erg/cm^2 − deg
PS/PDMS	6.1	6.1	0.0008
PDMS/PEO	10.8	10.2	0.0078
PS/PMMA	3.2	2.2	0.013

PS = polystyrene; PDMS = poly(dimethyl siloxane);
PEO = poly(ethylene oxide); PMMA = poly(methyl methacrylate)

Source: S. Wu, *Polymer Interface and Adhesion*, Marcel Dekker, New York, 1982, p. 126.

be positive. If

$$|\bar\gamma_a - \bar\gamma_b| < \bar\gamma_{ab} \qquad (12.4)$$

neither liquid can spread on the other (8). Figure 12.3 (9) shows that at concentrations of only about 1%, poly(dimethyl siloxane) covers 50–70% of the surface of a block copolymer-blend mixture with polycarbonate, as determined by ESCA. Noting the large difference in surface tensions (Table 11.6), this is to be expected. LeGrand and Gaines (10) found that the contact angles of all their blends of polycarbonate and poly(dimethyl siloxane) displayed contact angles very close to that of pure poly(dimethyl siloxane), providing important supporting information.

Figure 12.3 Studies by ESCA instrumentation reveal that surface segregation in polycarbonate – *blend* – polycarbonate – *block* – poly(dimethyl siloxane) segmented copolymer blends (9).

The surface tension of a crystalline polymer can be approximately calculated from (6)

$$\bar{\gamma}_c/\bar{\gamma}_a = (\rho_c/\rho_a)^n \tag{12.5}$$

where n is about 4, ρ represents the density, and c and a represent crystalline and amorphous, respectively. Since the density of crystalline polymers is higher than that in the corresponding amorphous state, the surface tension is higher.

For polymers undergoing the glass–rubber transition, the relationship can be written (6)

$$\left(\frac{d\bar{\gamma}}{dT}\right)_G = \frac{\alpha_G}{\alpha_R}\left(\frac{d\bar{\gamma}}{dT}\right)_R \tag{12.6}$$

where α represents the expansion coefficient. Since $\alpha_G < \alpha_R$, then

$$-\left(\frac{d\bar{\gamma}}{dT}\right)_G < -\left(\frac{d\bar{\gamma}}{dT}\right)_R \tag{12.6a}$$

While the surface tensions themselves are continuous across the transition, the temperature derivatives are not.

12.1.3 Polymers at an Interface

Consider the adsorption of a polymer on the surface of another substance; see Figure 12.4 (11). There are three parts to the adsorbed chain: the surface-held trains, the free-hanging loops, and the two tails, all in equilibrium. Where each mer in the surface-held trains may have only a small bonding energy, strong positive adsorption occurs, because many such bonds are formed. All must be released substantially simultaneously in order to debond the chain as a whole, creating low probabilities of desorption in many cases.

De Gennes pointed out that both attractive walls and repulsive walls must be considered (12). This is important if the polymer is in solution, and especially so if the surface prefers the solvent to the polymer. Figure 12.5 (12) illustrates the conditions at attractive versus repulsive walls. For the attractive wall situation, the concentration profile exhibits three regions: the proximal region, which is very sensitive to the details of the interactions, the central region, which exhibits self-similarity (see below), and the distal region, which is controlled by a few large loops and tails. Scaling analysis shows that the concentration, $\varphi(z)$, in the central region follows the relation-

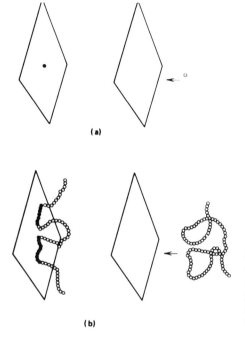

Figure 12.4 Small molecules are adsorbed differently than linear, flexible macromolecules (11). (*a*) A small molecule generally sits down on the surface in its entirety. (*b*) Polymer chains are often adsorbed only at one or more short sections.

ship between a and z,

$$\varphi(z) = (a/z)^{4/3} \tag{12.7}$$

where a represents the proximal distance (about the length of a mer), and z is the distance from the wall. The total thickness of the interface is governed by the size of the polymer chains and may be hundreds of angstroms wide.

The repulsive wall relationships are governed by the polymer chain correlation length, ξ; see Figure 12.5. Due to the decreases in entropy if the chain lies near the wall, the region next to the wall may contain nearly all solvent. This is called the depletion layer.

There are several important experimental results that arise because of the repulsive wall problem:

1. The effective pore size in GPC particles is smaller than the actual size.
2. During polymerization of latexes, the portion of the latex near the surface may be enriched in monomer relative to the interior.
3. Reaction rates of polymers with surfaces may be reduced.
4. The surfaces of polymer solutions may be substantially denuded of polymer.

(a) Attractive Wall

(b) Repulsive Wall

Figure 12.5   Concentration profiles at a wall are governed by the size of the chain and its bonding to the wall (12).

The central region of the polymer layer adsorbed onto an attractive wall may be modeled as a self-similar grid; see Figure 12.6 (12). The mesh size is determined by the overlap distance of the polymer chains, borrowing the concept from an analysis of the semidilute region; see Section 4.2. The mesh size is a decreasing function of the concentration,

$$\xi = a\varphi^{-3/4} \tag{12.8}$$

At a distance z from the wall, the local mesh size equals z, increasing with distance,

$$\xi(\varphi(z)) = z \tag{12.9}$$

There are a number of factors, both thermodynamic and kinetic, which control the concentration at the interface; see Table 12.3 (13). One of the more complicated kinetic aspects involves diffusion of a polymer chain through already adsorbed polymer layers. It is assumed diffusion follows reptation motions. However, the fraction of remaining available bonding sites becomes important when the chain actually reaches the surface.

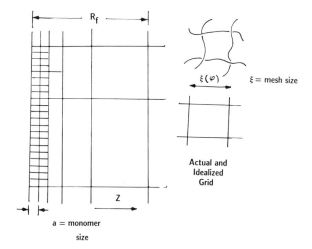

Figure 12.6 A self-similar polymer grid at an interface (12).

While the above discussion has assumed that each mer of the chain has an equal probability of bonding with the surface, this need not be so. The end groups may be more active, yielding an end-bonded chain. Or the chain may be a block copolymer, where one block is attracted to the wall and the other block is not. Of course, actual chemical bonds may be involved; in this case, the surface is described as being grafted.

Surface modification of polymers may allow a synergistic effect: a polymer may be given a low surface tension surface. In other cases, a reactive surface may be introduced. While the surface constitutes only a very small fraction of

Table 12.3   Factors controlling polymers at interfaces (13)

Thermodynamic	Kinetic
Equilibria	Diffusion of single chains
Competitive Adsorption	Evolution of structural rearrangements
Extent of spatial Entanglement	Segment-surface contact life
Number of bound segments	Lateral diffusion
	Diffusion through adsorbed layers
Adsorbed layer thickness	Replacement of one species by another
Patchiness and bare spots	Effect of dynamic entanglement

bulk polymer mass, it may be a very large fraction for finely divided materials such as latexes.

12.1.4 Composite Material Behavior

There are a number of practical problems associated with polymers at interfaces. For example, if the polymer is attracted to the interface in a composite material, it will become more dense in the vicinity of the interface. This results in an increase in the glass transition temperature, amounting to 10–20°C in some cases. For very finely divided fillers, a very large volume fraction of the polymer can be affected.

The development of crystallinity may differ at the interface with a filler as well, depending on several conditions. If the filler surface nucleates crystallization, the polymer at the interface may be more crystalline than the bulk. If the filler surface is neutral, then spherulite formation can be disrupted by the surface, and the degree of crystallinity at the surface can be lower than in the bulk. Similarly, the cross-link density of a polymer may be altered at the interface, being either higher or lower, depending on the chemistry of the filler surface.

Since polymers are often immiscible with each other, their surfaces tend to repel each other. This results in a rarification of matter at the interface, usually with poor mechanical behavior unless the interface is bonded through addition of a "compatibilizer," such as a block copolymer of the two polymers in question, or grafting between the two polymers.

12.2 ELECTRICAL BEHAVIOR OF POLYMERS

Most polymers are insulators. In fact, polyethylene and poly(vinyl chloride) are widely used as the insulating materials for electrical wiring, because they are highly insulating and weather resistant. However, to say a polymer is an insulator is a qualitative statement. To what extent does it conduct electricity? In a series of new developments, families of conducting polymers are being invented.

12.2.1 Basic Electrical Relationships

There are several basic relationships governing the electrical behavior of all materials. Ohm's law can be written

$$I = V/R \qquad (12.10)$$

where I is the current in amperes (A), V is the voltage in volts, and R is the resistance in ohms (Ω). The concept of the capacitance, C, involves storage of electrical charge,

$$C = Q/V \qquad (12.11)$$

where Q is the charge in coulombs (C) (14).

When an alternating voltage is applied to an imperfectly conducting material, a current flows that is displaced in time in such a way that it is out of phase with the voltage by an angle δ. The tangent of this angle plays a role similar to that described in Chapter 8 for mechanical behavior and is referred to as the dissipation factor. The power loss is given by

$$W_{\text{loss}} = 2\pi f C_p (\tan \delta) V^2 \qquad (12.12)$$

where W_{loss} is loss in watts, f is the frequency of the applied voltage, and C_p represents the parallel capacitance of the material. The relative dielectric constant (permittivity) is given by

$$\epsilon' = C_p / C_v \qquad (12.13)$$

where C_v is the capacitance of vacuum. The dielectric loss constant, ϵ'', sometimes called the loss factor, is given by

$$W_{\text{loss}} \cong \epsilon' \tan \delta = \epsilon'' \qquad (12.14)$$

These quantities can be used to describe the onset of molecular motion at the glass transition temperature; see Section 8.3. In the present setting, we are interested in determining the electrical behavior of polymers.

Two other quantities of interest are the resistivity, the resistance per unit distance, with units of $\Omega \cdot \text{cm}$, and its inverse, conductivity, with units of $\Omega^{-1} \cdot \text{cm}^{-1}$, or siemens per centimeter (S/cm).

12.2.2 Range of Polymer Electrical Behavior

The range of conductivities available for a range of polymers and other materials is given in Figure 12.7 (15). An insulator has conductivities in the range of 10^{-18} to 10^{-5} S/cm. A semiconductor is in the range of 10^{-7} to 10^{-3} S/cm, and a conductor is usually given as 10^{-3} to 10^6 S/cm. As seen in Figure 12.7, polystyrene, polyethylene, nylon, and a host of other ordinary polymers are insulators.

On the other end of the scale are the conducting and semiconducting polymers. Several types exist. For example, conducting fillers may be added, such as short metallic fibers that touch each other or carbon black. Alternately, an ionic polymer may be employed, or a salt may be added to the polymer. Conductivity in the latter systems depends on the moisture content of the polymer.

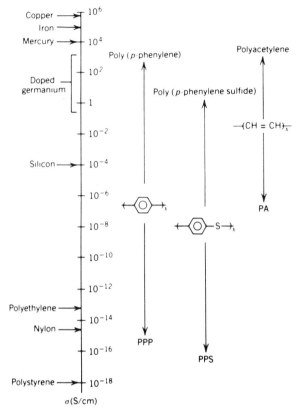

Figure 12.7 Conductivity scale comparing several polymers doped with AsF_5 with conventional materials (15).

12.2.3 Conducting Polymers

A much newer type is based on doped alternating double bond systems; see Table 12.4 (16) and Figure 12.7. However, conjugated organic polymers in their pure state are still electrical insulators. Chemical dopants include oxidative materials such as AsF_5 and I_2. Some newer doping materials include SbF_5, $AlCl_3$, Br_2, O_2 and a host of others. The identities of the anionic counterions derived from these dopants show a variety of structures such as $Sb_2F_{11}^-$, AsF_6^-, and I_3^-.

The basic mechanism of electronic conduction is illustrated in Figure 12.8 (15). The importance of π-bonds in electronic conduction must be emphasized, as the overlapping electronic clouds contribute to the conduction. Strong interactions among the π-electrons of the conjugated backbone are indicative of a highly delocalized electronic structure and a large valence

Table 12.4   Structures and conductivity of doped conjugated polymers (16)

Polymer	Structure	Typical Methods of Doping	Typical Conductivity s/cm
Polyacetylene		Electrochemical, chemical (AsF_5, I_2, Li, K)	$500-1.5 \times 10^3$
Polyphenylene		Chemical (AsF_5, Li, K)	500
Poly(phenylene sulfide)		Chemical (AsF_5)	1
Polypyrrole		Electrochemical	600
Polythiophene		Electrochemical	100
Poly(phenyl quinoline)		Electrochemical, chemical (sodium naphthalide)	50

bandwidth. For good conductivity, the ionization potential, IP, must be small. The electron affinity, EA, reflects the ease of addition of an electron to the polymer, especially through polymer doping. The bandgap, E_g, correlates with the optical-absorption threshold. The bandwidth, BW, correlates with carrier mobility; a large bandwidth suggests a high intrachain mobility, which favors high conductivity.

The mechanisms by which these polymers conduct electricity have been a source of controversy ever since conducting polymers were first discovered. At first, doping was assumed to remove electrons from the top of the valence band, a form of oxidation, or to add electrons to the bottom of the conduction band, a form of reduction. This model associates charge carriers with free spins, unpaired electrons. This results in theoretical calculations of conduction that are much too small (16). To account for spinless conductivity, the concept of transport via structural defects in the polymer chain was introduced. From a chemical viewpoint, defects of this nature include a

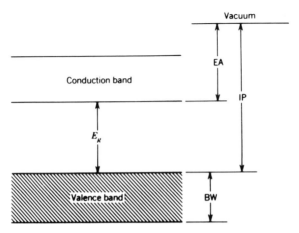

Figure 12.8 Polymer π-electron band structure. E_g represents the optical bandgap; BW is the bandwidth of the fully occupied valence band; EA is the electron affinity; and IP is the ionization potential (15).

radical cation for oxidation effects, or radical anion for the case of reduction. This is referred to as a polaron. Further oxidation or reduction results in the formation of a bipolaron. This can take place by the reaction of two polarons on the same chain to produce the bipolaron, a reaction calculated to be exothermic; see Figure 12.9 (15). In the bulk doped polymer, both intrachain and interchain electronic transport are important. Some applications, actual or proposed, include rechargeable batteries, $p - n$ junction devices, and electromagnetic (EMI) shielding.

Figure 12.9 Structures of *trans*-polyacetylene and poly(*p*-phenylene), showing reactions of two polarons (radical cations) to produce a bipolaron (dication) (15).

12.3 POLYMERS FOR NONLINEAR OPTICS

The development of nonlinear optic, NLO, materials constitutes one of the newest and most exciting interdisciplinary branches of research, demanding cooperation among polymer scientists, chemists, physicists, and materials scientists. Nonlinear optics is basically concerned with the interaction of optical frequency electromagnetic fields with materials, resulting in the alteration of the phase, frequency, or other propagation characteristics of the incident light. Some of the more interesting embodiments of second-order NLO include second harmonic generation, involving the doubling of the frequency of the incoming light; frequency mixing, where the frequency of two beams is either added or subtracted; and electrooptic effects, involving rotation of polarization and frequency and amplitude modulation (17). Third-order NLO involves three photons, for similar effects.

Nonlinear optical effects have been explored in a range of materials— inorganic, organic, and polymeric. The basic structural requirement for second-order NLO activity is a lack of an inversion center, which is derived principally from asymmetric molecular configurations and/or poling. Poling, the application of an electric field of high voltage across the material, orients all or a portion of the molecule, usually in the direction of the field. Third-order NLO behavior does not require asymmetry. Active materials include inorganic crystals such as lithium niobate and gallium arsenide (18, 19), organic crystals such as 2-methyl-4-nitroaniline, MNA (20, 21), polymers such as polydiacetylenes (22), and a plethora of polymers with liquid crystal side chains (17, 23–26). The organics and polymers with mesogenic side chains were found to be many times more active than lithium niobate (21), one of the best inorganic materials.

12.3.1 Theoretical Relationships

The starting point for nonlinear optics involves the constitutive relationship between the polarization induced in a molecule, P, and the electric field components, E, of the applied constant, low frequency, or optical fields (17). Ignoring magnetic dipoles and higher order multipoles, the induced polarization yields the approximation

$$P_i = \alpha_{ij}E_j + \beta_{ijk}E_jE_k + \gamma_{ijkl}E_jE_kE_l + \cdots \qquad (12.15)$$

The quantities P and E are vectors, and α, β, and γ are tensors. The subscripts arise through the introduction of Cartesian coordinates.

A similar expression can be written for the polarization induced in an ensemble of molecules, whether in the gas, liquid, or solid state. For this

Figure 12.10 Schematic portrayal of some of the more exciting experiments that compose the field of nonlinear optics. Note the frequency doubling in second harmonic generation, SHG (17).

case, the polarization P can be written

$$P_i = \chi_{ij}^{(1)} E_j + \chi_{ijk}^{(2)} E_j E_k + \chi_{ijkl} E_j E_k E_l + \cdots \tag{12.16}$$

where the coefficients $\chi_{ij}^{(1)}, \ldots$, are tensors describing the polarization induced in the ensemble. The quantity $\chi_{ijk}^{(2)}$ will yield a zero contribution to nonlinear polarization if the system is centrosymmetric. For nonzero values, some type of asymmetry must be induced in the medium. For polymeric materials, asymmetry is generally induced by the generation of a polar axis in the medium through electric field poling.

Several of the effects occurring through $\chi_{ijk}^{(2)}$ are shown in Figure 12.10 (17). The combination of two photons of frequency ω produces a single photon at 2ω in second harmonic generation. Basically, for second harmonic generation two photons must arrive at nearly the same time at a point in a nonsymmetric sample capable of absorbing them. Then the combined energy is released as one photon. The other effects in Figure 12.10 depend on other relationships. For example, the linear electrooptic effect follows from the interaction between an optical field and a direct current field in the nonlinear medium. This changes the propagation characteristics of the electromagnetic waves in the medium.

12.3.2 Experimental NLO Studies

When a polymer is subject to an intense sinusoidal electric field such as that due to an intense laser pulse, Fourier analysis of the polarization response can be shown to contain not only terms in the original frequency ω, but also terms in 2ω and 3ω (17). The intensity of the nonlinear response depends on the square of the the intensity of the incident beam for 2ω, and the third power for 3ω. For the second-order effects, the system must have some asymmetry, as discussed previously. For poling, this means both high voltage and a chemical organization that will retain the resulting polarization for extended periods of time. Polymeric systems investigated have been of three basic types:

1. Molecules exhibiting large NLO effects are dissolved or dispersed in the polymer, which then merely acts as a carrier (26). An example is 2-methyl-4-nitroaniline dissolved or dispersed in vinylidene fluoride copolymers (27).

2. A segment of the polymer chain backbone is the active portion. This method is clearly the most efficient, taking into account the weight of the system. If the polymer is poled above T_g, then cooled under poling, the orientation will have temporal longevity (28).

3. A special side-chain group can be attached to the polymer backbone. The advantage of side groups is their relatively easy orientation under electric fields. Some types of side groups can be bonded not only to the main chain but at a second point, providing cross-linking. This latter then holds the active portion of the chain in position for temporal longevity (29).

Typical responses for second-order effects are values in the range of 0.2–20×10^{-9} esu. Williams (17) remarked that uniquely useful applications for organic and polymeric materials will require a $\chi^{(2)}$ of 10^{-7} esu or greater. In addition, a variety of additional application-dependent properties and attributes must be present. These include uniform birefringence, minimized scattering losses, transparency, stability, and processibility. The relaxation time of the poled structures must be increased significantly.

NLO materials are now being investigated for devices for second-harmonic generators, modulators, shifters, amplifiers, laser technology, and interferometers. Important areas of development include waveguide devices such as optical switches and logic gates, communications and computational machines, and the like. One of the great advantages of electromagnetic waves instead of electrons in communications is the reduction in cross-talk. Effects in the picosecond time range have been investigated by Prasad and co-workers (30, 31) and Garito (32).

12.4 POLYMERS FOR EXTREME SERVICE CONDITIONS

Plastics today are no longer used only for inexpensive toys. Today's demanding application areas include aerospace, geothermal, and undersea environments. Polymers with sufficient thermal stability for construction of internal combustion engines are being considered. Because of their high strength to weight ratio, high modulus, thermal stability, and corrosion and oxidation resistance, they are becoming the materials of choice for many new products.

Table 12.5 Representative backbones for thermally stable polymers (33)

Name	Structure	Comments
Polybenzimidazole (PBI)		$T_m > 400°$ C, MW to 60,000, TGA(N_2) $500°$ C, soluble in polar aprotics, flame-resistant golden fiber
Aramids		High T_m, TGA(N_2) $450°$ C. Use temperatures of $\sim 370°$ C (short term), flame-resistant fabric
Polyimide		High T_m, TGA(N_2) up to $580°$ C insoluble, soluble when —CF_3 appears in backbone, films and foams
Polybenzoxazole		O-intractible, TGA(N_2) $500°$C S-TGA(air, 1% weight loss) $600°$C, both O and S used as adhesives

Poly(phenyl Quinoxaline)PPQ		$T_g \cong 370°C$, TGA(N_2) 500°C, MW to 300,000 SST adhesive, yellow films from m-cresol
Polyester		Use temperature of 325° C, fabrication by compression sintering, insoluble high flexural modulus, thermoplastic, machinable
Polypyrrone		TGA(air) 600° C, film former, high modulus, soluble in polar aprotics
Polyphosphazene (Inorganic polymer)		Elastic, soluble, T_g as low as $-84°C$, T_m up to 390° C, TGA(Ar) 410° C

An area of special interest involves high-temperature stability. Here, chemical moieties can be classified as "strong" or "weak" depending on their thermal stability (33, 34):

Weak	Strong
Alkyls	Aromatic
Amines	Ether
Alcohols	Sulfone
	Heterocyclic

This has resulted in a series of ladder or semiladder polymeric structures based on aromatics, as shown in Table 12.5 (33). The ladder structure allows individual bonds to break, without breaking the chain. It must be pointed out that aircraft skins heat up to several hundred degrees Celsius, through friction with the air. Materials employed must be stable for extended periods of time under these conditions. Frequently, these polymers are used in the form of fiber-based composites, to permit even higher strength retention at high temperatures.

12.5 MODERN ENGINEERING PLASTICS

One definition of an engineering plastic is one that has a high glass transition or a high melting temperature, for use at elevated temperatures. Other points of importance mention impact resistance, toughness, tensile strength, heat deflection temperature under load, creep, dimensional stability, and mold shrinkage.

Very few polymers are used commercially in the "pure" state. Some polyethylenes and polystyrenes are sold as homopolymers without any additives. However, several types of materials are added to most polymers to improve their properties:

1. Finely divided rubber is added to brittle plastics to toughen them.

2. Composites with glass, carbon, or boron fibers are made for high modulus and high strength.

3. Carbon black or silicas are added to rubber formulations to improve tear resistance and raise the modulus.

4. Various plasticizers may be added to lower the glass transition or reduce the amount of crystallinity. The main effect is to soften the final product.

5. Properties may be improved by adding silanes or other bonding agents to composites, to improve the bonding between polymer and other solid phases such as glass fibers.

6. Many types of polymers, both glassy and rubbery, are cross-linked to improve elastomer behavior or to control swelling.

7. Other items include fire retardants, colorants, and fillers to reduce price.

In each case, an understanding of how polymers behave and the effect of the various interactions allows the scientist and engineer to approach individual cases scientifically, rather than empirically.

12.6 CONCLUDING REMARKS

This chapter emphasizes new areas of research in physical polymer science, where the field is in its infancy. Polymer science as a whole is growing very rapidly. Concepts unheard of only a decade ago are occupying the attention of large teams of polymer scientists. As the time span from conception to commercialization continues to shrink, research effort is frequently increased accordingly. The student should not think of polymer science as completed and fully understood, but rather as just beginning.

REFERENCES

1. B. D. Ratner, *Ann. Biomed. Eng.*, **11**, 313(1983).
2. B. D. Ratner, S. C. Yoon, and N. B. Mateo, in *Polymer Surfaces and Interfaces*, W. J. Feast and H. S. Munro, Eds., Wiley, Chichester, England, 1987.
3. D. Campbell and J. R. White, *Polymer Characterization: Physical Techniques*, Chapman and Hall, New York, 1989, p. 330.
4. D. Briggs, in *Polymer Surfaces and Interfaces*, W. J. Feast and H. S. Munro, Eds., Wiley, Chichester, England, 1987.
5. A. W. Adamson, *Physical Chemistry of Surfaces*, 5th ed., Wiley-Interscience, New York, 1990.
6. S. Wu, *Polymer Interface and Adhesion*, Marcel Dekker, New York, 1982.
7. G. L. Gaines, Jr., *Polym. Eng. Sci.*, **12**, 1(1972).
8. G. L. Gaines, *Macromolecules*, **14**, 208(1981).
9. (a) I. Yilgor and J. E. McGrath, *Adv. Polym. Sci.*, **86**, 1(1988); (b) *Polysiloxane Copolymers/Anionic Polymerization*, Springer-Verlag, Berlin, 1988.
10. D. G. LeGrand and G. L. Gaines, Jr., *Polym. Prepr.*, **11**, 442(1970).
11. A. Silberberg, in *Encyclopedia of Polymer Science and Engineering*, 2nd ed., J. I. Kroschwitz, Ed., Wiley, New York, 1985.
12. P. G. de Gennes, *Adv. Colloid Interface Sci.*, **27**, 189(1987).
13. Based on M. Santore, Seminar, April 1991.
14. K. N. Mathes, in *Encyclopedia of Polymer Science and Engineering*, Vol. 5, J. I. Kroschwitz, Ed., Wiley, New York, 1986.

15. J. E. Frommer and R. R. Chance, in *Encylcopedia of Polymer Science and Engineering*, Vol. 5, J. I. Kroschwitz, Ed., Wiley, New York, 1986.

16. M. J. Bowden, in *Electronic and Photonic Applications of Polymers*, M. J. Bowden and S. R. Turner, Eds., Advances in Chemistry Series No. 218, American Chemical Society, Washington, DC, 1988.

17. D. Williams, in *Electronic and Photonic Applications of Polymers*, M. J. Bowden and S. R. Turner, Eds., Advances in Chemistry Series No. 218, American Chemical Society, Washington, DC, 1988.

18. D. J. Williams, Ed., *Nonlinear Optical Properties of Organic and Polymeric Materials*, ACS Symposium Series No. 283, American Chemical Society, Washington, DC, 1983.

19. C. Lee, D. Haas, H. T. Man, and V. Mechensky, *Photonics Spectra*, p. 171, April 1979.

20. B. F. Levine, C. G. Bethea, C. D. Thurmond, R. T. Lynch, and J. L. Bernstein, *J. Appl. Phys.*, **50**, 2523(1979).

21. B. I. Breene, J. Orenstein, and S. Schmitt-Rink, *Science*, **247**, 679(1990).

22. G. R. Meredith, J. G. VanDusen, and D. J. Williams, *Macromolecules*, **15** 1385(1982).

23. R. N. DeMartino, H. N. Yoon, J. R. Stamatoff, and A. Buckley, Eur. Pat. Applic. 0231770 (1987), to Celanese Corporation.

24. D. E. Stuetz, Eur. Pat. Applic. 017212 (1986).

25. N. A. Plate, R. V. Talroze, and V. P. Shibaev, in *Polymer Yearbook 3*, R. A. Pethrick and G. E. Zaikov, Eds., Harwood, London, 1986.

26. B. F. Levine, C. G. Bethea, C. D. Thurmond, R. T. Lynch, and J. L. Bernstein, *J. Appl. Phys.*, **50**, 2523(1979).

27. P. Pantelis and G. J. Davies, U.S. Pat. 4,748,074 (1988), to British Telecommunications.

28. B. I. Breene, J. Orenstein, and S. Schmitt-Rink, *Science*, **247**, 679(1990).

29. M. Eich, B. Reck, D. Y. Yoon, C. G. Wilson, and G. C. Bjorklund, *J. Appl. Phys.*, **66**, 3241(1989).

30. D. N. Rao, R. Burzynski, X. Mi, and P. N. Prasad, *Appl. Phys. Lett.*, **48**, 387(1986).

31. D. N. Rao, J. Swiatkiewicz, P. Chopra, S. K. Chosal, and P. N. Prasad, *Appl. Phys. Lett.*, **48**, 1187(1986).

32. A. F. Garito, presented at Pacifichem '89, Honolulu, Hawaii, December 1989.

33. P. E. Cassidy, in *Adhesives, Sealants, and Coatings for Space and Harsh Environments*, L. H. Lee, Ed., Plenum, New York, 1988.

34. P. E. Cassidy, *J. Chem. Ed.*, **58**, 951(1981).

GENERAL READING

A. W. Adamson, *Physical Chemistry of Surfaces*, 5th ed., Wiley, New York, 1990.

D. Avnir, *The Fractal Approach to Heterogeneous Chemistry: Surfaces, Colloids, Polymers*, Wiley, Chicester, England, 1989.

M. J. Bowden and S. R. Turner, Eds., *Electronic and Photonic Applications of Polymers*, Advances in Chemistry Series No. 218, American Chemical Society, Washington, DC, 1988.

W. J. Feast and H. S. Munro, Eds., *Polymer Surfaces and Interfaces*, Wiley, Chichester, England, 1987.

R. A. Hann and D. Bloor, Eds., *Organic Materials for Non-Linear Optics*, Royal Society of Chemistry, London, 1989.

A. J. Heeger, J. Orenstein, and D. R. Ulrich, Eds., *Nonlinear Optical Properties of Polymers*, Materials Research Society, Pittsburgh, 1988.

C. C. Ku and R. Liepins, *Electrical Properties of Polymers*, Hanser, Munich, 1987.

J. H. Lai, Ed., *Polymers for Electronic Applications*, CRC Press, Boca Raton, FL, 1989.

P. N. Prasad and D. J. Williams, *Nonlinear Optical Effects in Molecules and Polymers*, Wiley, New York, 1991.

R. B. Seymour, *Engineering Polymer Sourcebook*, McGraw-Hill, New York, 1990.

D. W. Van Krevelen, *Properties of Polymers*: *Their Correlation with Chemical Structure*; *Their Numerical Estimation and Prediction from Additive Group Contributions*, Elsevier, Amsterdam, 1990.

S. Wu, *Polymer Interface and Adhesion*, Marcel Dekker, New York, 1982.

STUDY PROBLEMS

1. One million molecular weight polystyrene chains are bound to a surface at one end only, at a density of one chain every 6400 Å^2. What is the minimum bonding energy between the polymer and the surface to hold the chains in place? [*Hint*: Use rubber elasticity theory for single chains.]

2. A particular polymer blend was found to have *negative* values for $-(d\bar{\gamma}/dT)$ for its interfacial tension. Does this polymer blend exhibit an upper or lower critical solution temperature? What is your reasoning?

3. Read any scientific or engineering paper concerned with conducting polymers published since 1990. What is the import of the paper? In what ways does it update the text?

INDEX

581